OPTICAL IMAGING AND ABERRATIONS

PART I

RAY GEOMETRICAL OPTICS

OPTICAL IMAGING AND ABERRATIONS

PART I

RAY GEOMETRICAL OPTICS

VIRENDRA N. MAHAJAN

THE AEROSPACE CORPORATION

AND

THE UNIVERSITY OF SOUTHERN CALIFORNIA

SPIE Optical Engineering Press

A Publication of SPIE—The International Society for Optical Engineering
Bellingham, Washington USA

Library of Congress Cataloging-in-Publication Data

Mahajan, Virendra N.
    Optical imaging and aberrations / Virendra N. Mahajan.
        p. cm.
    Includes bibliographical references and index.
    ISBN 0-8194-2515-X
    1. Aberrations. 2. Imaging systems. 3. Geometrical optics.
I. Title.
QC671.M36   1998
621.36—DC21   97-7721
                                                        CIP

Published by

SPIE—The International Society for Optical Engineering
P.O. Box 10
Bellingham, Washington 98227-0010 USA
Phone: 360/676-3290
Fax: 360/647-1445
Email: spie@spie.org
WWW: http://www.spie.org

Printed in the United States of America.

To my wife, **Shashi Prabha**
son, **Vinit Bharati**
and daughter, **Sangita Bharati**

FOREWORD

This book covers the subject of geometrical optics and aberrations in a consistent and fairly exhaustive manner. Vini Mahajan has explored the subject with a consistent and quite clear approach. He begins with the basic approaches to first order image formation, providing both analytic and graphical methods for locating the position and size of images. This understanding is extended to the important subject of the radiometry of images, and some associated basic relations.

The fact that aberrations provide intrinsic limitations to image quality has provided a source of fun and profit for generations of optical designers. This basic truth provides the motivation for the bulk of the content of this book. Mahajan approaches the subject from the effect of combined aberrations of various orders, and the relation between wave and ray aberrations. He then develops approaches to computing the amounts of the primary aberrations from constructional parameters of an optical system. The limits on aberration content for refracting systems are followed by an exposition of the aberrations arising in reflecting systems and perturbed optical systems.

The engineer interested in developing a wider understanding of the sources of aberrations and needing access to a consistent set of equations for analyzing the primary aberration content of systems will find the approach in this book to be quite valuable. The latter chapters provide an accessible set of equations that may be used in the initial design of the types of reflective and catadioptric lenses that find so much application in present day optical systems.

Anyone involved either deeply or occasionally in the field of optical engineering, design, and testing will find this book to be a valuable complement to the usual textbook that only briefly discusses the important subject of aberrations. The description of the sources of aberrations is useful to the designer. The discussion of symmetries of aberrations and how they relate to the widely used Zernike aberration terms is of great value to the optical test engineer. Working the large number of practical problems included in the text will develop a good understanding of the issues involved in the applications of aberration theory to real world issues.

Tucson, Arizona   R. R. Shannon
April 1998

TABLE OF CONTENTS

PART I. RAY GEOMETRICAL OPTICS

Preface xvii

Acknowledgments xxi

Symbols and Notation xxiii

CHAPTER 1: GAUSSIAN OPTICS 1

1.1   **Introduction** 3

1.2   **Foundations of Geometrical Optics** 5

       1.2.1   Fermat's Principle 5

       1.2.2   Laws of Geometrical Optics 8

       1.2.3   Optical Path Lengths of Neighboring Rays 10

       1.2.4   Malus-Dupin Theorem 11

       1.2.5   Hamilton's Point Characteristic Function and Direction of a Ray 13

1.3   **Gaussian Imaging** 14

       1.3.1   Introduction 14

       1.3.2   Sign Convention 14

       1.3.3   Spherical Refracting Surface 15

            1.3.3.1   Gaussian Imaging Equation 15

            1.3.3.2   Focal Lengths and Refracting Power 18

            1.3.3.3   Magnifications and Lagrange Invariant 19

            1.3.3.4   Graphical Imaging 22

            1.3.3.5   Newtonian Imaging Equation 24

       1.3.4   Thin Lens 24

            1.3.4.1   Gaussian Imaging Equation 24

            1.3.4.2   Focal Lengths and Refracting Power 25

            1.3.4.3   Undeviated Ray 26

            1.3.4.4   Magnifications and Lagrange Invariant 28

            1.3.4.5   Newtonian Imaging Equation 30

       1.3.5   Refracting Systems 31

            1.3.5.1   Cardinal Points and Planes 31

            1.3.5.2   Gaussian Imaging, Focal Lengths, and Magnifications 33

            1.3.5.3   Nodal Points 36

            1.3.5.4   Newtonian Imaging Equation 38

       1.3.6   Afocal Systems 38

       1.3.7   Spherical Reflecting Surface (Spherical Mirror) 42

            1.3.7.1   Gaussian Imaging Equation 42

            1.3.7.2   Focal Length and Reflecting Power 44

            1.3.7.3   Magnifications and Lagrange Invariant 46

            1.3.7.4   Graphical Imaging 49

            1.3.7.5   Newtonian Imaging Equation 52

1.4	**Paraxial Ray Tracing**		**52**
	1.4.1	Refracting Surface	52
	1.4.2	Thin Lens	54
	1.4.3	Two Thin Lenses	57
	1.4.4	Thick Lens	59
	1.4.5	Reflecting Surface (Mirror)	62
	1.4.6	Two-Mirror System	65
	1.4.7	Catadioptric System: Thin Lens-Mirror Combination	67
1.5	**Two-Ray Lagrange Invariant**		**69**
1.6	**Matrix Approach to Paraxial Ray Tracing and Gaussian Optics**		**73**
	1.6.1	Introduction	73
	1.6.2	System Matrix	73
	1.6.3	Conjugate Matrix	77
	1.6.4	System Matrix in Terms of Gaussian Parameters	81
	1.6.5	Gaussian Imaging Equations	81
References			**84**
Problems			**85**

CHAPTER 2: RADIOMETRY OF IMAGING 89

2.1	**Introduction**		**91**
2.2	**Stops, Pupils, and Vignetting**		**92**
	2.2.1	Introduction	92
	2.2.2	Aperture Stop, and Entrance and Exit Pupils	92
	2.2.3	Chief and Marginal Rays	94
	2.2.4	Vignetting	95
	2.2.5	Size of an Imaging Element	98
	2.2.6	Telecentric Aperture Stop	98
	2.2.7	Field Stop, and Entrance and Exit Windows	98
2.3	**Radiometry of Point Sources**		**100**
	2.3.1	Irradiance of a Surface	100
	2.3.2	Flux Incident on a Circular Aperture	103
2.4	**Radiometry of Extended Sources**		**104**
	2.4.1	Lambertian Surface	104
	2.4.2	Exitance of a Lambertian Surface	105
	2.4.3	Radiance of a Tube of Rays	106
	2.4.4	Irradiance by a Lambertian Surface Element	107
	2.4.5	Irradiance by a Lambertian Disc	108
2.5	**Radiometry of Point Object Imaging**		**112**
2.6	**Radiometry of Extended Object Imaging**		**114**
	2.6.1	Image Radiance	114
	2.6.2	Pupil Distortion	117
	2.6.3	Image Irradiance: Aperture Stop in Front of the System	118
	2.6.4	Image Irradiance: Aperture Stop in Back of the System	121

2.6.5   Telecentric Systems 123
2.6.6   Throughput 123
2.6.7   Condition for Uniform Image Irradiance 123
2.6.8   Concentric Systems 125

2.7   Photometry **126**
2.7.1   Photometric Quantities and Spectral Response of the Human Eye 126
2.7.2   Imaging by a Human Eye 127
2.7.3   Brightness of a Lambertian Surface 129
2.7.4   Observing Stars in the Daytime 130

Appendix: Radiance Theorem **134**

References **136**

Problems **137**

CHAPTER 3: OPTICAL ABERRATIONS 139

3.1   Introduction **141**

3.2   Wave and Ray Aberrations **142**
3.2.1   Definitions 142
3.2.2   Relationship Between Wave and Ray Aberrations 145

3.3   Defocus Aberration **148**

3.4   Wavefront Tilt **150**

3.5   Aberration Function of a Rotationally Symmetric System **152**
3.5.1   Rotational Invariants 152
3.5.2   Power-Series Expansion 155
        3.5.2.1   Explicit Dependence on Object Coordinates 156
        3.5.2.2   No Explicit Dependence on Object Coordinates 159
3.5.3   Zernike Circle-Polynomial Expansion 163
3.5.4   Relationships Between Coefficients of Power-Series and Zernike Polynomial Expansions 168

3.6   Observation of Aberrations **169**
3.6.1   Primary Aberrations 172
3.6.2   Interferograms 173

3.7   Conditions for Perfect Imaging **178**
3.7.1   Imaging of a 3-D Object 178
3.7.2   Imaging of a 2-D Transverse Object 181
3.7.3   Imaging of a 1-D Axial Object 183
3.7.4   Linear Coma and the Sine Condition 184
3.7.5   Optical Sine Theorem 186
3.7.6   Linear Coma and Offense Against the Sine Condition ... 188

Appendix A: Degree of Approximation in Eq. (3-11) **192**

Appendix B: Wave and Ray Aberrations: Alternative Definition and Derivation **194**

References **200**

Problems **201**

CHAPTER 4: GEOMETRICAL POINT-SPREAD FUNCTION **203**

4.1 **Introduction** 205

4.2 **Theory** 205

4.3 **Application to Primary Aberrations** 209

   4.3.1 Spherical Aberration 210
   4.3.2 Coma 217
   4.3.3 Astigmatism and Field Curvature 224
   4.3.4 Distortion 233

4.4 **Balanced Aberrations for Minimum RMS Spot Radius** 235

4.5 **Spot Diagrams** 236

4.6 **Summary of Results** 239

   4.6.1 Spherical Aberration 240
   4.6.2 Coma 240
   4.6.3 Astigmatism and Field Curvature 241
   4.6.4 Distortion 242
   4.6.5 Aberration Tolerance 242

References 243

Problems 244

**CHAPTER 5: CALCULATION OF PRIMARY ABERRATIONS:
               REFRACTING SYSTEMS** **245**

5.1 **Introduction** 247

5.2 **Spherical Refracting Surface with Aperture Stop at the Surface** 249

   5.2.1 On-Axis Point Object 249
   5.2.2 Off-Axis Point Object 252
      5.2.2.1 Aberrations with Respect to Petzval Image Point 253
      5.2.2.2 Aberrations with Respect to Gaussian Image Point 259

5.3 **Spherical Refracting Surface with Aperture Stop Not at the Surface** 261

   5.3.1 On-Axis Point Object 262
   5.3.2 Off-Axis Point Object 264

5.4 **Aplanatic Points of a Spherical Refracting Surface** .266

5.5 **Conic Refracting Surface** 271

   5.5.1 Sag of a Conic Surface 271
   5.5.2 On-Axis Point Object 275
   5.5.3 Off-Axis Point Object 278

5.6 **General Aspherical Refracting Surface** 281

5.7 **Series of Coaxial Refracting (and Reflecting) Surfaces** 281

   5.7.1 General Imaging System 282
   5.7.2 Petzval Curvature and Corresponding Field Curvature Wave Aberration .282
   5.7.3 Relationship Among Petzval Curvature, Field Curvature, and
          Astigmatism Wave Aberration Coefficients .287

5.8   **Aberration Function in Terms of Seidel Sums or Seidel Coefficients** **287**

5.9   **Effect of Change in Aperture Stop Position on the Aberration Function** **290**

    5.9.1   Change of Peak Aberration Coefficients 291

    5.9.2   Illustration of the Effect of Aperture-Stop Shift on Coma
          and Distortion 295

    5.9.3   Aberrations of a Spherical Refracting Surface with Aperture Stop Not at the
          Surface Obtained from Those with Stop at the Surface 297

5.10   **Thin Lens** **299**

    5.10.1   Imaging Relations 300

    5.10.2   Thin Lens with Spherical Surfaces and Aperture Stop at the Lens 301

    5.10.3   Petzval Surface 306

    5.10.4   Spherical Aberration and Coma 307

    5.10.5   Aplanatic Lens 310

    5.10.6   Thin Lens with Conic Surfaces 312

    5.10.7   Thin Lens with Aperture Stop Not at the Lens 313

5.11   **Field Flattener** **314**

    5.11.1   Imaging Relations 315

    5.11.2   Aberration Function 316

5.12   **Plane-Parallel Plate** **318**

    5.12.1   Introduction 318

    5.12.2   Imaging Relations 318

    5.12.3   Aberration Function 321

5.13   **Chromatic Aberrations** **323**

    5.13.1   Introduction 323

    5.13.2   Single Refracting Surface 323

    5.13.3   Thin Lens 327

    5.13.4   General System: Surface-by-Surface Approach 331

    5.13.5   General System: Use of Principal and Focal Points 336

    5.13.6   Chromatic Aberrations as Wave Aberrations 347

5.14   **Symmetrical Principle** **348**

5.15   **Pupil Aberrations and Conjugate-Shift Equations** **349**

    5.15.1   Introduction 349

    5.15.2   Pupil Aberrations 350

    5.15.3   Conjugate-Shift Equations 355

    5.15.4   Invariance of Image Aberrations 357

    5.15.5   Simultaneous Correction of Aberrations for Two or More
          Object Positions 358

References **360**

Problems **361**

CHAPTER 6: CALCULATION OF PRIMARY ABERRATIONS: REFLECTING AND CATADIOPTRIC SYSTEMS 365

6.1	**Introduction** 367	
6.2	**Conic Reflecting Surface** 367	
	6.2.1	Conic Surface 367
	6.2.2	Imaging Relations 370
	6.2.3	Aberration Function 370
6.3	**Petzval Surface** 375	
6.4	**Spherical Mirror** 377	
	6.4.1	Aberration Function and Aplanatic Points for Arbitrary Location of Aperture Stop 377
	6.4.2	Aperture Stop at the Mirror Surface 379
	6.4.3	Aperture Stop at the Center of Curvature of Mirror 381
6.5	**Paraboloidal Mirror** 384	
6.6	**Catadioptric Systems** 385	
	6.6.1	Introduction 385
	6.6.2	Schmidt Camera 385
	6.6.3	Bouwers-Maksutov Camera 394
6.7	**Beam Expander** 398	
	6.7.1	Introduction 398
	6.7.2	Gaussian Parameters 398
	6.7.3	Aberration Contributed by Primary Mirror 400
	6.7.4	Aberration Contributed by Secondary Mirror 401
	6.7.5	System Aberration 402
6.8	**Two-Mirror Astronomical Telescopes** .402	
	6.8.1	Introduction 402
	6.8.2	Gaussian Parameters 403
	6.8.3	Petzval Surface 408
	6.8.4	Aberration Contributed by Primary Mirror 408
	6.8.5	Aberration Contributed by Secondary Mirror 410
	6.8.6	System Aberration 412
	6.8.7	Classical Cassegrain and Gregorian Telescopes 413
	6.8.8	Aplanatic Cassegrain and Gregorian Telescopes 416
	6.8.9	Afocal Telescope 416
	6.8.10	Couder Anastigmatic Telescopes .417
	6.8.11	Schwarzschild Telescope 418
	6.8.12	Dall-Kirkham Telescope 420
6.9	**Astronomical Telescopes Using Aspheric Plates** 422	
	6.9.1	Introduction 422
	6.9.2	Aspheric Plate in a Diverging Object Beam 422
	6.9.3	Aspheric Plate in a Converging Image Beam 425
	6.9.4	Aspheric Plate and a Conic Mirror 426
	6.9.5	Aspheric Plate and a Two-Mirror Telescope 428

References **431**

Problems **432**

CHAPTER 7: CALCULATION OF PRIMARY ABERRATIONS: PERTURBED OPTICAL SYSTEMS **435**

7.1   **Introduction** **437**

7.2   **Aberrations of a Misaligned Surface** **438**

      7.2.1   Decentered Surface 438

      7.2.2   Tilted Surface 442

      7.2.3   Despaced Surface 444

7.3   **Aberrations of Perturbed Two-Mirror Telescopes** **445**

      7.3.1   Decentered Secondary Mirror 445

      7.3.2   Tilted Secondary Mirror 447

      7.3.3   Decentered and Tilted Secondary Mirror 448

      7.3.4   Despaced Secondary Mirror 451

7.4   **Fabrication Errors** **454**

      7.4.1   Refracting Surface 454

      7.4.2   Reflecting Surface 456

References **458**

Problems **459**

Bibliography **461**

Index **463**

PREFACE

The material presented here has been gathered from my lectures at the Electrical Engineering-Electrophysics Department of the University of Southern California, where I have been teaching a graduate course on optical imaging and aberrations since 1984. My objective for this course has been to provide the students with an understanding of how aberrations arise in optical systems and how they affect optical wave propagation and imaging based on both geometrical and physical optics. This book has been written with the same objective in mind. The emphasis of the text is on concepts, physical insight, and mathematical simplicity. Figures and drawings are given wherever appropriate to facilitate understanding and make the book reader friendly. An abbreviated version called *Aberration Theory Made Simple* was published by the SPIE Press in 1991 in their Tutorial Text Series (Vol. TT6). The current detailed version is divided into two parts just like the abbreviated one. In Part I of this text, which contains the first seven chapters, ray geometrical optics is discussed. In Part II, wave diffraction optics is discussed.

In Part I, Chapter 1 begins with the foundations of geometrical optics. Fermat's principle, the laws of geometrical optics, the Malus-Dupin theorem, and Hamilton's point characteristic function are described. Starting with a brief outline of the sign convention for object and image distances and their heights and ray angles, Gaussian imaging by a spherical refracting surface, a thin lens, an afocal system, and a spherical reflecting surface (mirror) are discussed. The cardinal points of an imaging system are defined and a paraxial ray-tracing procedure to determine them is described. It is emphasized that the results for a reflecting surface can be obtained from those for a refracting surface by substituting the refractive index associated with the reflected rays equal to the negative of the refractive index associated with the incident rays. A two-lens system, a two-mirror system, and a catadioptric system consisting of a lens and a mirror are considered as examples of imaging systems for which the focal length is explicitly determined using the ray-tracing equations. The sign convention used throughout the book is the Cartesian sign convention of analytical geometry. It is different from the one used in author's *Aberration Theory Made Simple* in some respects, which should be noted when making comparisons with the equations given there. A good understanding of this chapter is essential for performing Gaussian (or first-order) design and analysis of an optical imaging system. Given the radii of curvature and the positions of the surfaces of an optical system, and the refractive indices of the media around them, one can determine its cardinal points, and, in turn, the position and size of the image for a certain position and size of the object. However, there is no discussion in this chapter on the intensity of the image of a point object in terms of the object intensity, or the irradiance distribution of the image of an extended object in terms of the object radiance distribution.

The concepts of aperture stop, entrance and exit pupils, chief and marginal rays, sizes of imaging elements, and vignetting of rays are introduced in Chapter 2. Radiometry of point and extended sources and of point and extended object imaging are discussed next. The origins and limitations of the cosine-cube law of image intensity for point objects, and the cosine-fourth law of image irradiance for extended objects, are discussed in detail. It is pointed out that because of pupil distortion, integration must be performed across the aperture stop to calculate the total flux entering the system from an object element. For integrating across a pupil, its distortions determined by detailed ray tracing must be taken into account.

A brief discussion of photometry, which is a branch of radiometry involving the spectral response of the human eye, is also included.

Besides the position, size, and intensity or irradiance of an image, its quality, which depends on the aberrations of the system, is of paramount importance. In Chapter 3, the wave and ray aberrations are defined and a relationship between them is derived. Relationships between defocus wave aberration and longitudinal defocus, and wavefront tilt aberration and wavefront tilt angle, are described. The form of the aberration function of a rotationally symmetric system is derived, and its expansions in terms of a power series and Zernike circle polynomials are discussed. The relationships between the coefficients of the two expansions are given. It is shown that up to the fourth order in object and pupil coordinates, any system with an axis of rotational symmetry can have no more than five primary aberration terms, called Seidel aberrations. The form of the secondary (or Schwarzschild) and tertiary aberrations is discussed. How an aberration may be observed is described by discussing the interference patterns of the primary aberrations. The conditions under which an imaging system may form an aberration-free image are considered. In particular, the sine condition for coma-free imaging is discussed. It is not essential to understand all of the material in this chapter to understand the material in Chapters 5, 6, and 7, though it would be useful to read the first four sections and to know the form of the five primary aberrations of a rotationally symmetric system from Section 3.5.2.1.

In Chapter 4, the relationship between the ray and wave aberrations is utilized to discuss the geometrical point-spread functions and the ray spot diagrams for each of the five primary aberrations. The circle of least confusion is discussed for both spherical aberration and astigmatism, thereby introducing the concept of aberration balancing. The centroid, encircled power, and root-mean-square radius of an aberrated image spot are also discussed. The traditional examples of the image of a spoked wheel in the presence of astigmatism, and the image of a square grid in the presence of distortion, are explained. Thus, given the aberrations of a system, the quality of the image of a point object in terms of its size or the ray distribution can be determined using the material given in this chapter. Aberration tolerances can be obtained from the tolerable image spot sizes.

In order to determine the quality of an image formed by a certain system, its aberrations must be known. The remainder of Part I discusses how to calculate the aberrations of an optical system given the radii of curvature and positions of its surfaces, and the refractive indices of the media surrounding them. Of course, the task of a lens designer is to choose these parameters in a way that is practical yet meets his/her image quality objectives. Chapter 5 describes an approach for calculating the primary aberrations of a multisurface optical system with an axis of rotational symmetry. The theory is developed by starting with the simplest problem, namely, the aberrations of a spherical refracting surface with its aperture stop located at the surface. An on-axis point object is considered first, so that the only aberration that arises is spherical aberration. An off-axis point object is considered next, and expressions for field aberrations (coma, astigmatism, field curvature, and distortion) are obtained with respect to the Petzval image point. These are generalized next to obtain the aberrations with respect to the Gaussian image point. Only field curvature and distortion terms change as the image point is changed from Petzval to Gaussian. This completes the derivation of primary aberrations of a spherical refracting surface. The Gaussian imaging equations are obtained as a by-product of this derivation. The primary aberrations of a spherical refracting surface with an arbitrary location of the aperture stop are considered next and its aplanatic points are determined. The aberrations of a conic refracting surface and finally a general aspheric surface are obtained. Instead of starting with a derivation for the most complex case, namely, a general aspheric surface with a remote aperture stop, a step-by-step derivation of increasing complexity is given so that physical insight on the differences between different steps is not lost.

How the results given for a single refracting surface can be extended to obtain the aberrations of a multisurface system is described. The changes in the aberration function as a result of a change in the position of the aperture stop are discussed next. The aberration function is also considered in terms of Seidel sums and Seidel coefficients of an optical system. As applications of the theory, the aberrations of a thin lens and a plane-parallel plate are derived and discussed. The aplanatic and field-flattening lenses are also considered. Next, the chromatic aberrations of a refracting system are discussed in terms of the wavelength dependence of the position and magnification of an image formed by it. Finally, pupil aberrations are considered and conjugate-shift equations are obtained that relate the aberrations of the image of one object in terms of those of another.

The primary aberrations of reflecting and catadioptric systems are discussed in Chapter 6. As in the case of imaging relations, the aberration expressions for a reflecting surface may be obtained from those for a corresponding refracting surface by substituting the refractive index associated with the reflected rays equal to the negative of the refractive index associated with the incident rays. As examples of reflecting systems, expressions are obtained for the primary aberrations of a spherical mirror, paraboloidal mirror, a beam expander consisting of two confocal paraboloidal mirrors, and two-mirror astronomical telescopes. Schmidt and Bouwers-Maksutov cameras and telescopes with aspheric plates are discussed as examples of catadioptric systems.

Even if a practical design of a system has been chosen, its elements must be fabricated and assembled into a system. In Chapter 7, the last chapter of Part I of the book, the primary aberrations due to perturbations such as a decenter, a tilt, or a despace of the surface of a system are considered. When one or more of the imaging elements is decentered and/or tilted, a system loses its rotational symmetry. Hence, new aberrations arise which have different dependence on the object height but the same dependence on pupil coordinates as the aberrations of the unperturbed system. The expressions derived for the primary aberrations produced when a perturbation is introduced into the system are used to obtain the aberrations of misaligned two-mirror telescopes. Finally, the relationships between the fabrication errors of the surfaces and the corresponding aberrations or wavefront errors introduced by them are derived for both the refracting and reflecting surfaces.

Throughout the book, the primary aberrations of a system are emphasized since they are often the dominant aberrations in the early stages of the design of an optical system. Although expressions for higher-order aberrations have been given in the literature, their value in designing or analyzing optical systems has not been fully exploited or realized, mainly perhaps because of their complexity. The expressions for the primary aberrations of even simple systems such as a thin lens (made up of two surfaces with negligible thickness between them) or a two-mirror astronomical telescope are complex indeed. With the advent of computers and commercially available computerized ray-tracing and image-analysis programs (e.g., ZEMAX or CODE V), it is a simple matter to determine the aberrations of a system fully, not just its primary or secondary aberrations. However, it is this author's belief that it is essential to understand the primary aberrations of simple systems in order to be able to design systems that are more complex and provide high image quality. It is for this reason that full derivations and discussion of the expressions for the primary aberrations of simple systems are given. Key equations representing fundamental results are highlighted by putting a box around them. It is hoped that they will provide the reader with certain basic tools to develop new designs without endless surfing in a sea of potential designs.

Each chapter ends with a set of problems. These problems have been crafted carefully either as an extension of the theory given in the text, or, more often, as applications of the theory. They are an essential part of the book since only by working through such problems can the students appreciate the theory and validate their understanding of it.

In Part II, imaging based on diffraction is discussed. It starts with an introduction of the diffraction point-spread function and optical transfer function of a general imaging system. An understanding of diffraction effects is essential since geometrical point-spread functions in terms of the spot diagrams give at best a qualitative understanding of the image quality aspects of an imaging system, especially for high-quality systems. Optical systems with circular, annular, and Gaussian pupils are considered and aberration-free as well as aberrated images are discussed. Aberration tolerances based on the Strehl and Hopkins ratios of an image are obtained. The effect of random aberrations such as those introduced by atmospheric turbulence on the image formed by a system is also discussed.

ACKNOWLEDGMENTS

It is a great pleasure to acknowledge the generous support I have received over the years from my employer, The Aerospace Corporation, in preparing this book. My special thanks go to Mr. John Parsons for his continuous interest and encouragement in this endeavor. I also thank Dr. Bruce Gardner of The Aerospace Institute for providing support in preparing the figures in the book; Mr. John Hoyem for meticulously drawing the figures; and Ms. Carol Gibson and other staff for word processing and composition.

I am grateful to Dr. W. Swantner of BSC Optics for stimulating discussions on the subject of optical design and aberrations in practice. Occasionally, he verified my theory by way of numerical examples. My thanks to Yunsong Huang for his help in verifying the equations for pupil aberrations and conjugate shifts. I am also grateful to many friends and colleagues for reviewing the manuscript. Included among them are Dr. P. Mouroulis, Dr. D. Schroeder, Professor A. Walther, Dr. R. Buchroeder, and Mr. David Shafer. I took the advice of the first two in adopting the Cartesian sign convention. I had many helpful discussions with Dr. Mouroulis on some subtle and fine points. Of course, any shortcomings or errors in the book are my responsibility.

Many other people have helped me with this book; e.g., Dr. Rich Boucher and Captain Junichi Kamita for initially preparing some of the figures; and Mrs. Chizuko DeQueiroz for cover design. The Sanskrit verse on p. xxv was suggested by Professor Sally Sutherland of the University of California at Berkeley.

It has been a pleasure to work with Mr. Eric Pepper and other SPIE Press staff in bringing this book to its completion. My thanks to them for their cooperative spirit and quality support.

I dedicate this book to my wife and children, whom I cannot thank enough for their support to write it.

El Segundo, California   Virendra N. Mahajan
April 1998

SYMBOLS AND NOTATION

a	radius of exit pupil
a_i	aberration coefficient
A_i	peak aberration coefficient
AS	aperture stop
CR	chief ray
e	eccentricity
EnP	entrance pupil
ExP	exit pupil
f	focal length
F	focal ratio or f-number, focal point
GR	general ray
h	object height
h'	image height
H	principal point
K	power of a system
L	image distance from exit pupil
m	pupil-image magnification
M	object-image magnification
MR	marginal ray
n	refractive index
OA	optical axis
p	position factor
P	object point
P'	Gaussian image point
PSF	point-spread function
q	shape factor
R	radius of curvature of a surface or reference sphere
s	entrance pupil distance
s'	exit pupil distance
S	object distance
S'	image distance
t	thickness
V	Abbe number, spectral response
W	wave aberration
x, y	rectangular coordinates of a point
z	sag, object, or observation distance
z'	image distance
β	ray or field angle
ΔR	longitudinal defocus
r, θ	polar coordinates of a point
λ	optical wavelength
ξ, η	normalized rectangular coordinates
$\rho = \dfrac{r}{a}$	normalized radial coordinate in the pupil plane
σ_W	standard deviation of wave aberration
σ_F	standard deviation of figure errors
Φ	phase aberration
ψ	angular deviation of ray
$R_n^m(\rho)$	radial Zernike polynomial
$(-)\,x$	numerically negative quantity x

अनन्तरत्नप्रभवस्य यस्य हिमं न सौभाग्यविलोपि जातम् ।
एको हि दोषो गुणसन्निपाते निमज्जतीन्दोः किरणेष्विवाङ्कः ॥

Anantaratnaprabhavasya yasya himaṃ na saubhāgyavilopi jātam|

Eko hi doṣo gūṇasannipāte nimajjatindoḥ kiraṇesvivāṅkaḥ ‖

The snow does not diminish the beauty of the Himālayan mountains
which are the source of countless gems. Indeed, one flaw is lost
among a host of virtues, as the moon's dark spot is lost among its rays.

Kālidāsa *Kumārasambhava* 1.3

OPTICAL IMAGING AND ABERRATIONS

PART I

RAY GEOMETRICAL OPTICS

CHAPTER 1

GAUSSIAN OPTICS

1.1 Introduction **3**

1.2 Foundations of Geometrical Optics **5**

    1.2.1 Fermat's Principle 5

    1.2.2 Laws of Geometrical Optics 8

    1.2.3 Optical Path Lengths of Neighboring Rays 10

    1.2.4 Malus-Dupin Theorem 11

    1.2.5 Hamilton's Point Characteristic Function and Direction of a Ray 13

1.3 Gaussian Imaging **14**

    1.3.1 Introduction...14

    1.3.2 Sign Convention 14

    1.3.3 Spherical Refracting Surface 15

        1.3.3.1 Gaussian Imaging Equation 15

        1.3.3.2 Focal Lengths and Refracting Power 18

        1.3.3.3 Magnifications and Lagrange Invariant 19

        1.3.3.4 Graphical Imaging...22

        1.3.3.5 Newtonian Imaging Equation 24

    1.3.4 Thin Lens...24

        1.3.4.1 Gaussian Imaging Equation 24

        1.3.4.2 Focal Lengths and Refracting Power 25

        1.3.4.3 Undeviated Ray 26

        1.3.4.4 Magnifications and Lagrange Invariant 28

        1.3.4.5 Newtonian Imaging Equation 30

    1.3.5 Refracting Systems 31

        1.3.5.1 Cardinal Points and Planes 31

        1.3.5.2 Gaussian Imaging, Focal Lengths, and Magnifications 33

        1.3.5.3 Nodal Points 36

        1.3.5.4 Newtonian Imaging Equation 38

    1.3.6 Afocal Systems 38

    1.3.7 Spherical Reflecting Surface (Spherical Mirror)..................42

        1.3.7.1 Gaussian Imaging Equation 42

        1.3.7.2 Focal Length and Reflecting Power 44

        1.3.7.3 Magnifications and Lagrange Invariant 46

        1.3.7.4 Graphical Imaging...49

        1.3.7.5 Newtonian Imaging Equation 52

1.4 Paraxial Ray Tracing..**52**

    1.4.1 Refracting Surface 52

    1.4.2 Thin Lens...54

    1.4.3 Two Thin Lenses 57

    1.4.4 Thick Lens 59

1.4.5   Reflecting Surface (Mirror) 62

1.4.6   Two-Mirror System 65

1.4.7   Catadioptric System: Thin Lens-Mirror Combination 67

1.5   Two-Ray Lagrange Invariant **69**

1.6   Matrix Approach to Paraxial Ray Tracing and Gaussian Optics **73**

1.6.1   Introduction 73

1.6.2   System Matrix 73

1.6.3   Conjugate Matrix 77

1.6.4   System Matrix in Terms of Gaussian Parameters 81

1.6.5   Gaussian Imaging Equations 81

References **84**

Problems **85**

Chapter 1
Gaussian Optics

1.1 INTRODUCTION

In geometrical optics, light is assumed to consist of rays that propagate according to three laws: rectilinear propagation, refraction, and reflection. We begin this chapter with a statement of Fermat's principle and the derivation of these laws from it. We consider the refraction of two neighboring rays and show that their optical path lengths between planes that are perpendicular to one or both of them are equal to each other. The Malus-Dupin theorem, which states that rays are normal to a wavefront and remain so after refraction and/or reflection, is discussed. Hamilton's point characteristic function representing the optical path length of a ray from one point to another is introduced and a relationship between it and unit vectors along the ray through these points is obtained.

An optical imaging system consists of a series of refracting and/or reflecting surfaces that generally have a common axis of rotational symmetry called the *optical axis*. The surfaces bend light rays from an object according to the laws of geometrical optics to form its image. Gaussian optics or imaging, which is the subject of this chapter, relates the object distance and size to the image distance and size through the parameters of the imaging system such as the radii of curvature of the surfaces and refractive indices of the media between them. Throughout this book, a Cartesian sign convention is used for the object and image distances and their heights, the radii of curvature of surfaces, angles of incidence and refraction or reflection, and the slope angles of the rays. This convention offers the simplicity of few rules to remember, namely, those of a universally known right-handed coordinate system, regardless of whether an object or its image is real or virtual. Any quantities that are numerically negative are indicated with a parenthetical negative sign $(-)$ in the figures.

In Gaussian optics, the angle that a ray from a point object makes with the optical axis or a surface normal is treated as a small quantity so that its sine or tangent is replaced by the angle itself and, as a result, the law of refraction takes a simple form. This assumption or approximation is referred to as the *Gaussian* or the *paraxial* (meaning near the optical axis) *approximation*. Sometimes a distinction is made between Gaussian and paraxial optics in that paraxial optics is a limiting case of Gaussian optics in which the angles are infinitesimal quantities. The rays traced in this approximation are called *paraxial rays* and the corresponding method of ray tracing is referred to as *paraxial ray tracing*. The refraction or reflection of a ray incident on a surface takes place at a plane that is tangent to it and passes through its vertex. All of the rays diverging from a point object and propagating through the imaging system converge to a point called the *Gaussian image point*. In reality, of course, if the rays are traced according to the exact laws of geometrical optics (so-called *exact* or *finite ray tracing*) refracting at or reflecting from the surface at their points of incidence, they generally do not converge to an image point due to aberrations (discussed in the following chapters) of the system for the point object under consideration. The actual distribution of the rays in an image plane is discussed in Chapter 4.

We begin our discussion of Gaussian imaging with a brief introduction of the Cartesian sign convention for the distances and heights of object and image points, and the angles of incidence and refraction or reflection of the rays and their slope angles. The equations describing imaging by a spherical refracting surface are derived first. They are then used to derive the corresponding equations for a thin lens. The equations for a general multisurface imaging system are derived next. The cardinal points (principal, focal, and nodal points) of such a system are discussed. It is shown that simple imaging equations, similar to those for a single refracting surface, are obtained provided the object and image distances are measured from the respective principal points of the system. Both Gaussian and Newtonian forms of the imaging equations are given in each case. The imaging properties of afocal systems are described briefly. A parallel beam of light incident on such systems emerges from them as a parallel beam of light.

The imaging equations for a spherical reflecting surface are derived next, and it is shown that they can be obtained from those for a corresponding refracting surface by substituting the refractive index associated with the reflected rays equal to the negative of that associated with the incident rays. The Gaussian image of an object can also be determined graphically, and doing so is quite instructive and helpful in understanding the (Gaussian) imaging process. It is illustrated for both refracting and reflecting systems considered here. Once again, both Gaussian and Newtonian forms of the imaging equations are given.

Paraxial ray-tracing equations for refracting and reflecting surfaces are considered next, and they are used recursively to determine the principal and focal points of simple systems, e.g., a thin lens, two thin lenses, a thick lens, and two mirrors. These equations not only help determine the cardinal points of a system and the image location and magnification, but also the sizes of the imaging elements, vignetting of the object rays by these elements, and obscurations in mirror systems, as discussed in Chapter 2. The paraxial ray-tracing equations are used to develope a Lagrange invariant in terms of their hegihts from and slopes with the optical axis of a system.

Since the ray-tracing equations representing the transfer of a ray from one plane to another, and the refraction or reflection of a ray at a surface are linear in ray heights and slopes, they can be written in the form of 2×2 matrices. This is considered in the final section of the chapter. The system matrix representing the propagation of a ray from its point of incidence on to its point of emergence from a system is discussed. Similarly, the conjugate matrix representing the propagation of a ray from an object point to its Gaussian image point is discussed. This is followed by a matrix approach to Gaussian optics. Optical design codes perform ray tracing using matrices because of the simplicity of multiplication of 2×2 matrices, with the end result that a complete ray tracing operation involving any number of surfaces is described by a 2×2 matrix.

It should be noted that in Gaussian optics no distinction is made between a spherical and an aspheric (e.g., conic) surface. The vertex radius of curvature of an aspheric surface is used in the Gaussian imaging and paraxial ray-tracing equations. Thus, for a given object location and size, the image location and size for an aspheric surface of a certain vertex radius of curvature are the same as those for a spherical surface of the same radius of curvature. However, the quality of the images formed by these surfaces may be quite different due to the differences in their shapes. For example, a paraboloidal mirror focuses the rays from an axial point object at infinity to a point (see Section 4.5), but a corresponding spherical mirror does not (see Section 4.4). In practice, due to diffraction of light (discussed in Part II), even a paraboloidal mirror does not form a point image; the actual image is a bright spot surrounded by concentric dark and bright rings called the *Airy pattern*.

1.2 FOUNDATIONS OF GEOMETRICAL OPTICS

In this section, we discuss Fermat's principle and derive the three laws of geometrical optics from it. We consider refraction of two neighboring rays and show that their optical path lengths between planes that are perpendicular to one or both of them are equal to each other. We define an optical wavefront and show that rays which are normal to it remain so after refraction. Hamilton's point characteristic function representing the optical path length of a ray from one point to another is introduced, and relationships for the unit vectors along the ray through these points are obtained.

1.2.1 Fermat's Principle

In geometrical optics, light is assumed to consist of rays. Fermat's principle states that the time a ray takes in traveling from one point to another along its actual path is stationary with respect to small changes of that path. By definition, the refractive index of a medium is the ratio of the speed of light in vacuum and its corresponding value in the medium. Since the time taken by a ray is inversely proportional to the speed of light in a medium, which in turn is inversely proportional to its refractive index, the principle may also be stated as follows: The optical path length of a ray in traveling from one point to another along its actual path is stationary, where the optical path length is equal to the geometrical path length multiplied by the refractive index. The optical path length is stationary in the sense that any deviation of the path from the actual that is of first order in small quantities produces a deviation in the optical path length that is at least of second order in small quantities.

If we consider the actual and neighboring paths of a ray in going from a point P_1 to a point P_2 as indicated in Figure 1-1 so that the two paths deviate by no more than a small quantity ϵ, then the difference in their optical path lengths is given by

$$W(\epsilon) \; = \; \int_{P_1}^{P_2} n ds' - \int_{P_1}^{P_2} n ds \tag{1-1a}$$

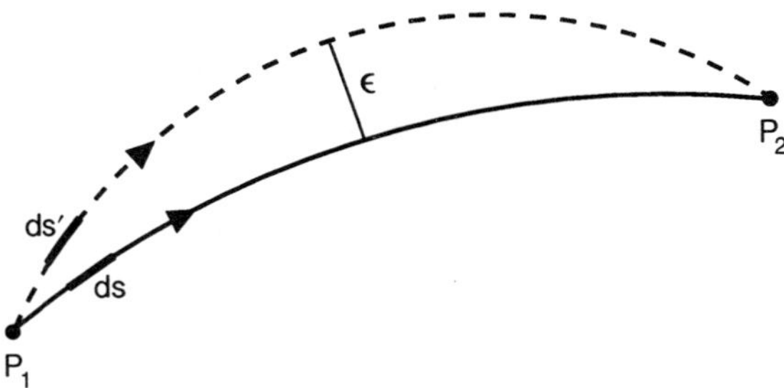

Figure 1-1. The actual and virtual paths of a ray in going from a point P_1 to a point P_2. The actual path is indicated by a solid line and the two paths deviate from each other by no more than a small quantity ϵ at any point along the path.

$$= O\left(\epsilon^2\right) \ , \tag{1-1b}$$

where ds and ds' are the differential elements of path length along the actual and neighboring virtual rays, respectively, n is the corresponding refractive index, and $O\left(\epsilon^2\right)$ indicates a function that depends on ϵ through ϵ^2 and/or higher powers of ϵ. It is clear from Eq. (1-1b) that

$$\lim_{\epsilon \to 0} \frac{\partial W}{\partial \epsilon} = 0 \ . \tag{1-2a}$$

Equation (1-2a) may also be written

$$\boxed{\delta \int_{P_1}^{P_2} n\, ds \ = \ 0} \ , \tag{1-2b}$$

where δ indicates a differential variation. Thus, up to the first order in ϵ, the two optical path lengths are equal.

The optical path length of an actual ray compared to those of the neighboring virtual rays may be a maximum or a minimum, or they may all be equal to each other. This may be seen from the properties of an ellipse (or ellipsoid) as illustrated in Figure 1-2. An ellipse has the property (see Figure 1-2a) that the sum of the distances of a point P on it from its geometrical focii F_1 and F_2 is independent of its location. Moreover, according to the law of reflection derived later in this section, the angles made by the lines F_1P and F_2P with the normal PN to the ellipse at P are equal. Thus, if we place a point source at the focus F_1 of an ellipsoidal mirror, all the rays from it pass through F_2 after reflection by the mirror, and their optical path lengths are equal to each other. Thus, for example, $\left[F_1PF_2\right]=\left[F_1QF_2\right]$, where the square brackets indicate an optical path length. However, for a plane mirror that is tangent to the ellipse at the point P, the optical path length $\left[F_1PF_2\right]$ of the actual ray will be a minimum compared with any neighboring optical path length such as $\left[F_1RF_2\right]$.

Similarly, if we consider a concave mirror shown dashed in Figure 1-2b so that it has a common tangent and therefore a common normal with the ellipse at the point. then the optical path length of the actual ray F_1PF_2 is maximum compared with the neighboring virtual (in the sense of ficticious) rays. We note, for example, that

$$[F_1RF_2] < [F_1QF_2] = [F_1PF_2] \ . \tag{1-3a}$$

Moreover, if we consider a convex mirror as in Figure 1-2c, having a common tangent with the ellipse at the point P, then the optical path length of the actual ray F_1PF_2 is minimum compared with the neighboring virtual rays. In this case,

$$[F_1RF_2] > [F_1QF_2] = [F_1PF_2] \ . \tag{1-3b}$$

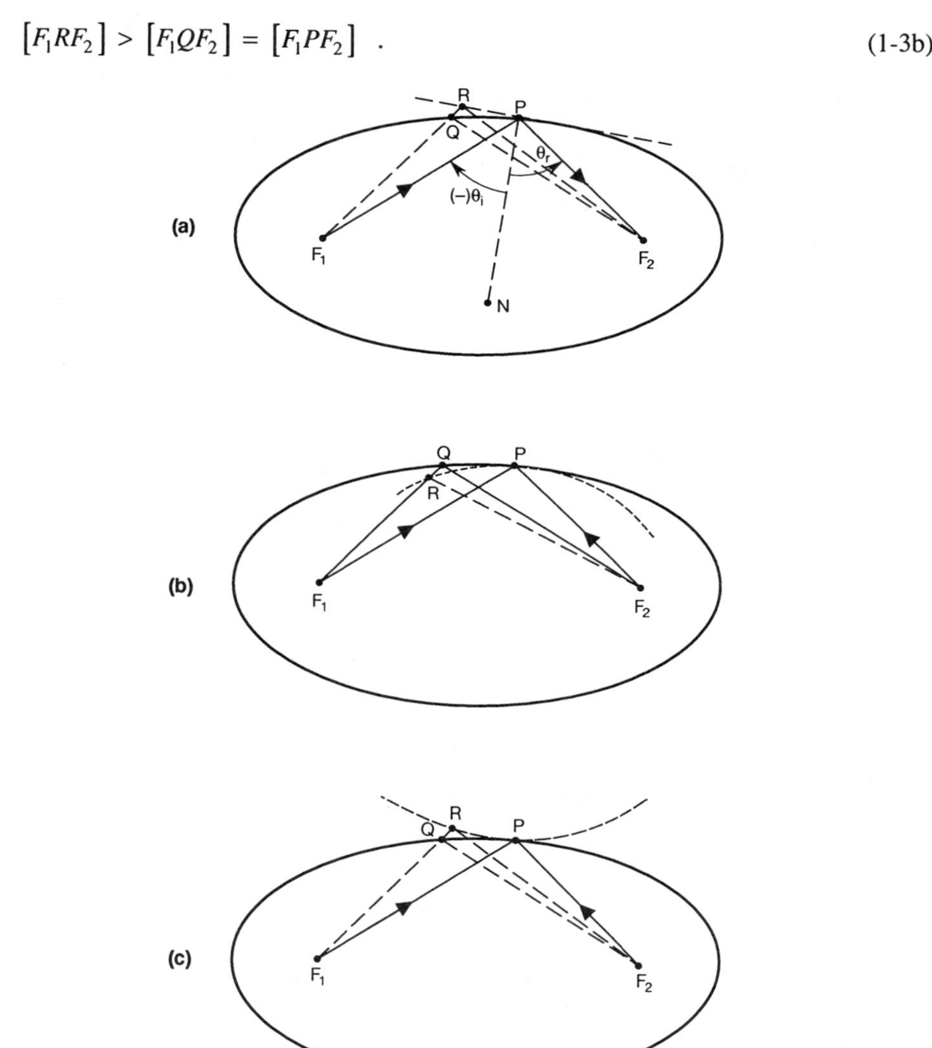

Figure 1-2. Stationarity of optical path length. (a) $[F_1PF_2]=[F_1QF_2]$ **for the elipsoidal mirror.** $[F_1PF_2]$ **is minimum for the plane mirror.** (b) $[F_1PF_2]$ **is maximum for the concave mirror.** (c) $[F_1PF_2]$ **is minimum for the convex mirror** .

1.2.2   Laws of Geometrical Optics

The laws of geometrical optics, namely, rectilinear propagation, refraction, and reflection, can be obtained from Fermat's principle. In this section, we give a simple derivation of these laws.

a.   Rectilinear Propagation

In a homogeneous medium, a light ray propagates in a straight line as indicated in Figure 1-3a. This law is self-evident since a ray propagating from one point to another in a straight line joining the two points propagates along the shortest optical path length. We note from Figure 1-3a that the difference in optical path lengths of a virtual (or ficticious) path $P_1 B P_2$ and the actual path $P_1 A P_2$ is given by

$$W(\epsilon) = n\left[\left(P_1 B + B P_2\right) - \left(P_1 A + A P_2\right)\right]$$

$$= n\left\{\left[\left(P_1 A\right)^2 + \epsilon^2\right]^{1/2} + \left[\left(A P_2\right)^2 + \epsilon^2\right]^{1/2} - \left(P_1 A + A P_2\right)\right\}$$

$$= O\left(\epsilon^2\right) \tag{1-4}$$

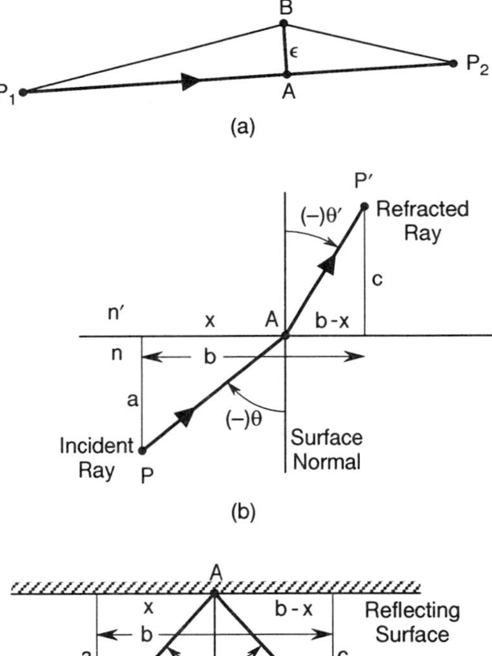

Figure 1-3. Laws of geometrical optics. (a) Rectilinear propagation. (b) Law of refraction. (c) Law of reflection.

as expected, where $\epsilon = AB$ is a small deviation of the virtual path from the actual and n is the refractive index of the homogeneous medium.

b. Refraction

At an interface between two media of refractive indices n and n', an incident light ray is refracted according to *Snell's law*,

$$\boxed{n'\sin\theta' = n\sin\theta \ ,} \tag{1-5}$$

where θ and θ' are the angles of incidence and refraction of the incident and refracted rays from the surface normal at the point of incidence (see Figure 1-3b). According to the Cartesian sign convention introduced later in Section 1.3.2, both of these angles are numerically negative. The incident ray, the refracted ray, and the surface normal are coplanar.

c. Reflection

At an interface between two media, an incident light ray is (also) reflected according to

$$\boxed{\theta' = -\theta \ ,} \tag{1-6}$$

where θ and θ' are the angles of incidence and reflection that the incident and reflected rays make with the surface normal at the point of incidence (see Figure 1-3c), respectively. According to the Cartesian sign convention outlined later in Section 1.3.2, the angle θ is numerically negative. The incident ray, the reflected ray, and the surface normal are coplanar.

d. Derivation of the Laws of Refraction and Reflection

To obtain the laws of refraction and reflection from Fermat's principle, we consider the optical path length of a ray in going from a point P_1 to another point P_2 after refraction at an interface between media of refractive indices n_1 and n_2 at a point A as in Figure 1-3b, or reflection as in Figure 1-3c. It is given by

$$[P_1AP_2] = n_1\left(a^2 + x^2\right)^{1/2} + n_2\left[(b-x)^2 + c^2\right]^{1/2} \quad \text{(Refraction)} \tag{1-7a}$$

or

$$[P_1AP_2] = \left(a^2 + x^2\right)^{1/2} + \left[(b-x)^2 + c^2\right]^{1/2} \ . \quad \text{(Reflection)} \tag{1-7b}$$

If we displace point A by a small amount along the interface, the value of x changes by that amount. According to Fermat's principle, the corresponding change in the optical path length in the limit of zero displacement is zero in the sense of calculus of variations, i.e., the derivative of the optical path length with respect to x is zero. Equating to zero the derivative of the right-hand side of Eqs. (1-7a) and (1-7b) with respect to x yields the laws of refraction and reflection, respectively. It should be noted that in Figures 1-3b and 1-3c, the incident ray, the refracted or reflected ray, and the surface normal are coplanar. A rigorous derivation of the laws of refraction and reflection from Fermat's principle is given by Klein.[1]

1.2.3   Optical Path Lengths of Neighboring Rays

We have seen from Fermat's principle that the optical path length of an actual ray from one point in space to another is equal to that of a neighboring virtual ray, at least up to the first order in their separation, e.g., $P_1A = P_1B$ to first order in ϵ in Figure 1-3a. Similarly, the optical path lengths of two actual but neighboring rays between planes that are perpendicular to one or both of them are equal to each other. We now consider refraction of two neighboring rays and show that their optical path lengths between planes that are perpendicular to one or both of them are equal at least up to the first order in their separation.

Consider a ray PQ incident on a spherical surface VQB of radius of curvature R with its center of curvature at C separating media of refractive indices n and n' as shown in Figure 1-4. The ray is refracted as a ray QP' so that the angles of incidence θ and refraction θ' are related to each other by Snell's law according to

$$n' \sin\theta' = n \sin\theta \quad .$$

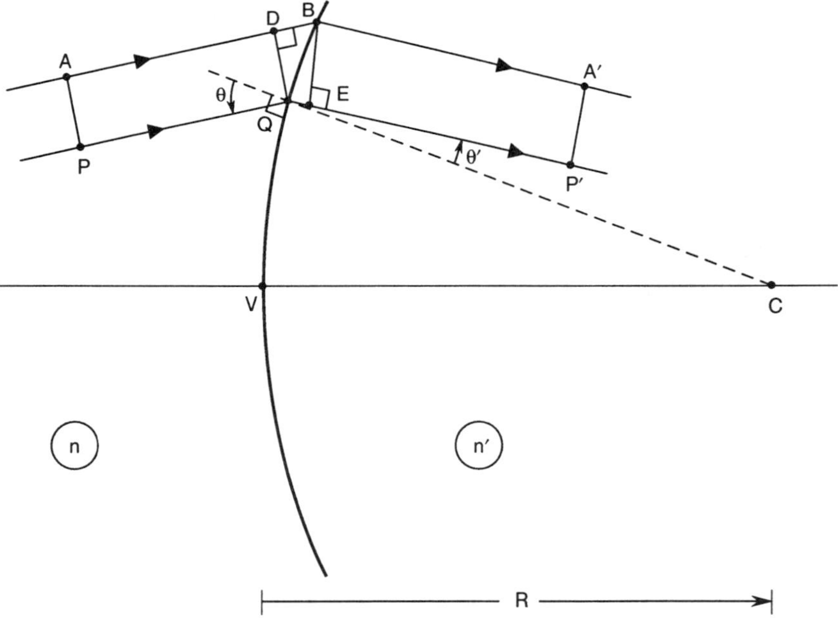

Figure 1-4. Two neighboring rays refracted by a spherical refracting surface of radius of curvature R separating media of refractive indices n and n'.

Another ray such as ABA' is considered to be a neighboring ray to the ray PQP' if the angle between their corresponding parts is small, and the distance between them is small compared with R. Let AP and $A'P'$ be perpendiculars to one of the rays. We now show that the optical path lengths $[PQP']$ and $[ABA']$ are equal to each other up to the first order in QB. Let QD be perpendicular to AB and BE be perpendicular to QP'. Since QB is small compared to R for neighboring rays, we may take QB to be perpendicular to the surface normal QC at Q. Thus, the angle DQB is equal to θ and the angle EQB is equal to $\pi/2 - \theta'$. Moreover, for neighboring rays, $PQ = AD$ and $EP' = BA'$. Hence, the difference in the two optical path lengths may be written

$$
\begin{aligned}
[PQP'] - [ABA'] &= [DB] - [QE] \\
&= QB\left(n\sin\theta - n'\sin\theta'\right) \\
&= 0 \quad .
\end{aligned}
$$

This result can be applied to each surface of an optical imaging system, yielding the result that the optical path lengths along two neighboring rays measured between planes that are perpendicular to one (or both) of them are equal.

1.2.4 Malus-Dupin Theorem

A surface passing through the end points of rays which have traveled equal optical path lengths from a point object is called an *optical wavefront*. The Malus-Dupin theorem states that a set of rays that are orthogonal to a wavefront remain orthogonal to a wavefront after refraction by a refracting surface. In a uniform medium, the wavefronts of the rays emanating from a point object are spherical and the rays are orthogonal to them at the points of their intersection. By using Fermat's principle, we show that they remain orthogonal to a wavefront after refraction by a refracting surface.

Let W be a spherical wavefront of rays emanating from a point object P. When these rays are refracted by a surface separating media of refractive indices n and n' so that they all travel equal optical path lengths as illustrated in Figure 1-5, a wavefront W' is obtained. By definition of the wavefronts, we have

$$[AVA'] = [BQB'] \quad , \tag{1-8}$$

where V and Q are the points of incidence of two neighboring rays PAV and PBQ. From Fermat's principle, the optical path length $[AQA']$ of the virtual ray AQA' may be written

$$[AQA'] = [AVA'] + O\!\left(\epsilon^2\right) \quad . \tag{1-9a}$$

where $\epsilon = VQ$ is a small quantity. Substituting Eq. (1-9a) into Eq. (1-8), we obtain

$$[AQA'] = [BQB'] + O\!\left(\epsilon^2\right) \quad . \tag{1-9b}$$

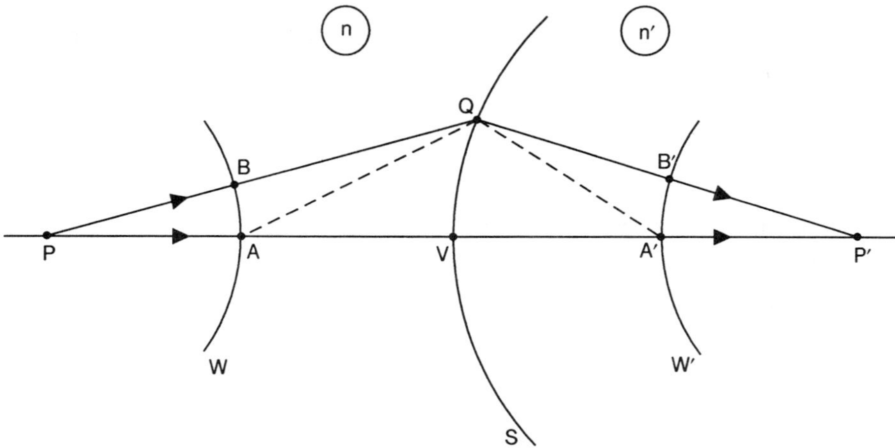

Figure 1-5. Refraction of a spherical wavefront W by a surface S separating media of refractive indices n and n', showing that rays such as BQ that are perpendicular to the wavefront W remain perpendicular to the wavefront W' after refraction.

Since BQ is perpendicular to the wavefront W at the point B,

$$[AQ] = [BQ] + O(\epsilon^2) \ ,$$

(1-10a)

where AB is of the same order of magnitude as VQ. Subtracting Eq. (1-10a) from Eq. (1-9b), we obtain

$$[QA'] = [QB'] + O(\epsilon^2) \ ,$$

(1-10b)

or the ray QB' is perpendicular to the wavefront W' at the point B'. If the wavefront W' is refracted by another refracting surface, the refracted rays and the wavefront produced by it can again be shown to be orthogonal to each other.

It should be noted that although the incident wavefront W is spherical with its center of curvature at P, the refracted wavefront W' may or may not be spherical depending on the shape of the refracting surface S. If W' is spherical with its center of curvature at P', then S is called a *Cartesian surface* and the points P and P' are called a *Cartesian pair* or *perfect conjugates*. Examples of Cartesian refracting surfaces are considered in Section 5.4 and Problems 5.1 and 5.2, and Cartesian reflecting surfaces are discussed in Section 6.2.1. In practice, however, we are interested in forming images of extended objects or point sources in a finite region of the object space. The task of a lens designer is to design systems with as few surfaces as possible yet that yield wavefronts in the image space that are close to being spherical over as wide a region of the object space as possible. As discussed in Section 3.2, the deviations of a wavefront from being spherical are called *wave aberrations* and the distances of the points of intersection of the rays from P' in an image plane are called *ray aberrations*.

1.2.5 Hamilton's Point Characteristic Function and Direction of a Ray

If we consider two points P and P' in media of refractive indices n and n', respectively, with 3-D position vectors $\vec{r}$ and $\vec{r}'$, the optical path length of a ray from P to P', called *Hamilton's point characteristic function*, may be written $V(P, P')$ or $V(\vec{r}, \vec{r}')$. Let $\hat{u}'$ be a unit vector along the ray through P'. Consider, as indicated in Figure 1-6, a neighboring point P'' with a position vector $\vec{r}' + \delta\vec{r}'$, where $\delta\vec{r}'$ is the vector $\overrightarrow{P'P''}$. The characteristic function for the points along a ray PBP'' is given by $V(\vec{r}, \vec{r}' + \delta\vec{r}')$. Regarding V as a function of $\vec{r}'$, we may write

$$V(\vec{r}, \vec{r}' + \delta\vec{r}') - V(\vec{r}, \vec{r}') = \delta\vec{r}' \cdot \nabla' V \quad , \tag{1-11}$$

where Δ' represents a gradient operator with respect to $\vec{r}'$. The left-hand side of Eq. (1-11) represents the optical path length of the ray segment $P'A$ (or BP'') lying between two wavefronts W' and W'' of rays from P passing through P' and P'', respectively. Now, $[P'A]$ is equal to $n'\hat{u}' \cdot \delta\vec{r}'$. Thus,

$$n'\hat{u}' \cdot \delta\vec{r}' = \delta\vec{r}' \cdot \nabla' V \quad ,$$

or

$$\boxed{n'\hat{u}' = \nabla' V} \quad . \tag{1-12a}$$

Thus, the unit vector along a ray through P' is given by the gradient of $V(P, P')$ with respect to its position vector $\vec{r}'$. Similarly, by considering a ray from P' to P (and a point in the neighborhood of P) and then reversing the ray path, we find that

$$\boxed{n\hat{u} = -\nabla V} \quad , \tag{1-12b}$$

where $\hat{u}$ is a unit vector along the ray (from P to P') through P.

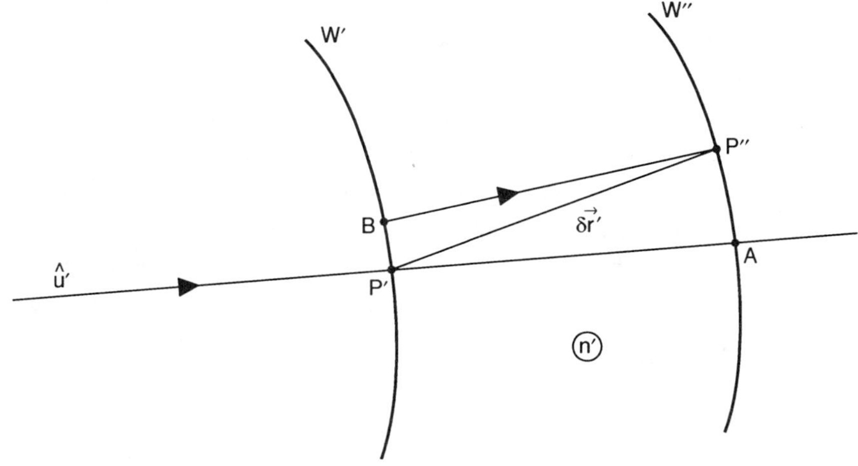

Figure 1-6. Optical path of a ray with unit vector $\hat{u}'$ in a medium of refractive index n'. W' and W'' are the wavefronts passing through two points P' and A on its path.

1.3 GAUSSIAN IMAGING

1.3.1 Introduction

An optical imaging system consists of a series of refracting and/or reflecting surfaces that refract and/or reflect the light rays from an object to form its image. Generally, the surfaces have a common axis of rotational symmetry referred to as the *optical axis*. Such a system is called a *centered* or a *rotationally symmetric system*. For light rays and surface normals to refracting and reflecting surfaces making small angles with the optical axis, Gauss gave an extremely useful approximation to the exact theory. In this approximation, the sines and tangents of the angles of the rays with the optical axis are replaced by the angles, and any diagonal distances are approximated by the corresponding axial distances. The image of an object obtained according to geometrical optics in the Gaussian approximation is called the *Gaussian image*. (The aberrations of the system are neglected in calculating this image.) Rays making small angles with the optical axis are called *paraxial*. Because of the rotational symmetry, only rays lying in the plane containing the optical axis and the point object under consideration need to be considered. Such a plane is called the *tangential* (or *meridional*) *plane* and rays lying in this plane are called *tangential* (or *meridional*) *rays*. Those rays that intersect this plane are called *skew rays*. The object is assumed to be on the left-hand side of the system so that initially light travels from left to right to form an image of the object.

We begin our discussion of Gaussian imaging with a brief introduction of the Cartesian sign convention for the distances and heights of object and image points, and the angles of incidence and refraction or reflection of the rays and their slope angles. We first derive imaging equations for a spherical refracting surface. Both Gaussian and Newtonian forms of the imaging equations are given. These are used to obtain the corresponding equations for a thin lens. The imaging equations for a multisurface refracting system are derived next. The principal, focal, and nodal points, collectively called the *cardinal points*, of such systems are discussed. It is shown that simple imaging equations, similar to those for a single refracting surface, are obtained provided the object and image distances are measured from the respective principal points of the system in the Gaussian form and from the focal points in the Newtonian form of the imaging equations. The concept of *Lagrange invariant* is discussed in each case. Afocal systems, i.e., those for which a parallel beam of light incident on them emerges as a parallel beam of light, or the object and its image both lie at infinity, are also discussed. Finally, the imaging equations for a spherical reflecting surface are derived. It is shown that they can be obtained from those for a corresponding refracting surface by simple substitutions of the refractive indices and taking into account the differences in sign convention for the two types of surfaces.

1.3.2 Sign Convention

Although there is no universally accepted standard sign convention, we will use the Cartesian sign convention. It has the advantage that there are no special rules to remember other than those of a right-handed Cartesian coordinate system. Our sign convention is the same as that used by Mouroulis and Macdonald,[2] but it is slightly different in its implementation from those of Born and Wolf,[3] Welford,[4] and Schroeder.[5] It is different from the sign convention used by the author in his *Aberration Theory Made Simple*, and used, for example, by Jenkins and White,[6] Klein and Furtak,[7] and Hecht and Zajac.[8] For convenience we list the rules of our sign convention as they apply to the quantities we will encounter in our discussion of Gaussian optics.

1. Light is incident on a system from left to right.

2. Distances to the right of and above (left of and below) a reference point are positive (negative).

3. The radius of curvature of a surface is treated as the distance of its center of curvature from its vertex. Thus, it is positive (negative) when the center of curvature lies to the right (left) of the vertex.

4. The acute angle of a ray from the optical axis or from the surface normal is positive (negative) if it is counterclockwise (clockwise).

5. When light travels from right to left, as when it is reflected by an odd number of mirrors, then the refractive index and the spacing between two adjacent surfaces are given a negative sign.

Throughout the book, any quantities that are numerically negative are indicated in the figures by a parenthetical negative sign $(-)$.

1.3.3 Spherical Refracting Surface

In this section, we derive equations that describe the imaging of an object by a spherical refracting surface. The concept of Lagrange invariance is introduced, showing that a large transverse magnification of an image is accompanied by a small angular magnification of the rays so that the product of the two magnifications is a constant.

1.3.3.1 Gaussian Imaging Equation

As indicated in Figure 1-7, consider a spherical refracting surface of a radius of curvature R separating media of refractive indices n and n'. The line VC joining its vertex V and center of curvature C defines its optical axis OA. Since the mirror is spherical, it does not have a unique vertex. However, for a given size mirror, its central point defines its vertex.

We first consider the imaging of an axial point object P_0 lying at a distance S from V. An object ray P_0Q incident at a point Q on the surface at a height x from the optical axis is refracted as a ray QP_0' intersecting the optical axis at a point P_0' at a distance S' from V. Let the angles of incidence and refraction (i.e., the angles of the incident and refracted rays from the surface normal QC at the point of incidence Q) be θ and θ',

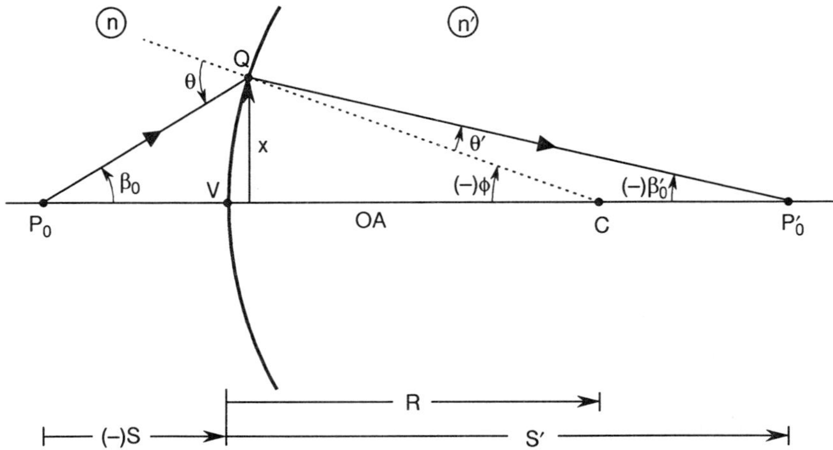

Figure 1-7. Gaussian imaging by a convex spherical refracting surface of radius of curvature R separating media of refractive indices n and n', where $n' > n$. VC is the optical axis of the surface, where V is the vertex of the surface and C is its center of curvature. The axial point object P_0 lies at a (numerically negative) distance S and its image P_0' lies at distance S' from V. The angles θ and θ' are the angles of the incident and refracted rays P_0Q and $P_0'Q$, respectively, from the surface normal at the point Q. They are both numerically negative. The slope angles of these rays from the optical axis are β_0 and β_0'. The angle β_0' is numerically negative. The angle ϕ of the surface normal at Q from the optical axis is also numerically negative.

respectively. Similarly, let the slope angles of these rays from the optical axis be β_0 and β_0', respectively.

In the Gaussian approximation of Snell's law (i.e., for small angles), the angles of incidence and refraction are related to each other according to

$$n'\theta' = n\theta \quad . \tag{1-13}$$

The rays propagating according to this approximation are called *paraxial rays*. From the triangle P_0CQ, we note that

$$\theta = \beta_0 - \phi \quad , \tag{1-14}$$

where the angle ϕ of the surface normal from the optical axis is numerically negative. Similarly, from triangle $CP_0'Q$, we note that

$$\theta' = \beta_0' - \phi \quad , \tag{1-15}$$

where β_0' is numerically negative. Substituting Eqs. (1-14) and (1-15) into Eq. (1-13) and noting that

$$\beta_0 = -x/S \quad , \tag{1-16}$$

$$\beta_0' = -x/S' \quad , \tag{1-17}$$

and

$$\phi = -x/R \ , \tag{1-18}$$

where the object distance S is numerically negative since P_0 lies to the left of V, we find that

$$\frac{n'}{S'} - \frac{n}{S} = \frac{n'-n}{R} \ . \tag{1-19}$$

We note that Eq. (1-19) is independent of the value of x, i.e., the height of the point of incidence Q of the ray. Thus, in the Gaussian approximation, all rays incident on the surface pass through P_0' after refraction by it. In reality, the refracted rays intersect the axis at slightly different points in the vicinity of P_0', which leads to aberrations as discussed in Chapter 5. Equation (1-19) is called the *Gaussian imaging equation.* It gives the position of the image point for a given position of the object point. The reference point for the object and image distances is the vertex V of the refracting surface. A point object such as P_0 and its corresponding Gaussian image point P_0' are called *conjugate points*.

In Figure 1-7, the object is real and its distance S from the refracting surface is numerically negative. If a beam of rays converging to a point on the right-hand side of the vertex V is incident on the refracting surface, the point object is considered *virtual* and its distance S from the surface is numerically positive. If, however, a real point object lies on the right-hand side of V at a distance S (as, for example, when considering the image of one element of a system by another that precedes it), so that light initially travels from right to left, then the object lies in a medium of refractive index n' and the image lies in a medium of index n. Since light is traveling backwards (from right to left), the signs of the refractive indices are reversed, i.e., they become negative quantities, and Eq. (1-19) is replaced by

$$\frac{n'}{S} - \frac{n}{S'} = \frac{n'-n}{R} \ , \tag{1-19$'$}$$

where S is numerically positive and a numerically positive value of S' implies that the image point lies on the right-hand side of V. Note that changing the sign of the refractive indices does not change the equation. An alternative and perhaps a simpler approach is to treat the real object on the right-hand side of V as an image and determine its conjugate object, giving the actual image. Thus, we let S' equal to the numerically positive object distance in Eq. (1-19) and determine the value of S. A numerically positive value of S implies that the actual image lies on the right-hand side of V. Of course, the system could be mentally rotated by 180 degrees about a vertical line passing through V, use an equation such as (1-19) in which S is numerically negative in a medium of index n', reverse the sign of R, determine S' in a medium of index n, and rotate the system back to its original configuration.

1.3.3.2 Focal Lengths and Refracting Power

If an object lies at infinity, i.e., if $S = -\infty$, then the corresponding image distance $S' \equiv VF' = f'$ (see Figure 1-8a), where f' is called the *image-space focal length* of the refracting surface. The point F' is called the *image-space focal point* of the surface. Rays incident on the surface parallel to its optical axis are focused at F' after refraction by it. Similarly, the object distance $S \equiv VF = f$ (see Figure 1-8b) for which the image lies at infinity (i.e., $S' = \infty$) is called the *object-space focal length*, where F is called the *object-space focal point*. Rays originating at F and incident on the surface are made parallel by it. Of course, if rays parallel to the optical axis are incident on the surface from right to left, they will be focused at F after refraction by it. The planes passing through the focal points F and F' that are perpendicular to the optical axis are called object-space and image-space focal planes, respectively. It should be evident from Figure 1-8 that the focal points F and F' are not conjugate points.

By their definitions, the image-space and object-space focal lengths of the refracting surface, obtained from Eq. (1-19), are given by

$$\boxed{f' = \frac{n'}{n'-n}R}$$

$$(1\text{-}20)$$

and

$$f = -\frac{n}{n'-n}R \ ,$$

$$(1\text{-}21)$$

respectively. If f' is numerically positive as in Figure 1-8a, then f is numerically negative (since F lies to the left of V) as in Figure 1-8b. The focal points F and F' lie on the opposite sides of the vertex V at different distances from each other.

The quantity on the right-hand side of Eq. (1-19) is called the *refracting power K* of the surface. It is a measure of the ability of the refracting surface to focus a parallel beam incident on it in as short a distance as possible. Its reciprocal is called the *equivalent* or *effective focal length f_e* of the surface. Thus, we may write

$$\boxed{K = \frac{n'-n}{R} = \frac{1}{f_e}}\ .$$

$$(1\text{-}22)$$

The power K and the equivalent focal length f_e are positive if $n'-n$ and R have the same sign. Such a surface is called a *positive* or a *converging surface*. Similarly, K and f_e are negative if $n'-n$ and R have opposite signs. Such a surface is called a *negative* or a *diverging surface*. We also note that $f_e = f'$ if $n' = 1$, i.e., the equivalent focal length represents the image-space focal length when the refractive index n' of the image space is unity. In terms of the refracting power and focal lengths, Eq. (1-19) may be written

$$\boxed{\frac{n'}{S'}-\frac{n}{S} = K = \frac{1}{f_e} = \frac{n'}{f'} = -\frac{n}{f}}\ .$$

$$(1\text{-}23)$$

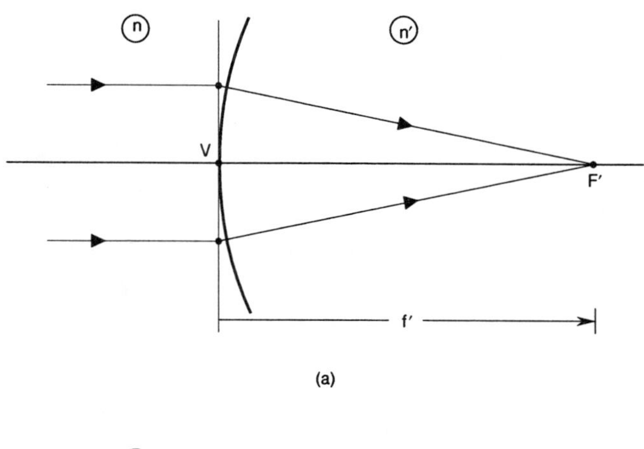

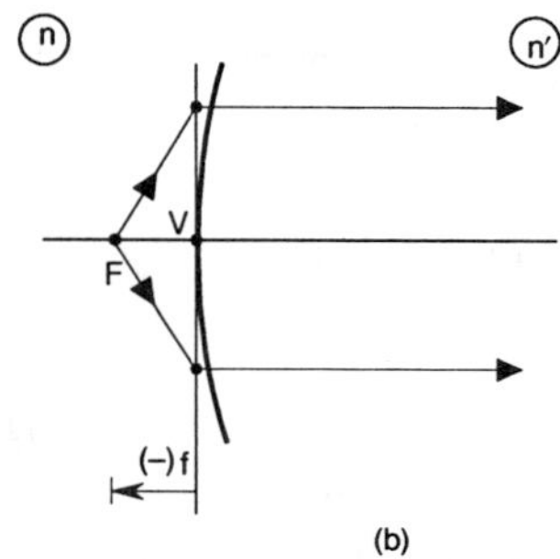

Figure 1-8. Focal points of a refracting surface. (a) F' is the image-space focal point. (b) F is the object-space focal point.

1.3.3.3 Magnifications and Lagrange Invariant

Now we consider the imaging of an off-axis point object P lying at a height h from the optical axis in the object plane passing through P_0, as illustrated in Figure 1-9. The incident and the refracted rays PV and VP', respectively, are shown in the figure passing through the vertex V. The image lies at the point P' where the refracted ray VP' intersects the image plane passing through P_0'. Both the object and the image planes are mutually parallel and perpendicular to the optical axis. It is evident from the figure that the angles of incidence and refraction from the surface normal at V, i.e., from the optical axis, are given by

$$\theta = h/S \tag{1-24a}$$

and

$$\theta' = h'/S \ , \tag{1-24b}$$

respectively. Note that θ, θ', and h' are all numerically negative. Substituting Eqs. (1-24) into the Snell's law equation (1-13), we find that the *transverse magnification* of the image is given by

$$\boxed{M_t = \frac{h'}{h} = \frac{nS'}{n'S}} \ . \tag{1-25a}$$

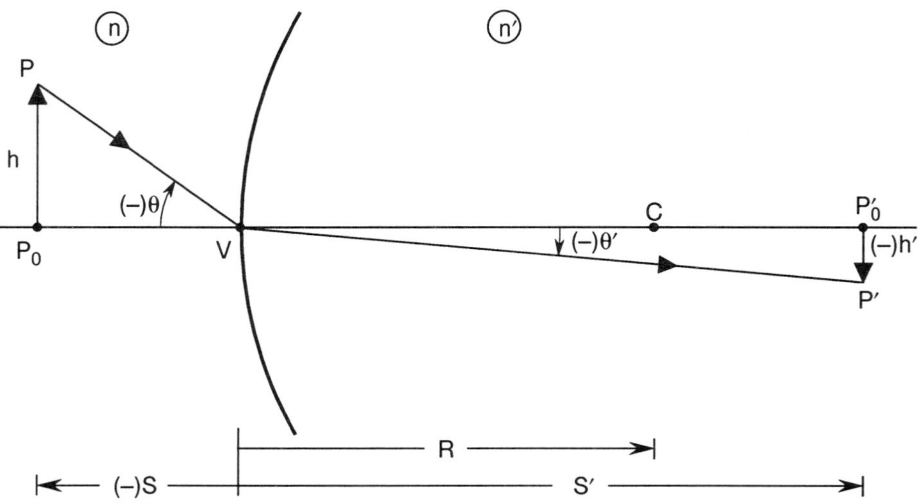

Figure 1-9. Gaussian imaging of an off-axis point object *P* lying at a height *h* from the optical axis. The image lies at *P′* at a height *h′*.

For an object lying at infinity ($S = -\infty$), $M_t = 0$. For an object lying between infinity and the object-space focal plane, $M_t < 0$, i.e., the image is inverted. As the object approaches the object-space focal plane, $M_t \to \infty$. For an object lying between the object-space focal plane and the surface, $M_t > 0$, or the image is virtual (in that the rays appear to diverge from it rather than converge upon it) and erect. As the object approaches the surface, the image also approaches it with $M_t = 1$.

The *ray angular magnification*, representing the ratio of the angular divergence of the rays from P_0 and their angular convergence to P_0' (see Figure 1-7) is given by

$$\boxed{M_\beta \ = \ \beta_0' / \beta_0 \ = \ S / S' \ .}$$
$$(1\text{-}26)$$

Note that M_β is not the ratio of the angular sizes θ' and θ of the image $P_0'P'$ and the object P_0P, respectively, subtended at V in Figure 1-9. From Eqs. (1-25a) and (1-26), we find that the product of the transverse and angular magnifications is given by

$$\boxed{M_t M_\beta \ = \ n / n' \ ,}$$
$$(1\text{-}27)$$

which depends only on the refractive indices of the object and image spaces. In particular, it does not depend on the object and image distances. It has the consequence that a large transverse magnification of the image can be obtained only with a correspondingly small angular magnification of the rays, i.e., by having a much smaller angular divergence of the rays at the image than at the object. From the definitions of the magnifications, namely, Eqs. (1-25a) and (1-26), Eq. (1-27) can also be written

$$\boxed{n' h' \beta_0' \ = \ n h \beta_0 \ ,}$$
$$(1\text{-}28)$$

showing that the quantity $nh\beta_0$ does not change upon refraction (see Figure 1-10). This quantity is called the *Lagrange* (or the *Smith-Helmholtz*) *invariant*. (As stated following Eq. (1-69), the square of the Lagrange invariant is proportional to the flux entering or exiting from an imaging system and the Lagrange invariance of Eq. (1-28) is related to the conservation of energy in the imaging process.) From Eq. (1-28), the transverse magnification of the image can also be written

$$\boxed{M_t \;=\; \frac{n\beta_0}{n'\beta_0'}}\; . \tag{1-25b}$$

i.e., it can be obtained from the slope angles of the incident and refracted rays for an axial object point.

For a small change ΔS in the object distance, let the corresponding change in the image distance be $\Delta S'$, as illustrated in Figure 1-11. The ratio $\Delta S'/\Delta S$ is called the *longitudinal magnification M_l*. Differentiating both sides of Eq. (1-19), we find that

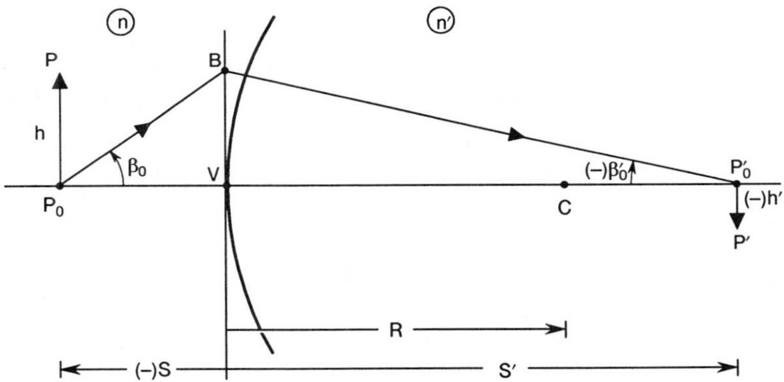

Figure 1-10. Lagrange invariant $nh\beta_0$ of a refracting surface.

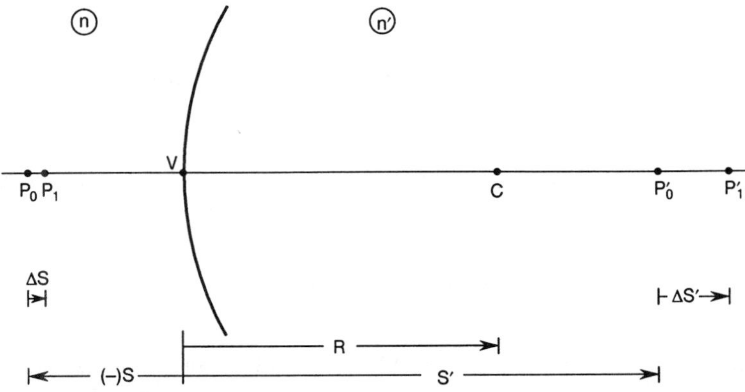

Figure 1-11. Imaging of an object lying along the optical axis and illustration of longitudinal magnification. As the object is moved closer to the surface from P_0 to P_1 by an amount ΔS, the image moves farther from P_0' to P_1' by an amount $\Delta S'$ as indicated.

$$M_l \; = \; \Delta S'/\Delta S \; = \; (n/n')(S'/S)^2 \; = \; (n'/n)M_t^2 \; = \; M_t \, / \, M_\beta \quad . \tag{1-29}$$

Whether the transverse magnification M_t is positive or negative, the longitudinal magnification M_l is always positive, indicating, for example, that if the object distance S increases (from a larger negative value to a smaller one), i.e., if the object moves closer to the refracting surface, then the image distance S' also increases (from a smaller positive value to a larger one), i.e., the image moves farther from the surface. Thus the image moves in the same direction as the object. Since the value of M_t varies with the position of the object, M_l also varies with it. Hence, Eq.(1-29) is valid only for infinitesimal values of ΔS. For finite values, it can be shown that $M_l = (n'/n)M_0 M_1$, where M_0 and M_1 are the transverse magnifications of the images lying in planes passing through P_0' and P_1', respectively. Thus, for example, the image of a cube is a truncated pyramid which is approximately a rectangular parallelepiped if the cube is infinitesimal. It is shown in Section 1.3.6 that for an afocal telescope, M_t and, therefore, M_l also are independent of the position of the object. It should be evident that the longitudinal magnification also represents the magnification of the image of a small axial object.

In Eq. (1-29), the refracting surface is assumed to be fixed in position and $\Delta S'$ represents the displacement of the image corresponding to a displacement ΔS of the object. However, if the object is fixed and the refracting surface is displaced by an amount Δ, then the corresponding displacement of the image is given by $\left(1 - n'M_t^2/n\right)\Delta$.

1.3.3.4 Graphical Imaging

The location of the Gaussian image P' can be determined graphically as the point of intersection of any two of the following three conveniently drawn rays from the point object P, as illustrated in Figure 1-12.

1. Ray 1 incident parallel to the optical axis passes through the image-space focal point F' after refraction.

2. Ray 2 incident in the direction of the center of curvature C of the refracting surface is refracted by it without any deviation. This is because the angle of incidence of the ray is zero; hence, the angle of the refracted ray is also zero.

3. Ray 3 incident passing through the object-space focal point F is refracted parallel to the optical axis.

Extension of one or more of these rays may be necessary for them to intersect with each other. Moreover, in Gaussian optics, which is based on the paraxial rays, any refraction (or reflection) at a surface takes place at a plane that is tangent to it at its vertex as shown, for example, in Figures 1-10 and 1-12. The tangent plane is sometimes called the *paraxial refracting surface*.

The Gaussian image P_0' of an on-axis point object P_0 can be determined independently (rather than as the point of intersection of the optical axis and the line that is perpendicular to it and passes through P') as follows: Consider a ray P_0E incident on the surface as shown in Figure 1-13. A hypothetical ray incident parallel to it and passing through C intersects the image-space focal plane at a point D. The refracted ray corresponding to the incident ray P_0E passes through the point D and intersects the optical axis at the Gaussian image point P_0'. The point D may also be determined by considering a hypothetical parallel ray passing through the object-space focal point F. It is refracted as a ray parallel to the optical axis intersecting the focal plane at the point.

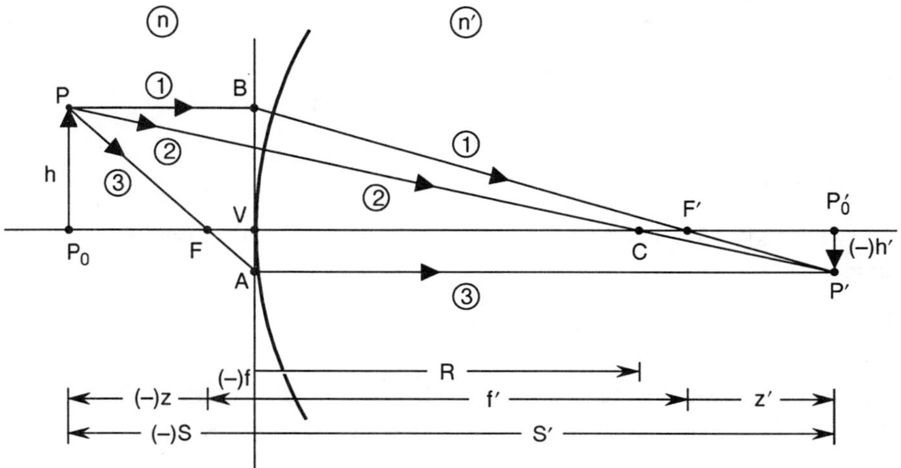

Figure 1-12. Graphical Gaussian imaging of an object P_0P by a spherical refracting surface of radius of curvature R separating media of refractive indices n and n'. $P_0'P'$ is the Gaussian image.

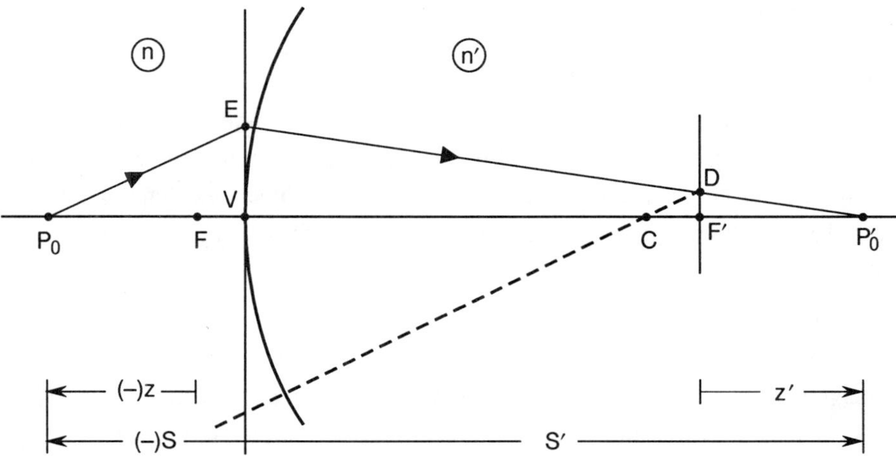

Figure 1-13. Graphical Gaussian imaging of an axial point object P_0 by a spherical refracting surface .

1.3.3.5 Newtonian Imaging Equation

In the Gaussian imaging equation (1-19), the object and image distances S and S', respectively, are measured from the vertex V of the refracting surface. In the Newtonian imaging equation, they are measured from the respective focal points F and F'. Thus, let z and z' be the object and image distances from the focal points F and F', respectively, as indicated in Figure 1-12. From similar triangles P_0FP and FVA in this figure, we note that the transverse magnification may be written

$$\boxed{M_t \ = \ h'/h = \ -f/z \quad ,}\tag{1-30}$$

where z (like f) is numerically negative since P_0 lies to the left of F. Similarly, from similar triangles $VF'B$ and $F'P_0'P'$, it may also be written

$$\boxed{M_t \ = \ -z'/f' \quad .}\tag{1-31}$$

Equating the right-hand sides of these equations, we obtain the *Newtonian imaging equation*:

$$\boxed{zz' = ff' \quad .}\tag{1-32}$$

Differentiating both sides of Eq. (1-32) and using Eqs. (1-30), (1-31), 1-20), and (1-21), we obtain Eq. (1-29) relating the longitudinal and transverse magnifications.

1.3.4 Thin Lens

A thin len*s* consists of two refracting surfaces such that their separation is negligible compared to its focal length and the object and image distances. Now we use the results of the last section to derive the imaging equations for such a lens. We also show that a ray incident toward the center of a thin lens emerges from it undeviated and undisplaced after refraction by it.

1.3.4.1 Gaussian Imaging Equation

Consider a lens made of a material with a refractive index n_l in a medium of refractive index n_m, as illustrated in Figure 1-14. Let the radii of curvature of the refracting surfaces be R_1 and R_2. Consider an axial point object P_0 lying at a distance S_1 from the lens. Its image P_0' formed by the first surface lies at a distance S_1' which, according to Eq. (1-19), is given by

$$\frac{n_l}{S_1'} - \frac{n_m}{S_1} \ = \ \frac{n_l - n_m}{R_1} \quad .\tag{1-33}$$

This image is the *virtual object* for the second surface lying at a distance $S_2 = S_1'$. (An object is virtual when the rays associated with it appear to converge to it rather than actually diverge from it.) The image P_0'' of the object P_0' formed by the second surface lies at a distance S_2' which, according to Eq. (1-19), is given by

$$\frac{n_m}{S_2'} - \frac{n_l}{S_1'} \ = \ \frac{n_m - n_l}{R_2} \quad .\tag{1-34}$$

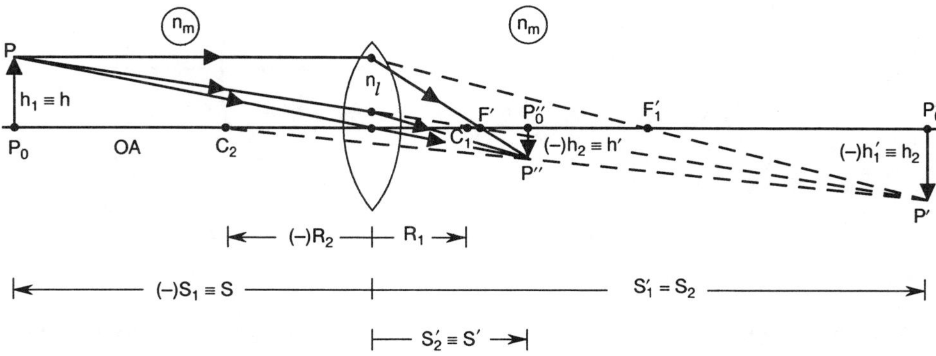

Figure 1-14. Paraxial imaging by a thin lens of refractive index n_l in a medium of refractive index n_m. The radii of curvature of the lens surfaces are R_1 and R_2.

Adding Eqs. (1-33) and (1-34), we obtain

$$\frac{1}{S'} - \frac{1}{S} = \frac{n_l - n_m}{n_m}\left(\frac{1}{R_1} - \frac{1}{R_2}\right) , \tag{1-35}$$

where we have let $S_1 = S$ and $S'_2 = S'$ be the object and final image distances, as indicated in Figure 1-14.

1.3.4.2 Focal Lengths and Refracting Power

When the object lies at infinity, i.e., when $S = -\infty$, then $S' = f'$, where f' given by

$$\frac{1}{f'} = \frac{n_l - n_m}{n_m}\left(\frac{1}{R_1} - \frac{1}{R_2}\right) \tag{1-36}$$

is called the *image-space focal length* of the lens. Thus, a ray incident on the lens parallel to its optical axis passes through the *image-space focal point F'* upon refraction, as illustrated in Figure 1-15a. Similarly, when $S = f = -f'$, then $S' = \infty$ and a ray from the *object-space focal point F* incident on the lens emerges from it parallel to its optical axis upon refraction, as illustrated in Figure 1-15b. The focal points F and F' lie on the opposite sides of the lens at equal distances from it. This would not be the case if the refractive indices of the object and image spaces of the lens were different.

The right-hand side of Eq. (1-35) represents the *refracting power K* of the lens. Its reciprocal is called the *equivalent* or *effective focal length f_e* of the lens. Thus, we may write

$$K = \frac{n_l - n_m}{n_m}\left(\frac{1}{R_1} - \frac{1}{R_2}\right) = \frac{1}{f_e} = \frac{1}{f'} . \tag{1-37}$$

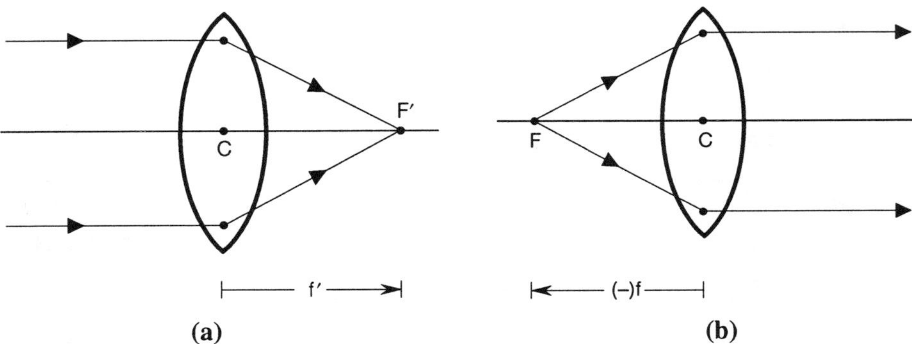

Figure 1-15. Focal points of a thin lens. (a) F' is the image-space focal point. (b) F is the object-space focal point. C is the center of the lens.

We note that the refracting power of the lens is equal to the sum of the refracting powers K_1 and K_2 of its two surfaces, i.e.,

$$K = K_1 + K_2 \quad , \tag{1-38}$$

where

$$K_1 = \frac{n_l - n_m}{R_1} \tag{1-39}$$

and

$$K_2 = \frac{n_m - n_l}{R_2} \quad . \tag{1-40}$$

Generally, the lens is used in air, i.e., $n_m = 1$, in which case Eq. (1-37) reduces to

$$\boxed{K = (n-1)\left(\frac{1}{R_1} - \frac{1}{R_2}\right) = \frac{1}{f_e} = \frac{1}{f'} \quad ,} \tag{1-41}$$

where $n = n_l$ is the refractive index of the lens. Equation (1-41) is called the *lens maker's formula*. In terms of the refracting power and the focal lengths of the lens, Eq. (1-35) for imaging may be written

$$\boxed{\frac{1}{S'} - \frac{1}{S} = K = \frac{1}{f_e} = \frac{1}{f'} = -\frac{1}{f} \quad .} \tag{1-42}$$

A lens with a positive value of K or f_e or f' is called a *converging* or a *positive lens*. Similarly, a lens with a negative value of K or f_e or f' is called a *diverging* or a *negative lens*.

1.3.4.3 Undeviated Ray

We now show that a ray passing through the center of a thin lens remains undeviated and undisplaced upon refraction. As indicated in Figure 1-16, the center O of a lens is defined to be that point on the optical axis for which

$$\frac{R_1}{R_2} = \frac{OC_1}{OC_2} \quad , \tag{1-43}$$

where C_1 and C_2 are the centers of curvature of its surfaces. An axial ray passes through the lens center undeviated since it is perpendicular to both surfaces of the lens (the angles of incidence and refraction are zero at each of the two surfaces).

Consider a ray traversing the path AB through the lens center O. We now show that such a ray emerges from the lens in the same direction as that of its incidence. Let the angle of incidence of this ray be θ_m. Then, according to Snell's law for paraxial rays,

$$n_m \theta_m = n_l \theta_l \quad , \tag{1-44}$$

where θ_l is the angle of refraction of the ray at the first surface. Let θ_l' and θ_m' be the angles of incidence and refraction of the ray at the second surface. Once again, according to Snell's law,

$$n_l \theta_l' = n_m \theta_m' \quad . \tag{1-45}$$

Now $AC_1 = R_1$ and $BC_2 = R_2$; hence, Eq. (1-43) may be written

$$\frac{AC_1}{BC_2} = \frac{OC_1}{OC_2} \quad , \tag{1-46}$$

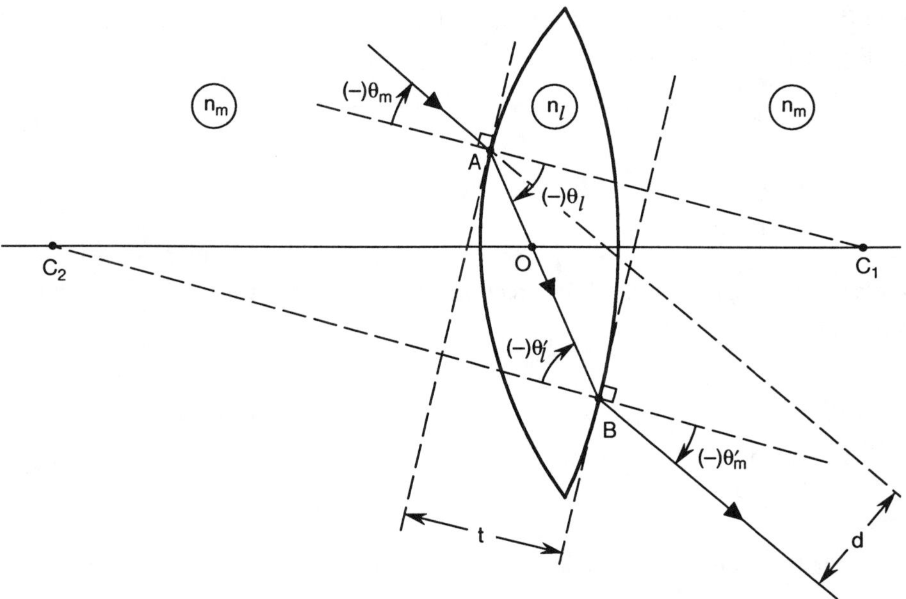

Figure 1-16. Input and output segments of a ray passing through the optical center O of a lens are parallel. In the case of a thin lens, a ray passing through its center remains undeviated and undisplaced. All of the four angles in the figure are numerically negative.

showing that the triangles OC_1A and C_2OB are similar. Hence, AC_1 is parallel to C_2B, and, therefore, $\theta_l = \theta_l'$. As a consequence, we find from Eqs. (1-44) and (1-45) that $\theta_m = \theta_m'$, implying that the incident and the emergent rays are parallel to each other. These rays when extended intersect the optical axis at the nodal points of the lens, which are discussed in Section 1.3.5.3.

Since the surface normals AC_1 and BC_2 at the points A and B are parallel, the tangent planes at these points are also parallel. Hence, the lens acts as a plane-parallel plate for the ray under consideration, with the consequence that the lateral displacement of the ray is given by

$$
\begin{aligned}
d &= AB\sin(\theta_m - \theta_l) \\[2mm]
&= \frac{t}{\cos\theta_l}\sin(\theta_m - \theta_l) \quad,
\end{aligned}
\tag{1-47}
$$

where t is the distance between the tangent planes at the points A and B. Since, by definition, t is negligible for a thin lens, the displacement d of the ray is also negligible. Hence, a ray incident in the direction of the center O of a thin lens remains undeviated and undisplaced after refraction by it.

1.3.4.4 Magnifications and Lagrange Invariant

The transverse magnification of the image formed by the lens can be obtained by applying Eq. (1-25) to the images formed by its two surfaces. Thus, using Eq.(1-25), the magnification of the inverted image $P_0'P'$ of the object P_0P formed by the first surface (see Figure 1-14) is given by

$$
M_1 = h_1' / h_1
\tag{1-48a}
$$

$$
= \frac{n_m S_1'}{n_l S_1} \quad.
\tag{1-48b}
$$

Similarly, the magnification of the erect image $P_0''P''$ of the object $P_0'P'$ formed by the second surface is given by

$$
M_2 = h_2' / h_2 = h_2' / h_1'
\tag{1-49a}
$$

$$
= \frac{n_l S_2'}{n_m S_1'} \quad.
\tag{1-49b}
$$

Hence, the transverse magnification of the $P_0''P''$ of the object P_0P formed by the lens as a whole is given by

$$
\begin{aligned}
M_t &= M_1 M_2 \\[1mm]
&= h_2' / h_1 \\[1mm]
&= S_2' / S_1 \quad,
\end{aligned}
\tag{1-50}
$$

or

$$M_t = h'/h = S'/S \ ,$$

(1-51a)

where we have let $h = h_1$ and $h' = h'_2$ be the object and final image heights, respectively.

The Gaussian image of a point object can be located graphically, as illustrated in Figure 1-17, in the same manner as in the case of a refracting surface except that a ray through the center of curvature of the surface is replaced by one through the center of the lens. Figure 1-17 is similar to Figure 1-14 except that the two-step imaging (one for each surface) has been replaced by a single-step imaging. The transverse magnification given by Eq. (1-51) is immediately evident from the similar triangles P_0CP and P'_0CP' in Figure 1-17.

The *angular magnification* of a ray bundle diverging from the axial point object P_0 and converging toward its image P'_0 (see Figure 1-18) is given by

$$M_\beta = \beta'_0/\beta_0 = S/S' \ .$$

(1-52)

From Eqs. (1-51a) and (1-52), we find that the product of the transverse magnification of the image and the angular magnification of the ray bundle for a thin lens in a medium of refractive index n_m is given by

$$M_t M_\beta = 1 \ .$$

(1-53)

From the definitions of the magnifications, Eq. (1-53) can also be written

$$h'\beta'_0 = h\beta_0 \ ,$$

(1-54)

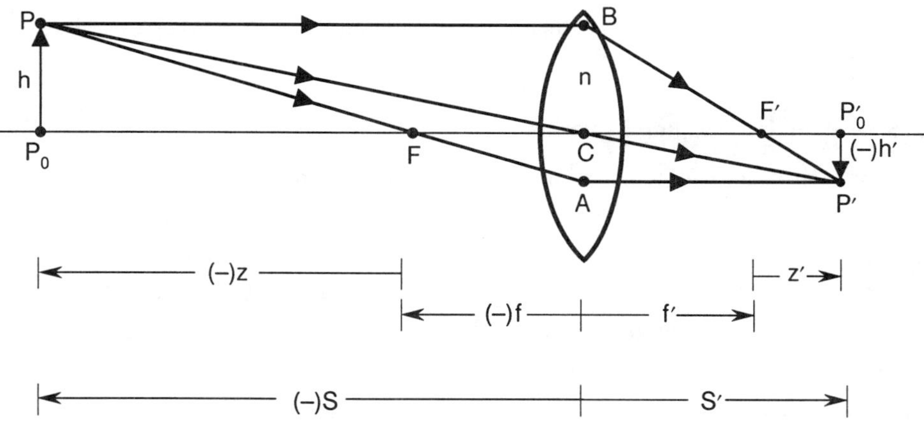

Figure 1-17. Paraxial imaging by a lens of refractive index n and focal length f'. Compared with Figure 1-14, the two-step imaging (one for each surface) has been replaced by a single-step imaging.

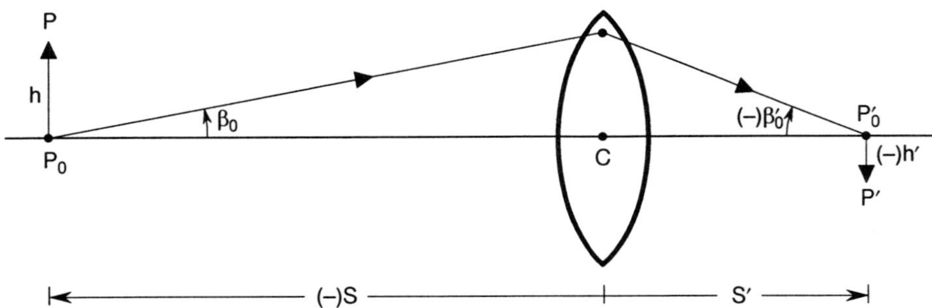

Figure 1-18. Lagrange invariant $h\beta_0$ of a thin lens.

showing that the quantity $h\beta_0$ is invariant upon refraction by the lens. It is the *Lagrange invariant* of the lens illustrated schematically in Figure 1-18. From Eq. (1-54), the transverse magnification of the image can also be written

$$M_t = \beta_0/\beta_0' \quad , \tag{1-51b}$$

i.e., it is given by the ratio of the slope angles of the incident and refracted rays for an axial point object.

Differentiating both sides of Eq. (1-35), we obtain the longitudinal magnification of the image,

$$M_l = \Delta S'/\Delta S = \left(S'/S\right)^2 = M_t^2 = M_t/M_\beta \quad . \tag{1-55}$$

The comments made following Eq. (1-29) apply to Eq. (1-55) as well. In Eq. (1-55), the lens is assumed to be fixed in position and $\Delta S'$ represents the displacement of the image corresponding to a displacement ΔS of the object. However, if the object is fixed and the lens is displaced by an amount Δ, then the corresponding displacement of the image is given by $\left(1 - M_t^2\right)\Delta$.

1.3.4.5 Newtonian Imaging Equation

In the Gaussian imaging equation (1-35), the object and image distances S and S', respectively, are measured from the lens center. In the corresponding *Newtonian imaging equation*, they are measured from the respective focal points. Thus, as indicated in Figure 1-17, let z and z' be the object and image distances from the focal points F and F', respectively. From similar triangles P_0FP and FCA, we note that the transverse magnification of the image can be written

$$M_t = h'/h = -f/z \quad . \tag{1-56}$$

Similarly, from similar triangles $CF'B$ and $P_0'F'P'$, it may also be written

$$h'/h = -z'/f' \quad . \tag{1-57}$$

The negative sign on the right-hand sides of Eqs. (1-56) and (1-57) has been introduced since M_t in Figure 1-17 is numerically negative due to h' being numerically negative. From Eqs. (1-56) and (1-57), we obtain

$$\boxed{z\,z' = f\,f' = -f'^2} \qquad (1\text{-}58)$$

which is the *Newtonian imaging equation*.

1.3.5 Refracting Systems

In this section, we discuss the cardinal points of a general imaging system and derive equations that govern its Gaussian imaging properties. A thick lens consists of two refracting surfaces whose separation along the optical axis is nonnegligible. Such lenses are necessary, for example, to obtain short focal lengths. A compound lens is made up of several thick and/or thin lenses. Such lenses are necessary if the aberrations, i.e., ray deviations from their Gaussian behavior, of single thin lenses are to be corrected. The Gaussian imaging properties of a complex lens system can be obtained by repeating the application of imaging by a single refracting surface, as we did in the case of imaging by a thin lens in Section 1.3.4. However, we now show that in many ways, Gaussian imaging equations for any imaging system can be reduced to those for a single refracting surface.

1.3.5.1 Cardinal Points and Planes

A general imaging system is characterized by *six cardinal points*: two *principal points*, two *focal points* and two *nodal points*. The planes normal to the optical axis and passing through these points are called *principal planes*, *focal planes*, and *nodal planes*, respectively. The location of the principal and focal points is sufficient to describe Gaussian imaging by the system. As illustrated in Figure 1-19a, the *image-space focal point F'* of a system is defined as the point through which rays incident parallel to its optical axis from the left pass after refraction by it. The rays converging toward F' when extended backward intersect the incident parallel rays in a plane called the *image-space principal plane*. This plane intersects the optical axis at a point H' called the *image-space principal point*. The rays behave as if all of their deviation takes place at the principal plane. The distance $H'F'$ of the focal point F' from the principal point H' is called the *image-space focal length f'*.

The *object-space focal point F* shown in Figure 1-19b is defined as the axial point such that the rays originating from it and incident on the system emerge from the system parallel to its optical axis after refraction by it. The rays originating from F when extended forward intersect the emergent parallel rays in a plane called the *object-space principal plane*. This plane intersects the optical axis at a point H called the *object-space principal point*. The rays behave as if all of their deviation takes place at the principal plane. The distance HF of the focal point F from the principal point H is called the *object-space focal length f*.

The *principal planes are planes of unit transverse magnification.* As illustrated in Figure 1-20, consider a system whose focal points are F and F'. A ray 1 incident in the direction AQ parallel to the optical axis emerges from the system passing through F', and the extensions of the incident and emergent rays intersect at a point Q'. A second ray 2 incident on the system passing through F emerges from it in the direction $Q'A'$ parallel to the optical axis and the extensions of the incident and emergent rays intersect at a point Q. Thus, the two rays initially directed toward Q emerge in directions that intersect at Q'.

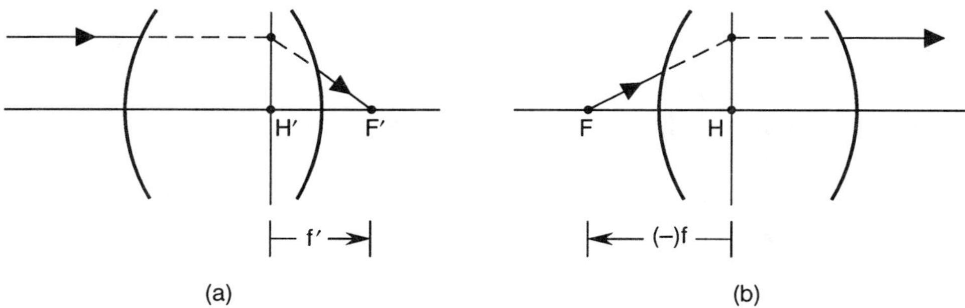

Figure 1-19. Principal and focal points of an imaging system. (a) The image-space focal point F' and principal point H'. (b) The object-space focal point F and principal point H.

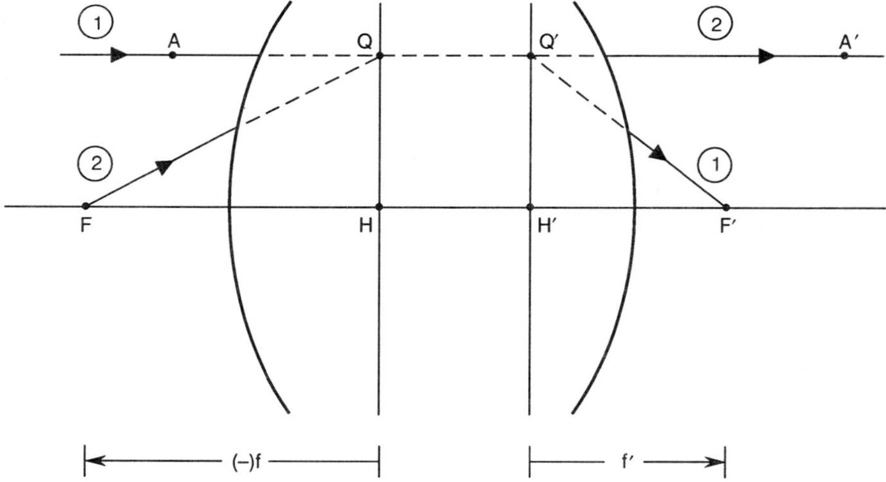

Figure 1-20. Unit transverse magnification of principal planes.

Hence, Q' is an image of Q and vice versa, i.e., Q and Q' are conjugate points. Similarly, the principal planes HQ and $H'Q'$ are conjugate planes. Since $HQ = H'Q'$, they are conjugate planes of unit (positive) transverse magnification.

Any imaging system is associated with object and image spaces. Its *object space* is the space that contains all the physical objects lying to its left and all the points that are conjugate to any physical objects lying to its right. Similarly, its *image space* is the space that contains all the physical objects lying to its right and all the points that are conjugate to any physical objects lying to its left. Of course the two spaces are conjugate of each other. Now, every object is associated with an image, which may be real or virtual lying on either side of the system. Hence, both the object and image spaces extend from infinity on the left of the system to infinity on its right. Thus, the two spaces are superimposed on each other. A distinction is made between the two spaces by considering rays before entering the system as lying in its object space and those emerging from it as lying in its image space. Sometimes, a distinction is made between a real and a virtual space. The portion of the object space lying on the left of a system is called its *real object space* and the portion of the image space lying on its right is called its *real image space*. The remaining portions are correspondingly called *virtual object* and *image spaces*.

1.3.5.2 Gaussian Imaging, Focal Lengths, and Magnifications

Figure 1-21 illustrates Gaussian imaging by a general optical system. Given the cardinal points of a system, the Gaussian image of a point object formed by it can be determined graphically in the same manner as in the case of a refracting surface except that the center of curvature of the surface is replaced by the nodal points of the system discussed later (see Figure 1-24). Moreover, because of the conjugate property of the principal planes, a ray incident in the direction of H emerges as if coming from H'. If n and n' are the refractive indices of the object and image spaces of the system, then repeated application of Eq. (1-27) for a refracting surface yields for a system

$$\boxed{M_t M_\beta = n/n'} \tag{1-59}$$

Since $M_t = 1$ for the principal planes, the angular magnification of a ray 3 incident in the direction of H making an angle β with the optical axis and appearing to emerge from H' making an angle β' is given by

$$M_\beta = \beta'/\beta \tag{1-60}$$
$$= n/n' \; .$$

Thus, for the principal planes,

$$n'\beta' = n\beta \; . \tag{1-61}$$

Note that both β and β' are numerically negative in Figure 1-21.

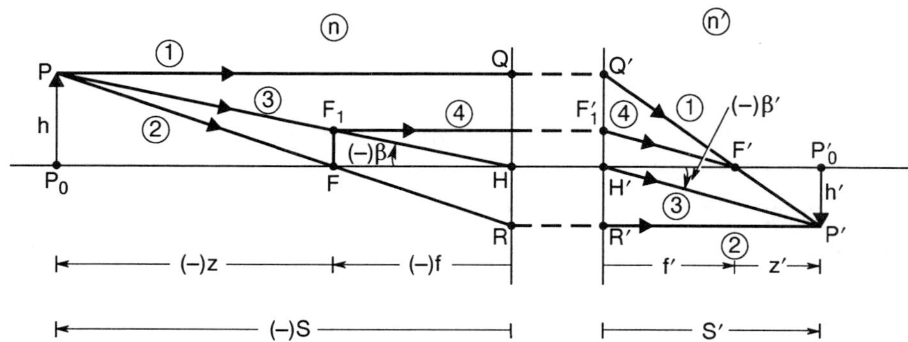

Figure 1-21. Imaging by a general optical imaging system.

Now consider a ray 4 such that it and ray 3 leave the left focal plane from the same point F_1. The image of F_1 is formed at infinity, i.e., the emergent rays $F_1'F'$ and $H'P'$ are parallel to each other and ray 4 passes through F' after refraction by the system. From the triangle FHF_1, we note that

$$\beta = FF_1/f \ . \tag{1-62a}$$

Similarly, since $FF_1 = H'F_1'$, we find from triangle $H'F'F_1'$ that

$$\beta' = -FF_1/f' \ , \tag{1-62b}$$

where we have introduced a negative sign on the right-hand side because β' is numerically negative. Substituting Eqs. (1-62) into Eq. (1-61) we obtain

$$\boxed{\frac{n'}{f'} = -\frac{n}{f}} \ . \tag{1-63}$$

The *transverse magnification* of the image $P_0'P'$ of the object P_0P is given by

$$\boxed{M_t = \frac{h'}{h} = \frac{\beta'S'}{\beta S} = \frac{nS'}{n'S}} \ . \tag{1-64}$$

Considering similar triangles P_0FP and FHR, we find that

$$\frac{h'}{h} = -\frac{f}{S-f} \ . \tag{1-65}$$

where the negative sign is due to h' being numerically negative. Note that f and S are both numerically negative. Comparing Eqs. (1-64) and (1-65), we obtain

$$\boxed{\frac{n'}{S'} - \frac{n}{S} = \frac{n'}{f'} = -\frac{n}{f}} \ . \tag{1-66}$$

The ratio of the image (object)-space refractive index and the image (object)-space focal length is called the *refracting power K* of the system. Its reciprocal represents the equivalent focal length f_e. Hence, Eq. (1-66) may be written

$$\boxed{\frac{n'}{S'} - \frac{n}{S} = \frac{n'}{f'} = -\frac{n}{f} = K = \frac{1}{f_e}} \quad . \tag{1-67}$$

When the focal length is expressed in meters, the unit of power is called a *diopter* and it is denoted by D.

The angular magnification of a ray bundle diverging from the axial point object P_0 and converging to its image point P_0' after refraction by the system, as illustrated in Figure 1-22, is given by

$$\boxed{M_\beta = \beta_0' / \beta_0 = S / S'} \quad , \tag{1-68}$$

where we have used the fact that $HQ = H'Q'$. From Eqs. (1-64) and (1-68), we obtain Eq. (1-59). From the definitions of the magnifications, Eq. (1-59) may also be written

$$\boxed{n'h'\beta_0' = nh\beta_0} \quad , \tag{1-69}$$

thus demonstrating the *Lagrange invariance* for the entire system. It is a paraxial approximation of the optical sine theorem discussed in Section 3.7. It can be shown that for a small object of radius h centered on the optical axis with an axial marginal ray slope angle β_0, the flux entering the system is proportional to $(nh\beta_0)^2$ (see Figure 2-2 and Problem 2.7). Similarly, the flux emerging from the system is proportional to $(n'h'\beta_0')^2$, where β_0' is the slope angle of the axial marginal ray in the image space. Hence, the Lagrange invariance of Eq. (1-69) is related to the conservation of energy in the imaging process. From Eq. (1-69), the transverse magnification of the image can also be written

$$\boxed{M_t = \frac{n\beta_0}{n'\beta_0'}} \quad , \tag{1-70}$$

i.e., it can be obtained from the slope angles of an axial incident ray and the corresponding refracted ray in the image space of the system.

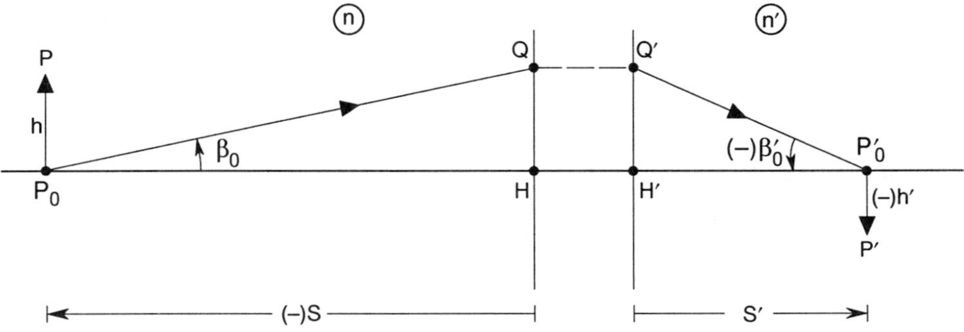

Figure 1-22. Lagrange invariant $nh\beta_0$ of an optical imaging system.

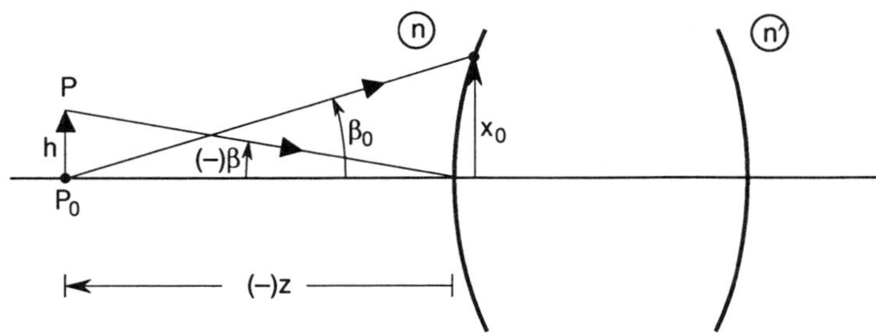

Figure 1-23. The Lagrange invariant $nh\beta_0 \rightarrow -nx\beta$ for an object lying at infinity at an angle β from the optical axis of a system. The object lies at a very large distance z; hence, it does not matter whether the reference point for this distance is the object-space principal point or the vertex of the first surface of the system.

When an object lies at infinity at a certain angle β from the optical axis, then $h \rightarrow \infty$ and $\beta_0 \rightarrow 0$, but the product $h\beta_0$ remains finite. Consider, as illustrated in Figure 1-23, an object lying at a very large distance z from an optical system. The location of the reference point for measuring the object distance does not matter since the distance is very large. An axial ray making an angle β_0 with its optical axis is incident on it at a height $x = -z\beta_0$. For an off-axis point object at a height h at an angle β from the optical axis, we have $h = \beta z$. Thus, $h\beta_0 = -x\beta$. For an object lying at infinity, $z \rightarrow -\infty$, $\beta_0 \rightarrow 0$ (i. e., the axial ray is incident parallel to the optical axis at a height x) and $h \rightarrow \infty$, but $h\beta_0$ is finite and equal to $-x\beta$. Hence, the object-space Lagrange invariant $nh\beta_0$ approaches $-nx\beta$. This expression is utilized in Section 1.3.6, where an afocal system working at infinite conjugates is discussed. It is rederived in Section 1.5 from a two-ray Lagrange invariant.

Differentiating Eq. (1-66), we find that the longitudinal magnification of the image is given by

$$\boxed{M_l = \Delta S'/\Delta S = (n/n')(S'/S)^2 = (n'/n)M_t^2 = M_t / M_\beta} \qquad (1\text{-}71)$$

The comments made following Eq. (1-29) apply to Eq. (1-71) also. Of course, Eq. (1-71) for a multisurface system is a generalization of Eq. (1-29) for the single-surface system.

1.3.5.3 Nodal Points

The *nodal points N and N'* indicated in Figure 1-24 *correspond to unit ray angular magnification*, i.e., a ray incident in the direction of N emerges parallel to it as if coming from N'. From the parallelogram $AA'NN'$, we note that

$$NN' = AA' = HH' , \qquad (1\text{-}72)$$

i.e., the distance between the nodal points is equal to the distance between the principal points. If we consider a second ray FB parallel to the first but passing through F, it

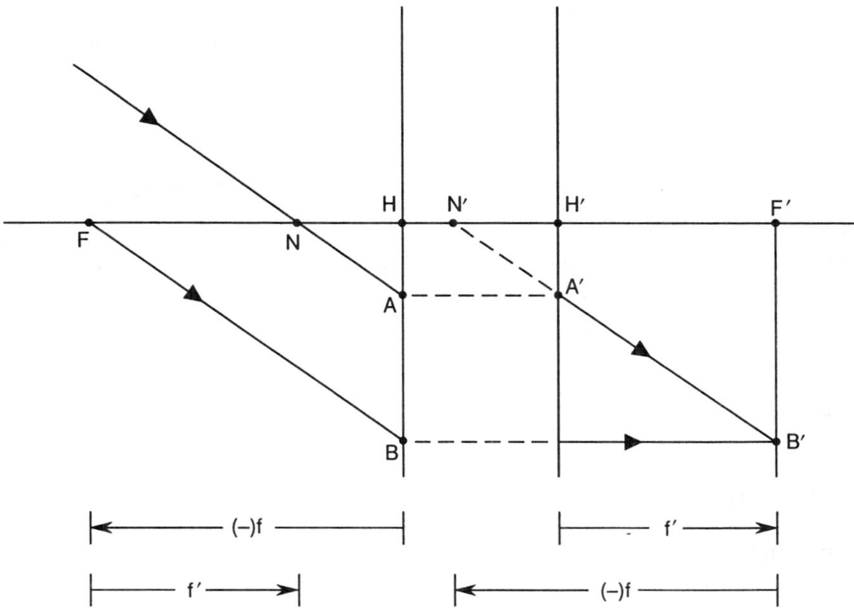

Figure 1-24. Unit angular magnification of nodal points N and N' of an optical imaging system.

emerges parallel to the optical axis in the direction BB'. From the congruent triangles HFB and $F'N'B'$, we find that

$$\boxed{F'N' = HF = f} \tag{1-73}$$

Also

$$F'H' + H'N' = HN + NF \quad . \tag{1-74}$$

But

$$H'N' = H'H - N'H$$

$$= N'N - N'H$$

$$= HN \quad . \tag{1-75}$$

Substituting Eq. (1-75) into Eq. (1-74), we obtain

$$F'H' = NF \quad ,$$

or

$$\boxed{FN = H'F' = f'} \tag{1-76}$$

Letting $M_\beta = 1$ in Eq. (1-59), we note that the nodal planes are conjugate planes with a transverse magnification of n/n'. This may also be seen directly by noting from Figure

1-24 that $S \equiv HN = (f + f')$ and $S' \equiv H'N' = (f + f')$ and substituting into Eq. (1-64). Note that if $n = n'$, then $f = -f'$, and, therefore, N and H coincide, and N' and H' coincide.

1.3.5.4 Newtonian Imaging Equation

If we measure the object and image distances z and z' from the focal points F and F', respectively, as illustrated in Figure 1-18a, we find from similar triangles P_0FP and FHR, and similar triangles $H'F'Q'$ and $F'P_0'P$ that

$$\boxed{M_t = h'/h = -f/z = -z'/f'} \quad . \tag{1-77}$$

Hence,

$$\boxed{zz' = ff'} \quad , \tag{1-78}$$

which is the *Newtonian imaging equation.*

Comparing the imaging equations for a general optical system with those for a single refracting surface, we find that they are indeed similar to each other. The only significant difference is that now the object and image distances are measured from the principal points H and H', respectively (instead of the surface vertex in the case of a single surface). It should be evident that the principal and nodal points of a thin lens coincide at its center. For a single refracting (or reflecting) surface, the principal points coincide with its vertex and its nodal points coincide with its center of curvature. The Newtonian imaging equation for a general system is the same as for a single refracting surface or a thin lens since the principal points are not utilized in this equation.

1.3.6 Afocal Systems

An *afocal* (or without focus, or without focal length) optical imaging system is one that forms the image at infinity of an object at infinity, i.e., a parallel beam of light incident on such a system emerges from it as a parallel beam. Since an emerging ray does not intersect the optical axis or the corresponding incident ray, the concepts of focal points, principal points, and therefore, focal lengths lose their meanings for such a system. One may say that the corresponding principal and focal points lie at infinity on opposite sides of the system and that the focal length is infinity as well, or the focusing power $K = 0$. However, it does not affect the discussion in this section.

A simple example of an afocal system is a refracting astronomical telescope, which is a confocal (or common focus) two-lens system, as illustrated in Figure 1-25 (a reflecting afocal telescope is discussed in Section 6.7). In Figure 1-25a, both lenses are positive and the telescope is called *Keplerian.* In Figure 1-25b, the first lens is positive but the second is negative and the telescope is called *Galilean.* It is easily seen that the object-space focal point F_1 of the first lens and the image-space focal point F_2' of the second lens are conjugates (these focal points are not shown in Figure 1-25). A parallel beam of light incident on the system is focused at the common focus by the first lens and

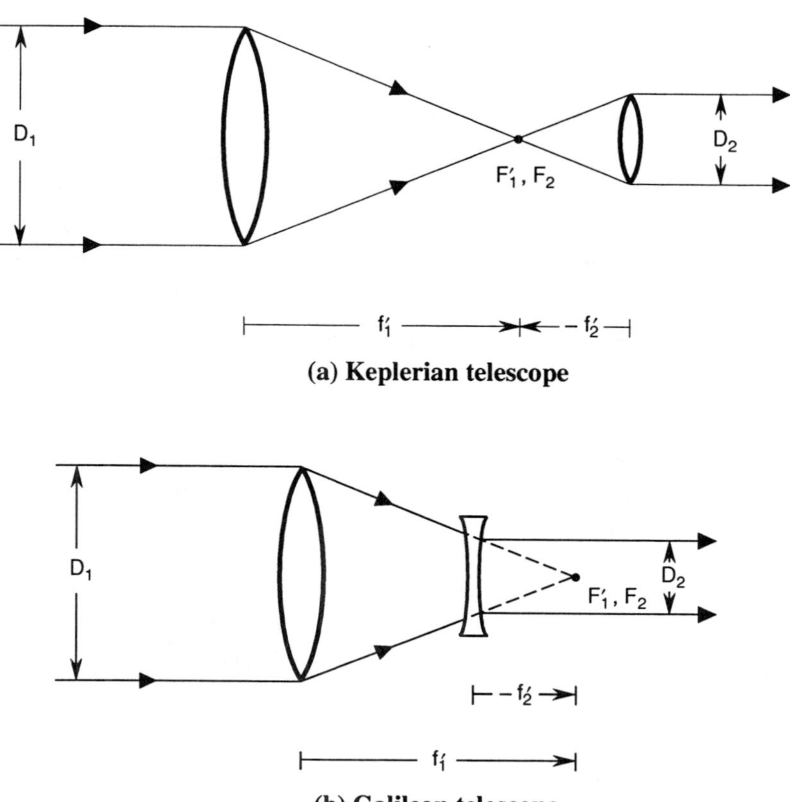

(a) Keplerian telescope

(b) Galilean telescope

Figure 1-25. Afocal system consisting of two lenses with a common focus. (a) A Keplerian telescope has positive lenses, i.e., f_1' and f_2' are both numerically positive. (b) A Galilean telescope has a positive first lens but a negative second lens, i.e., f_1' is numerically positive but f_2' is numerically negative. F_1' is the image-space focal point of the first lens and F_2 is the object-space focal point of the second lens.

emerges as a parallel beam from the second. If the first lens has a longer focal length than that of the second, the system may also be used as a beam reducer. Similarly, if the second lens has a longer focal length, then the system can be used as a beam expander. It is easy to see from the figure that the beam expansion ratio D_2/D_1 is given by $|f_2'/f_1'|$, where D and f' are the diameter and the image-space focal length of a lens.

In Section 1.3.5.2, we discussed the Lagrange invariant for an object lying at infinity. For an afocal system working at infinite conjugates, as illustrated in Figure 1-26a, the Lagrange invariant equation (1-69) becomes

$$n'x_0'\beta' = nx_0\beta \quad , \tag{1-79}$$

where n and n' are the refractive indices of the object and image spaces, x_0 and x_0' are the heights of the axial conjugate rays (i.e., rays parallel to the axis), and β and β' are the slope angles of the off-axis conjugate rays from the optical axis. Hence, the ray angular magnification is given by

$$\frac{\beta'}{\beta} = \frac{nx_0}{n'x_0'} \ .$$

$$(1\text{-}80)$$

Now we consider how afocal systems form images of objects located at finite distances. Consider, for example, the imaging of an object P_0P of height h by an afocal system, as illustrated in Figure 1-26b. Its image $P_0'P'$ has a height of h' which can be obtained by determining the image formed successively by each surface of the system. To obtain an analogue of the imaging equation (1-67), we consider an object Q_0Q at a distance S from P_0P. The distance S' of the image $Q_0'Q'$ from the image $P_0'P'$ can be determined as follows. Since the system is afocal, the transverse magnification h'/h is independent of the position of the object, as may be seen from the figure. For the afocal systems of Figure 1-25, it is equal to f_2'/f_1'. Numerically, it is negative for the Keplerian telescope and positive for the Galilean. If we consider a ray P_0Q incident on the system at an angle β_0, the corresponding angle β_0' of the emerging ray may be obtained from the Lagrange invariant equation (1-69). Hence, the image distance S' is given by

$$S' = h'/\beta_0'$$

$$= \frac{n'h'^2}{nh\beta_0} \ .$$

$$(1\text{-}81)$$

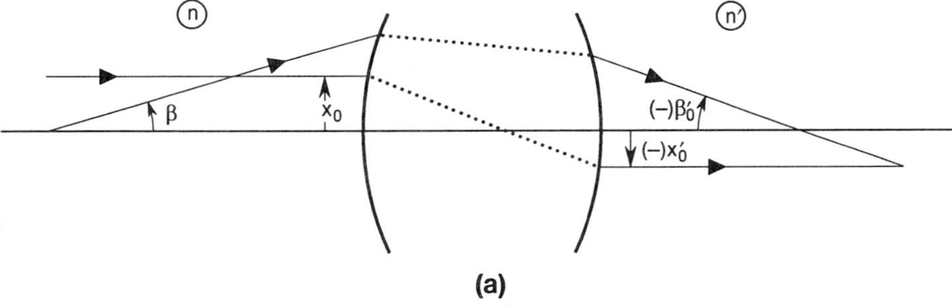

(a)

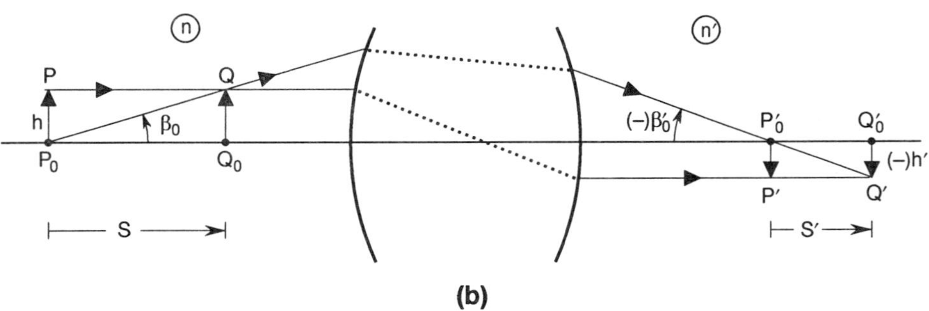

(b)

Figure 1-26. (a) Lagrange invariant of an afocal system for infinite conjugates. (b) Finite conjugate imaging by an afocal system. Conceptually, the system is assumed to be a multisurface one. Hence, a dotted line in the figure does not represent a ray but merely a line joining its point of incidence on and its point of emergence from the system.

Or, since $h = \beta_0 S$, the ratio of the conjugate distances is given by

$$\boxed{\frac{S'}{S} = \frac{n'h'^2}{nh^2}} \ . \tag{1-82}$$

This ratio, representing the longitudinal magnification of the system, is independent of the position of the reference conjugate points P_0 and P_0'. Since the transverse magnification is independent of the object position, the longitudinal magnification S'/S is also constant. In Figure 1-25, the image of a nearby object can be determined by using the Newtonian imaging equation (1-78) recursively to the two lenses. Now if the object position changes by a distance S, then the image position changes by a distance $S' = M_t^2 S$, where M_t is the transverse magnification of the image.

Besides its use as a telescope or a beam expander, an afocal system can also be used to change the effective focal length and field of view (which is defined in Section 2.2.6) of another system by inserting it in the collimated region of the other system. Consider, for example, a system with an image-space focal length f' and an angular field of view β. Figure 1-27a shows how it can be combined with an afocal system to form *a telephoto system*. The first lens in the figure is positive and the second is negative. Thus, f_1' is numerically positive but f_2' is numerically negative, where $\left| f_2' \right| < f_1'$. The combined system has a longer effective focal length $f_e' = f'/M_t$, where $M_t = -f_2'/f_1'$ is the transverse magnification of the afocal system and $M_t < 1$. We note from the figure that the angular magnification is $M_\beta = \beta'/\beta_e = 1/M_t$, and $M_\beta > 1$ or $\beta_e < \beta'$. Now, the size of the image of a distant object formed by an optical system depends linearly on its focal length. Hence, such an image is increased in size by the use of the afocal system, i.e., the combined system is a telephoto system. Letting $\beta' = \beta$, we find that the effective field of view of the combined system is reduced to β_e, which is smaller than β' by a factor of $1/M_t$. Note that to avoid vignetting (discussed in Section 2.2.3) by the afocal system, β_e must be $\leq D_2/f_1'$, where D_2 is the diameter of the beam emerging from the afocal system. It should be noted that adding an afocal system to an imaging system is not the only way to achieve the telephoto effect. A positive and a negative lens of suitable focal lengths also form a telephoto system, as illustrated by Problem 1.12.

When the afocal system is used in reverse so that the first lens is negative ($f_1' < 0$) and the second lens is positive ($f_2' > 0$) as in Figure 1-27b, the effective focal length of the combined system is reduced to $f_e' = f'/M_t$, where $M_t = -f_2'/f_1'$ and $M_t > 1$. The angular magnification of the afocal system is $M_\beta = \beta'/\beta_e = -1/M_t$, and $\left| M_\beta \right| < 1$ or $\beta_e > \left| \beta' \right|$. Note that β' is numerically negative in Figure 1-27b. Letting $\left| \beta' \right| = \beta$, we find that the effective field of view β_e of the combined system is larger than β, or that the combined system is a *wide-angle system*.

Afocal systems are used in a wide variety of applications such as periscopes, binoculars, range finders, and as relay lenses.[9]

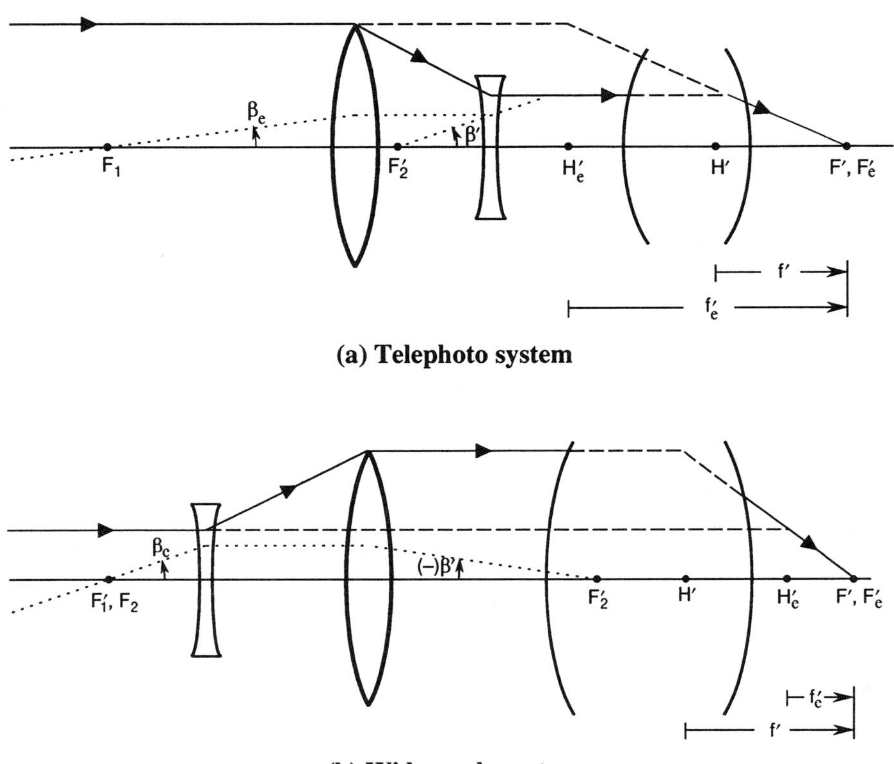

(a) Telephoto system

(b) Wide-angle system

Figure 1-27. (a) Afocal system attached to an imaging system giving a long effective focal length and thereby a telephoto combined system. (b) Afocal system attached to an imaging system giving a short effective focal length and thereby a wide angle combined system. F' and H' are the image-space focal point and principal point of the imaging system. Similarly, F'_e and H'_e are the object-space focal point and principal point of the combined system.

1.3.7 Spherical Reflecting Surface (Spherical Mirror)

Now we consider the imaging properties of a spherical reflecting surface. We derive the equations for imaging, magnification, and Lagrange invariant. We show that the equations for a reflecting surface can be obtained from the corresponding equations for a refracting surface by substituting the refractive index associated with the reflected rays equal to the negative of that associated with the incident rays.

1.3.7.1 Gaussian Imaging Equation

We now consider the imaging properties of a spherical reflecting surface of radius of curvature R as illustrated in Figure 1-28. The line VC joining its vertex V and its center of curvature C defines its optical axis. Consider an axial point object P_0 at a distance S.

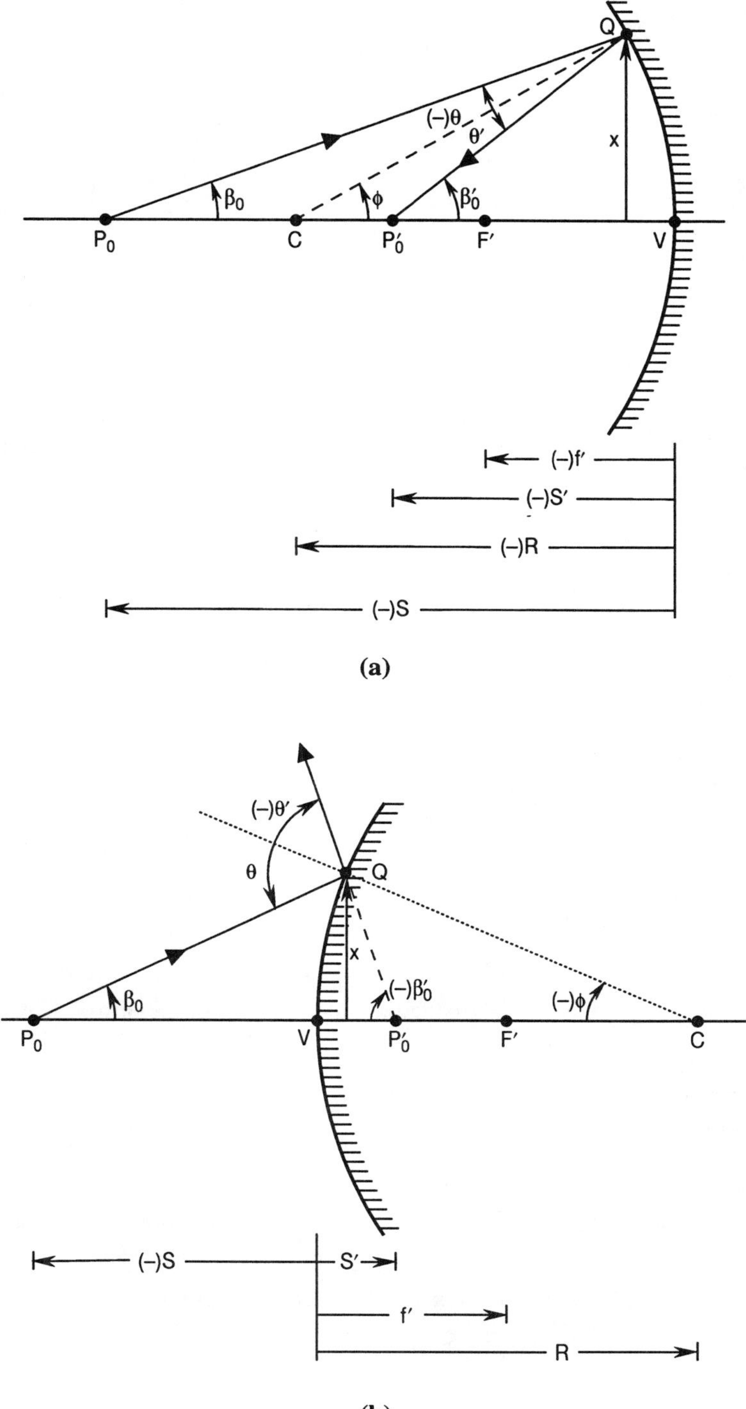

Figure 1-28. Gaussian imaging of an axial point object P_0 by a spherical reflecting surface of radius of curvature R. (a) Concave mirror forms a real Gaussian image P_0'. (b) Convex mirror forms a virtual Gaussian image.

Let the slope angles of the incident and reflected rays from the optical axis be β_0 and β_0', respectively. According to the law of reflection

$$\theta' = -\theta \ . \tag{1-83}$$

From triangle P_0CQ we note that

$$\phi = \beta_0 - \theta \ . \tag{1-84}$$

Similarly, from triangle $CP_0'Q$, we note that

$$\beta_0' = \phi + \theta' \ . \tag{1-85}$$

Substituting for θ and θ' from Eqs. (1-84) and (1-85) into Eq. (1-83) and noting that

$$\phi = -x/R \ , \tag{1-86}$$

$$\beta_0 = -x/S \ , \tag{1-87}$$

and

$$\beta_0' = -x/S' \ , \tag{1-88}$$

we obtain

$$\boxed{\frac{1}{S'} + \frac{1}{S} = \frac{2}{R}} \ . \tag{1-89}$$

1.3.7.2 Focal Length and Reflecting Power

When the object lies at infinity, i.e., when $S = -\infty$, the corresponding image distance $S' \equiv VF' = f'$, where f' is called the *focal length* of the mirror. Thus, the focal length of the mirror is given by

$$\boxed{f' = R/2} \ . \tag{1-90}$$

The rays incident parallel to the optical axis come to focus after reflection by the mirror at the point F', which lies halfway between V and C. It is evident that a mirror has only one focal point. The object- and image-space focal points are coincident just as the two spaces are coincident. Thus, if a point source is placed at the focal point F', its rays incident on the mirror become parallel after reflection by it. Substituting Eq. (1-90) into Eq. (1-89), we obtain

$$\boxed{\frac{1}{S'} + \frac{1}{S} = \frac{1}{f'}} \ . \tag{1-91}$$

The focal point F' of a mirror is illustrated in Figure 1-29 for both a concave and a convex mirror. It is real in the case of a concave mirror but is virtual in the case of a convex mirror. We note that Eq. (1-89) is independent of the refractive index of the medium in which the rays are incident or reflected. Hence, it is independent of the direction of propagation of the rays. The focal length f' is numerically negative for a concave mirror but positive for a convex mirror.

For object rays propagating from left to right, the rays on the first mirror (not necessarily the first imaging element) of a system will be incident propagating from left to right. In a medium of refractive index n, the incident rays will be associated with a refractive index n, but the reflected rays will be associated with a refractive index $n' = -n$. Any refracting imaging elements following the mirror will be assigned refractive indices with a negative value of their actual refractive indices since the rays on

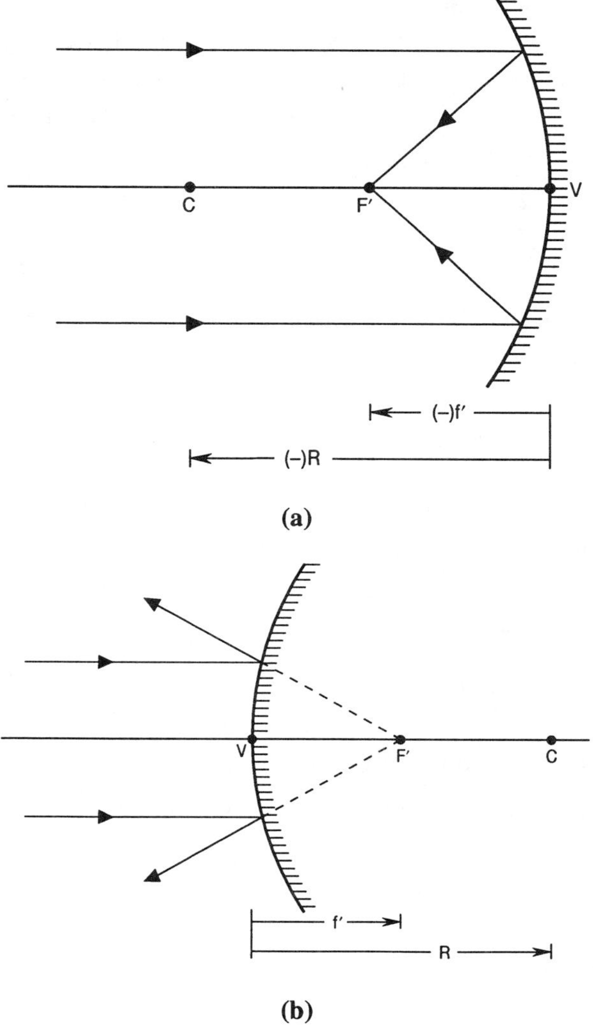

Figure 1-29. The focal point F' of a mirror. It lies halfway between the vertex V and the center of curvature C of the mirror. (a) Concave mirror. (b) Convex mirror.

them are incident propagating from right to left. When reflected by a second mirror in the system, these rays will propagate from left to right and will be associated with a refractive index $n'_2 = n$. Hence, we define the *reflecting power K* and the *equivalent focal length* f_e of a mirror according to

$$K = \frac{1}{f_e} = \frac{2n'}{R} \ , \tag{1-92}$$

where n' is the refractive index associated with the rays reflected by it. Thus, if the first mirror in a system is concave, it has a negative value of R, a negative value of n' in Eq. (1-92), and positive values of K and f_e. Similarly, a second concave mirror in a system will have a positive value of R, a positive value of n' in Eq. (1-92), and positive values of K and f_e. Hence, *a concave mirror is always a converging or a positive imaging element* regardless of the direction of the rays incident on it. Similarly, *a convex mirror* has negative values of K and f_e, i.e., it *is always a diverging or a negative imaging element* regardless of the direction of the rays incident on it. In air, $n' = -1$ for a first mirror and its reflecting power and equivalent focal length are given by

$$K_1 = \frac{1}{f_{e1}} = -\frac{2}{R_1} \ , \tag{1-93}$$

where R_1 is its radius of curvature. Similarly, since $n' = 1$ for a second mirror, its reflecting power and equivalent focal length are given by

$$K_2 = \frac{1}{f_{e2}} = \frac{2}{R_2} \ , \tag{1-94}$$

where R_2 is its radius of curvature. Continuing in this manner, we find that the reflecting power K_j and equivalent focal length f_{ej} of a jth mirror in air of radius of curvature R_j in a system is given by

$$K_j = \frac{1}{f_{ej}} = (-1)^j \frac{2}{R_j} \ . \tag{1-95}$$

1.3.7.3 Magnifications and Lagrange Invariant

Now we consider the imaging of an off-axis point object P at height h from the optical axis in the object plane passing through P_0 as illustrated in Figure 1-30. A ray PV incident at the vertex V of the mirror is reflected as a ray VP' intersecting the image plane passing through P'_0 at the point P', which locates the image point at a height h'. It is evident from the figure that

$$\theta = h / S \tag{1-96}$$

and

$$\theta' = h' / S' \ . \tag{1-97}$$

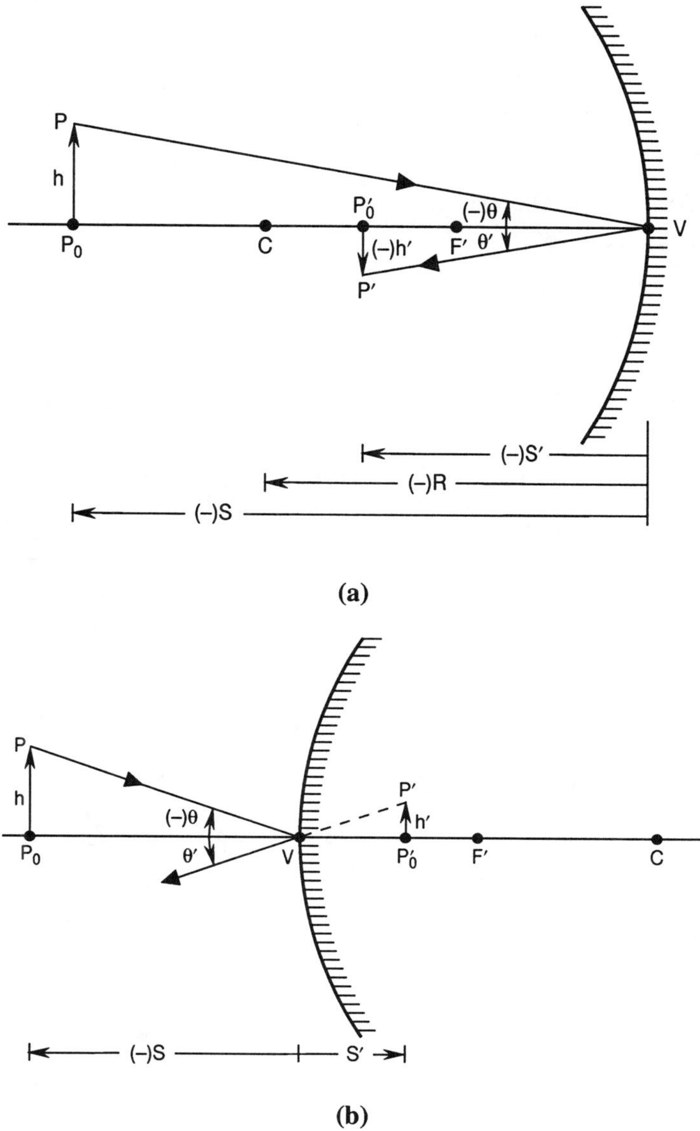

(a)

(b)

Figure 1-30. Gaussian imaging of an off-axis point object *P* at height *h*. (a) Concave mirror forms an inverted image at *P′* at a height *h′*. (b) Convex mirror forms a virtual and erect image.

Substituting Eqs. (1-96) and (1-97) into Eq. (1-83), we find that the *transverse magnification* of the image is given by

$$\boxed{M_t = h'/h = -S'/S \ .}$$

$$(1\text{-}98a)$$

The image formed by a concave mirror is inverted as in Figure 1-30a, but that by a convex mirror is erect as in Figure 1-30b. Accordingly, the magnification is negative in Figure 1-30a and positive in Figure 1-30b.

The ray *angular magnification* representing the ratio of the angular divergence of the rays from P_0 and the angular convergence of these rays to P_0' as in Figure 1-31, is given by

$$M_\beta = \beta_0' / \beta_0 = S / S' \ . \tag{1-99}$$

From Eqs. (1-98a) and (1-99), we obtain

$$M_t M_\beta = -1 \ . \tag{1-100}$$

Equation (1-100) may also be written

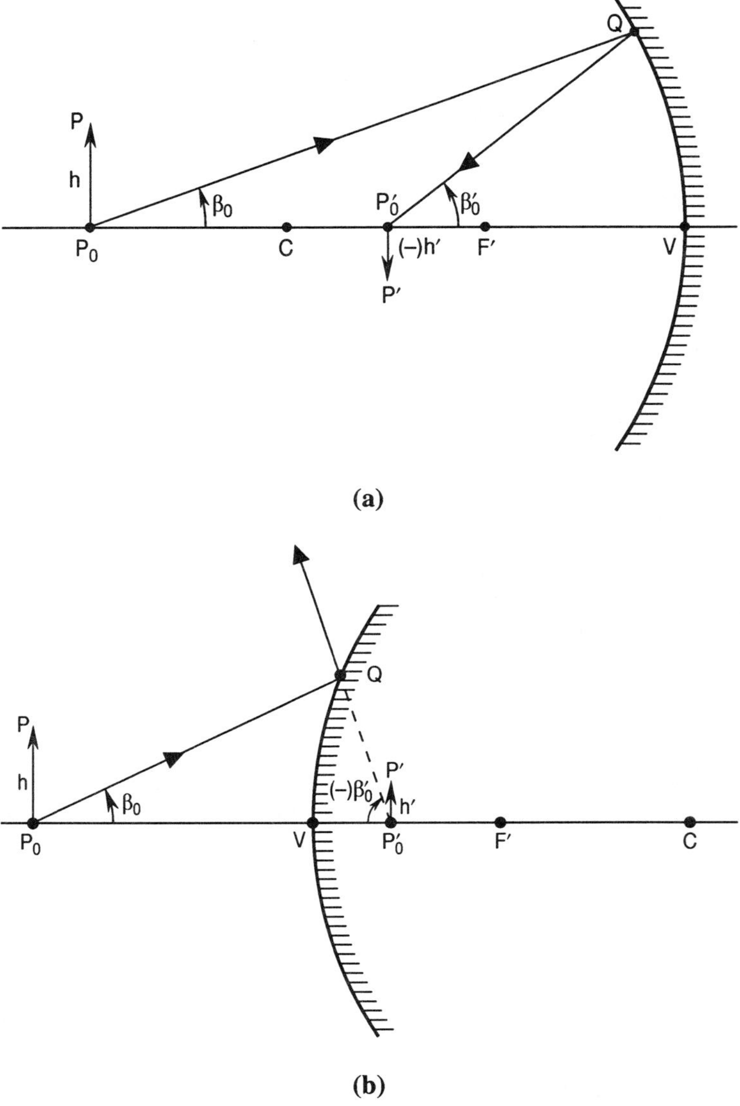

(a)

(b)

Figure 1-31. Lagrange invariant $nh\beta_0$ of a mirror. (a) Concave mirror. (b) Convex mirror.

$$\boxed{h'\beta_0' = -h\beta_0 \ ,} \tag{1-101}$$

representing the Lagrange invariance for the mirror. The Lagrange invariant is $nh\beta_0$, where $n = 1$. From Eq. (1-101), the transverse magnification of the image can also be written

$$\boxed{M_t = -\beta_0/\beta_0' \ ,} \tag{1-98b}$$

i.e., it can also be obtained from the slope angles of the axial incident ray and the corresponding reflected ray.

Differentiating Eq. (1-89), we obtain the longitudinal magnification M_l of the image in terms of its transverse magnification M_t according to

$$\boxed{M_l = \Delta S'/\Delta S = -(S/S')^2 = -M_t^2 \ .} \tag{1-102}$$

Equation (1-102) shows that whether M_t is positive or negative, M_l is always negative. Thus, for example, if the object distance increases, the image distance decreases. For a real object, an increase in the object distance takes place (from a larger negative value to a smaller one) by moving the object closer to the mirror. In Figure 1-30a, a decrease in the image distance (from a smaller negative value to a larger one) implies that the image moves away from the mirror. Similarly, in Figure 1-30b, a decrease in the image distance (from a larger positive value to a smaller one) implies that the image moves closer to the mirror. Thus the image moves in a direction opposite to that of the object. This is true for a system with an odd number of mirrors, as may be seen from Eq. (1-71) by letting $n'/n = -1$. The opposite is true if the number of mirrors is even since then $n'/n = 1$.

In Eq. (1-102), the mirror is assumed to be fixed in position and $\Delta S'$ represents the displacement of the image corresponding to a displacement ΔS of the object. However, if the object is fixed and the mirror is displaced by an amount Δ, then the corresponding displacement of the image is given by $\left(1 + M_t^2\right)\Delta$.

Comparing Eq. (1-89) with Eq. (1-19), we note that the imaging properties of a spherical reflecting surface can be obtained from those of a spherical refracting surface if we let $n = 1$ since the medium between the object and the mirror is air and $n' = -1$, representing a reflected ray propagating backward. Similarly, the expression for the focal length, reflecting power, magnifications, and Lagrange invariant for a mirror can be obtained from the corresponding expressions for a refracting surface by letting $n' = -n$, where $n = 1$.

1.3.7.4 Graphical Imaging

The graphical construction of the image for a reflecting surface is similar to that for a refracting surface, except that the former has only one focal point. It is illustrated in Figure 1-32 for a concave and a convex mirror. It should be remembered that in Gaussian optics, which is based on paraxial rays, any reflection at a surface takes place at a plane that is tangent to it at its vertex, as illustrated in Figure 1-32.

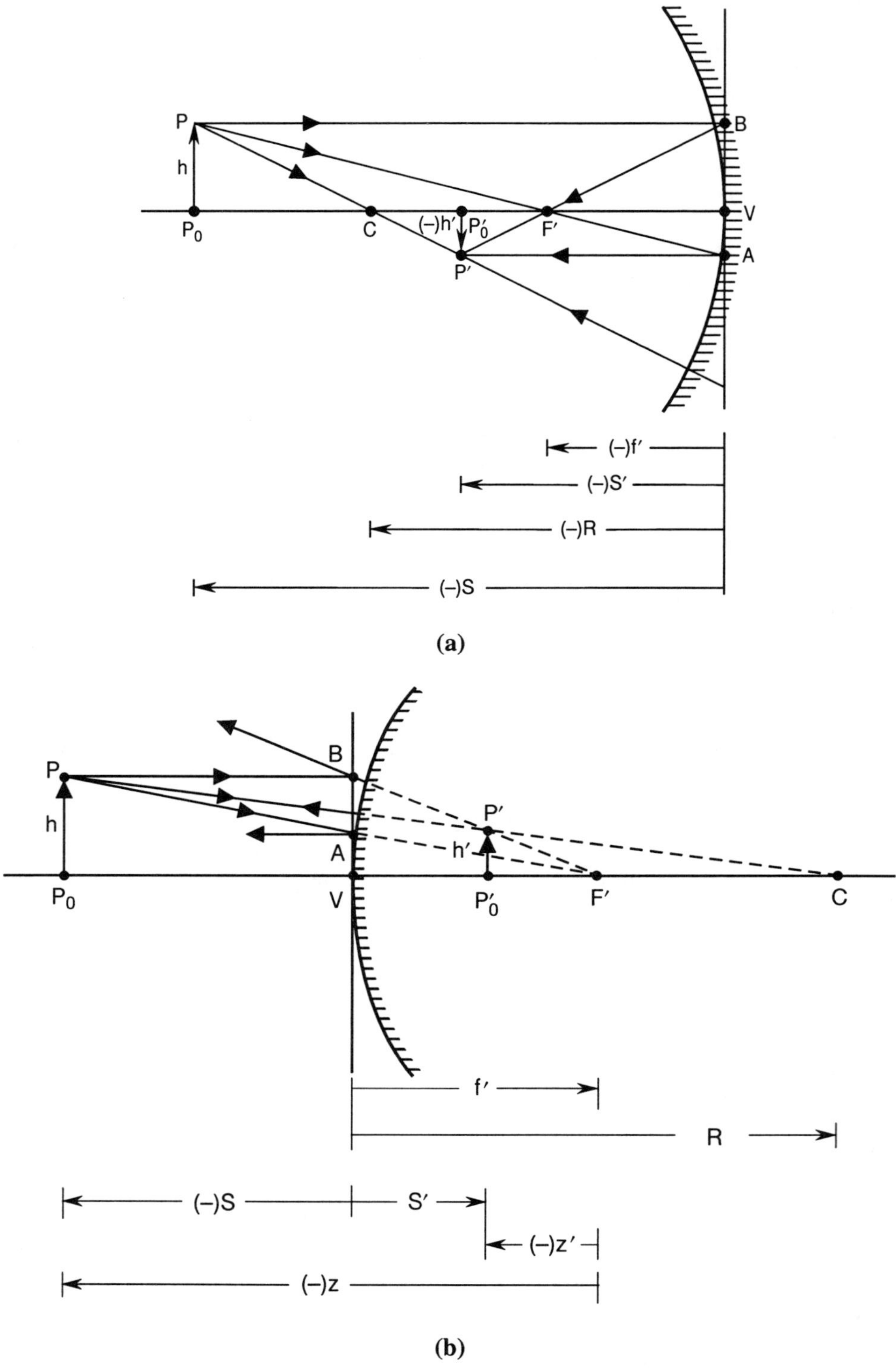

Figure 1-32. (a) Paraxial imaging of a real object P_0P of height h. (a) Concave mirror forms a real and inverted image $P_0'P'$ of height h'. (b) Convex mirror forms a virtual and erect image.

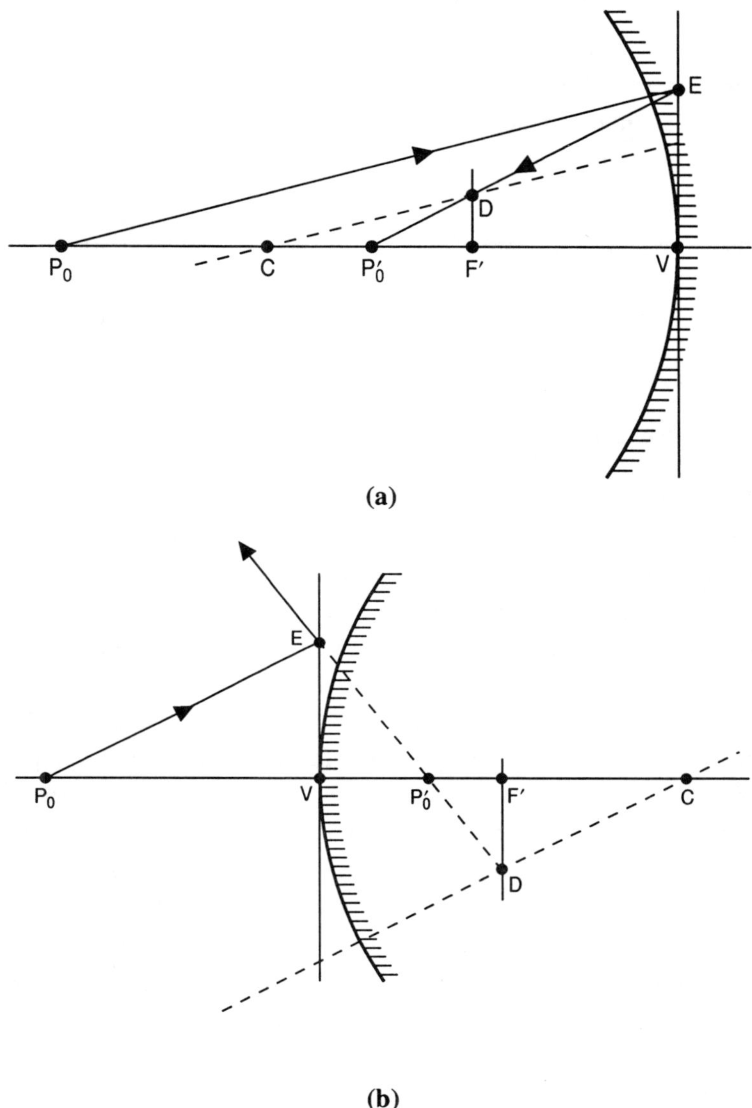

(a)

(b)

Figure 1-33. Graphical Gaussian imaging by a spherical reflecting surface of an axial point object P_0. (a) Concave mirror. (b) Convex mirror.

The Gaussian image P_0' of an on-axis point object P_0 can be determined independently (rather than as the point of intersection of the optical axis and the line that is perpendicular to it and passes through P') as follows: Consider a ray P_0E incident on the surface as shown in Figure 1-33. A hypothetical ray incident parallel to it and passing through C intersects the focal plane at a point D. The reflected ray corresponding to the incident ray P_0E passes through the point D and intersects the optical axis at the Gaussian image point P_0'. The point D may also be determined by considering a hypothetical parallel ray passing through the focal point F'. It is refracted as a ray parallel to the optical axis intersecting the focal plane at the point D.

1.3.7.5 Newtonian Imaging Equation

Equation (1-91) is the *Gaussian imaging equation*. If we measure the object and image distances z and z', respectively, from the focal point F', as indicated in Figure 1-32, then from similar triangles $VF'B$ and $P_0'F'P'$, and similar triangles $P_0F'P$ and $VF'A$, we find that the transverse magnification of the image is given by

$$\boxed{M_t = h'/h = z'/f' = f'/z} \ . \tag{1-103}$$

Hence,

$$\boxed{z\,z' = f'^2} \ , \tag{1-104}$$

which is the *Newtonian imaging equation.*

1.4 PARAXIAL RAY TRACING

The cardinal points of an imaging system consisting of a series of coaxial refracting and/or reflecting surfaces can be determined by repeated application of the paraxial ray-tracing equations for refraction or reflection at a surface. Starting at an object point, a ray undergoes rectilinear propagation to the first surface of the system; it is reflected or refracted at the surface depending on whether it is a refracting or a reflecting surface; it undergoes rectilinear propagation again until it reaches the next surface, and the process repeats until the ray reaches the image plane. In this section, we develop ray-tracing equations for a refracting surface, a thin lens, and a reflecting surface. These equations are used recursively to determine the focal length of a combination of two lenses and a combination of two mirrors. The ray-tracing equations are used, not only for determining the Gaussian properties of a system, but also for determining the sizes of the imaging elements and stops, vignetting of the rays, and obscurations in mirror systems. For the exact (or finite as opposed to infinitesimal for paraxial) ray tracing, the reader may refer to Welford.[4]

1.4.1 Refracting Surface

We now derive the ray-tracing equations for a refracting surface. As indicated in Figure 1-34, consider a spherical refracting surface of radius of curvature R_1 separating media of refractive indices n_0 and n_1. An object ray A_0A_1 from a point object A_0 incident at a point A_1 on the refracting surface is refracted as a ray A_1A_2. Let x_0, x_1, and x_2 be the heights of the points A_0, A_1 and A_2, respectively, from the optical axis VC, where V is the vertex and C is the center of curvature of the surface. Let t_0 and t_1 be the axial distances of A_0 and A_2 from the vertex V.

It is evident from Figure 1-34 that for paraxial rays, rectilinear propagation from A_0 to A_1 gives

$$\boxed{x_1 = x_0 + t_0\beta_0} \ , \tag{1-105}$$

where β_0 is the slope angle of the incident ray A_0A_1 from the optical axis. Equation (1-105) is called the *transfer ray-tracing equation*. According to Snell's law, refraction of the ray at A_1 gives

$$n_1\theta_1 = n_0\theta_0 \quad , \tag{1-106}$$

where θ_0 and θ_1 are the angles of incidence and refraction (i.e., the angles of the incident and refracted rays from the surface normal at point A_1), respectively.

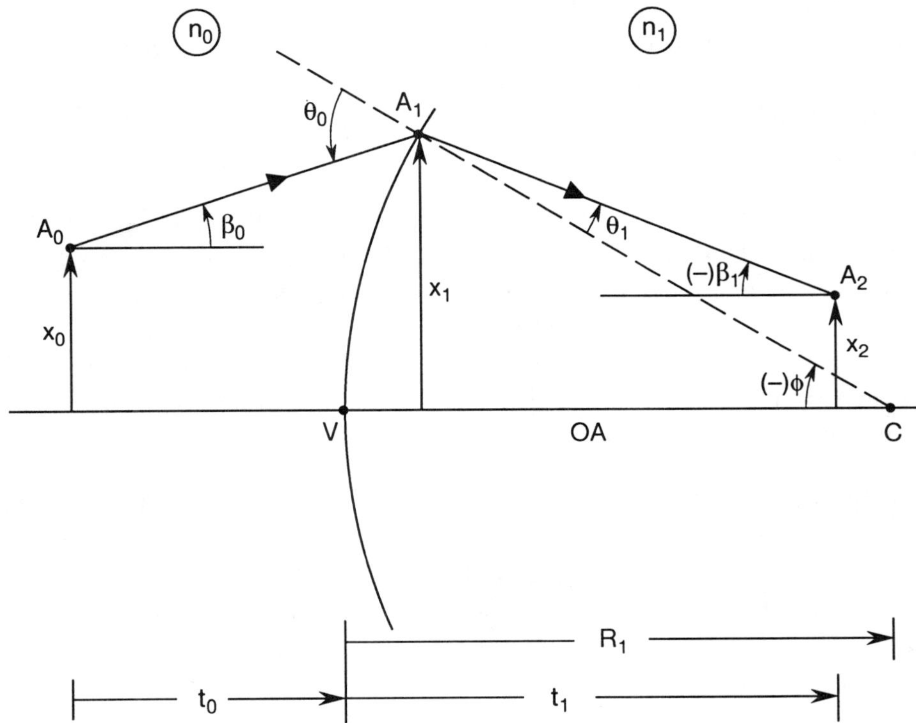

Figure 1-34. Ray tracing by a spherical refracting surface of radius of curvature R separating media of refractive indices n_0 and n_1.

We note from Figure 1-34 that

$$\phi = \beta_1 - \theta_1 \tag{1-107}$$

and

$$\theta_0 = \beta_0 - \phi \quad , \tag{1-108}$$

where

$$\phi = -x_1/R_1 \quad . \tag{1-109}$$

Substituting for θ_0, θ_1, and ϕ from Eqs. (1-107), (1-108) and (1-109) into Eq. (1-106), we obtain

$$\boxed{n_1\beta_1 \;=\; n_0\beta_0 + (n - n')\frac{x_1}{R_1}} \;\; .$$

(1-110)

Equation (1-110) is called the *refraction ray-tracing equation*. The rectilinear propagation of the refracted ray from A_1 to A_2 gives

$$x_2 \;=\; x_1 + t_1\beta_1 \;\; ,$$

(1-111)

where β_1 is numerically negative. If the next surface lies at a distance t_1 from the first, then A_2 determines the point of incidence on it. In that case the ray $A_1 A_2$ is refracted at the point of incidence A_2 by the second surface according to an equation similar to Eq. (1-110) and the ray propagates rectilinearly until it reaches the next surface. Using Eqs. (1-105) and (1-110) recursively, the ray can be propagated to the image plane of a multisurface system.

The ray tracing of a refracting surface is illustrated schematically in Figure 1-35a. A ray starts at a height x_0 from the optical axis at a distance t_0 from the refracting surface. It is incident on the surface with a slope angle β_0. The starting point is indicated by coordinates (x_0, β_0). The ray is incident on the surface at a height x_1 and is refracted with a slope β_1. The point of incidence is indicated by (x_1, β_1), where x_1 and β_1 are given by Eqs. (1-105) and (1-110), respectively. The height of a point x_2 on the refracted ray at a distance t_1 from the refracting surface is given by Eq. (1-111).

As a simple example, we use the ray-tracing equations to determine the focal length of the refracting surface. If we let $\beta_0 = 0$, corresponding to a ray incident parallel to the optical axis of the system, as in Figure 1-35b, and let $x_2 = 0$, corresponding to the point of intersection of the refracted ray with the optical axis, then the corresponding value t_1 gives the focal length f_1'. Letting $\beta_0 = 0$ and $x_2 = 0$ in Eqs. (1-105), (1-110), and (1-111), we find that

$$f_1' \;=\; \frac{n_1}{n_1 - n_0}\, R_1 \;\; ,$$

(1-112)

which is in agreement with Eq. (1-20).

Next we apply Eqs. (1-105) and (1-110) recursively to obtain the ray-tracing equations for a thin lens and the imaging properties of a thick lens.

1.4.2 Thin Lens

In the case of a thin lens, the refraction of an incident ray takes place at its two surfaces, which have a negligible spacing between them. It is illustrated schematically in Figure 1-36a. It starts at point (x_0, β_0) in a medium of refractive index n_0 at a distance t_0 from the first surface. The point of incidence is (x_1, β_1) on the first surface and (x_2, β_2) on the second. The lens has a refractive index n and a thickness t_1 which is negligible. The ray ends at a height x_3 from the optical axis at a distance t_2 from the second surface in a medium of refractive index n_0.

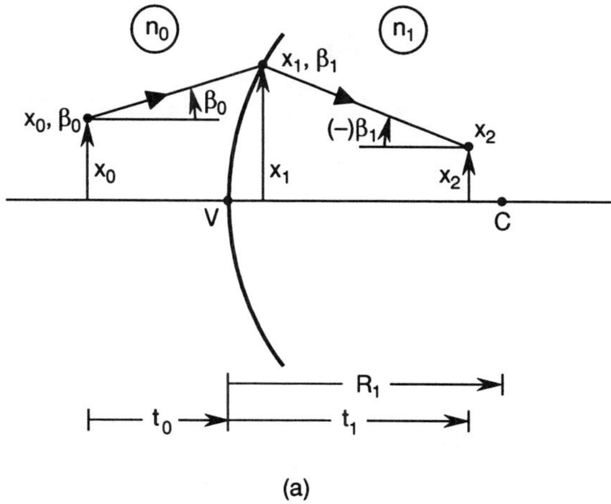

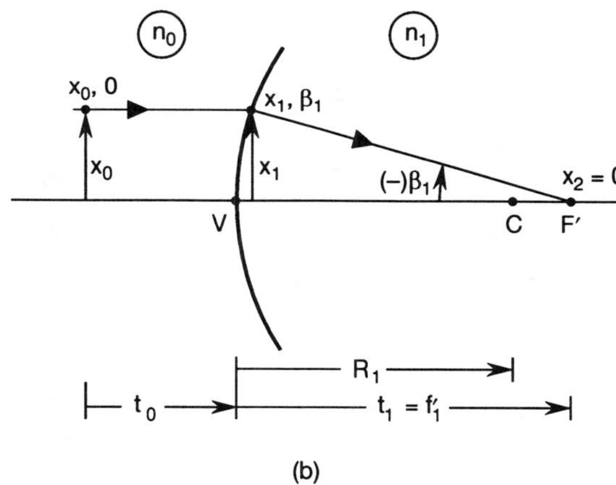

Figure 1-35. Paraxial ray tracing of a spherical refracting surface. (a) General case. (b) Determination of focal point F'.

Now, we apply Eqs. (1-105) and (1-110) recursively to obtain the focal length of a thin lens of a refractive index n and spherical surfaces of radii of curvature R_1 and R_2. The paraxial ray-tracing equations for the lens may be written as follows (see Figure 1-36b):

$$x_1 = x_0 + t_0 \beta_0 \tag{1-113a}$$

$$= x_0 \text{ for a ray incident parallel to the optical axis,} \tag{1-113b}$$

$$n_1 \beta_1 = n_0 \beta_0 + (n_0 - n_1)\frac{x_1}{R_1} \tag{1-114a}$$

$$= (n_0 - n_1)\frac{x_0}{R_1} \ , \tag{1-114b}$$

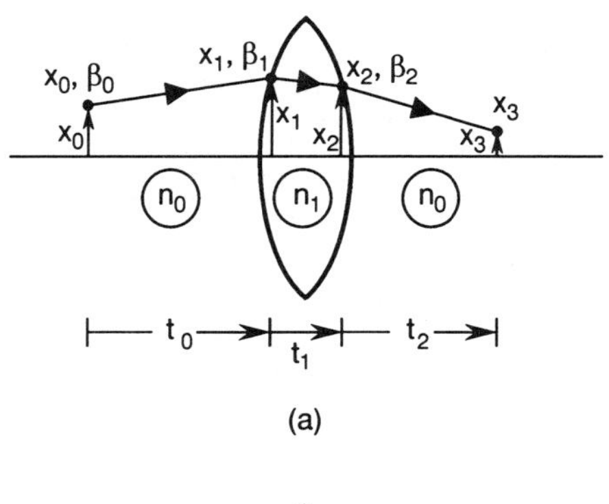

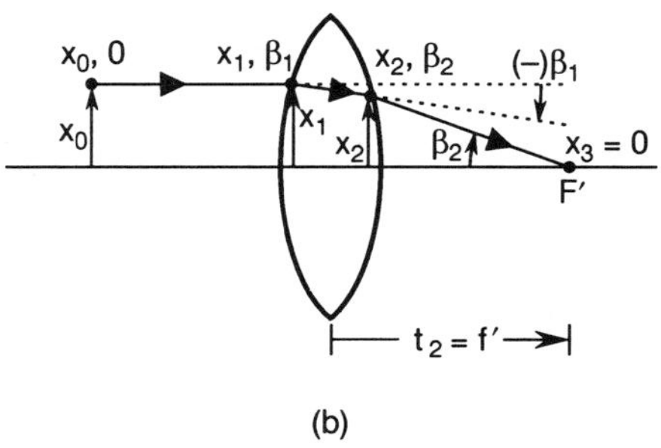

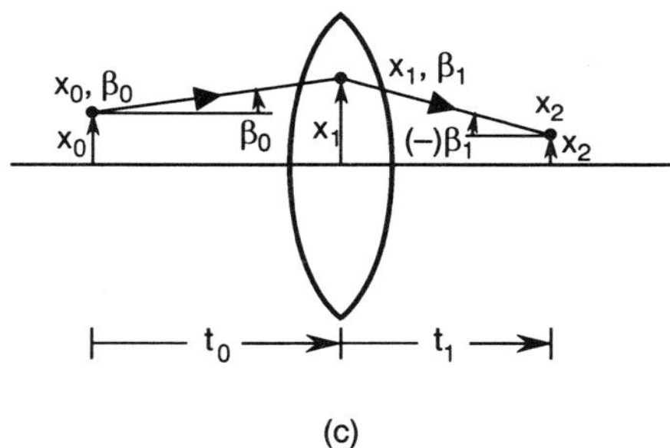

Figure 1-36.   Paraxial ray tracing of a thin lens. (a) General case. (b) Object at infinity. (c) Simplified paraxial ray tracing of a thin lens. Lens thickness t_1 in (c) is neglected.

$$x_2 = x_1 + t_1 \beta_1$$

$$= x_1 \text{ since we neglect } t_1$$

$$= x_0 \quad , \tag{1-115}$$

$$n_2 \beta_2 = n_1 \beta_1 + (n_1 - n_2)\frac{x_2}{R_2} \quad , \tag{1-116a}$$

or, since $n_2 = n_0$,

$$n_0 \beta_2 = (n_0 - n_1)\left(\frac{1}{R_1} - \frac{1}{R_2}\right)x_0 \quad , \tag{1-116b}$$

$$x_3 = x_2 + t_2 \beta_2 \tag{1-117}$$

$$= x_1 + t_2 \beta_2$$

$$= x_0 \left[1 + \frac{t_2}{n_0} (n_0 - n_1) \frac{1}{R_1} - \frac{1}{R_2} \right]$$

$$= 0 \text{ for the right focal point,}$$

and

$$\frac{1}{f'} \equiv \frac{1}{t_2} = \frac{n_1 - n_0}{n_0} \left(\frac{1}{R_1} - \frac{1}{R_2} \right) \ . \tag{1-118}$$

Except for the notation, Eq. (1-118) is the same as Eq. (1-36). Substituting Eqs. (1-114a) and (1-115) into Eq. (1-116a) and noting that $n_2 = n_0$, we obtain

$$\beta_2 = \beta_0 - \frac{x_1}{f'} \ . \tag{1-119}$$

Referring to Figure 1-36c, where a ray incident on the lens is shown refracted by it, the ray-tracing Eqs. (1-113a), (1-119), and (1-117) for a thin lens of image-space focal length f_1' may be written

$$\boxed{x_1 = x_0 + t_0 \beta_0 \ ,} \tag{1-120}$$

$$\beta_1 = \beta_0 - \frac{x_1}{f_1'} \ , \tag{1-121}$$

and

$$x_2 = x_1 + t_1 \beta_1 \ , \tag{1-122}$$

respectively, in one step (rather than in two as in Figures 1-36a and 1-36b). It should be evident that the principal and nodal points of a thin lens coincide at its center.

1.4.3 Two Thin Lenses

We now consider a combination of two thin lenses spaced a certain distance apart and determine its focal length as well as its principal and focal points. Consider, as shown in Figure 1-37, two thin lenses L_1 and L_2 of image-space focal lengths f_1' and f_2' separated by a distance t_1. Using Eqs. (1-120) and (1-121) recursively, we can obtain the focal points and the principal points of the combined imaging system as follows:

$$x_1 = x_0 + t_0 \beta_0$$

$$= x_0 \text{ for a ray incident parallel to the optical axis,}$$

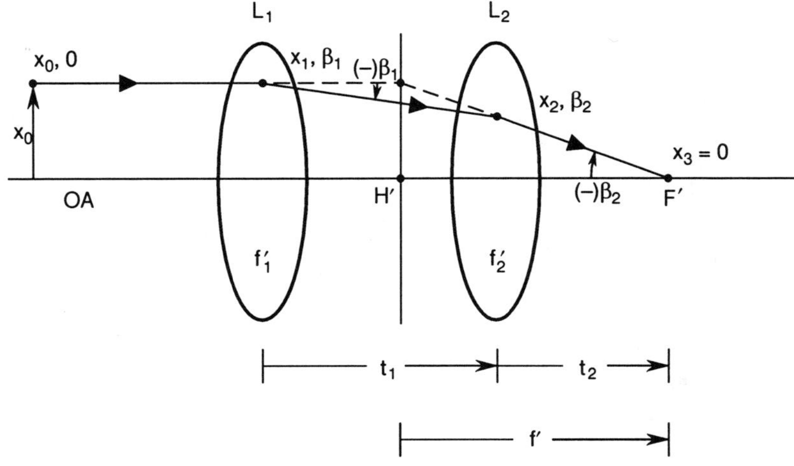

Figure 1-37. Paraxial ray tracing of a two-lens system.

$$\beta_1 = \beta_0 - \frac{x_1}{f_1'}$$

$$= -\frac{x_0}{f_1'} \quad ,$$

$$x_2 = x_1 + t_1\beta_1$$

$$= x_0\left(1 - \frac{t_1}{f_1'}\right) \quad ,$$

$$\beta_2 = \beta_1 - \frac{x_2}{f_2'}$$

$$= -x_0\left(\frac{1}{f_1'} + \frac{1}{f_2'} - \frac{t_1}{f_1'f_2'}\right) \quad ,$$

$$\frac{1}{f'} = -\frac{\beta_2}{x_0}$$

or

$$\boxed{\frac{1}{f'} = \frac{1}{f_1'} + \frac{1}{f_2'} - \frac{t_1}{f_1'f_2'}} \quad , \tag{1-123}$$

$$x_3 = x_2 + t_2\beta_2$$

$$= 0 \text{ for the right focal point,}$$

and

$$t_2 = -\frac{x_2}{\beta_2}$$

or

$$t_2 = f'\left(1 - \frac{t_1}{f_1'}\right) \ . \tag{1-124}$$

The quantity t_2 locates the image-space focal point F'. A positive value of t_2 implies that F' lies to the right of the center of lens L_2. A distance such as t_2 of the focal point F' from the vertex of the last element of a system in the sense of light propagation is called its *image-space focal distance*. The principal point H' is located by noting that $H'F' = f'$. It lies to the left of F' for a positive value of f'. Since the refractive index of the image space is unity, f' is also equal to the equivalent focal length of the system. Accordingly, the refracting power of the system is $K = 1/f'$. The object-space focal point F and principal point H can be determined in a similar manner by considering a ray incident parallel to the axis from right to left. We find that F lies at a distance $f\left(1 - t_1/f_2'\right)$ from lens L_1, where $f = -f'$ since the lenses are in air. A negative value in this case implies that the focal point F lies to the left of the center of lens L_1. Such a distance of the focal point F from the vertex of the first element of a system is called its *object-space focal distance*.

It is easy to see from Eqs. (1-123) and (1-124) that if $t_1 = f_1' + f_2'$, as in Figure 1-25, then $f' \rightarrow \infty$ and, therefore, $t_2 \rightarrow \infty$. Thus, the system is afocal and the focal point F' lies at infinity on the right-hand side of the system. The principal point H' lies to the left-hand side of the lens L_2 at a distance $f' - t_2 = f' t_1 / f_1' \rightarrow \infty$, i.e., it lies at infinity on the left-hand side of the system. Similarly, we can show that the principal point H and the focal point F lie at infinity on the right-hand and left-hand sides of the system, respectively. If $t_1 < f_1' + f_2'$, then the system has a positive focal length. If, however, $t_1 > f_1' + f_2'$, then the system has a negative focal length.

1.4.4   Thick Lens

Now, we consider a thick lens of refractive index n, thickness t and surfaces with radii of curvature R_1 and R_2 and determine its focal length by repeated application of the ray-tracing equations (1-105) and (1-110) for transfer and refraction at a refracting surface, respectively. With reference to Figure 1-38, and noting that $n_0 = 1$, $n_1 = n$, $n_2 = 1$, we proceed as follows by considering a ray incident on the lens from left to right parallel to its axis so that $\beta_0 = 0$:

$$x_1 = x_0 + t_0 \beta_0$$

$$= x_0 \ , \quad \text{(for a ray incident parallel to the optical axis)}$$

$$n_1 \beta_1 = n_0 \beta_0 + (n_0 - n_1)\frac{x_1}{R_1} \ ,$$

or

$$n\beta_1 = (1 - n)\frac{x_0}{R_1} \ ,$$

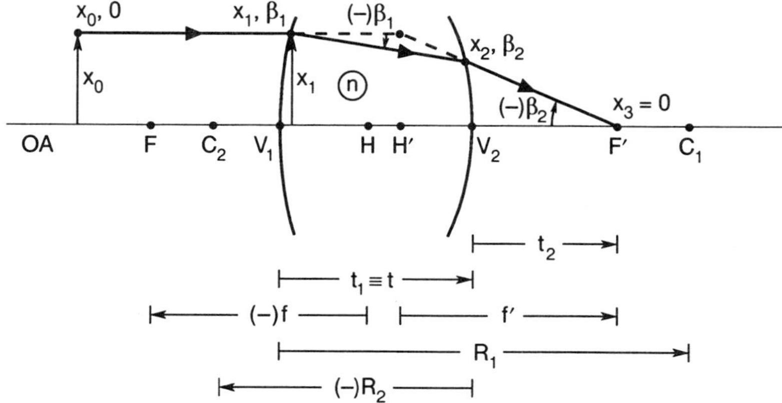

Figure 1-38. Paraxial ray tracing of a thick lens of refractive index n and thickness t. C_1 and C_2 are the centers of curvature of the surfaces of the lens with vertices V_1 and V_2 and radii of curvature R_1 and R_2, respectively.

$$x_2 = x_1 + t_1\beta_1$$

$$= \left(1 - t\frac{n-1}{nR_1}\right)x_0 \quad,$$

$$n_2\beta_2 = n_1\beta_1 + (n_1 - n_2)\frac{x_2}{R_2} \quad,$$

or

$$\beta_2 = \left[\frac{1-n}{R_1} + \frac{n-1}{R_2}\left(1 - t\frac{n-1}{nR_1}\right)\right]x_0 \quad.$$

and

$$\frac{1}{f'} = -\frac{\beta_2}{x_0}$$

or

$$\boxed{\frac{1}{f'} = (n-1)\left(\frac{1}{R_1} - \frac{1}{R_2}\right) + \frac{t(n-1)^2}{nR_1R_2}} \quad. \tag{1-125}$$

Since the medium surrounding the lens is air, the refractive index of the image space is unity. Hence, f' is also the equivalent focal length of the lens. In the case of a thin lens, the thickness t is assumed to be small. Therefore, neglecting the second term on the right-hand side of Eq. (1-125), we obtain Eq. (1-41) for the focal length of a thin lens.

Similarly, letting $x_2 = x_1$ and substituting for β_1 in terms of β_0 and x_1 into the equation for β_2, we obtain Eq. (1-121).

The image-space focal point F' is located by letting $x_3 = 0$. Thus,

$$x_3 = x_2 + t_2\,\beta_2$$
$$= 0 \ ,$$

or

$$t_2 = -\frac{x_2}{\beta_2}$$

or

$$t_2 = f'\left(1 - t\,\frac{n-1}{nR_1}\right) \ . \tag{1-126}$$

The quantity $t_2 \equiv V_2 F'$, where V_2 is the vertex of the second surface, locates the focal point F' and represents the *back focal distance*. A positive value of t_2 implies that the focal point F' lies to the right of V_2. The principal point H' is located by noting that $H'F' = f'$. A positive value of f' implies a converging or a positive lens which, in turn, implies that H' lies to the left of F'. It lies at a distance

$$V_2 H' = t_2 - f'$$
$$= -tf'\,\frac{n-1}{nR_1} \tag{1-127}$$

from V_2 and a negative value indicates that H' lies to the left of V_2.

The object-space focal point F and the principal point H can be determined in a similar manner by considering a ray incident parallel to the axis from right to left. Thus, we can show that the distance of the focal point F from the vertex V_1 of the first surface is given by

$$V_1 F = f\left(1 + t\,\frac{n-1}{nR_2}\right) \ , \tag{1-128}$$

where $f = -f'$ is the object-space focal length of the lens. The distance of the principal point H from the vertex V_1 is given by

$$V_1 H = -tf'\,\frac{n-1}{nR_2} \ , \tag{1-129}$$

and a positive value implies that H lies to the right of V_1. The distance of H' from H is given by

$$HH' = t - \left(V_1 H + H' V_2 \right)$$

$$= t \left[1 - f' \frac{n-1}{n} \left(\frac{1}{R_1} - \frac{1}{R_2} \right) \right] \tag{130a}$$

$$\simeq \frac{n-1}{n} t \tag{130b}$$

$$= t/3 \quad , \tag{130c}$$

where we have used the thin lens formula for the focal length in obtaining Eq. (1-130b) and $n = 1.5$ in further obtaining Eq. (1-130c). Thus, unless the lens is very thick, the separation of its principal points is approximately equal to one third of its thickness independent of its radii of curvature.

As illustrated in Figure 1-39, it is interesting to explore the variation in the positions of the principal and focal points of a thick lens with increasing thickness. In this figure, the magnitudes of the radii of curvature of its two surfaces are assumed to be equal. Figure 1-39a shows the thin-lens approximation of a thick lens. Accordingly, the principal points coincide. In practice, there is some spacing between them, as indicated in Figure 1-38b. The principal points coincide if the two surfaces of the thick lens are concentric, as shown in Figure 1-39c. According to Eq. (1-126), the image-space focal point F' lies at the back vertex V_2, as in Figure 1-39d, if $t = n R_1 / (n-1)$. Its focal length in that case is given by $f' = R_1 / (n-1)$, which is independent of the value of R_2, showing that the second refracting surface has no effect on the image formed at its vertex. Similarly, according to Eq. (1-128), if $t = - n R_2 / (n-1)$, the object-space focal point F lies at the front vertex V_1 and the focal length of the lens is given by $f' = - R_2 / (n-1)$, independent of the value of R_1.

If the thickness is increased to $t = n \left(R_1 - R_2 \right) / (n-1)$, the focal length approaches infinity, i.e., the lens becomes *afocal*. Since $R_1 = - R_2$ in the figure, F lies at V_1 when F' lies at V_2. As illustrated in Figure 1-39e, parallel rays incident on the lens are focused inside it and emerge from it as parallel rays. The corresponding principal and focal points lie at infinity on the opposite sides of the lens. If the thickness of the lens is increased further, the principal points lie farther from the respective vertices than the corresponding focal points, as shown in Figure 1-39f, thus giving it a negative image-space focal length f' even though its shape is biconvex.

1.4.5 Reflecting Surface (Mirror)

We now derive the ray-tracing equations for a reflecting surface. Consider a spherical reflecting surface of radius of curvature R_1 with a vertex V and center of curvature C, as illustrated in Figure 1-40. An object ray $A_0 A_1$ from a point object A_0 incident on the surface at a point A_1 with a slope angle β_0 is reflected as a ray $A_1 A_2$ so that the magnitudes of the angles of incidence θ and reflection θ' from the surface

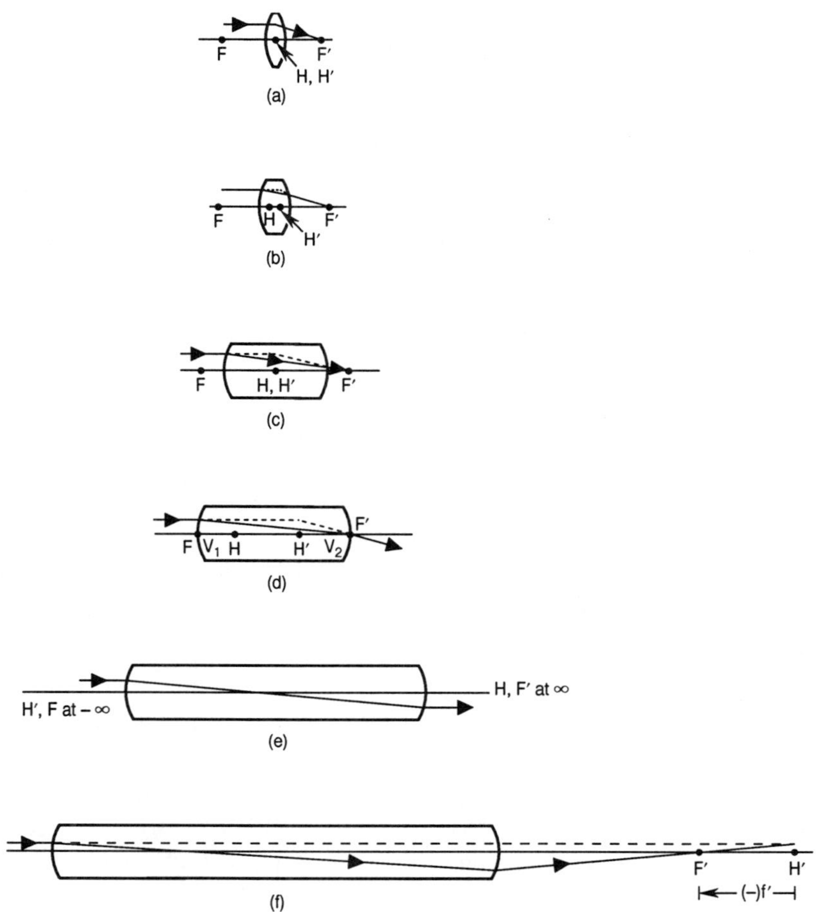

Figure 1-39. The principal and focal points of a thick lens of increasing thickness. The magnitudes of the radii of curvature of its two surfaces are assumed to be equal in the figure. (a) Thin lens. (b) Thick lens. (c) Concentric lens. (d) Thick lens such that the image-space focal point F' lies at the back vertex V_2. (e) Afocal thick lens. (f) Convex thick lens with a negative image-space focal length f'.

A_0, A_1, and A_2, respectively, from the optical axis VC of the mirror. Also, let t_0 and t_1 be the axial distances of points A_0 and A_2 from the vertex V. We note from the figure that for paraxial rays, rectilinear propagation from A_0 to A_1 gives

$$\boxed{x_1 = x_0 + t_0\beta_0}\ .\qquad\qquad (1\text{-}131)$$

According to the law of reflection

$$\theta' = -\theta\ .\qquad\qquad (1\text{-}132)$$

We note from Figure 1-40 that

$$\beta_0 - \beta_1 = \theta - \theta'\ ,\qquad\qquad (1\text{-}133)$$

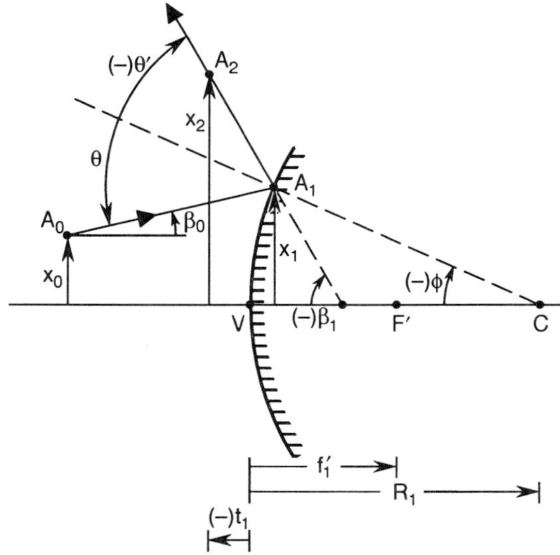

Figure 1-40. Paraxial ray tracing of a convex spherical mirror of radius of curvature R_1 with center of curvature C and vertex V.

$$\beta_1 = \phi - \theta \ , \tag{1-134}$$

normal $A_1 C$ are equal to each other. Let x_0, x_1, and x_2 be the heights of points and

$$\phi = -\frac{x_1}{R_1} \ . \tag{1-135}$$

Note that β_1, θ', and ϕ are all numerically negative angles in the figure. Substituting for θ, θ', and ϕ from Eqs. (1-133) through (1-135) into Eq. (1-132), we obtain

$$\boxed{\beta_1 = -\beta_0 - \frac{2x_1}{R_1} \ .} \tag{1-136}$$

Equation (1-136) is called the *reflection ray-tracing equation.* Rectilinear propagation of the reflected ray from A_1 to A_2 gives

$$x_2 = x_1 + t_1 \beta_1 \ . \tag{1-137}$$

Note that t_1 is numerically negative in this equation since the rays are propagating from right to left in going from A_1 to A_2. Hence, the quantity $t_1\beta_1$ is numerically positive.

The ray tracing of a reflecting surface is illustrated schematically in Figure 1-41a. A ray starts at a height x_0 from the optical axis at a distance t_0 from the refracting surface. It is incident on the surface with a slope angle β_0. The starting point is indicated by coordinates (x_0, β_0). The ray is incident on the surface at a height x_1 and is reflected with a slope β_1. The point of incidence is indicated by (x_1, β_1), where x_1 and β_1 are given by Eqs. (1-131) and (1-136), respectively. The height x_2 of a point A_2 on the reflected ray at a distance t_1 from the refracting surface is given by Eq. (1-137).

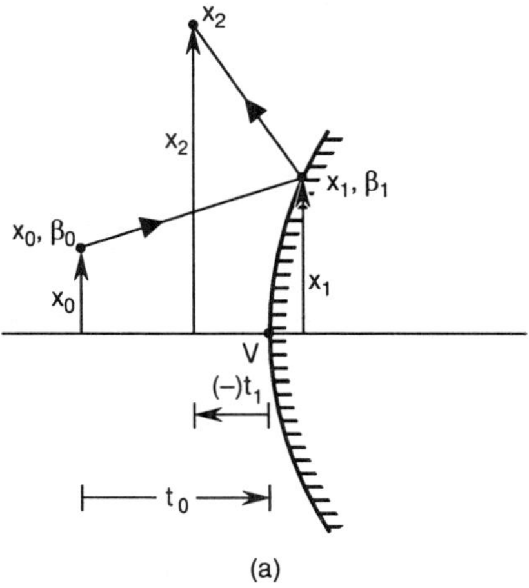

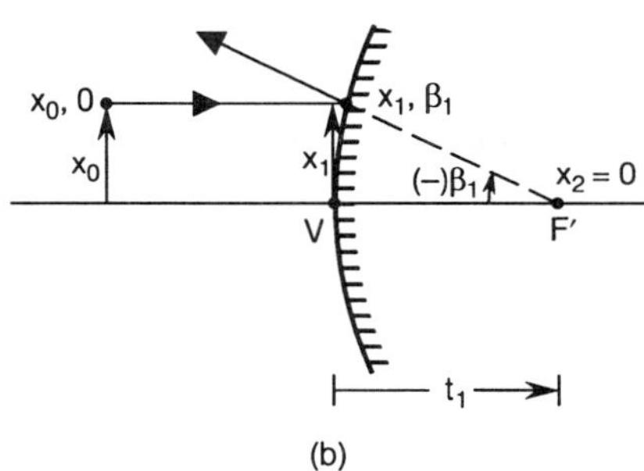

Figure 1-41. Paraxial ray tracing of a reflecting surface. (a) General case. (b) Determination of focal point.

If we let $\beta_0 = 0$, corresponding to a ray incident parallel to the optical axis, and let $x_2 = 0$, corresponding to the intersection of the reflected ray with the optical axis, as in Figure 1-41b, then the corresponding value of t_1 gives the focal length of the mirror. Letting $\beta_0 = 0$ in Eqs. (1-131) and (1-136), and $x_2 = 0$ in Eq. (1-137), we find that the focal length of the mirror is given by

$$f_1' = \frac{R_1}{2} \ ,$$
(1-138)

which is in agreement with Eq. (1-90). The reflecting power of the mirror is given by

$$K \equiv n'/f' = -1/f_1' \ .$$
(1-139)

1.4.6 Two-Mirror System

We, now consider, as shown in Figure 1-42, an imaging system consisting of two mirrors M_1 and M_2, of radii of curvature R_1 and R_2 separated by a distance t_1, and determine its focal length and its principal and focal points. Starting with a ray incident parallel to the optical axis, we apply Eqs. (1-131) and (1-136) recursively as follows:

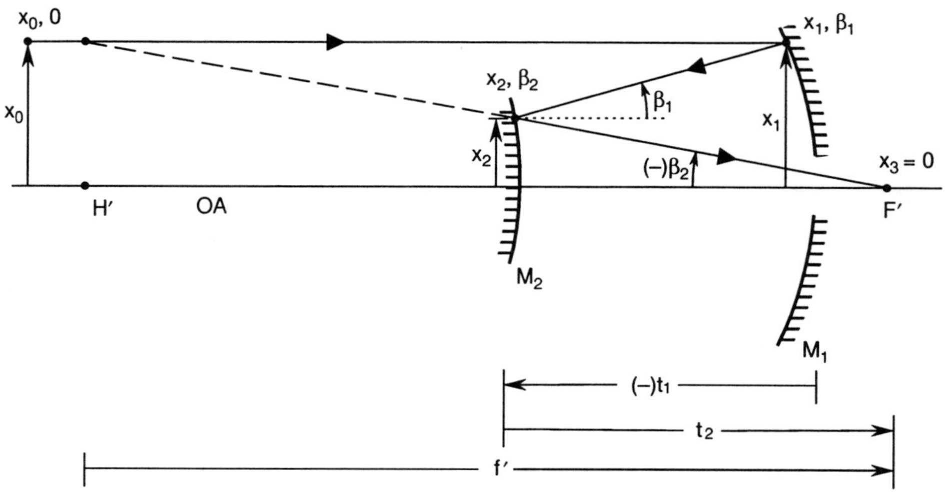

Figure 1-42. Paraxial ray tracing of a two-mirror system to determine its focal point F' and principal point H'.

$$x_1 = x_0 + t_0\beta_0$$

$$x_2 = x_1 + t_1\beta_1$$

$$= x_0\left(1 - \frac{2t_1}{R_1}\right) \quad ,$$

$$\beta_2 = -\beta_1 - \frac{2x_2}{R_2}$$

$$= 2x_0\left[\frac{1}{R_1} - \frac{1}{R_2}\left(1 - \frac{2t_1}{R_1}\right)\right] \quad ,$$

$$\frac{1}{f'} = -\frac{\beta_2}{x_0}$$

or

$$\boxed{\frac{1}{f'} = -2\left(\frac{1}{R_1} - \frac{1}{R_2} + \frac{t_1}{R_1 R_2}\right) \quad ,} \tag{1-140}$$

$$x_3 = x_2 + t_2\beta_2$$

$$= 0 \text{ for a focal point,}$$

$$t_2 = -\frac{x_2}{\beta_2}$$

or

$$\boxed{t_2 = f'\left(1 - \frac{2t_1}{R_1}\right)} \ .$$

(1-141)

The quantity t_2 locates the focal point F' and represents the *image-space focal distance* of the system. The principal point H' is located by noting that $H'F = f'$. A positive value of f' implies that H' lies to the left of F' at a distance f' from it. Similarly, by considering a ray incident parallel to the optical axis from right to left, the location of the focal point F and the principal point H can be determined. We find that F lies at a distance $f(1 - 2t_1/R_2)$ from M_1, where $f = -f'$ is the object-space focal length of the system and a negative value of t_1 has been maintained. This distance is the *object-space focal distance* of the system.

Letting $f_1' = R_1/2$ and $f_2' = R_2/2$ denote the focal lengths of the mirrors, Eq. (1-140) for the focal length of the system can be written

$$\frac{1}{f'} = -\frac{1}{f_1'} + \frac{1}{f_2'} - \frac{t_1}{f_1'f_2'} \ .$$

(1-142)

In terms of the equivalent focal lengths of the mirrors, $f_{e1} = -R_1/2$ and $f_{e2} = R_2/2$ defined by Eqs. (1-93) and (1-94), it can also be written

$$\boxed{\frac{1}{f'} = \frac{1}{f_{e1}} + \frac{1}{f_{e2}} - \frac{t}{f_{e1}f_{e2}}} \ ,$$

(1-143)

where we have let t be the magnitude (so that it is numerically positive) of the spacing between the mirrors. We note that $f' \to \infty$ if $t = f_{e1} + f_{e2}$, i.e., the system becomes afocal if the mirrors are confocal. An example of an afocal system is discussed in Section 6.7, where two confocal paraboloidal mirrors are considered as an astigmatic beam expander. If the spacing between the mirrors is less, then the system has a positive focal length and it is called a *Cassegrain telescope*. If it is larger, then the focal length of the system is negative and it is called a *Gregorian telescope*. These telescopes are discussed in Section 6.8, where the conic shapes of the mirrors are considered to reduce their aberrations. It should be evident from Figure 1-42 that the central portion of a bundle of rays incident on the primary mirror M_1 is blocked by the secondary mirror M_2. Hence, the image-forming beam is hollow on the inside; it is said to be *centrally obscured* in the case of an axial object lying at infinity.

1.4.7 Catadioptric System: Thin Lens-Mirror Combination

Finally, we consider a catadioptric system consisting of a thin lens of focal length f_l' and a concave mirror of focal length f_m' separated by a distance t, as illustrated in

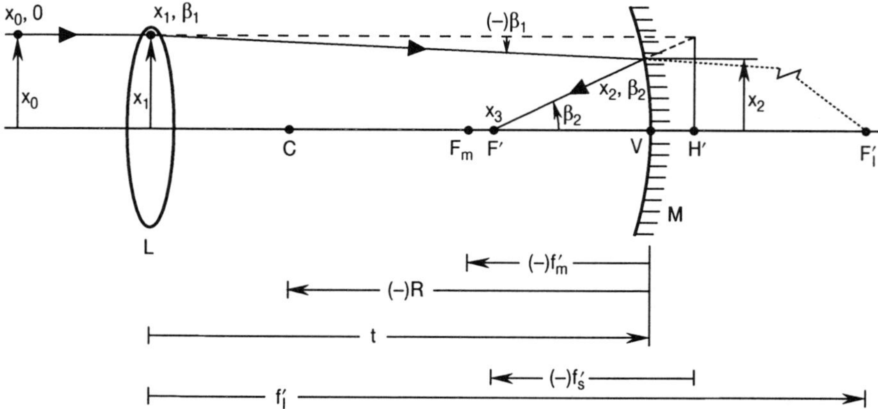

Figure 1-43. Catadioptric system consisting of a thin lens L of image-space focal length f_l' and a mirror M of radius of curvature R separated by a distance t. The dotted line is a continuation of the ray refracted by the lens and intersects the optical axis at the image-space focal point F_l' of the lens.

Figure 1-43, and determine its focal length. Applying the ray-tracing Eqs. (1-120) and (1-121) for a thin lens and Eqs.(1-131) and (1-136) for a mirror, we obtain the focal length f_s' of the system as follows:

$$x_1 = x_0 + t_0\beta_0$$

$$= x_0 \text{ for a ray incident parallel to the optical axis,}$$

$$\beta_1 = \beta_0 - \frac{x_1}{f_l'}$$

$$= -\frac{x_0}{f_l'} \quad ,$$

$$x_2 = x_1 + t_1\beta_1$$

$$= x_0\left(1 - \frac{t}{f_l'}\right) \quad ,$$

$$\beta_2 = -\beta_1 - \frac{x_2}{f_m'}$$

$$= x_0\left[\frac{1}{f_l'} - \frac{1}{f_m'}\left(1 - \frac{t}{f_l'}\right)\right]$$

$$= -\frac{x_0}{f_s'} \quad ,$$

where the negative sign in the last step accounts for the fact that β_2 is numerically positive but f_s' is numerically negative in the figure. Thus, the focal length of the system is given by

$$\frac{1}{f'_s} = -\left(\frac{1}{f'_l} - \frac{1}{f'_m} + \frac{t}{f'_l f'_m}\right) .$$

(1-144)

The focusing power K_s and the equivalent focal length f_e of the system are given by

$$K_s \equiv \frac{n'_s}{f'_s} = -\frac{1}{f'_s} = \frac{1}{f_e} ,$$

(1-145)

where $n'_s = -1$ is the refractive index of the image space of the system. The distance t_2 of the focal point F' from the vertex V of the mirror is given by

$$x_3 = x_2 + t_2 \beta_2$$

$$= 0 \text{ for a focal point.}$$

Hence,

$$t_2 = -\frac{x_2}{\beta_2} ,$$

or

$$t_2 = f'_s\left(1 - \frac{t}{f'_l}\right) .$$

(1-146)

If the lens is placed at the center of curvature C of the mirror, then $t = -R = -2f'_m$, where R is the radius of curvature of the mirror, and Eqs. (1-144) and (1-146) reduce to

$$\frac{1}{f'_s} = \frac{1}{f'_l} + \frac{1}{f'_m} .$$

(1-147)

and

$$t_2 = f'_s\left(1 + \frac{2f'_m}{f'_l}\right) .$$

(1-148)

Note that there is only one principal point and one focal point. Hence, the reference point for both object and image distances is either the principal point H' or the focal point F' depending on whether the Gaussian or the Newtonian imaging equation is used. Equation (1-147) will be used in Section 6.6.3, where we discuss the Bouwers-Maksutov camera.

1.5 TWO-RAY LAGRANGE INVARIANT

In earlier sections we have shown that the Lagrange invariant, which is the product of the slope angle of a ray from an axial point object, object height, and the refractive index of the object space, is invariant upon refraction or reflection by a surface, and hence for a system consisting of any number of such surfaces. Now we consider this invariant in terms of the heights and slopes of two arbitrary rays incident on the system.

We show how this invariant reduces to that for finite or infinite conjugates. We also show that the slope and the height of any other ray incident on the system can be obtained anywhere in space as a linear combination of the slopes and heights of the other two in that space.

Consider, as illustrated in Figure 1-44, two linearly independent rays (so that one is not a scaled version of the other) incident at heights x_0 and x with slope angles β_0 and β on a refracting surface of radius of curvature R separating media of refractive indices n and n'. From Eq. (1-110), the slope angles β_0' and β' of the corresponding refracted rays are given by

$$n'\beta_0' \;=\; n\beta_0 + (n-n')\frac{x_0}{R} \tag{1-149}$$

and

$$n'\beta' \;=\; n\beta + (n-n')\frac{x}{R} \;\;. \tag{1-150}$$

From Eqs. (1-149) and (1-150), we find that

$$n'\left(\beta_0'x - \beta'x_0\right) \;=\; n\left(\beta_0 x - \beta x_0\right) \;\;, \tag{1-151}$$

showing that the quantity $n\left(\beta_0 x - \beta x_0\right)$, called the *two-ray Lagrange invariant*, is invariant upon refraction of the rays. If we let x_0' and x' be the heights of the rays in a plane at a distance t from the refracting surface, we find from Eq. (1-105) that

$$x_0' = x_0 + t\beta_0' \tag{1-152}$$

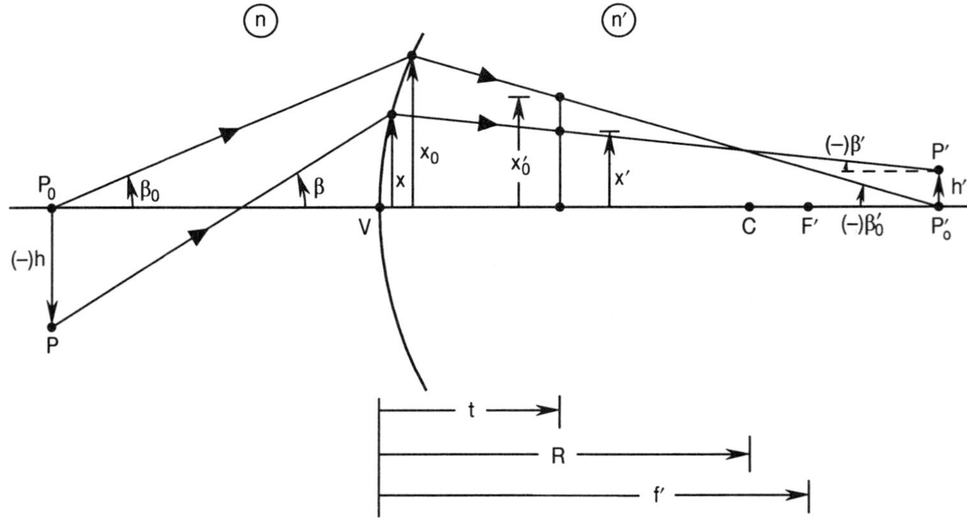

Figure 1-44. Lagrange invariant of two rays incident on a refracting surface of radius of curvature R separating media of refractive indices n and n'.

and

$$x' = x + t\beta' \quad . \tag{1-153}$$

From Eqs. (1-152) and (1-153), we find that

$$n'\left(\beta_0' x' - \beta' x_0'\right) = n'\left(\beta_0' x - \beta' x_0\right) \quad . \tag{1-154}$$

Thus, the quantity $n'\left(\beta_0' x - \beta' x_0\right)$ remains invariant upon transfer of the rays from one plane to another. Hence, the two-ray Lagrange invariant remains the same throughout the optical system including the object and image spaces. From Eqs. (1-151) and (1-154), we may write

$$\boxed{n\left(\beta_0 x - \beta x_0\right) = n'\left(\beta_0' x' - \beta' x_0'\right)} \quad , \tag{1-155}$$

showing the equality of the Lagrange invariant in object and image spaces.

If we consider the Lagrange invariant in two conjugate planes passing through P_0 and P_0' so that $x_0 = 0$, $x = h$, $x_0' = 0$, and $x' = h'$, as in Figure 1-10, we find that Eq. (1-155) reduces to Eq. (1-28). If one of the conjugates lies at infinity, we let $\beta_0 = 0$ and find that the object space-Lagrange invariant reduces to $-nx_0\beta$, as discussed in Section 1.3.5.2. For the refracting surface, it is easy to show that the various expressions for the Lagrange invariant are equal to each other. For example, in Figure 1-45, $-nx_0\beta$ and $n'\left(\beta_0' x - \beta' x_0\right)$ at the surface, and $n'h'\beta_0'$ at the image plane, are all equal to $nf'\beta_0'\beta$,

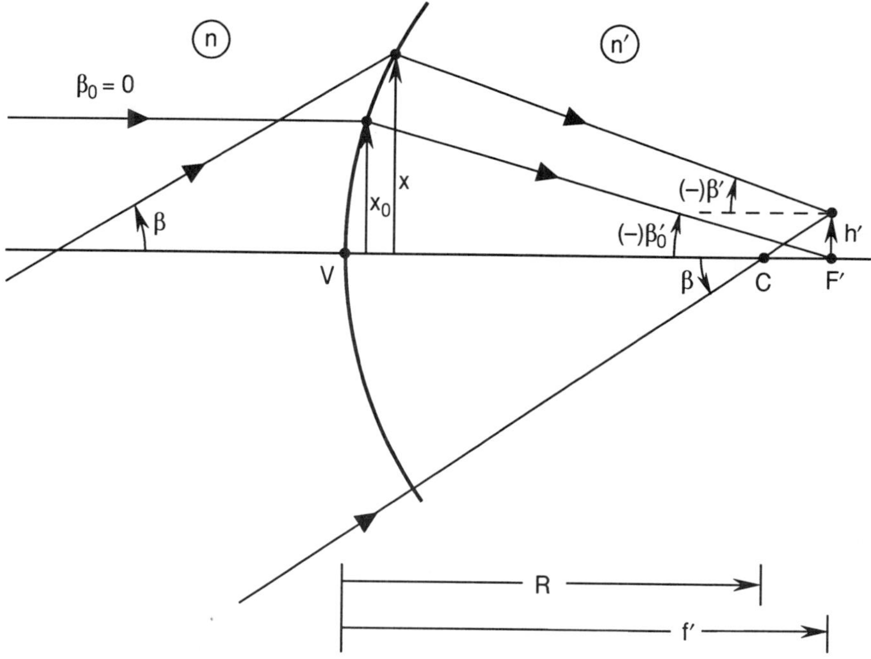

Figure 1-45. Lagrange invariant of a refracting surface for an object at infinity.

where f' is the image-space focal length of the refracting surface. If the system is afocal, as in Figure 1-26a, then β_0 and β_0' are both equal to zero, and the image-space Lagrange invariant reduces to $-n'x_0'\beta'$, thus yielding Eq. (1-79).

Now we show that if the slope and height of any third ray are known in a certain space, they can be obtained in any other space, without tracing it, as a linear combination of the values of the two rays in that other space. Let the slope of the rays in a certain space be β_1, β_2, and β_3. Let their heights in a certain plane in that space be x_1, x_2, and x_3, respectively. The two-ray Lagrange invariant in this plane for the rays may be written

$$L_{12} = n(\beta_1 x_2 - \beta_2 x_1) \ , \tag{1-156a}$$

$$L_{13} = n(\beta_1 x_3 - \beta_3 x_1) \ , \tag{1-156b}$$

and

$$L_{23} = n(\beta_2 x_3 - \beta_3 x_2) \ . \tag{1-156c}$$

Using primes for the corresponding quantities in another space, the Lagrange invariant in terms of the quantities in this space may be written

$$L_{12} = n'(\beta_1' x_2' - \beta_2' x_1') \ , \tag{1-157a}$$

$$L_{13} = n'(\beta_1' x_3' - \beta_3' x_1') \ , \tag{1-157b}$$

and

$$L_{23} = n'(\beta_2' x_3' - \beta_3' x_2') \ , \tag{1-157c}$$

respectively. From these equations we find that the height and the slope of the third ray in the other space are given by

$$x_3' = \frac{L_{13} x_2' - L_{23} x_1'}{L_{12}} \tag{1-158}$$

and

$$\beta_3' = \frac{L_{13}\beta_2' - L_{23}\beta_1'}{L_{12}} \ . \tag{1-159}$$

As discussed later in Section 2.2.3, the two object rays that are traced in practice are the marginal ray from an on-axis point object and the chief ray from a point on the edge of the object. The first ray passes through the edges of the entrance and exit pupils and so determines their sizes. It also passes through the center of the image and thereby locates it. The second ray passes through the centers of the entrance and exit pupils and, therefore, determines their locations. It also passe through the edge of the image and thus determines its size.

1.6 MATRIX APPROACH TO PARAXIAL RAY TRACING AND GAUSSIAN OPTICS

1.6.1 Introduction

We have seen that the transfer, refraction, and reflection ray-tracing equations are linear in ray heights and slopes. Hence, they can be written in the form of a matrix equation[9-13]. Each operation, e.g., the transfer of a ray from one surface to another, or its refraction at or reflection from a surface, is represented by a 2×2 matrix. Accordingly, a series of such operations is represented by the product of 2×2 matrices resulting in a 2×2 matrix for the whole series. In this section we discuss the system matrix representing the tracing of a ray from its point of incidence on to its point of emergence from the system. We also discuss a conjugate matrix representing the tracing of a ray

from an object point to its Gaussian image point. The relationships between the elements of the system matrix and the Gaussian parameters of the system are given. Matrix approach to Gaussian optics in terms of the imaging and magnification equations, and cardinal points is also described. The case of a mirror as an imaging element is not considered separately, since it can treated as a special case of a refracting surface by letting the refractive index associated with the reflected rays equal to the negative of the refractive associated with the incident rays.

1.6.2 System Matrix

Consider, as an example, the propagation of a ray from a point A_0 in a medium of refracting index n_0 to a point A_3 in a medium of refractive index n_3, through an optical system made up of two refracting surfaces of radius of curvature R_1 and R_2 separated by a distance t_1, as shown in Figure 1-46. The first surface separates media of refractive indices n_1 and n_1', where $n_1 = n_0$. Similarly, the second surface separates media of refractive indices n_2 and n_2', where $n_2 = n_1'$. Consider a ray $A_0 A_1$ incident on the first surface at a point A_1 with a slope β_0. Let x_0 and x_1 be the heights of the points A_0 and A_1, respectively, from the optical axis. From the transfer Eq. (1-105), the height and slope of the ray at a point slightly to the left of A_1 may be written

$$x_1 = x_0 + t_0 \beta_0 \qquad (1-160)$$

and

$$n_1 \beta_1 = n_0 \beta_0 \quad , \qquad (1-161)$$

where t_0 is the axial distance of the surface from A_0 and $n_1 = n_0$. We note that in a transfer operation, the height of the ray changes, but its slope does not. Equations (1-160) and (1-161) can be written in a matrix form

$$\begin{pmatrix} x_1 \\ n_1 \beta_1 \end{pmatrix} = T_{01} \begin{pmatrix} x_0 \\ n_0 \beta_0 \end{pmatrix} \quad , \qquad (1-162)$$

where

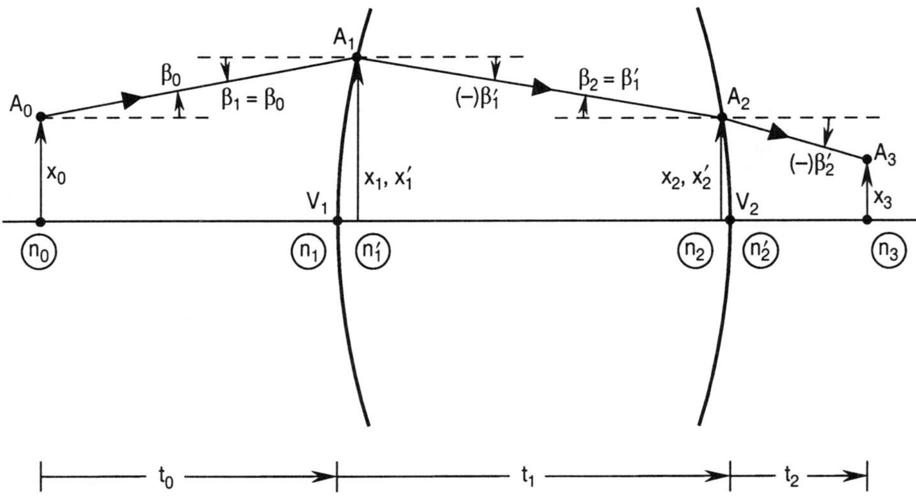

Figure 1-46. Propagation of a ray through an optical system consisting of two refracting surfaces.

$$T_{01} = \begin{pmatrix} 1 & t_0/n_0 \\ 0 & 1 \end{pmatrix} \tag{1-163}$$

represents the *transfer matrix* of the ray in propagating from the point A_0 to the point A_1.

The ray is refracted by the surface at the point A_1. Let the slope of the refracted ray be β_1'. If we represent the height of the ray at a point just to the right of A_1 by x_1', then we may write for the height and slope of the ray at the this point [see Eq. (1-110)]

$$x_1' = x_1 \tag{1-164}$$

and

$$n_1'\beta_1' = n_1\beta_1 + \frac{n_1 - n_1'}{R_1} \quad , \tag{1-165}$$

where n_1 and n_1' are the refractive indices of the media separated by the first surface. We note that in a refraction operation, the ray height does not change but its slope does. Equations (1-164) and (1-165) can be written in a matrix form

$$\begin{pmatrix} x_1' \\ n_1'\beta_1' \end{pmatrix} = R_1 \begin{pmatrix} x_1 \\ n_1\beta_1 \end{pmatrix} \quad , \tag{1-166}$$

where

$$R_1 = \begin{pmatrix} 1 & 0 \\ -K_1 & 1 \end{pmatrix} \tag{1-167}$$

represents the *refraction matrix* for the refraction of the ray at the first surface and

$$K_1 = \frac{n_1' - n_1}{R_1} \tag{1-168}$$

is the refracting power of this surface.

The propagation of the ray from the point A_1 to a point A_2 on the second refracting surface is again a transfer operation. The height x_2 and slope β_2 of the ray at a point just to the left of A_2 are given by

$$x_2 = x_1' + t_1 \beta_1' \tag{1-169}$$

and

$$n_2 \beta_2 = n_1' \beta_1' \quad , \tag{1-170}$$

respectively, where $n_2 = n_1'$ and, therefore, $\beta_2 = \beta_1'$. Equations (1-169) and (1-170) can be written in a matrix form

$$\begin{pmatrix} x_2 \\ n_2 \beta_2 \end{pmatrix} = T_{12} \begin{pmatrix} x_1' \\ n_1' \beta_1' \end{pmatrix} \quad , \tag{1-171}$$

where

$$T_{12} = \begin{pmatrix} 1 & t_1/n_1' \\ 0 & 1 \end{pmatrix} \tag{1-172}$$

represents the transfer matrix of the ray in propagating from the point A_1 to the point A_2.

Next the ray is refracted at the point A_2 by the second refracting surface. Let the slope of the refracted ray be β_2'. The height x_2' and slope β_2' of the ray at a point just to the right of A_2 are given by

$$x_2' = x_2 \tag{1-173}$$

and

$$n_2' \beta_2' = n_2 \beta_2 + \frac{n_2 - n_2'}{R_2} \quad , \tag{1-174}$$

where $n_2 = n_1'$ and n_2' are the refractive indices of the media separated by the second surface. Equations (1-173) and (1-174) can be written in a matrix form

$$\begin{pmatrix} x_2' \\ n_2' \beta_2' \end{pmatrix} = R_2 \begin{pmatrix} x_2 \\ n_2 \beta_2 \end{pmatrix} \quad , \tag{1-175}$$

where

$$R_2 = \begin{pmatrix} 1 & 0 \\ -K_2 & 1 \end{pmatrix} \tag{1-176}$$

represents the refraction matrix of the ray at the second surface and

$$K_2 = \frac{n'_2 - n_2}{R_2} \tag{1-177}$$

is the refracting power of this surface.

Finally, the propagation of the ray from the point A_2 to the point A_3 is again a transfer operation. The height x'_3 and slope β_3 of the ray at the point A_3 are given by

$$x_3 = x'_2 + t_2\beta'_2 \tag{1-178}$$

and

$$n_3\beta_3 = n'_2\beta'_2 \quad , \tag{1-179}$$

respectively, where t_2 is the axial distance of the point A_3 from the second surface and $n_3 = n'_2$. Equations (1-178) and (1-179) can be written in a matrix form

$$\begin{pmatrix} x_3 \\ n_3\beta_3 \end{pmatrix} = T_{23}\begin{pmatrix} x'_2 \\ n'_2\beta'_2 \end{pmatrix} \quad , \tag{1-180}$$

where

$$T_{23} = \begin{pmatrix} 1 & t_2/n'_2 \\ 0 & 1 \end{pmatrix} \tag{1-181}$$

represents the transfer matrix of the ray in propagating from the point A_2 to the point A_3.

From Eqs. (1-162), (1-166), (1-171), (1-175), and (1-180), the propagation of the ray from the point A_0 to the point A_3 through the optical system consisting of two surfaces can be described by

$$\begin{pmatrix} x_3 \\ n_3\beta_3 \end{pmatrix} = T_{23}MT_{01}\begin{pmatrix} x_0 \\ n_0\beta_0 \end{pmatrix} \quad , \tag{1-182}$$

where

$$M = R_2T_{12}R_1 \tag{1-183}$$

represents the *system matrix* for propagation of the ray from a point just to the left of the point A_1 to a point just to the right of the point A_2. We note that the progression of a ray, as it propagates undergoing refractions and transfers from the left to the right in a system, is represented by a sequence of the matrix operations ordered from right to left. The ordering of the matrices in this manner is essential since two matrices do not commute (unless one of them is diagonal). We also note that the determinant of each matrix is unity. Hence, the determinant of the product of any number of such matrices is also unity. Thus, for example, if

$$M = \begin{pmatrix} A & B \\ C & D \end{pmatrix} \tag{1-184}$$

is the system matrix of an optical system, then its determinant

$$AD - BC = 1 \quad . \tag{1-185}$$

Hence, only three of the four elements of the system matrix are independent of each other. By repeated application of the refraction and transfer operations, the system matrix of any optical system can be determined, just as we did above for a system with two refracting surfaces.

1.6.3 Conjugate Matrix

To relate the matrix elements A, B, C, and D of the system matrix to the Gaussian parameters of a system, we consider the propagation of a ray from a point object P at a height x in the object plane to its Gaussian conjugate point P' at a height x' in the

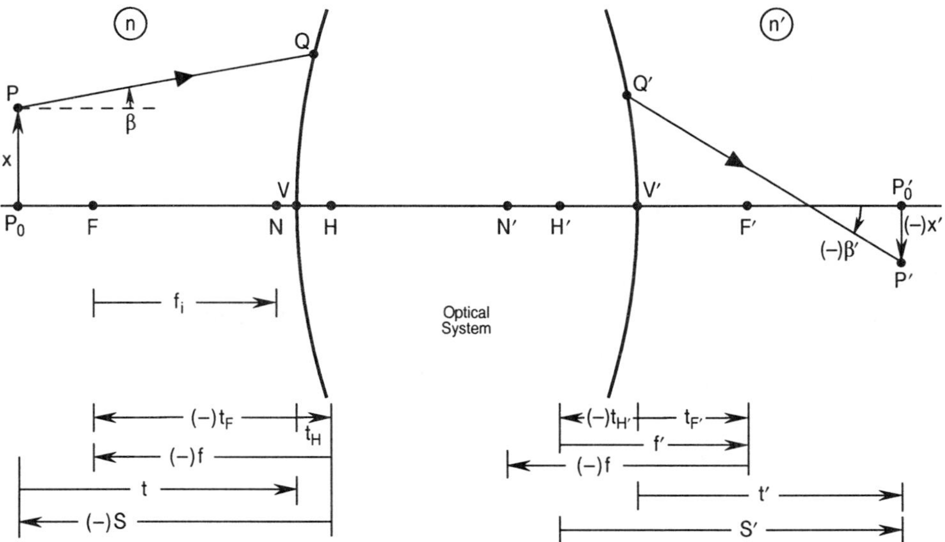

Figure 1-47. Gaussian imaging of a point object P at a height x by an optical system. The image lies at P' at a height x'. H and H' are the object- and image-space principal points of the system. Similarly, F and F', and N and N' are the object- and image-space focal and nodal points, respectively. The refractive indices of the object and image spaces are n and n', repectively. The object lies in a plane at a distance S from H and the image lies at a distance S' from H'.

image plane, as illustrated in Figure 1-47. Let t be the distance of the first imaging surface of the system from the object plane and t' be the distance of the image plane from the last imaging surface. Consider an object ray with a slope β. Let the slope of the corresponding ray in the image space be β'. The height and slope of the ray at the image point P' are related to the height and slope of the ray at the object point P according to

$$\begin{pmatrix} x' \\ n'\beta' \end{pmatrix} = N_{PP'} \begin{pmatrix} x \\ n\beta \end{pmatrix} \quad , \tag{1-186}$$

where n and n' are the refractive indices of the object and image spaces, respectively, and

$$N_{PP'} = \begin{pmatrix} N_{11} & N_{12} \\ N_{21} & N_{22} \end{pmatrix} \tag{1-187}$$

$$= \begin{pmatrix} 1 & t'/n' \\ 0 & 1 \end{pmatrix} \begin{pmatrix} A & B \\ C & D \end{pmatrix} \begin{pmatrix} 1 & t/n \\ 0 & 1 \end{pmatrix} \tag{1-188}$$

$$= \begin{pmatrix} A + C(t'/n') & A(t/n) + B + (t'/n')\big[C(t/n) + D\big] \\ C & C(t/n) + D \end{pmatrix} \tag{1-189}$$

is the conjugate matrix representing the propagation of the ray from the object point P to the image point P'. From Eqs. (1-186) and (1-187), it is evident that

$$x' = N_{11}x + N_{12}n\beta \tag{1-190}$$

and

$$n'\beta' = N_{21}x + N_{22}n\beta \quad . \tag{1-191}$$

Since P' is the image of P, all rays originating at P and transmitted by the system must pass through P'. Therefore, x' in Eq. (1-190) must be independent of β, thus yielding $N_{12} = 0$. Substituting $N_{12} = 0$ in Eqs. (1-189) and (1-190) yields the image location

$$\frac{t'}{n'} = -\frac{A(t/n) + B}{C(t/n) + D} \tag{1-192}$$

and

$$x' = N_{11}x \quad . \tag{1-193}$$

Moreover, the conjugate matrix of Eq. (1-189) reduces to

$$N_{PP'} = \begin{pmatrix} A + C(t'/n') & 0 \\ C & C(t/n) + D \end{pmatrix} \quad . \tag{1-194}$$

Since

$$M_t = x'/x \tag{1-195}$$

is the transverse magnification of the image, Eq. (1-193) yields

$$\begin{aligned} N_{11} &= M_t \\ &= A + C\left(t'/n'\right) \quad . \end{aligned} \tag{1-196}$$

Moreover, since the determinant of any ray-tracing matrix is unity,

$$\begin{aligned} N_{22} &= 1/N_{11} = 1/M_t \\ &= C(t/n) + D \quad . \end{aligned} \tag{1-197}$$

Hence, the conjugate matrix may be written

$$N_{PP'} = \begin{pmatrix} M_t & 0 \\ C & 1/M_t \end{pmatrix} \quad . \tag{1-198}$$

Now by definition, the principal planes correspond to $M_t = 1$. Hence, the conjugate matrix for the principal planes becomes

$$N_{HH'} = \begin{pmatrix} 1 & 0 \\ C & 1 \end{pmatrix} \quad , \tag{1-199}$$

where H and H' are the object-space and image-space principal points of the system. Comparing Eqs. (1-194) and (1-199), we find that the locations of the principal planes are given by

$$\boxed{t_{H'} = n'(1-A)/C} \tag{1-200}$$

and

$$\boxed{t_H = n(D-1)/C \quad .} \tag{1-201}$$

Note that $t_{H'}$ is the distance of the image-space principal plane from the last surface of the system and t_H is the distance of the object-space principal plane from the first surface of the system. Since t is the distnace of the first surface from the object plane (rather than the distance of the object plane from the first surface), we substituted $-t_H$ for t in obtaining Eq. (1-201).

If parallel rays are incident on the system, they all come to a focus in the image-space focal plane of the system, i.e., rays with different values of x but the same value of β in the object space have the same value of x' in the image space. Hence, Eq. (1-190) yields $N_{11} = 0$ and Eq. (1-189), in turn, implies that the image-space focal distance is given by

$$\boxed{t_{F'} = -n'A/C \quad .} \tag{1-202}$$

Similarly, if rays with different slopes from a point in the object-space focal plane are incident on the system, they all emerge from the system parallel to each other. Thus, β' must be independent of β in Eq. (1-191), implying that $N_{22} = 0$. Hence, Eq. (1-194) yields that the object-space focal distance is given by

$$\boxed{t_F = nD/C} \quad .$$

(1-203)

As in the case of t_H, since t_F is the distance of the object-space focal point F from the first surface, $-t_F$ was substituted for t in obtaining Eq. (1-203). By definition, the image-space focal length is the distance of the focal plane from the principal plane in this space. Thus, it is given by

$$\boxed{\begin{aligned} f' &= t_{F'} - t_{H'} \\ &= -n'/C \end{aligned}} \quad ,$$

(1-204)

where we have made use of Eqs. (1-200) and (1-202). The object-space focal length is the distance of the focal plane from the principal plane in this space, i.e., it is given by

$$\boxed{\begin{aligned} f &= t_F - t_H \\ &= n/C \end{aligned}} \quad ,$$

(1-205)

where we have made use of Eqs. (1-201) and (1-203). By definition, the refracting power of a system is given by

$$K = \frac{n'}{f'} = -\frac{n}{f} \quad .$$

(1-206)

Hence, comparing Eq. (1-206) with Eq. (1-204) or (1-205), we obtain

$$C = -K \quad .$$

(1-207)

Substituting Eq. (1-207) into Eq. (1-198), we finally obtain the conjugate matrix for general imaging

$$\boxed{N_{PP'} = \begin{pmatrix} M_t & 0 \\ -K & 1/M_t \end{pmatrix}} \quad .$$

(1-208)

Similarly, substituting Eq. (1-207) into Eq. (1-199), we obtain the conjugate matrix for the principal planes

$$N_{HH'} = \begin{pmatrix} 1 & 0 \\ -K & 1 \end{pmatrix} \quad .$$

(1-209)

Of course, Eq. (1-208) reduces to Eq. (1-209) for the principal planes by letting $M_t = 1$.

As an example, consider a thick lens of refractive index n, thckness t, and surfaces with radii of curvature R_1 and R_2 surrounded by air. Its system matrix can be obtained by substituting Eqs. (1-167), (1-172), and (1-176) into Eq. (1-183) and letting $n_1 = 1$, $n_1' = n = n_2$, $n_2' = 1$, and $t_1 = t$. Thus, we find that

$$M = \begin{pmatrix} 1 - K_1(t/n) & t/n \\ -K_1 - K_2 + K_1 K_2(t/n) & 1 - K_2(t/n) \end{pmatrix} . \tag{1-210}$$

Hence, Eq. (1-204), (1-205), and (1-207) yield the focal lengths f and f' of the lens. Similarly, Eqs. (1-200) through (1-203) yield Eqs. (1-126) through (1-128), where $V_2 H' = t_{H'}$, $V_1 H = t_H$, $V_2 F' = t_{F'}$, and $V_1 F = t_F$.

1.6.4 System Matrix in Terms of Gaussian Parameters

If the Gaussian parameters of a system are known, the system matrix can be constructed from them. Substituting Eq. (1-207) into Eq. (1-200) and (1-201), we obtain the matrix elements A and D in the form

$$A = 1 + K t_{H'}/n' \tag{1-211}$$

and

$$D = 1 - K t_H/n \ . \tag{1-212}$$

Substituting Eqs. (1-207), (1-211), and (1-212) into Eq. (1-185), we obtain

$$B = \frac{t_H}{n} - \frac{t_{H'}}{n'} + \frac{K t_H t_{H'}}{nn'} \ . \tag{1-213}$$

Hence, Eq. (1-184) for the system matrix can be written in terms of the Gaussian parameters of the system in the form

$$M = \begin{pmatrix} 1 + K\left(t_{H'}/n'\right) & \dfrac{t_H}{n} - \dfrac{t_{H'}}{n'} + \dfrac{K t_H t_{H'}}{nn'} \\ -K & 1 - K\left(t_H/n\right) \end{pmatrix} . \tag{1-214}$$

It can be shown that the diagonal elements of the system matrix of a system interchange when the system is reversed (by rotating it by 180 degrees about an axis normal to its optical axis). It can also be shown that the diagonal elements of the system matrix of a symmetrical system (so that it does not change when reversed) are equal to each other (see Problem 1-18).

1.6.5 Gaussian Imaging Equations

If $S = -t$ is the distance of an object from the object-space principal point H, then the distance $S' = t'$ of the image from the image-space principal point H' can be obtained by substituting for the system matrix the conjugate matrix for the principal planes given by Eq. (1-209) into Eq. (1-188) and comparing the result obtained with Eq.

(1-208). The negative sign in $S = -t$ accounts for the fact that a thickness is considered positive when a ray propagates through it from left to right, but the object distance S is numerically negative when the object lies to the left of H. Thus, we may write

$$\begin{pmatrix} 1 & S'/n' \\ 0 & 1 \end{pmatrix} \begin{pmatrix} 1 & 0 \\ -K & 1 \end{pmatrix} \begin{pmatrix} 1 & -S/n \\ 0 & 1 \end{pmatrix} = \begin{pmatrix} M_t & 0 \\ -K & 1/M_t \end{pmatrix} \; , \tag{1-215}$$

or

$$\begin{pmatrix} 1 - K\dfrac{S'}{n'} & -\dfrac{S}{n} + \dfrac{S'}{n'}\left(1 + K\dfrac{S}{n}\right) \\ -K & 1 + K\dfrac{S}{n} \end{pmatrix} = \begin{pmatrix} M_t & 0 \\ -K & 1/M_t \end{pmatrix} \; . \tag{1-216}$$

Hence, comparing the (1,2) and (1,1) matrix elements on both sides of Eq. (1-216), we obtain

$$\frac{n'}{S'} - \frac{n}{S} = K \tag{1-217}$$

and

$$M_t = \frac{nS'}{n'S} \; . \tag{1-218}$$

Equations (1-217) and (1-218) are the Gaussian imaging and magnification equations, respectively, discussed in Section 1.3.5.

From Eq. (1-191), the angular magnification M_β of a ray originating at an axial point object $(x = 0)$ is given by

$$M_\beta = \beta'/\beta = (n/n')N_{22} = n/n'M_t \; , \tag{1-219}$$

where we have let $N_{22} = 1/M_t$ from Eq. (1-208). Substituting Eq. (1-195) into Eq. (1-219), we obtain the Lagrange invariant equation

$$n'x'\beta' = nx\beta \; . \tag{1-220}$$

Except for notation, Eq. (1-220) is the same as Eq. (1-69). See Figure 1-22 also.

Since by definition, the nodal points N and N' are the conjugate points with an angular magnification of unity, Eq. (1-219) shows that they also lie in conjugate planes with a transverse magnification of n/n'. Hence, Eq. (1-218) shows that their distances HN and $H'N'$ from their respective principal points are equal. For the nodal points, letting $S = S'$ in Eq. (1-217) and utilizing Eq. (1-206), we obtain

$$S = S' = f + f' \; . \tag{1-221}$$

In a similar manner the Newtonian imaging equations of Section 1.3.5.4 (see Problem 1.19) and Eq. (1-82) for the longitudinal magnification of an afocal system can be obtained (see Problem 1.21).

REFERENCES

1.   M. V. Klein, *Optics*, Wiley, New York (1970).

2.   P. Mouroulis and J. Macdonald, *Geometrical Optics and Optical Design*, Oxford, New York (1997).

3.   M. Born and E. Wolf, *Principles of Optics*, Pergamon, New York (1985).

4.   W. T. Welford, *Aberrations of the Symmetrical Optical System*, Academic Press, New York (1974).

5.   D. J. Schroeder, *Astronomical Optics*, Academic Press, New York (1987).

6.   F. A. Jenkins and H. E. White, *Fundamentals of Optics*, 4th ed., McGraw Hill, New York, (1976).

7.   M. V. Klein and T. E. Furtak, *Optics*, Wiley, New York (1988).

8.   E. Hecht and A. Zajac, *Optics*, Addison–Wesley, Reading, Massachusetts (1973).

9.   W. B. Wetherell, "Afocal lenses," in *Applied Optics and Optical Engineering*, eds. R. R. Shannon and J. C. Wyant, Vol. X, pp. 109–192, Academic Press, Orlando, Florida (1987); also, "Afocal systems," in *Handbook of Optics*, ed. M. Bass, Vol I, 2.1–2.23, McGraw-Hill, New York (1995).

10.   K. Halback, "Matrix representation of Gaussian optics," *Am. J. Phys.* **32**, 90–108 (1964).

11.   D. C. Sinclair, "The specifications of optical systems by paraxial transfer matrices," in *Applications of Geometrical Optics*, ed. W. J. Smith, SPIE Proc. **39**, 141–149 (1973).

12.   A. Gerrard and J. M. Burch, *Introduction to Matrix Methods in Optics*, Wiley, New York (1975).

13.   W. Brouwer, *Matrix Methods in Optical Instrument Design*, W. A. Benjamin, New York (1964).

14.   W. Blaker, *Geometric Optics: The Matrix Theory*, Marcel Dekker, New York (1971).

PROBLEMS

Illustrate each problem by a diagram.

1.1 Show that the minimum separation between the real conjugates of a positive thin lens of focal length f' is $4f'$. A pair of real conjugates can be obtained for two positions of the lens. Show that if t is the spacing between the real conjugates and d is the distance between the positions of the lens, then the focal length of the lens can be determined as $f' = \left(t^2 - d^2\right)/4t$.

1.2 A *thin lens* with a focal length of 10 cm is located at a distance of 3 cm in front of a concave spherical mirror of radius of curvature 20 cm. (a) Determine the focal point and the principal point of the system. (b) Repeat the problem when the lens is in contact with the mirror.

1.3 A *thick lens* has a refractive index of 1.5. Its surfaces have radii of curvature of 10 cm and -25 cm. If the second surface is silvered and the lens is 2 cm thick, locate the focal point and the principal point of the system.

1.4 A *Mangin mirror* consists of a thin negative miniscus lens with a silvered back surface. Show that, if R_1 and R_2 are the radii of curvature of the lens and n is its refractive index, the focal length of the Mangin mirror is given by $f_s'^{-1} = 2nR_2^{-1} - 2(n-1)R_1^{-1}$.

1.5 Consider a thick *equiconvex lens* with radii of curvature $R_1 = 4\,\text{cm}$ and $R_2 = -4\,\text{cm}$, and refractive index $n = 1.5$. Calculate its focal length and sketch its principal and focal points if its thickness is 0.3 cm, 2 cm, 8 cm, 12 cm, 24 cm, or 36 cm.

1.6. Consider a *plane-parallel* plate of thickness t and refractive index n. (a) Considering it as a limiting case of a thick lens with surfaces having infinite radii of curvature, determine its principal and focal points. (b) Calculate the location and size of the image of an object lying at a distance S_o from its front surface. (c) Determine the location and size of the image of its front surface formed by its back surface. (d) Sketch the various quantities determined for $t = 1$ cm, $n = 1.5$, and $S_o = 3\,\text{cm}$.

1.7 The *human eye* may be represented in a simplified form as follows:

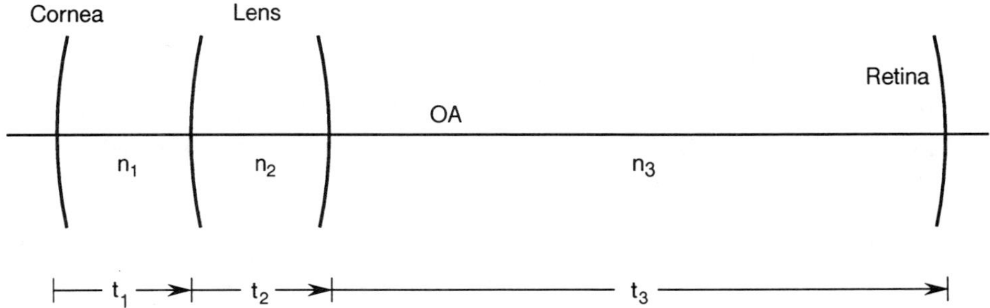

$$C_1 = 0.1282051 \qquad n_1 = 1.336 \qquad t_1 = 3.6$$

$$C_2 = 0.10 \qquad n_2 = 1.413 \qquad t_2 = 3.6$$

$$C_3 = 0.16667 \qquad n_3 = 1.336$$

(a) Determine t_3. (b) Determine the six cardinal points and show them on the axis. (c) Determine the cardinal points for an underwater swimmer. Indicate the changes from (b). Note that C_i is the *curvature* of a surface, i.e., it is the reciprocal of its radius of curvature. (Hint: One focal point is on the retina. Refractive index of water is 1.336.) Note: t is in units of mm and C is in units of mm^{-1}.

1.8   In a *nearsighted eye*, the focal point F' lies in front of the retina. Assume that the eye can be approximated as shown in the figure below such that F' is 23 mm from the cornea instead of 24.387 mm as in a normal eye. (a) Determine the prescription of a corrective lens placed 15 mm in front of the cornea that makes F' lie on the retina. (b) Repeat the calculation for a *contact lens*.

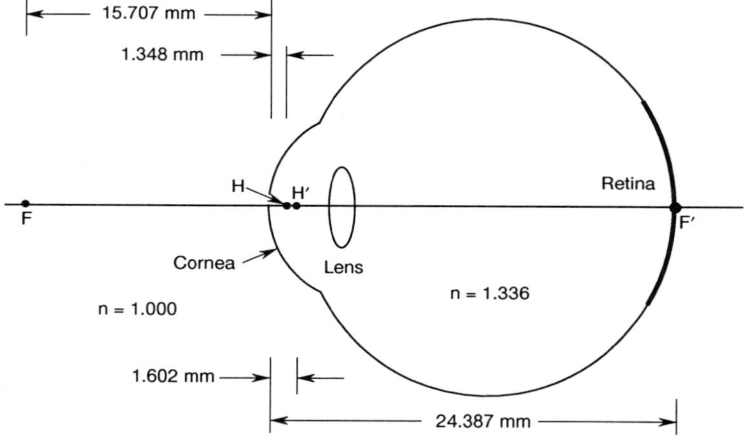

1.9   Two thin lenses of focal lengths $\pm f'$ are placed a distance f' apart. (a) Determine the focal points, principal points, and the focal length of the system. How does the order of the lenses affect the result? (b) Repeat the problem when the lenses have focal lengths of f' and $-f'/6$ and are placed a distance $2f'/3$ apart.

1.10   Consider a system of two thin lenses of focal lengths f_1' and f_2' spaced a distance t apart. (a) Determine its cardinal points if $f_1' = 2f_2'$ and $t = 0.5f_2'$, f_2', $1.5f_2'$ (*Huygens eyepiece*), $2f_2'$, and $3f_2'$ (*astronomical telescope*). (b) Repeat the problem if $f_1' = -2f_2'$ and $t = 0.5f_2'$, $-f_2'$ (*Galilean telescope*), and $-1.5f_2'$ (*telephoto lens*). Let $f_1' = 10\,\text{cm}$.

1.11   Consider an *afocal system* consisting of two lenses of equal focal lengths f' placed $2f'$ apart (a) Determine the transverse and longitudinal magnifications of the image of a nearby object. (b) Determine the space between the object and its image. Show that the position and size of the image do not change as the system is moved along its axis. (c) How are the imaging properties of the system affected if a third lens of focal length is placed at the common focal point of the first two. (d) As an example, consider $f' = 10$ cm and an object placed at a distance of 30 cm from the first lens.

1.12   The size of the image of a distant object depends on the focal length of the imaging system. A *telephoto lens* consisting of a positive lens and a negative lens is used to obtain a large image such that the back focal distance is kept small. (a) Design a telephoto lens with a focal length of 20 cm and a back focal distance of 4 cm. Let the focal length of the positive lens be 4 cm. (b) Determine the focal length and the back focal distance of the lens when it is reversed. Show that the reversed lens works as a *wide-angle lens*.

1.13   Consider a lens of refractive index n and thickness t with its two surfaces having *equal radii of curvature R*. (a) Show that the distance between its principal points is also equal to t. (b) Determine its principal and focal points for $n = 1.5$, $t = 2$ cm, and $R = 10$ cm.

1.14   Consider a *concentric lens* of refractive index n with its two surfaces having radii of curvature R_1 and R_2. Show that such a lens behaves as a negative thin lens placed at the common center of curvature of its two surfaces with a focal length that is n times the focal length of a thin lens of the same refractive index and surfaces with the same radii of curvature. Determine its principal and focal points for $n = 1.5$, $R_1 = 10$ cm, and $R_2 = 8$ cm.

1.15   Consider a *glass sphere* of radius of curvature R and refractive index n. This is an example of a concentric lens with its two surfaces having radii of curvature that are equal in magnitude but opposite in sign. (a) Determine its cardinal points. (b) Determine the position and relative size of the image of an object placed at a distance of 6 cm from its surface for $R = 3$ cm and $n = 1.5$. (c) For what value of refractive index is the image of an object at infinity found at the back surface of the lens?

1.16   For the *glass sphere* considered in Problem 1.15, determine the apparent position and relative size of a flower (a) imbedded at its center, (b) placed at a distance of R/n from its center and observed from the other side of the center. This problem illustrates the concept of a *contact magnifier*. A typical lens magnifier produces a magnified (virtual) image of an object placed between it and its front focal plane. A hemispherical or hyperhemispherical contact lens magnifier produces a magnified (virtual) image of an object placed in contact with its planar surface. It is shown in Section 5.4 that the images produced by these contact magnifiers are aplanatic, i.e. they are free of spherical aberration and coma. The image formed by the *aplanatic hyperhemispherical magnifier,* also called *Amici lens*, is free of astigmatism and all orders of spherical aberration and not just the primary (See

Problem 5.2). The contact magnifiers can be used in reverse as in *immersed detectors* where the image is focused on the detector, which is in contact with the planar surface of the hemispherical or the hyperhemispherical lens. The image on the detector is smaller in size by the magnification of the lens determined in parts (a) and (b). See R. C. Jones, "Immersed radiation detectors," *Appl. Opt.* **1**, 607–613 (1962).

1.17   By considering the ray-tracing equations given in Section 1.4, determine the system matrix for (a) refraction at a plane boundary separating media of refractive indices n and n', (b) a thin lens of focal length f', (c) reflection from a plane mirror, and (d) reflection from a mirror of radius of curvature R. Illustrate each case schematically by a diagram. Show that the system matrices obtained are in agreement with the corresponding matrices obtained by using Eq. (1-214).

1.18   (a) Show that the power of a system changes when it is reversed (by rotating it by 180 degrees about an axis normal to its optical axis) unless the refractive indices of its object and image spaces are equal. Consider a thin lens of refractive index 1.5 and radii of curvature 10 cm and -15 cm with air in its object space and water in its image space. Calculate its focal lengths and power. Now reverse the lens and repeat the calculations. (b) Show that the diagonal elements of the system matrix of a system with equal refractive indices for its object and image spaces interchange when the system is *reversed*. Also show that these elements are equal to each other if such a system is *symmetrical* (so that it does not change when reversed).

1.19   Given the conjugate matrix for the propagation of a ray from the object-space principal plane to the image-space principal plane [see Eq. (1-209)], obtain the matrix representing the propagation of a ray from the object-space focal point to the image-space focal point. Considering an object lying at distance z from the object-space focal plane, show that the image lies at a distance z' given by Eq. (1-78). Also show that the magnification of the image is given by Eq. (1-77).

1.20   Consider an *afocal system* forming the image of an object as in Figure 1-26b. Starting with its conjugate matrix, show that the longitudinal and transverse magnifications of the image are related to each other according to Eq. (1-82).

1.21   (a) Consider a *glass hemisphere* of radius of curvature R and refractive index n. Determine its focal length and principal points using the matrix approach. Illustrate by a diagram for $n = 1.5$ and $R = 2$ cm. Determine the image of an object lying at a distance of 6 cm from the vertex of the hemisphere. (b) Now, consider two hemispherical lenses with their plane surfaces facing each other separated by a distance t. Determine the focal length and the principal points of the system. Illustrate by a diagram for $t = 3$ cm.

CHAPTER 2

RADIOMETRY OF IMAGING

2.1   Introduction **91**

2.2   Stops, Pupils, and Vignetting **92**

    2.2.1   Introduction 92

    2.2.2   Aperture Stop, and Entrance and Exit Pupils 92

    2.2.3   Chief and Marginal Rays 94

    2.2.4   Vignetting 95

    2.2.5   Size of an Imaging Element 98

    2.2.6   Telecentric Aperture Stop 98

    2.2.7   Field Stop, and Entrance and Exit Windows 98

2.3   Radiometry of Point Sources **100**

    2.3.1   Irradiance of a Surface 100

    2.3.2   Flux Incident on a Circular Aperture 103

2.4   Radiometry of Extended Sources **104**

    2.4.1   Lambertian Surface 104

    2.4.2   Exitance of a Lambertian Surface 105

    2.4.3   Radiance of a Tube of Rays 106

    2.4.4   Irradiance by a Lambertian Surface Element 107

    2.4.5   Irradiance by a Lambertian Disc 108

2.5   Radiometry of Point Object Imaging **112**

2.6   Radiometry of Extended Object Imaging **114**

    2.6.1   Image Radiance 114

    2.6.2   Pupil Distortion...117

    2.6.3   Image Irradiance: Aperture Stop in Front of the System 118

    2.6.4   Image Irradiance: Aperture Stop in Back of the System 121

    2.6.5   Telecentric Systems 123

    2.6.6   Throughput 123

    2.6.7   Condition for Uniform Image Irradiance 123

    2.6.8   Concentric Systems 125

2.7   Photometry **126**

    2.7.1   Photometric Quantities and Spectral Response of the Human Eye.........126

    2.7.2   Imaging by a Human Eye 127

    2.7.3   Brightness of a Lambertian Surface 129

    2.7.4   Observing Stars in the Daytime 130

Appendix: Radiance Theorem **134**

References **136**

Problems **137**

Chapter 2
Radiometry of Imaging

2.1 INTRODUCTION

In Chapter 1, we showed how the position and size of the Gaussian image of an object formed by an imaging system can be determined from the radii of curvature of its surfaces and the refractive indices of the media around them. However, we did not consider the sizes of the imaging elements or the apertures in the system. Accordingly, no effort was made there to determine the cone of object rays that enters or exits from the imaging system. Such calculations are essential for the determination of the image intensity in terms of the object intensity, or the image irradiance in terms of the object radiance

We begin this chapter by introducing the concept of an *aperture stop* and its images, the *entrance* and *exit pupils* of an imaging system. The light cone from a point object that enters the system is limited by the entrance pupil. Similarly, the light cone that exits from the system and converges onto the image point is limited by the exit pupil. Certain specicosine-nd *marginal rays*, are defined. *Vignetting* or blocking of the rays from an off-axis point object by the aperture stop and/or other elements of the system, thus changing the effective shape of the stop and pupils, is explained. A *telecentric stop* is defined and its advantages are briefly discussed. The *field stop* and its images, the *entrance* and *exit windows* and *angular field of view* of a system are also described.

The field of *radiometry* deals with the determination of the amount of light radiated by a source per unit area per unit solid angle, or falling on a surface per unit area.[1] To discuss the *radiometry of imaging*, we start with the radiometry of point and extended sources. Terms such as *intensity* of a point source, *radiance* of an extended source, *irradiance* of a surface, and a *Lambertian source* are introduced. The *inverse square, cosine, and cosine-third laws of irradiance* for a *point source* are explained. Similarly, the *cosine law of intensity* and *cosine-fourth law of irradiance* for an *extended source* are discussed. In particular, the irradiance of a surface by a Lambertian disc source is discussed. The results thus obtained are used next to discuss the radiometry of optical imaging.

The radiometry of point-object imaging is discussed first and a relationship between the intensities of a point object and its point image is derived. This is followed by the radiometry of extended-object imaging. An invariant relation between the radiances of an object and its image is derived, and the irradiance of the image is discussed. Pupil distortion is discussed, showing that the transverse magnification between the entrance and exit pupils varies with the location of an area element on the pupil. The total flux in an image element may be determined by integrating across the entrance pupil or the exit pupil. If the aperture stop is located in front of the imaging system so that it is also the entrance pupil, then the exit pupil is distorted and the entrance pupil is more convenient for determining the total flux. However, if the aperture stop lies in the back of the system so that it is also the exit pupil, then the entrance pupil is distorted and the exit pupil is more convenient. When the aperture stop lies inside the system, then integration may be performed across the entrance or the exit pupil, provided the area of the pupil (i.e., the region of integration) is determined (e.g., by ray tracing,) for the location of the object element under consideration. Dividing the total flux calculated for a certain image element by its area yields its irradiance. The cosine-fourth law of image irradiance is derived and the range of its validity is discussed. It is shown that an imaging system with barrel distortion gives a uniform-irradiance image of a uniformly radiating object. The irradiance distribution of the images formed by systems that are telecentric or concentric is also discussed.

A brief discussion of *photometry*, the branch of radiometry limited to human observations in the visible region of the electromagnetic spectrum, is given. Photometry is different from the rest of radiometry in that the spectral response of the human eye is taken into account to determine the results of any observation. The brightness of a Lambertian surface is discussed, showing that it appears equally bright at all distances along all directions of observation. It is also shown why stars may be observed during daytime with the aid of a telescope.

2.2 STOPS , PUPILS , AND VIGNETTING

2.2.1 Introduction

In this section, we define and discuss how to determine the aperture stop and entrance and exit pupils of an optical system. The chief and marginal rays are defined as the object rays that pass through the center and edge of the aperture stop, respectively. The chief ray from the edge of an object locates the pupils and determines the image size. Similarly, the marginal ray from the axial point object locates the image plane and determines the sizes of the pupils. The minimum size of an imaging element (e.g., a lens or a mirror) required to avoid vignetting of rays is determined by the intersection of the marginal ray from the edge of an object with the element and can be obtained by adding the magnitudes of the heights of intersection of the edge chief ray and the axial marginal ray with the element. A telecentric aperture stop is discussed which offers the advantage of increased defocus error tolerance to the size or the shape of an image. Finally, a field stop is defined whose images in its object and image spaces, called the entrance and exit windows, define the angular fields of view in those spaces, respectively.

2.2.2 Aperture Stop, and Entrance and Exit Pupils

Figures 2-1a and 2-1b show an optical system consisting of two thin lenses with an optical axis OA forming the Gaussian images of an on-axis point object P_0 and an off-axis point object P at P_0' and P', respectively. Not all of the rays emanating from an object point and incident on the system are transmitted by it; some of them are blocked by one or another element of the system. An aperture in the system that physically limits the

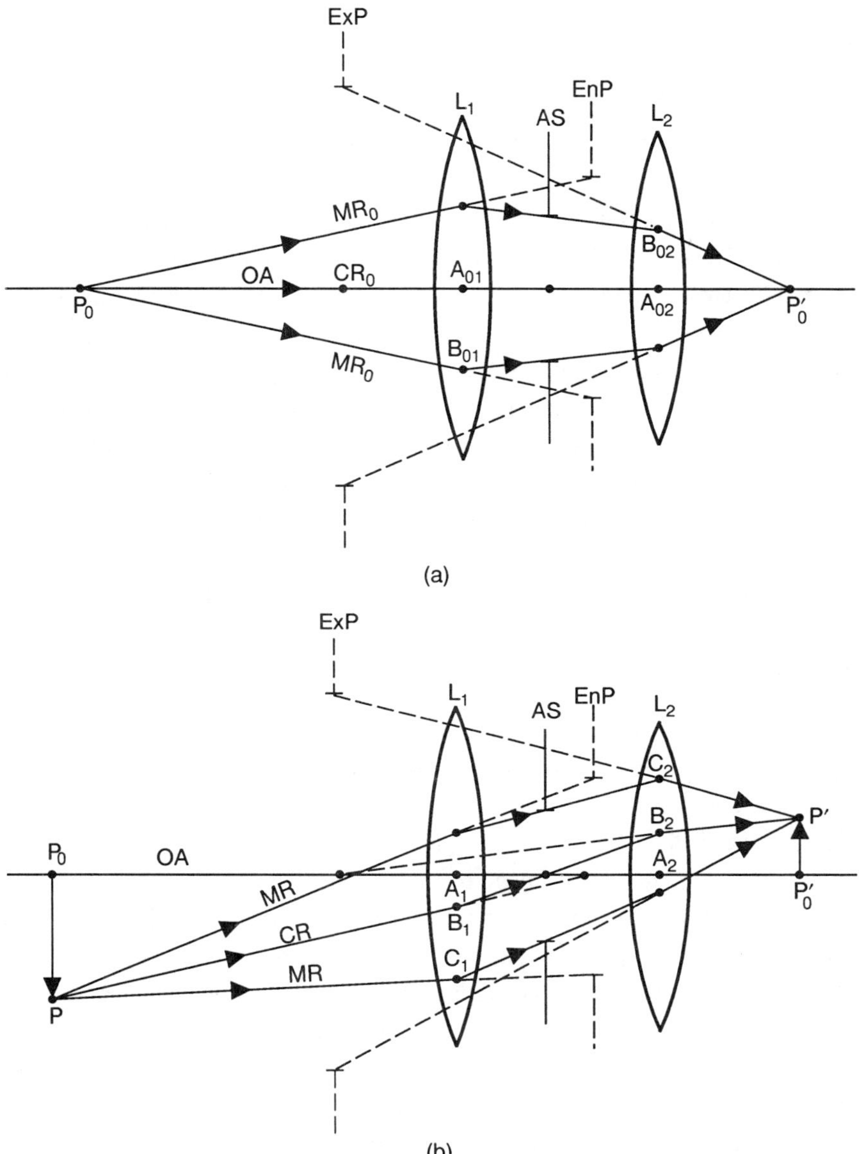

Figure 2-1. (a) Imaging of an on-axis point object P_0 by an optical imaging system consisting of two lenses L_1 and L_2. *OA* is the optical axis. The Gaussian image is at P'_0. *AS* is the aperture stop; its image by L_1 is the entrance pupil *EnP*, and its image by L_2 is the exit pupil *ExP*. CR_0 is the axial chief ray, and MR_0 is the axial marginal ray. (b) Imaging of an off-axis point object P. The Gaussian image is at P'. *CR* is the off-axis chief ray, *MR* is the off-axis marginal ray.

solid angle of the transmitted rays from a point object the most is called its *aperture stop* (*AS*). For an extended (i.e., a non point) object, it is customary to consider the aperture stop as the limiting aperture for an axial point object, and determine the vignetting or blocking of some rays by this stop and other elements of the system for off-axis object points. The object is assumed to be placed to the left of the system so that, initially, light travels from left to right.

The image of the stop by surfaces of the system that precede it in the sense of light propagation, i.e., by those that lie between it and the object, is called the *entrance pupil* (*EnP*). When observed from the object side, the entrance pupil appears to limit the rays entering the system to form the image of the object. Similarly, the image of the aperture stop by surfaces that follow it, i.e., by those that lie between it and the image, is called the *exit pupil* (*ExP*). The object rays reaching its image appear to be limited by the exit pupil. Since the entrance and exit pupils are images of the aperture stop formed by the system elements that precede and follow it, respectively, the two pupils are conjugates of each other for the whole system, i.e., if one pupil is considered as the object, the other is its image formed by the system. In Figure 2, *AS* is the aperture stop and its images by   the lenses L_1 and L_2 are the entrance and exit pupils *EnP* and *ExP*, respectively. From the definition of the object and image spaces given in Section 1.3.5.1, we note that the aperture stop lies in the image space of lens L_1 and the object space of lens L_2. Similarly, the entrance pupil lies in the (virtual) object space of lens L_1, and the exit pupil lies in the (virtual) image space of lens L_2. Moreover, the entrance pupil lies in the (virtual) object space and the exit pupil lies in the (virtual) image space of the two-lens system.

The aperture stop of a multielement imaging system may be determined by forming the image of each element and aperture by the imaging elements that precede it and determining the smallest image as seen by the axial point of the object for the system. The smallest image is the entrance pupil of the system, and the corresponding element or aperture is its aperture stop. The image of the entrance pupil by the whole system or, equivalently, the image of the aperture stop by the elements that follow it, is the exit pupil of the system. Alternatively, the aperture stop may be determined directly by tracing a ray from the axial point object and calculating the ratio of the ray height at and radius (semidiameter) of each element and aperture in the system. The element or aperture with the highest ratio is the aperture stop. If the angle of the chosen ray is increased, each ratio increases by a proportional amount until it reaches a value of unity for the aperture stop. Any further increase in the angle of the ray will lead to its vignetting by the aperture stop.

It is possible that the stop of a system may also be its entrance and/or exit pupil. For example, a stop placed to the left of a lens is also its entrance pupil. Similarly, a stop placed to the right of a lens is also its exit pupil. Finally, a stop placed at a single thin lens is both its entrance and exit pupils.

2.2.3   Chief and Marginal Rays

An object ray passing through the center of the aperture stop and appearing to pass through the centers of the entrance and exit pupils is called the *chief* (or the *principal*) *ray* (*CR*). An object ray passing through the edge of the aperture stop and appearing to pass through the edges of the entrance and exit pupils is called a *marginal*

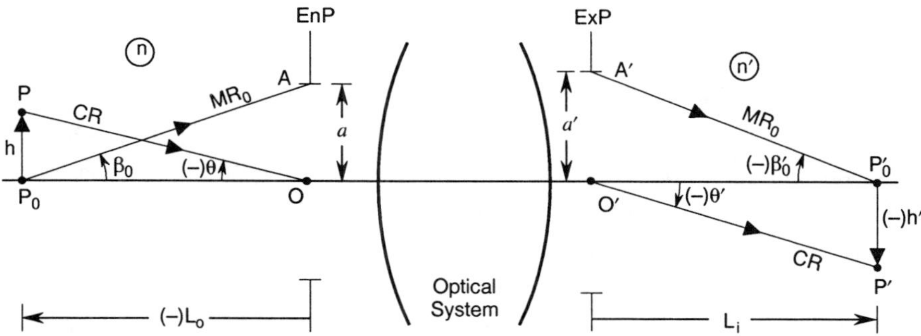

Figure 2-2. Schematic diagram of a system and its entrance and exit pupils *EnP* and *ExP*, respectively, showing the marginal ray $P_0A\cdots A'P_0'$ from the the axial point object P_0 and the chief ray $PO\cdots O'P'$ from the off-axis point object P'.

ray (*MR*). The rays lying between the center and the edge of the aperture and, therefore, appearing to lie between the center and edge of the entrance and exit pupils, are called *zonal rays*. As illustrated in Figure 2-2, the axial marginal ray $P_0A\cdots A'P_0'$ determines the sizes of the entrance and exit pupils, and the location of the axial image point. Similarly, the off-axis chief ray $PO\cdots O'P'$ from the edge of the object determines the locations of the entrance and exit pupils, and the size of the image. Using the Lagrange invariant Eq. (1-69), we find that the angles of the marginal and chief rays are related to each other according to

$$\boxed{nh\beta_0 = -na\theta = -n'a'\theta' = n'h'\beta_0' \quad ,} \tag{2-1}$$

where n and n' are the refractive indices of the object and image spaces, h and h' are the object and image heights, θ and θ' are the chief ray angles (both numerically negative) in the object and image spaces, and a and a' are the radii of the entrance and exit pupils, respectively. Moreover, we have used the fact that the object and image distances from the entrance and exit pupils, respectively, are given by

$$L_0 = h/\theta = -a/\beta_0 \quad , \tag{2-2a}$$

and

$$L_i = h'/\theta' = -a'/\beta_0' \quad . \tag{2-2b}$$

The quantities in Eq. (2-1) represent from left to right, the two-ray Lagrange invariant (discussed in Section 1.5) in the planes of the object, entrance pupil, exit pupil, and the image.

2.2.4 Vignetting

The vignetting of the rays from an off-axis point object by the system may be determined by projecting the images of all elements and apertures (by the preceding elements) on the entrance pupil using the point object as the center of projection. The common area of these projections represents the *effective entrance pupil* of the system for the point object under consideration. Its images formed by the elements that precede it and by the entire system are the *effective aperture stop* and the *effective exit pupil* of the system. An alternative but equally valid approach to determining the vignetting of rays is to project the images of all elements (e.g., images of A and L_1 by L_2 in Figure 2-3c) on the exit pupil using the Gaussian image point as the center of projection. The common area of these projections on the exit pupil represents the effective exit pupil. The exit pupil, which is the image of A by L_2, is not shown in Figure 2-3. The images of the common area by the elements that follow it (looking at them from the image point) and by the entire system are the effective aperture stop and the effective entrance pupil, respectively.

In Figure 2-1a, the lenses are quite large compared with the aperture stop; therefore, they do not in any way limit the ray bundle from the object point P_0 transmitted by the system. AS is indeed the aperture stop since it does limit the ray bundle. Similarly, we note from Figure 2-1b that for any point on the object $P_0 P$, there is no vignetting of the aperture stop, i.e., any ray that is not blocked by the aperture stop is also not blocked by either of the two lenses. Thus, for a circular aperture stop, the entrance and exit pupils are also circular. We note that the cone of light rays from an axial point object illuminates the lenses symmetrically, but the one from the off-axis point object illuminates them eccentrically. We also note that different portions of the lenses are used for different point objects. The same region of an imaging element is used for different point objects only when the aperture stop is located at the element.

However, consider Figure 2-3a, which also shows a system consisting of two lenses L_1 and L_2 with an aperture A placed between them. The images of A and L_2 by L_1 are indicated as A', and L_2'. We note that A is the aperture stop of the system for only those objects that have their axial points lying between P_1 and P_2, where P_1 and P_2 are the points of intersection of the lines joining the upper edges of L_1 and A', and A' and L_2', respectively, with the optical axis. For these objects, A' subtends the smallest angle (at an axial point) among L_1, A', and L_2'. It is, therefore, the entrance pupil of the system. For objects lying to the left of P_1, L_1 subtends the smallest angle. Hence, it $\left(L_1\right)$ is the aperture stop of the system for such objects, in which case it is also the entrance pupil of the system. For objects lying to the right of P_2, L_2' subtends the smallest angle. Therefore, for these objects, L_2 is the aperture stop and the exit pupil of the system, and L_2' is its entrance pupil.

To illustrate vignetting, we consider an object such as $P_0 P$. It is evident from the foregoing that as indicated in Figure 2-3b, A is the aperture stop AS, and A' is the entrance pupil EnP. For the axial point object P_0, the projections of L_1 and L_2' on the entrance pupil are indicated in the figure and illustrated on its right-hand side. It is evident that EnP is smaller than the projections of L_1 and L_2', and there is no vignetting as expected. As stated earlier, for a circular aperture stop, the entrance and exit pupils are also circular.

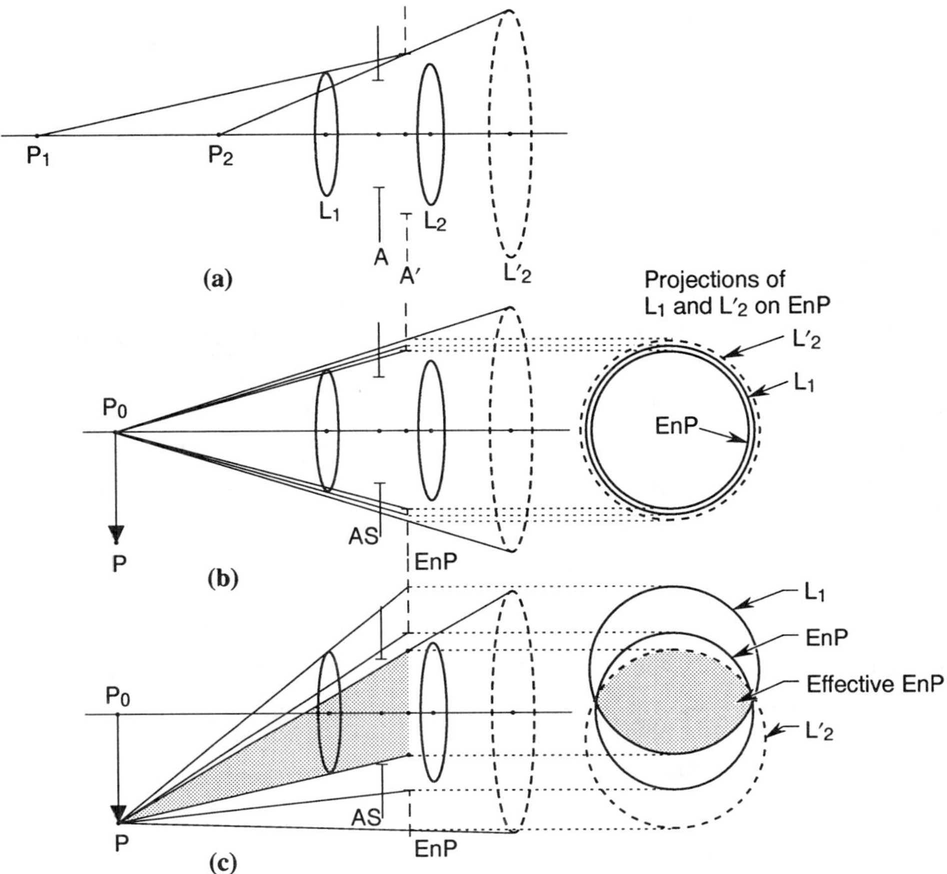

Figure 2-3. Aperture stop of a system and its vignetting. A' and L_2' are the images of A and L_2 by L_1. **(a) Determination of aperture stop. (b) Diagram showing no vignetting for an on-axis point object** P_0. **(c) Vignetting diagram for an off-axis point object** P. **The circles on the right-hand side of the figure show projections of** L_1 **and** L_2' **on** EnP **with the point object under consideration as the center of projection.**

Figure 2-3c shows the projections of L_1 and L_2' on EnP as viewed from an off-axis point object P. These projections, illustrated as eccentric circles on the right-hand side of the figure, are shown to be circular only as an approximation of the actual ellipses. The ray bundle originating at P and transmitted by the system is shown shaded in the figure. It is clear that the upper marginal ray (sometimes called the *upper rim ray*) is limited by L_2 and the lower marginal ray (sometimes called the *lower rim ray*) is limited by L_1; i.e., the upper portion of the ray bundle from P is blocked by L_2, and its lower portion is blocked by L_1. Thus, there is vignetting of the aperture stop and the *effective aperture stop*, and the corresponding entrance and exit pupils are no longer circular. The shape of the *effective entrance pupil* is shown shaded in the figure as the region of EnP that is common with the projections of L_1 and L_2' on it. Its Gaussian images by L_1 and L_2 give the shapes of the effective aperture stop and exit pupil, respectively. The consequence of the variation of the shape of the entrance pupil with the location of point object P lies not only in the loss of light in its image but also in the distribution of the image light (since it depends on the shape of the pupil). Diagrams such as those shown on the right-hand side of Figures 2-3b and 2-3c illustrating the shape of the pupil for a point object are called *vignetting diagrams*.

2.2.5   Size of an Imaging Element

To avoid vignetting for a certain field of view, the size of of an imaging element in a system, e.g., a lens or a mirror, can be determined by tracing the marginal ray from a point on the edge of the object and making the size of the element large enough that this ray is not obstructed by it. The approximate size of an element can be obtained by adding the magnitudes of the heights of the chief ray from an edge point object and the marginal ray from the axial point object. This is because the angle between the chief and marginal rays is approximately independent of the location of the point object in the object plane. Thus, we do not have to trace the marginal ray from the edge of the object. For example, in Figure 2-1, the radius of lens L_1 required to avoid vignetting of rays from the point object P is $\left|A_1 C_1\right|$ or $\left|A_1 B_1 + B_1 C_1\right|$. However, $\left|B_1 C_1\right|$ is approximately equal to $\left|A_{01} B_{01}\right|$. Hence, the lens radius is given by the sum of the magnitudes of the heights of the axial marginal ray and the edge chief ray on the lens. Similarly, the radius of lens L_2 is given by $A_2 C_2$ or $A_2 B_2 + B_2 C_2$, where $B_2 C_2$ is approximately equal to $A_{02} B_{02}$.

2.2.6   Telecentric Aperture Stop

If the aperture stop lies in the front focal plane of a system, as in Figure 2-4a, then the exit pupil lies at infinity, any chief ray in the image space lies parallel to the optical axis, and the system is said to be *telecentric on the image side*. If, however, it lies in its back focal plane, then the entrance pupil lies at infinity, any chief ray in the object space lies parallel to the optical axis, and the system is said to be *telecentric on the object side*. If the system is *afocal* (i.e., one which forms the image at infinity of an object at infinity, as discussed in Section 1.3.6) and if the aperture stop is placed in an intermediate focal plane, then both the entrance and exit pupils lie at infinity, and the system is said to be *telecentric on both object and image sides*. However, a system cannot be telecentric on the object side if the object lies at infinity, for then the aperture stop will lie in the image plane where it cannot control the cross section of the focused beams. A telecentric stop on the image side, for example, has the advantage that the size or the shape of an image is insensitive to small focus errors, as may be seen from Figure 2-5. In Figure 2-4a, the height of the image center does not change with defocus, i.e., P' and P'' are at the same height. However, in Figure 2-4b, where the aperture stop does not lie in the front focal plane, a small defocus changes the height of the image center, as may be seen from the fact that P'' is at a slightly larger height than P', i.e., $h'' > h'$.

2.2.7   Field Stop, and Entrance and Exit Windows

Whereas the aperture stop limits the solid angle of the rays from a point on an object transmitted by the system, there is another stop called the *field stop* which limits the

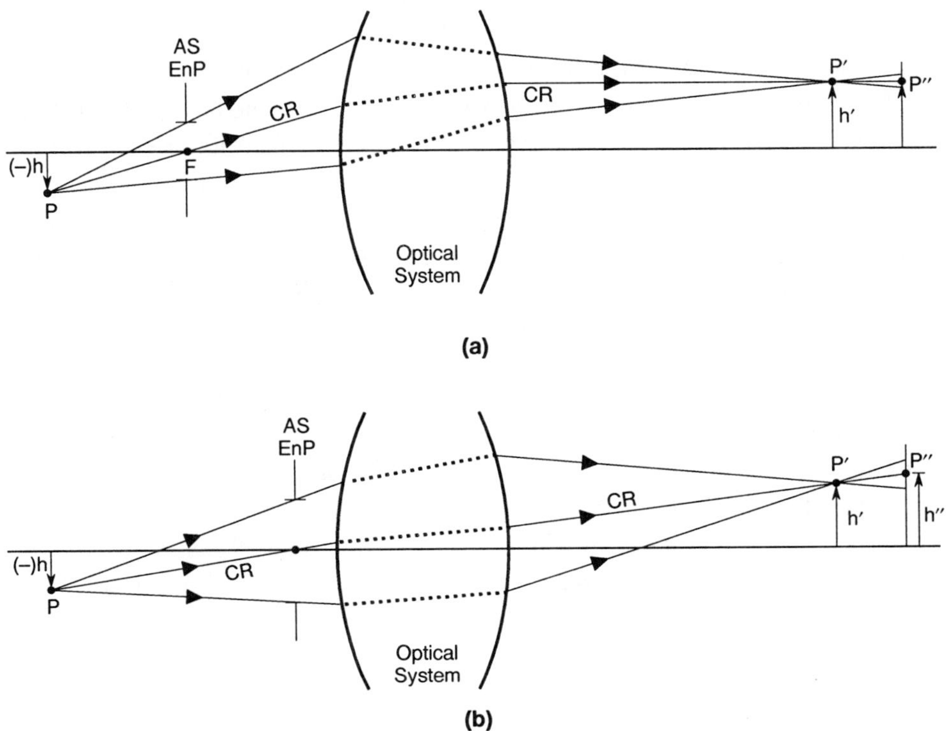

(a)

(b)

Figure 2-4. (a) Telecentric aperture stop on the image side. (b) Nontelecentric aperture stop. A dotted line shown within the system here and in Figure 2-5 does not represent a ray but merely a line joining its points of incidence on and emergence from the system.

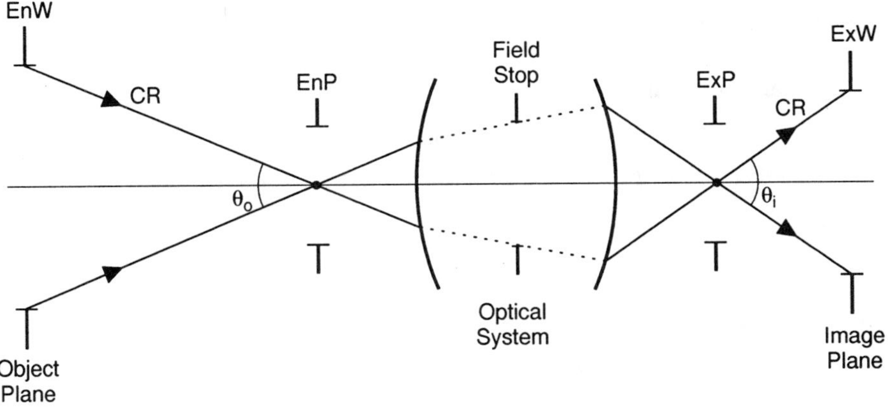

Figure 2-5. Field stop, entrance and exit windows, and field of view of a system. The field stop is assumed to lie at an intermediate image of the object.

solid angle of the transmitted chief rays from the object, as illustrated in Figure 2-5. The image of the field stop by the imaging elements that precede it is called the *entrance window EnW*, and its image by the elements that follow it is called the *exit window ExW*. The field stop is placed at a real image of the object. The image may be an intermediate or the final one. Accordingly, the entrance and exit windows lie in the object and image planes, respectively. The entrance window defines the object field that is actually imaged. Simple examples of field stops are the rectangular diaphragm or the plate holder for the film in a camera or for a slide in a slide projector. The field stop of a system is determined by finding the image of each aperture and element by the imaging elements that precede it and determining the image that subtends the smallest angle at the center of the entrance pupil. This image is the entrance window, and the physical stop corresponding to it is the field stop.

The angle θ_o subtended by the entrance window at the center of the entrance pupil defines the *angular field of view of the system in object space*. Similarly, the angle θ_i subtended by the exit window at the center of the exit pupil is the *angular field of view of the system in image space*. According to Eq. (2-1), their ratio θ_o/θ_i is equal to the magnification of the exit pupil when the refractive indices of the object and image spaces are equal.

It should be noted that whereas the position and the size of the aperture stop determine the quality and the amount of light in the final image (by virtue of blocking rays with large aberrations), the field stop only determines the portion of the object that is imaged. Additional stops and baffles are placed in optical systems to block stray light from reaching the final image area. An example is a stop called a *Lyot stop* (or a *cold stop* when used in an infrared system) placed at a real image of the aperture stop.

2.3 RADIOMETRY OF POINT SOURCES

We first discuss the radiometry of point sources. In particular, we calculate the irradiance (i. e., the flux per unit area) of a surface and the total flux incident on a circular aperture when it is irradiated by a point source.

2.3.1 Irradiance of a Surface

If a point source radiates a flux dF (in watts or W) in a certain direction into a solid angle $d\Omega$, then its *intensity I* (in watts/steradian or W/sr) in that direction is given by

$$I = \frac{dF}{d\Omega} \ .$$

(2-3)

A point source is said to radiate uniformly or isotropically if its intensity is the same in all directions. The total flux F emitted by such a point source is

$$F = 4\pi I \ .$$

(2-4)

For a nonuniform point source emitting a total flux of F, $F/4\pi$ may be called the *mean intensity*.

If a flux dF from a point source irradiates a surface element of area dS, the flux incident on the surface per unit area is called the *irradiance E* (in watts/square meter or W/m^2) of the surface, i.e., the irradiance of the surface is given by

$$E = \frac{dF}{dS} \; . \tag{2-5}$$

If the surface lies at a distance R as in Figure 2-6a, then the solid angle subtended by it at the point source is given by

$$d\Omega = \frac{dS}{R^2} \; . \tag{2-6}$$

Hence, the flux incident on it is given by

$$dF = I \, d\Omega$$
$$= I \, dS/R^2 \; . \tag{2-7}$$

Its irradiance, according to Eq. (2-5), is given by

$$\boxed{E = I/R^2 \; .} \tag{2-8}$$

Equation (2-8) represents the *inverse-square law of irradiance*; namely, the irradiance of a surface by a point source lying on its surface normal is inversely proportional to the square of its distance from the radiating source.

If the irradiated surface element is not normal to the line joining its center and the point source as in Figure 2-6b, then the solid angle subtended by it at the point source is given by

$$d\Omega = dS \cos\theta/R^2 \; , \tag{2-9}$$

where θ is the angle between the surface normal and the line joining the surface center and the point source. The quantity $dS \cos\theta$ is called the *projected area* of the surface in the direction of the point source. The flux incident on the surface is given by

$$dF = I \, d\Omega$$
$$= I \, dS \cos\theta/R^2 \; . \tag{2-10}$$

Its irradiance is given by

$$E = \frac{dF}{dS}$$
$$= I \frac{d\Omega}{dS}$$
$$= I \cos\theta/R^2 \; . \tag{2-11}$$

Equation (2-11) represents the inverse-square law and the *cosine law of irradiance*. Thus, the irradiance of a surface element whose normal makes an angle θ with the line joining its center and the point source is proportional to $\cos\theta$.

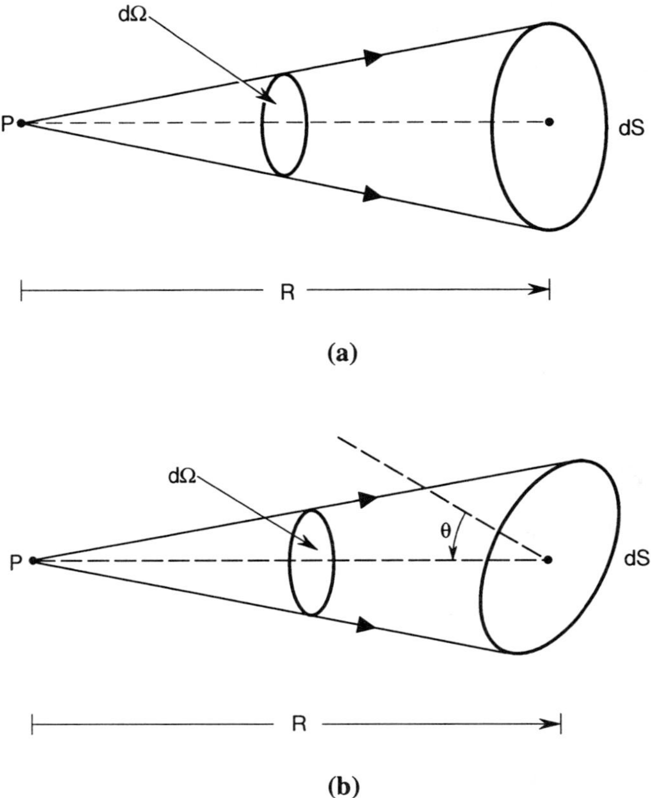

(a)

(b)

Figure 2-6. Irradiance of a surface element irradiated by a point source. The line joining the point source and the center of the surface is normal to the surface in (a) and makes an angle θ with the normal in (b).

Now consider a uniform point source irradiating a plane surface lying at a distance R as illustrated in Figure 2-7. The solid angle subtended by a surface element of area dS inclined at an angle θ with the direction joining its center and the point source is given by

$$d\Omega = dS\cos\theta / (R/\cos\theta)^2$$
$$= dS\cos^3\theta / R^2 \quad . \tag{2-12}$$

Hence, its irradiance is given by

$$E_\theta = I\frac{d\Omega}{dS}$$
$$= I\cos^3\theta / R^2 \tag{2-13}$$

$$= E_0\cos^3\theta \quad , \tag{2-14}$$

where

$$E_0 = I/R^2 \tag{2-15}$$

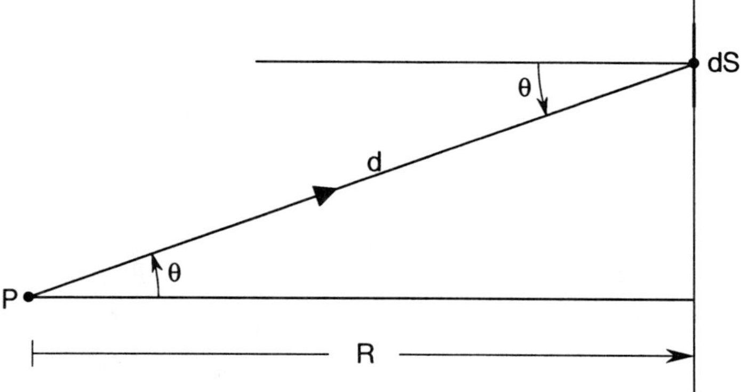

Figure 2-7. Irradiance of a surface by a point source.

is the irradiance at an axial point of the surface lying on its normal passing through the
point source. Equation (2-14) represents the *cosine-third law of irradiance by a point
source*. If we write it in the form

$$E_0 \; = \; \left(I/d^2\right)\cos\theta \; \; ,$$
(2-16)

where

$$d \; = \; R/\cos\theta$$
(2-17)

is the distance of the surface element from the point source, we note that the cosine-third
law of irradiance represents a combination of inverse-square and cosine laws of
irradiance.

2.3.2   Flux Incident on a Circular Aperture

Now we determine the flux incident on a circular aperture of radius a from a point
source P of intensity I lying at a distance R on its axis (see Figure 2-8). The flux incident
on an annular element of the aperture of radius r and width dr is given by

$$
\begin{aligned}
dF \; &= \; \frac{I\cos\theta}{d^2}\,dS \\[2mm]
&= \; IR\,\frac{2\pi\,rdr}{\left(R^2+r^2\right)^{3/2}} \; \; ,
\end{aligned}
$$
(2-18)

where $dS = 2\pi\,rdr$ is the area of the annular element, θ is the angle between its surface
normal and the line joining it with the point source, and d is its distance from the point
source. The total flux incident on the aperture is given by

$$
\begin{aligned}
F \; &= \; \int dF \\[2mm]
&= \; 2\pi\,IR\,\int_0^a \frac{rdr}{\left(R^2+r^2\right)^{3/2}}
\end{aligned}
$$

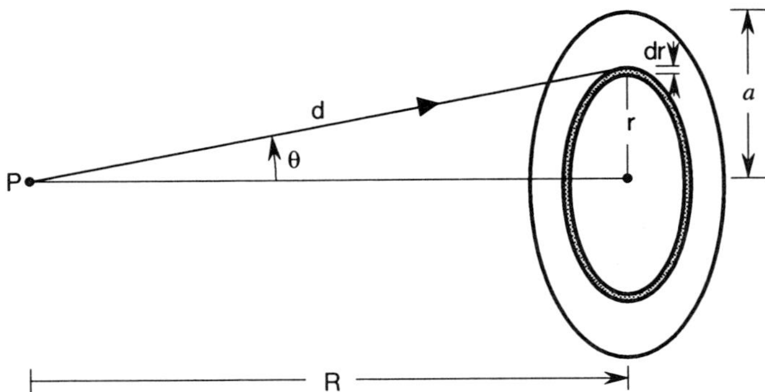

Figure 2-8. Point source irradiating an aperture.

$$= 2\pi IR \left[\frac{1}{R} - \frac{1}{\left(R^2 + a^2\right)^{1/2}} \right] \; . \tag{2-19}$$

When $a << R$, Eq. (2-19) reduces to

$$F = \frac{\pi a^2 I}{R^2}$$

$$= \frac{IS}{R^2} \; , \tag{2-20}$$

where $S = \pi a^2$ is the area of the aperture. For a distant point source, S/R^2 is the solid angle subtended by the aperture, or I/R^2 is the uniform irradiance on it, and hence Eq. (2-20).

2.4 RADIOMETRY OF EXTENDED SOURCES

Next, we discuss the radiometery of extended sources. In particular, we define a Lambertian surface, calculate its exitance, and show that the radiance of a tube of rays propagating in a given medium is invariant. We also discuss the cosine-fourth law of irradiance by a Lambertian surface element and determine the irradiance of a surface by a Lambertian disc.

2.4.1 Lambertian Surface

Unless an irradiated surface is highly polished, it will reflect or radiate over a wide range of directions like a self-radiating object. Its intensity at any point per unit projected area in a certain direction is called its *radiance* (in watts/square meter steradian or W/m^2 sr) at that point in the direction under consideration. Thus, if an area element dS has an intensity dI in a direction inclined to its normal at an angle θ (see Figure 2-9), the projected area is $dS \cos\theta$ and its radiance B is given by

$$B = \frac{dI}{dS \cos\theta} \; . \tag{2-21}$$

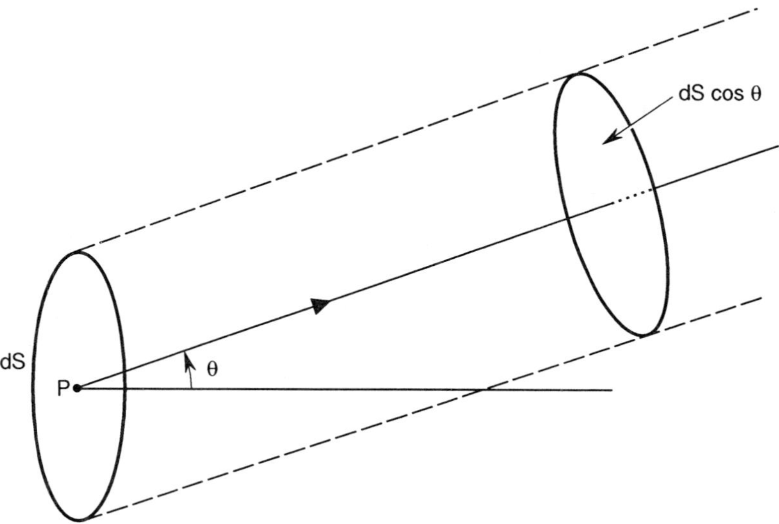

Figure 2-9. Radiance of a surface.

Generally, the radiance of self-radiating and reradiating surfaces does not vary strongly with the direction of radiation. According to Eq. (2-21), the radiance of a surface element is independent of the direction of radiation if its intensity is proportional to $\cos\theta$. Such a surface is said to obey *Lambert's cosine law of intensity*. A surface that radiates uniformly in all directions is called a *Lambertian surface* or a *uniform diffuser*, depending on whether it is self-radiating or reradiating.

2.4.2 Exitance of a Lambertian Surface

The flux radiated by a unit area of a radiating surface is called its *exitance*. The exitance of a Lambertian surface of area dS and radiance B may be determined as follows. The flux emitted by it in a direction making an angle θ with the surface normal into a solid angle $d\Omega$ (see Figure 2-10) is given by

$$dF = B\,dS\cos\theta\,d\Omega \ , \tag{2-22}$$

where $dS\cos\theta$ is the projected area of the surface normal to the direction under consideration and

$$d\Omega = 2\pi\sin\theta\,d\theta \tag{2-23}$$

is equal to the area of an annulus of angular width $d\theta$ on a unit sphere centered at the surface center. The flux radiated by the surface into a cone of half-angle θ_0 is given by

$$
\begin{aligned}
F &= \int dF \\
  &= 2\pi\,B\,dS\int_0^{\theta_0}\sin\theta\cos\theta\,d\theta \\
  &= \pi\,B\,dS\sin^2\theta_0 \ \ .
\end{aligned}
\tag{2-24}
$$

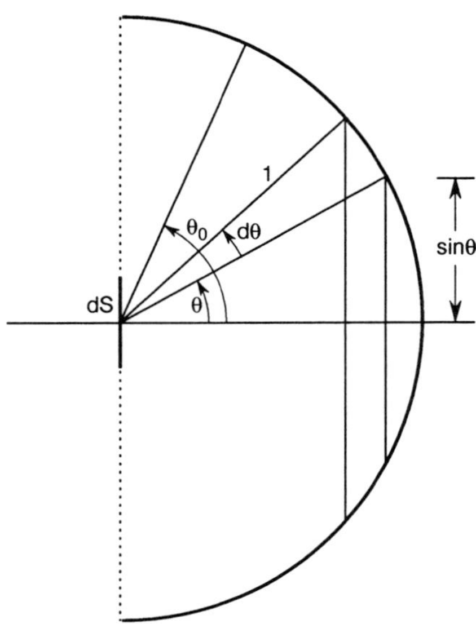

Figure 2-10. Exitance of a Lambertian surface.

Note that we brought B outside the integral sign since it is independent of θ for a Lambertian radiating surface. Letting $\theta_0 = \pi/2$, we obtain the total flux radiated by the surface into a hemisphere, i.e.,

$$F = \pi B \, dS \ . \tag{2-25}$$

Hence, the exitance of the surface is given by

$$M = F \ / dS$$

$$ = \pi B \ . \tag{2-26}$$

Since a hemisphere subtends a solid angle of 2π [as may be seen by integrating Eq. (2-23) from 0 to $\pi/2$], one may expect a factor of 2 on the right-hand sides of Eqs. (2-25) and (2-26) for a uniformly radiating (Lambertian) surface. However, that is not the case owing to the $\cos\theta$ factor in the projected area in Eq. (2-22).

2.4.3   Radiance of a Tube of Rays

Consider, as indicated in Figure 2-11, a Lambertian surface element dS_1 of radiance B_1 irradiating a surface element dS_2 at a distance R such that the normals to the two surface elements make angles θ_1 and θ_2, respectively, with the line joining their centers. According to Eq. (2-21), the intensity of dS_1 in the direction of dS_2 is $B_1 dS_1 \cos\theta_1$. It represents the flux radiated by dS_1 per unit solid angle in the direction of dS_2. The solid angle subtended by dS_2 at dS_1 is given by $dS_2 \cos\theta_2/R^2$. Hence, the total flux received by dS_2 from dS_1 is given by

$$dF_2 = B_1 \, dS_1 \, dS_2 \cos\theta_1 \cos\theta_2 / R^2 \ , \tag{2-27}$$

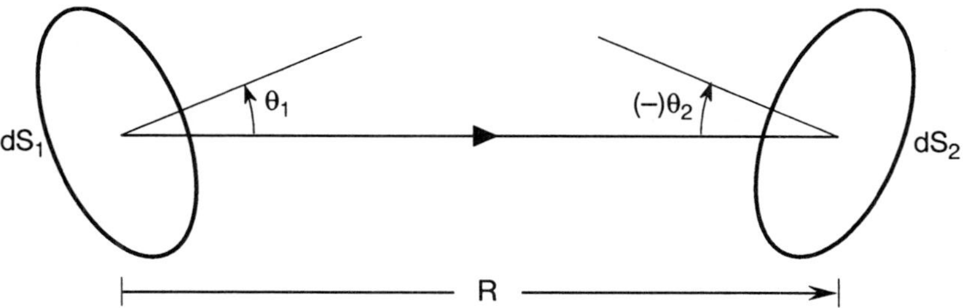

Figure 2-11. Irradiance of a surface element by a Lambertian surface element.

and the irradiance of dS_2 is given by

$$E = \frac{dF_2}{dS_2}$$

$$= B_1 \, dS_1 \, \cos\theta_1 \, \cos\theta_2 / R^2 \quad . \tag{2-28}$$

If the roles of the two surface elements are reversed, i.e., if dS_2 is a radiator of radiance B_2 and dS_1 is irradiated by it, then the intensity of dS_2 in the direction of dS_1 is $B_2 \, dS_2 \cos\theta_2$. It represents the flux radiated by dS_2 per unit solid angle in the direction of dS_1. The solid angle subtended by dS_1 at dS_2 is given by $dS_1 \cos\theta_1 / R^2$. Hence, the total flux received by dS_1 from dS_2 is given by

$$dF_1 = B_2 \, dS_1 \, dS_2 \, \cos\theta_1 \, \cos\theta_2 / R^2 \quad . \tag{2-29}$$

The flux given by Eq. (2-27) can also be interpreted as the flux received by the element dS_1 from the element dS_2. From conservation of energy, $dF_1 = dF_2$. Hence, equating the right-hand sides of Eqs. (2-27) and (2-29) shows that

$$\boxed{B_1 = B_2} \quad , \tag{2-30}$$

i.e., *the radiance along a tube of rays connecting the two surface elements is invariant.*

2.4.4   Irradiance by a Lambertian Surface Element

Now we determine the irradiance on a screen at a distance R from a Lambertian surface element of radiance B, the two being parallel to each other as shown in Figure 2-12. The intensity of dS_1 in the direction of dS_2 is $B dS_1 \cos\theta$. The solid angle subtended by dS_2 at dS_1 is given by $dS_2 \cos\theta / (R/\cos\theta)^2$. The total flux received by dS_2 is given by

$$dF_2 = B \, dS_1 \, dS_2 \, \cos^4\theta / R^2 \quad , \tag{2-31}$$

and its irradiance is given by

$$E_0 = \frac{dF_2}{dS_2}$$

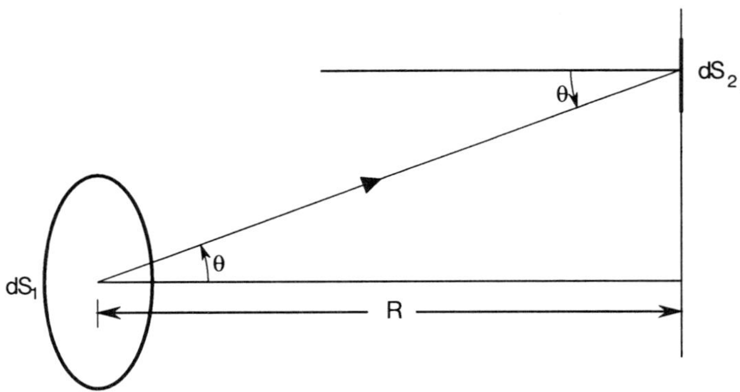

Figure 2-12. Irradiance of a surface by a Lambertian surface element.

$$= B\,dS_1 \cos^4\theta/R^2$$

$$= E_0 \cos^4\theta \quad, \tag{2-32}$$

where

$$E_0 = B\,dS_1/R^2 \tag{2-33}$$

is the irradiance at the axial point of the screen. Equation (2-32) represents the *cosine-fourth law of irradiance by an extended source.*

2.4.5   Irradiance by a Lambertian Disc

We now derive an expression for the irradiance on a screen due to a Lambertian disc[2,3] source of radius a and radiance B in a parallel plane at a distance R. Consider an area element $dS_1 = r_1 dr_1 d\theta_1$ centered at a point P_1 with coordinates (x_1, y_1) or (r_1, θ_1) with a radial width of dr_1 and an angular width of $d\theta_1$, as indicated in Figure 2-13. The flux radiated by it per unit solid angle in a direction making an angle β with the axis of the disc is given by $B\,dS_1 \cos\beta$. The solid angle subtended by an area element dS_2 centered at a point P_2 with coordinates (x_2, y_2) or (r_2, θ_2) is given by $dS_2 \cos\beta/(P_1 P_2)^2$ or $dS_2 \cos^3\beta/R^2$. Hence, the irradiance at the point P_2 due to the element dS_1 of the disc is given by

$$dE(r_2) = B\,dS_1 \cos^4\beta/R^2 \quad . \tag{2-34}$$

We note from Figure 2-13 that

$$\cos\beta = R/P_1 P_2$$

$$= R\Big/\big[R^2 + (x_1 - x_2)^2 + (y_1 - y_2)^2\big]^{1/2}$$

$$= R\Big/\big[R^2 + r_1^2 + r_2^2 - 2r_1 r_2 \cos(\theta_1 - \theta_2)\big]^{1/2} \quad . \tag{2-35}$$

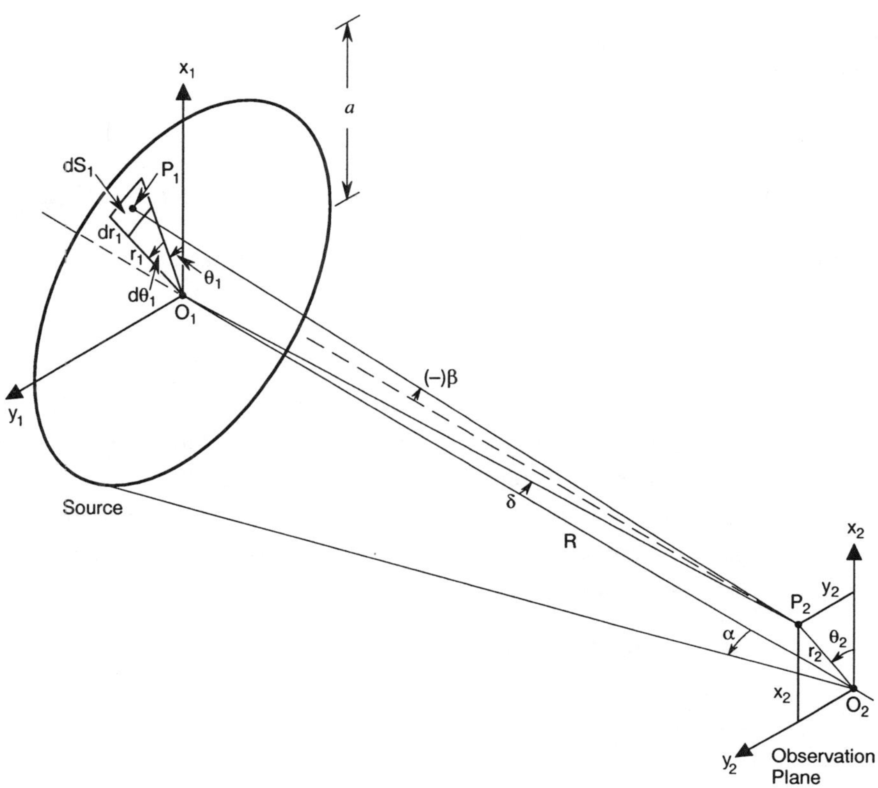

Figure 2-13. Irradiance of a surface by a Lambertian disc. The dashed line passing through P_2 is parallel to the axis O_1O_2 of the disc.

The irradiance at P_2 due to the whole disc is obtained by integrating across the disc, i.e.,

$$E(r_2) = \int dE$$

$$= BR^2 \int_0^a \int_0^{2\pi} \left[R^2 + r_1^2 + r_2^2 - 2r_1 r_2 \cos(\theta_1 - \theta_2) \right]^{-2} r_1 \, dr_1 \, d\theta_1$$

$$= (\pi B/2) \left\{ 1 - \left[1 + \frac{4a^2 R^2}{\left(R^2 + r_2^2 - a^2 \right)^2} \right]^{-1/2} \right\} . \tag{2-36}$$

Note that it does not depend on θ_2; that is, it is radially symmetric about the axial point O_2 on the screen. Equation (2-36) may also be written in the form

$$E(\delta) = (\pi B/2) \left\{ 1 - \left[1 + \frac{4 \tan^2\alpha \cos^4\delta}{\left(1 - \tan^2\alpha \cos^2\delta \right)^2} \right]^{-1/2} \right\} , \tag{2-37}$$

where α is the half-angle of the cone subtended by the disc at the screen and δ is the angle O_1P_2 makes with the axis O_1O_2 of the disc. The angles α and δ are given by

$$\tan\alpha = a/R \tag{2-38}$$

and

$$\tan\delta \;=\; r_2/R \;\;.$$

(2-39)

respectively.

The irradiance at an axial point of the disc is obtained by letting $r_2 = 0$. Thus,

$$E(0) \;=\; \pi B a^2 / \!\left(a^2 + R^2\right)$$

(2-40a)

$$=\; \pi B \sin^2\alpha \;\;.$$

(2-40b)

Multiplying both sides of Eq. (2-37) by dS_2 and comparing the result obtained with Eq. (2-24), we find, as expected, that the same amount of flux is transmitted from a source to a receiver if their roles are interchanged. When $R \ll a$, $E(0) \to \pi B$, which is numerically equal to the exitance of the disc [see Eq. (2-26)]. Thus, when the source is very large compared with the distance of the receiver, the axial irradiance is independent of the distance between the two.

When $R \gg a$, Eqs. (2-36) and (2-37) reduce to

$$E(r_2) \;=\; \frac{\pi a^2 B R^2}{\left(R^2 + r_2^2\right)^2}$$

$$=\; \pi B \left(\frac{a}{R}\right)^{\!2} \frac{R^4}{\left(R^2 + r_2^2\right)^2}$$

(2-41)

and

$$E(\delta) \;=\; \pi B \tan^2\alpha \, \cos^4\delta$$

(2-42)

respectively. The axial irradiance in this case is given by

$$E(0) \;=\; \pi B (a/R)^2$$

(2-43a)

$$=\; I/R^2 \;\;,$$

(2-43b)

where $I = \pi a^2 B$ is the intensity of the disc along its axis. Thus, along its axis but away from it, the disc behaves like a point source of intensity $\pi a^2 B$. The actual value of axial irradiance given by Eq. (2-40a) is smaller than the approximate value given by Eq. (2-43a). The difference between the two values decreases as R/a increases. For example, for $R/a \geq 5$, the difference is $\leq 4\%$.

Figure 2-14a shows how the axial irradiance decreases according to Eq. (2-40a) as R increases. We note that when $R/a \geq 5$, the disc behaves like a point source according to Eq. (2-43a). The irradiance distribution in a plane at a distance R from the disc at a point at a distance r_2 from its axis, as given by Eq. (2-36), is shown in Figure 2-14b for different values of R/a. Each curve in the figure is normalized by the corresponding axial value. The irradiance as a function of r_2 falls off less rapidly than that according to $\cos^4\delta$ as given by Eq. (2-39). The difference between the two decreases as R/a increases; the curves for $R/a \geq 10$ and the $\cos^4\delta$ curve are indistinguishable from each other in Figure 2-14b.

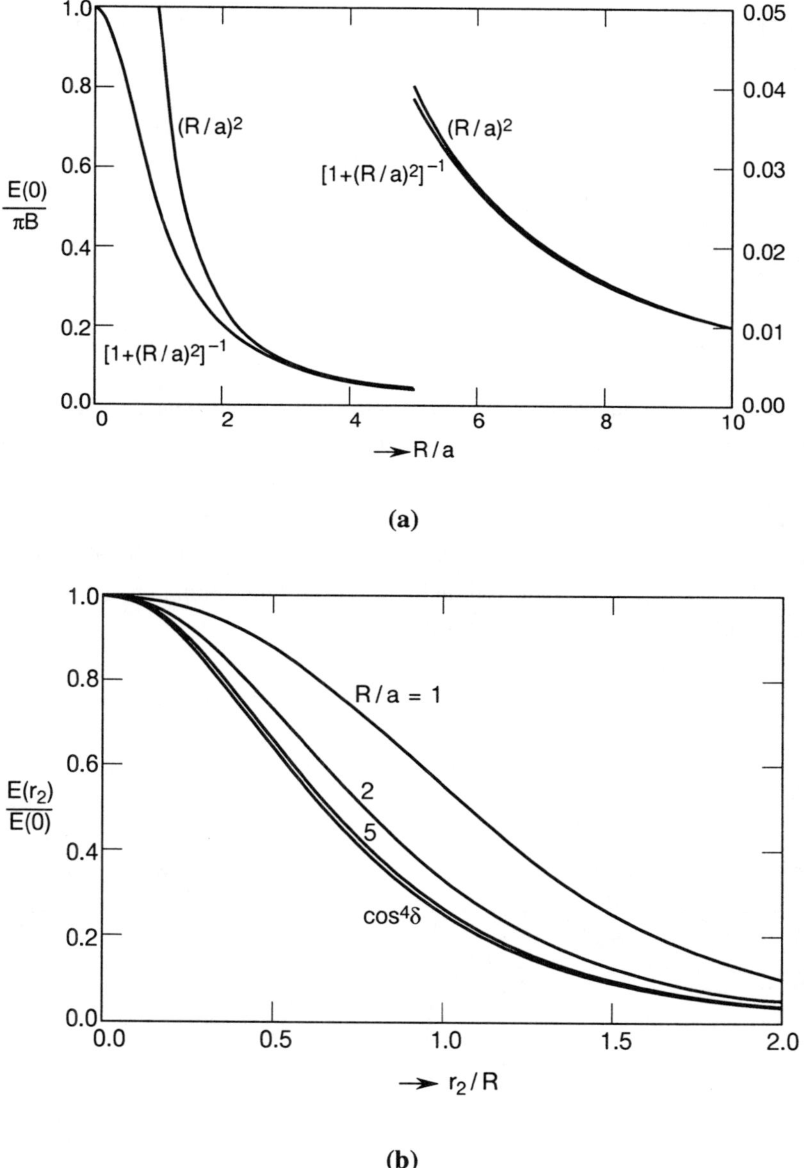

Figure 2-14. Irradiance distribution on a surface by a Lambertian disc of radius a and radiance B. (a) Axial irradiance at a distance R. The vertical scale on the right-hand side is for the right-hand side curves. (b) Irradiance in a plane at a distance R normalized by the axial irradiance. r_2 is the distance of a point in the plane from the axis of the disc.

2.5 RADIOMETRY OF POINT OBJECT IMAGING

Now we discuss the radiometry of point object imaging, i.e., determine the intensity of the image point in terms of the intensity of the object point and the parameters of the system. Consider, as indicated in Figure 2-15, a point object P lying in a plane at a distance L_o from the entrance pupil of an optical imaging system. Its Gaussian image lies at P' in a plane at a distance L_i from the exit pupil of the system. If the object is a uniform point source of intensity I_o, the flux incident on the entrance pupil of area S_{en} is given by

$$F = I_o\,\Omega(P)\ ,\tag{2-44}$$

where

$$\Omega(P) = \frac{S_{en}\cos\theta}{(PO)^2}$$
$$= \left(S_{en}/L_o^2\right)\cos^3\theta\tag{2-45}$$

is the solid angle subtended by the entrance pupil at the point object P. Here, PO is the distance between the point object P and the center O of the entrance pupil and θ is the angle the chief ray makes with the optical axis of the system in object space. It is assumed here that the dimensions of the entrance pupil are small enough compared to its distance from the object plane that the variation of the angle θ with the location of an area element on the pupil can be neglected and, therefore, integration across the pupil is not required. Equation (2-44) may also be written

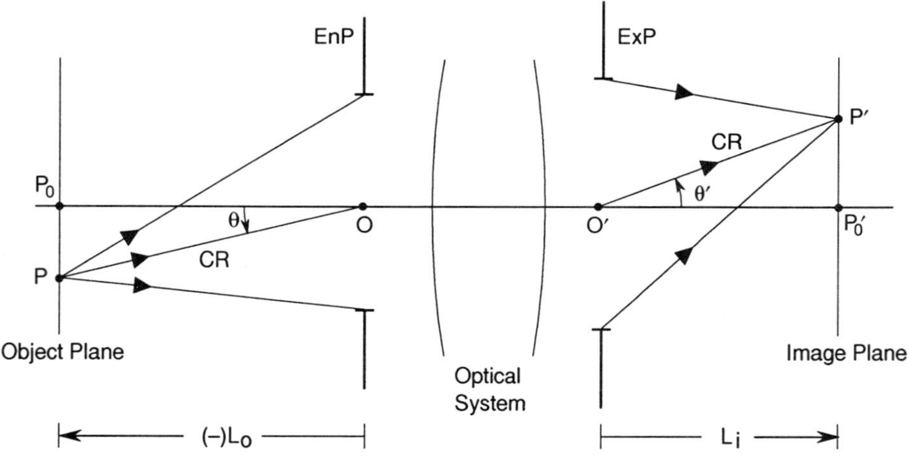

Figure 2-15. Radiometry of point object imaging. A point object P lies in the object plane at a distance L_o from the entrance pupil EnP of the system. Its Gaussian image P' lies in the image plane at a distance L_i from the exit pupil ExP of the system. The chief ray CR makes an angle θ in the object space and θ' in the image space of the system.

$$F = I_o \, \Omega(P_0) \cos^3 \theta \quad , \tag{2-46}$$

where

$$\Omega(P_0) = S_{en}/L_o^2 \tag{2-47}$$

is the solid angle subtended by the entrance pupil at the axial point object P_0.

In the absence of any transmission losses in the system, the flux F emerges from the exit pupil and focuses on the image point P'. If I_i is the intensity of the image point, then the flux emerging from the exit pupil is given by

$$F' = I_i \, \Omega'(P') \quad , \tag{2-48}$$

where

$$\Omega'(P') = \frac{S_{ex} \cos \theta'}{(O'P')^2}$$

$$= \left(S_{ex}/L_i^2 \right) \cos^3 \theta' \tag{2-49}$$

is the solid angle subtended by the exit pupil at the image point. Here, $O'P'$ is the distance between the center O' of the exit pupil and the image point P', S_{ex} is the area of the exit pupil, and θ' is the angle the chief ray makes with the optical axis in image space. Equation (2-49) may also be written

$$\Omega'(P') = \Omega'(P_0') \cos^3 \theta' \quad , \tag{2-50}$$

where

$$\Omega'(P_0') = S_{ex}/L_i^2 \tag{2-51}$$

is the solid angle subtended by the exit pupil at the axial image point P_0'. As in the case of the entrance pupil, the dimensions of the exit pupil are assumed to be small enough compared with its distance from the image plane that the variation of the angle θ' with the location of an area element on the exit pupil can be neglected and, therefore, integration across the pupil is not required. Hence, Eq. (2-48) may be written

$$F' = I_i \, \Omega'(P_0') \cos^3 \theta' \quad . \tag{2-52}$$

From conservation of energy, $F = F'$. Hence, equating the right-hand sides of Eqs. (2-48) and (2-52), we obtain the intensity of the image point

$$\boxed{ I_i = \frac{I_o \, \Omega(P_0) \cos^3 \theta}{\Omega'(P_0') \cos^3 \theta'} } \tag{2-53}$$

It should be noted that the image point is a uniform point source only within the solid angle $\Omega'(P')$ since (according to geometrical optics) there is no radiation outside it.

For an axial point object, both θ and θ' approach zero and Eq. (2-53) reduces to

$$I_i = \frac{I_o \, \Omega(P_0)}{\Omega'(P_0')} \quad . \tag{2-54}$$

If D_{en} and D_{ex} are the diameters of the entrance and exit pupils, then Eq. (2-54) may also be written

$$I_i = \frac{I_o F_{ex}^2}{F_{en}^2} \quad , \tag{2-55}$$

where $F_{en} = |L_o|/D_{en}$ and $F_{ex} = L_i/D_{ex}$ are the focal ratios of the optical beams entering and exiting from the system.

2.6.   RADIOMETRY OF EXTENDED OBJECT IMAGING

Finally, we discuss the radiometry of extended object imaging.[4-9] We derive an invariant property of the radiance of rays when they are refracted or reflected and use it to obtain the irradiance of the image of a Lambertian object formed by an optical system. We show that the iradiance in the image plane decreases as the fourth power of the cosine of the chief ray angle in image space. This decrease is shown to be compensated in a system with barrel distortion to yield a uniform-irradiance image. (Distortion is introduced in Chapter 3 and barrel distortion is discussed in Chapter 4.)

2.6.1   Image Radiance

We have already shown that the radiance along a tube of rays is invariant [see Eq. (2-30]. Although we did not state it explicitly, it was assumed that the rays were propagating in a given medium and, therefore, without any refraction or reflection. Now we consider how their radiance changes when they are refracted or reflected.[4]

Consider an elementary beam of solid angle $d\Omega$ incident in a direction (θ, ϕ) on an interface separating media of refractive indices n and n' as shown in Figure 2-16. The solid angle $d\Omega$ is given by

$$d\Omega = \frac{(r\,d\theta)(r\sin\theta\,d\phi)}{r^2}$$

$$= \sin\theta\,d\theta\,d\phi \quad . \tag{2-56}$$

It represents the area on a unit sphere lying between the angles θ and $\theta + d\theta$, and ϕ and $\phi + d\phi$, as may be seen from the figure. If B is the radiance of the beam, the flux incident on an elementary area dS is given by

$$dF = B\,dS\cos\theta\,d\Omega \quad . \tag{2-57}$$

The beam is refracted at the interface in the direction (θ', ϕ), where θ' is given by Snell's law according to

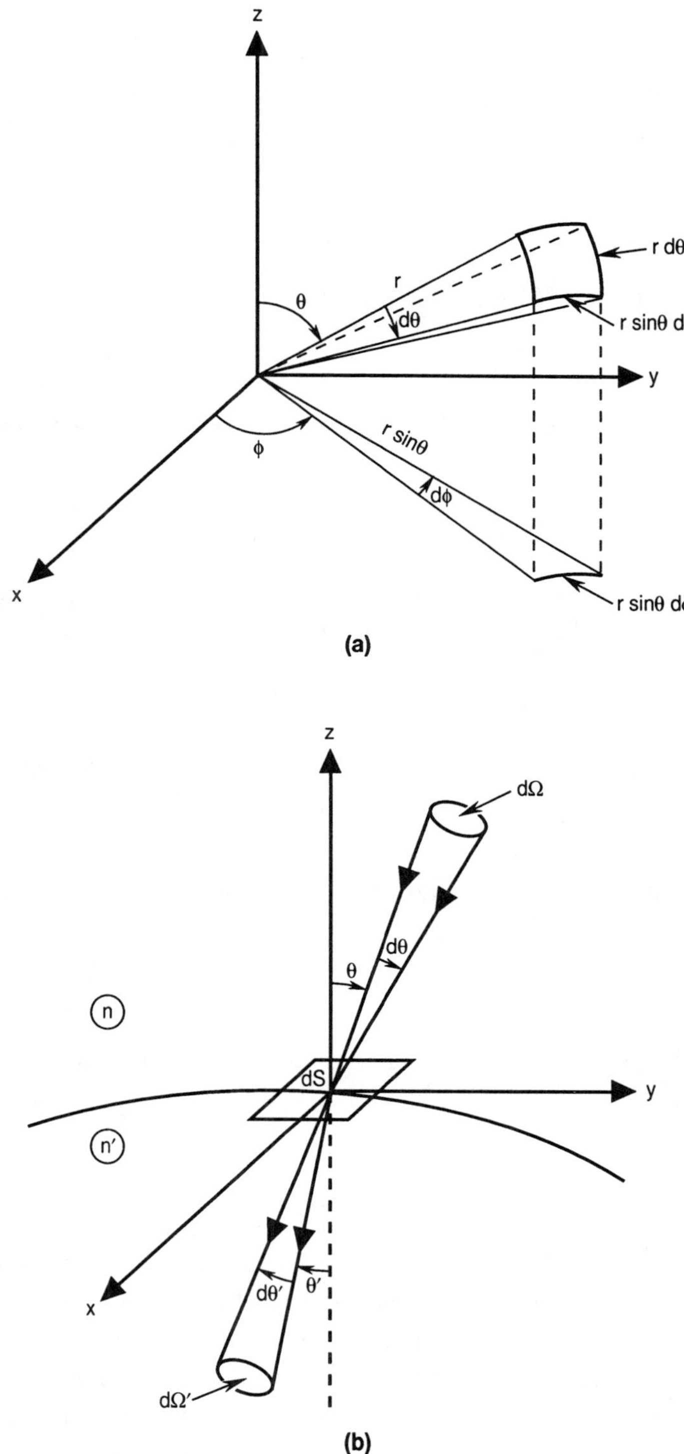

Figure 2-16. (a) Solid angle of an elementary beam in polar coordinates (r, θ, ϕ) and (b) its change from $d\Omega$ to $d\Omega'$ upon refraction at an interface separating media of refractive indices n and n'.

$$n' \sin \theta' = n \sin \theta \quad . \tag{2-58}$$

The azimuthal angle ϕ does not change upon refraction by virtue of the fact that the incident ray, the refracted ray, and the surface normal are coplanar. The solid angle of the refracted beam is given by

$$d\Omega' = \sin \theta' \, d\theta' \, d\phi \quad . \tag{2-59}$$

Differentiating both sides of Eq. (2-58), we obtain

$$n' \cos \theta' \, d\theta' = n \cos \theta \, d\theta \quad . \tag{2-60}$$

Substituting Eqs. (2-58) and (2-60) into Eq. (2-56) and comparing it with Eq. (2-59), we find that

$$n'^2 \cos \theta' \, d\Omega' = n^2 \cos \theta \, d\Omega \quad , \tag{2-61}$$

i.e., the quantity $n^2 \cos \theta \, d\Omega$ is invariant upon refraction. The two solid angles are different from each other because the rays bounding $d\Omega$ are refracted by slightly different amounts due to their slightly different angles of incidence. For a reflecting surface, they are equal since then $\theta' = \theta$.

If B' is the radiance of the refracted beam, the flux contained in it is given by

$$dF' = B' \, dS \cos \theta' \, d\Omega' \quad . \tag{2-62}$$

In the absence of any transmission loss, the incident flux is equal to the refracted flux, i.e.,

$$dF' = dF \quad . \tag{2-63}$$

Hence, equating the right-hand sides of Eqs. (2-57) and (2-62), and substituting Eq. (2-61), we obtain

$$\boxed{\frac{B'}{n'^2} = \frac{B}{n^2}} \quad . \tag{2-64}$$

Thus, when the rays are refracted by a surface, the quantity B/n^2 associated with them is invariant. We refer to this invariance as the *radiance theorem*. When the rays are reflected by a lossless surface, their radiance is invariant since $n' = -n$ in that case. Since the entrance pupil of an optical imaging system lies in its object space, the radiance of rays at the entrance pupil is equal to the object radiance. Similarly, since the exit pupil lies in the image space, the radiance of rays at the exit pupil is equal to the image radiance. In optical imaging by a multisurface system, if the refractive indices n and n' of the object and image spaces are equal (in practice, they are often both equal to unity), then the radiance of an image element is equal to the radiance of the corresponding object element. Taking into account the loss of energy at a refracting or a reflecting surface, we conclude that image radiance can at most be equal to the object radiance, i.e., $B' \leq B$ when $n = n'$.

A general proof of invariance of B/n^2 using the Hamilton's point characteristic function is given in the Appendix. The invariance for a small object centered on the optical axis can be obtained by using the Lagrange invariance of Eq. (1-70) in the small-angle approximation (see Problem 2.7). For large angles, it can be obtained by using the *sine condition* discussed in Section 3.7.4. A *generalized Lagrange invariant* is also introduced in the Appendix.

2.6.2 Pupil Distortion

To determine the irradiance distribution of the image of a planar Lambertian object formed by an optical system, we proceed as follows. Consider an object element dS of radiance B centered at an off-axis point object P imaged as an image element dS' of radiance B' centered at P' as shown in Figure 2-17. The flux radiated by the object element per unit solid angle in a direction making an angle γ with the optical axis (which is perpendicular to the mutually parallel planes of the entrance pupil and the object) is given by $B\,dS\cos\gamma$. The solid angle subtended by an area element dS_{en} of the entrance pupil *EnP* centered on the direction under consideration is given by $dS_{en}\cos^3\gamma/L_o^2$, where L_o is the (numerically negative) distance of the object plane from the plane of the entrance pupil. Hence, the flux incident on dS_{en} is given by

$$dF = B\,dS\left(dS_{en}/L_o^2\right)\cos^4\gamma \quad . \tag{2-65}$$

In the absence of transmission losses in the system, this flux reaches a corresponding area dS_{ex} in the exit pupil *ExP*, emerges from it, and converges on the image element dS'. The flux reaching the image element can also be obtained as follows. The intensity of dS_{ex} in the direction of dS' making an angle γ' with the optical axis (which is

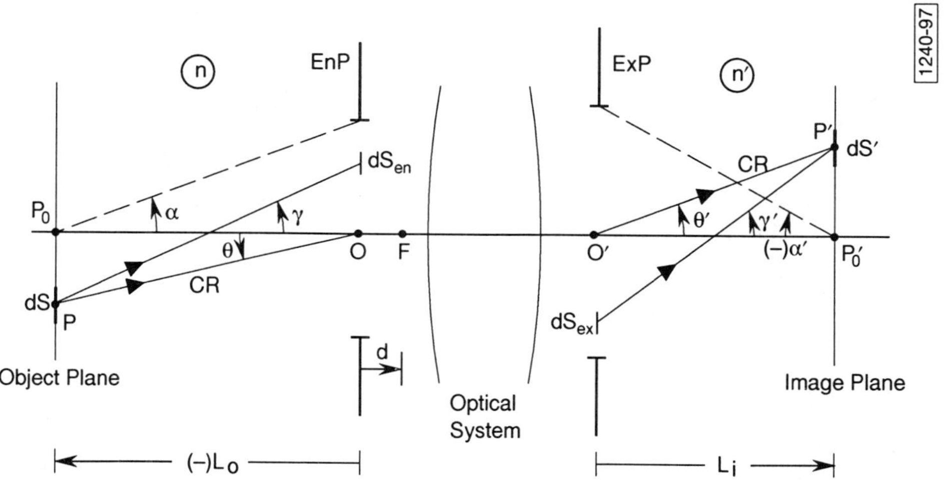

Figure 2-17. Radiometry of extended object imaging.

perpendicular to the mutually parallel planes of the exit pupil and the image) is given by $B'dS_{ex}\cos\gamma'$. The solid angle subtended by dS' at the exit pupil is equal to $dS'\cos^3\gamma'/L_i^2$, where L_i is the distance of the image plane from the plane of the exit pupil. Hence, the flux emerging from the area element dS_{ex} of the exit pupil and converging on dS' is given by

$$dF' = B'dS'\left(dS_{ex}/L_i^2\right)\cos^4\gamma' \quad . \tag{2-66}$$

Since $dF = dF'$, comparing Eqs. (2-65) and (2-66), we obtain

$$\boxed{\frac{dS_{ex}}{dS_{en}} = \left(\frac{n}{n'}\right)^2 \frac{L_i^2}{L_o^2}\frac{dS}{dS'}\frac{\cos^4\gamma}{\cos^4\gamma'}} \quad , \tag{2-67}$$

where we have substituted $B' = B\left(n'/n\right)^2$. Equation (2-67) shows that for a fixed position of dS (and therefore of dS'), the areal magnification dS_{ex}/dS_{en} of the pupil varies with the location of dS_{en} according to $\cos^4\gamma/\cos^4\gamma'$, signifying *pupil distortion*.

It also shows that if the position of dS_{en} (and therefore of dS_{ex}) is fixed, including when they are centered on the optical axis, and if the image is distortion free so that its areal magnification dS'/dS is independent of the location of dS in the object plane, then the areal pupil magnification dS_{ex}/dS_{en} depends on the location of dS. The total flux in the image element may be determined by integrating Eq. (2-65) across the entrance pupil or Eq. (2-66) across the exit pupil. If the aperture stop is located in front of the imaging system so that it is also the entrance pupil, then the exit pupil is distorted and Eq. (2-65) is more convenient for determining the total flux. However, if the aperture stop lies in the back of the system so that it is also the exit pupil, then the entrance pupil is distorted and Eq. (2-66) is more convenient. When the aperture stop lies inside the system, then either Eq. (2-65) or Eq. (2-66) may be integrated provided the area of the corresponding pupil (i.e., the region of integration) is determined, e.g., by ray tracing, for the location of the object element under consideration. Dividing the total flux calculated for a certain image element by its area yields its irradiance.

2.6.3   Image Irradiance: Aperture Stop in Front of the System

If the aperture stop lies in the object space so that it is also the entrance pupil, as in astronomical telescopes, then the flux F entering the optical system may be obtained by integrating Eq. (2-65) across it, i.e.,

$$F = \left(BdS/L_o^2\right)\int_{EnP}\cos^4\gamma\, dS_{en} \quad . \tag{2-68}$$

The irradiance at the image element is given by

$$E = F/dS'$$
$$= \left(B/M^2 L_o^2\right)\int_{EnP}\cos^4\gamma\, dS_{en} \quad , \tag{2-69}$$

where

$$M = (dS'/dS)^{1/2}$$ (2-70)

is the magnification of the image element. The calculation of the irradiance is equivalent to calculating it for a disc with a radius equal to that of the entrance pupil and a radiance of B/M^2 in a plane at a distance L_o from it. Hence, the irradiance is given by Eq. (2-36), where $R = |L_o|$ and $r_2 = |L_o|\tan\theta$, i.e.,

$$E(\theta) = \frac{\pi B}{2M^2}\left\{1 - \left[1 + 4\cos^4\theta\,\frac{\tan^2\alpha}{\left(1 - \tan^2\alpha\cos^2\theta\right)^2}\right]^{-1/2}\right\},$$ (2-71)

where $|\alpha| = \tan^{-1}\left(a_{en}/|L_o|\right)$ is the semiangle of the cone subtended by the entrance pupil at the axial object point P_0' and θ is the angle of the chief ray in the object space. The angle $2|\alpha|$ is called the *angular aperture* of the light cone incident on the system. Letting $\theta = 0$, we obtain the irradiance at the axial image point P_0' for an object element centered at the axial point object P_0:

$$E(0) = \pi\left(B/M^2\right)\sin^2\alpha \quad .$$ (2-72)

The quantity $n\sin\alpha$ is called the *numerical aperture* of the light cone incident on the system.

The variation of irradiance with θ normalized by its axial value is shown in Figure 2-18, which has been adapted from Figure 2-14. It is evident from the figure that

$$E(\theta) \geq E(0)\cos^4\theta \quad ,$$ (2-73)

equality holding more and more closely as the f-number $|L_o|/2a_{en}$ of the object light cone increases. Thus for small values of $|L_o|/2a_{en}$, the decrease of the image irradiance with θ is not as rapid as $\cos^4\theta$. For $|L_o|/2a_{en} \geq 5$, the decrease is practically the same as $\cos^4\theta$.

For *small entrance pupils*, the angle γ for any area element of *EnP* is approximately equal to the chief ray angle θ in the object space and the integral in Eq. (2-69) simply reduces to $S_{en}\cos^4\theta$. Hence, Eq. (2-69) for the image irradiance may be written

$$E(\theta) = E(0)\cos^4\theta \quad ,$$ (2-74)

where

$$E(0) = \left(B/M^2\right)\left(S_{en}/L_o^2\right)$$

$$= \left(B/M^2\right)\Omega(P_0)$$ (2-75a)

$$= \left(\pi B/4M^2\right)/F_{en}^2$$ (2-75b)

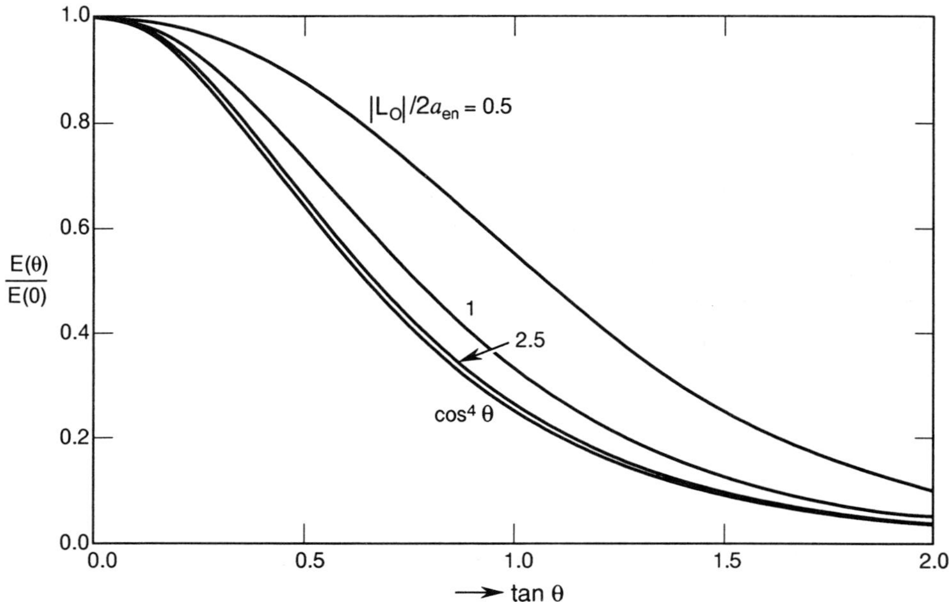

Figure 2-18. Irradiance distribution of the image of a Lambertian object adapted from Figure 2-14b.

is the axial irradiance. Equation (2-74) represents the *cosine-fourth law in the object space, showing that the irradiance of the image of a Lambertian object decreases as the fourth power of the cosine of the chief ray angle* θ *in the object space.* This result may also be obtained from Eq. (2-71) by letting the angle α be small so that $\tan\alpha \simeq \alpha$. Of course, there may be additional decrease due to vignetting.

If d is the distance of the object-space focal point from the entrance pupil and f is the object-space focal length of the system, then Eq. (1-77) gives

$$M = -f/(L_o - d) \quad . \tag{2-76}$$

For an object at infinity, as in astronomical observations, $L_o M \to -f$. Moreover, in that case, all the rays are parallel to the chief ray, i.e., the angle γ has the same value equal to θ for any area element of the entrance pupil. Hence, Eq. (2-69) reduces to

$$E(\theta) = \left(B/f^2\right) S_{en} \cos^4\theta$$

$$= \left(\pi B n'^2 / 4n^2 F_\infty^2\right)\cos^4\theta \quad , \tag{2-77}$$

where

$$\boxed{F_\infty = f'/D_{en}} \tag{2-78}$$

is the focal ratio of the image-forming light cone for an object lying at infinity. F_∞ is called the *f-number* or the *relative aperture* of the system. If D_{en} is increased by a certain factor so that the system collects more light, the image irradiance does not change if the image-space focal length f' is also increased by the same factor. The amount of light collected increases as D_{en}^2 and the image area increases as f'^2 so that the irradiance does not change unless the f-number also changes. A lens with a small f-number is said to be *fast* since it requires a shorter exposure time. Its *speed* is inversely proportional to the square of its f-number.

The focal ratio of the image-forming light cone for finite conjugates can be related to F_∞ as follows. From Eqs. (1-63) and (1-66), the image distance S' of P_0' can be written

$$S' = f'(1 - M) \quad . \tag{2-79}$$

Similarly, the image distance s' of the exit pupil can be written in terms of the pupil magnification $m = D_{ex}/D_{en}$. Hence, we may write

$$\begin{aligned}
F_{ex} &= L_i/D_{ex} \\
&= (S' - s')/D_{ex} \\
&= F_\infty(1 - M/m) \quad .
\end{aligned} \tag{2-80}$$

We note that $F_{ex} \to F_\infty$ as $M \to 0$, i.e., as the object moves to infinity.

2.6.4   Image Irradiance: Aperture Stop in Back of the System

When the aperture stop lies in the image space of the system so that it is also its exit pupil, then integrating Eq. (2-66) across it, we obtain the total flux in the image:

$$F' = \left(B'dS'/L_i^2\right) \int_{ExP} \cos^4 \gamma' \, dS_{ex} \quad . \tag{2-81}$$

Accordingly, the irradiance of the image element is given by

$$\begin{aligned}
E &= F'/dS' \\
&= \left(B'/L_i^2\right) \int_{ExP} \cos^4 \gamma' \, dS_{ex} \quad .
\end{aligned} \tag{2-82}$$

For an exit pupil of radius a_{ex} and radiance B', the irradiance of the image element can be determined in exactly the same manner as the irradiance due to a circular disc. Thus, it is given by Eq. (2-36), where $R = L_i$ and $r_2 = L_i \tan\theta'$, and B is replaced by B' or $B(n'/n)^2$, i.e.,

$$E(\theta') = \frac{\pi B}{2}\left(\frac{n'}{n}\right)^2 \left\{1 - \left[1 + 4\cos^4\theta' \, \frac{\tan^2\alpha'}{\left(1 - \tan^2\alpha'\cos^2\theta'\right)^2}\right]^{-1/2}\right\} \quad . \tag{2-83}$$

Here, $|\alpha'| = \tan^{-1}(a_{ex}/L_i)$ is the semiangle of the cone subtended by the exit pupil at the axial image point P_0' and θ' is the chief ray angle in the image space. The angle $2|\alpha'|$ is called the angular aperture of the image-forming light cone exiting from the system. Letting $\theta' = 0$ in Eq. (2-83), we obtain the irradiance at the axial image point P_0', i.e.,

$$\boxed{E(0) = \pi B(n'/n)^2 \sin^2 \alpha'} \qquad (2\text{-}84)$$

The quantity $n' \sin|\alpha'|$ is called the *numerical aperture* of the image-forming light cone exiting from the system.

If the object lies at infinity, then the image point P_0' coincides with the image-space focal point F'. If, in addition, the system obeys the sine condition [discussed in Section 3.7; see Eq. (2-91) also], then

$$D_{en} = -2f' \sin \alpha_\infty \quad , \qquad (2\text{-}85)$$

or

$$F_\infty = n'/2NA_\infty' \quad , \qquad (2\text{-}86)$$

where α_∞ is the corresponding semiangle of the image-forming light cone and

$$NA_\infty' = n' \sin|\alpha_\infty| \qquad (2\text{-}87)$$

is the corresponding image-space numerical aperture.

The angular aperture, the f-number, and the numerical aperture, all give a measure of the light gathering capability of an optical system in the sense that the image illumination depends on them. It is customary to use the f-number of the image-forming light cone for systems such as cameras imaging objects lying at large distances. The term numerical aperture is used when imaging objects at short distances, as in microscopes.

The variation of irradiance normalized by its axial value is the same as shown in Figure 2-18 provided we replace θ by θ', $|L_o|/a_{en}$ by L_i/a_{ex}, and use Eq. (2-84) for $E(0)$ instead of Eq. (2-72). In this case, we find that

$$\boxed{E(\theta') \geq E(0)\cos^4 \theta'} \qquad (2\text{-}88)$$

The decrease of the image irradiance with θ' is not as rapid as $\cos^4 \theta'$. When the f-number L_i/D_{ex} of the image-forming light cone approaches 5, the variation of irradiance with θ' is practically the same as $\cos^4 \theta'$.

For small exit pupils, i.e., for small values of a_{ex} and, therefore, large f-numbers, the angle γ' for any element on ExP is approximately equal to the chief ray angle θ' in the image space of the system, and the integral in Eq. (2-82) simply reduces to $S_{ex} \cos^4 \theta'$. Accordingly, Eq. (2-82) for the image irradiance distribution may be written

$$\boxed{E(\theta') = E(0)\cos^4 \theta'} \qquad (2\text{-}89)$$

where the axial irradiance is given by

$$\begin{aligned}
E(0) &= B(n'/n)^2 \left(S_{ex}/L_i^2\right) \\
&= B(n'/n)^2 \Omega(P_0') \quad .
\end{aligned} \qquad (2\text{-}90)$$

This is the *cosine-fourth law in the image space, showing that the irradiance of the image of a Lambertian object decreases as the fourth power of the cosine of the chief ray angle θ' in the image space*. This result may also be obtained from Eq. (2-86) for a small angle α' so that $\tan\alpha' \simeq \alpha'$.

2.6.5 Telecentric Systems

If the aperture stop lies in the object-space focal plane, then the system is *telecentric on the image side* and the chief ray angle θ' in the image space is zero for any position of the object element. The irradiance distribution of the image is given by Eq. (2-71) with an appropriate value of L_o. If in addition, the object lies at infinity, then the distribution is given by Eq. (2-77). Incidentally, if we use Eq. (2-82) to determine the image irradiance, we note that the exit pupil is infinite in size and lies at infinity. Moreover, the element of area dS_{ex} given by Eq. (2-67) must be used which, in turn, yields Eq. (2-69). Indeed it would be a mistake to consider Eq. (2-89) with $\theta' = 0$ as the correct equation for this case, leading to the incorrect result that the image irradiance is uniform as given by Eq. (2-87). Similarly, if the aperture stop lies in the image-space focal plane, then the system is *telecentric on the object side* and the chief ray angle θ in the object space is zero for any object element. The image irradiance distribution in this case is given by Eq. (2-86).

2.6.6 Throughput

If we consider the corresponding object and image elements centered on the optical axis at P_0 and P_0', respectively, then equating the axial image irradiances given by Eqs. (2-72) and (2-84), we obtain

$$\boxed{n'h'\sin\alpha' \ = \ nh\sin\alpha \ \ ,} \tag{2-91}$$

where we have substituted $M = h'/h$, h and h' being the heights of the object and image elements from the optical axis, respectively. Equation (2-91) is called the *sine condition* and represents the absence of coma as discussed in Section 3.7.4. Similarly, equating the axial irradiances from Eqs. (2-75a) and (2-90) in the small angle approximation, we obtain

$$n'^2 \, dS' \, \Omega'\left(P_0'\right) \ = \ n^2 \, dS\Omega\left(P_0\right) \ \ , \tag{2-92}$$

where we have used Eq. (2-70) for the magnification. Thus, the quantity $n^2 \, dS\Omega\left(P_0\right)$, called the *optical throughput*,[10] is an invariant. Note that if $n = n'$ (in practice, they are often both equal to unity), then $B = B'$ and the product of the area and the solid angle may simply be called the *throughput*. In that case, the throughput multiplied by the radiance gives the flux passing through the system.

2.6.7 Condition for Uniform Image Irradiance

In Eq. (2-69) we assumed that the image magnification M was independent of the position of the object element. We now consider a system with *distortion* so that the value of M depends on the position of the object element. If we consider an area element dS in

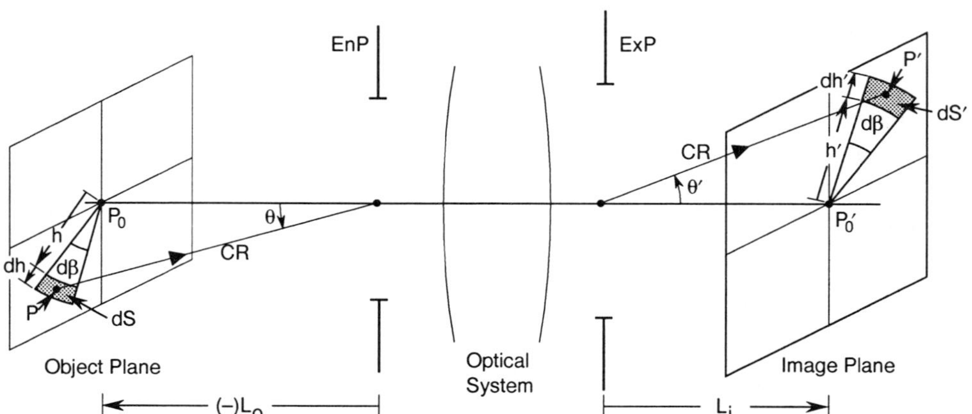

Figure 2-19. Imaging of an area element by an optical system with barrel distortion.

the object plane at a height h with a width dh subtending an angle $d\beta$ at the center of the image plane, as illustrated in Figure 2-19, then

$$dS = h\, d\beta\, dh \quad .\tag{2-93}$$

If the corresponding image is at a height h_i' with a width dh_i' subtending an angle $d\beta$ at the center of the image plane, then

$$dS' = h_i'\, d\beta\, dh_i' \quad .\tag{2-94}$$

Hence, the *areal magnification* of the image element is given by

$$\frac{dS'}{dS} = \frac{h_i' dh_i'}{h\, dh} \quad .\tag{2-95}$$

If M is the paraxial image magnification and h' is the corresponding image height if the magnification is independent of h, then

$$h = L_o \tan\theta \tag{2-96}$$

and

$$h' = L_o M \tan\theta \quad .\tag{2-97}$$

Note that h is numerically negative and h' is numerically positive in Figure 2-19. Accordingly, the magnification M is numerically negative. If we assume that the actual image height is given by

$$h_i' = L_o M \sin\theta \quad ,\tag{2-98}$$

then Eq. (2-95) may be written

$$\frac{dS'}{dS} = \frac{h'\,dh'}{h\,dh}\,\frac{h_i'\,dh_i'}{h'\,dh'}$$

$$= M^2 \cos^4\theta \quad . \tag{2-99}$$

Hence, Eq. (2-69) is replaced by

$$E = \left(B/L_o^2\right)(dS/dS')\int_{EnP}\cos^4\gamma\,dS_{en}$$

$$= \left(B/M^2L_o^2\right)\cos^{-4}\theta\int_{EnP}\cos^4\gamma\,dS_{en} \quad . \tag{2-100}$$

Comparing Eqs. (2-69) and (2-100), Eq. (2-73) yields

$$\boxed{E(\theta) \geq E(0) \quad .} \tag{2-101}$$

Thus, the irradiance at image points away from the center is somewhat higher than at the center for object light cone f-numbers ≤ 5 when the image height is given by Eq. (2-98). The difference between the actual and the paraxial image heights is given by

$$\boxed{h_i' - h' = L_o M(\sin\theta - \tan\theta) \quad ,} \tag{2-102}$$

which is numerically negative. This difference represents *barrel distortion* (distortion is introduced in Chapter 3 and barrel distortion is discussed in Chapter 4). It may be noted that we are able to introduce the $\cos^{-4}\theta$ factor into Eq. (2-69) by virtue of Eq. (2-99) and thereby make the image relatively uniform. This is not possible when the aperture stop lies in the back of the system as may be seen from Eq. (2-82). However, barrel distortion can improve the irradiance at off-axis points.

For an object at infinity, all of the rays incident on the entrance pupil are parallel with $\gamma = \theta$, $L_o M \to -f$, and Eq. (2-100) reduces to

$$E = BS_{en}/f^2$$

$$= n'^2 BS_{en}/n^2 f'^2 \quad . \tag{2-103}$$

In this case, the cosine-fourth decrease in image irradiance is compensated and an image of uniform irradiance is obtained.

2.6.8 Concentric Systems

In a *concentric system* such as a *Schmidt* or a *Bouwers-Maksutov camera* (discussed in Section 6.6), the aperture stop and the entrance and exit pupils all lie at its common center of curvature and the image is formed on a spherical surface concentric with the system. The chief ray incident through the common center passes undeviated and the chief ray angles in the object and image spaces are equal. The image element area dS' is normal to the line joining its center and the center of the exit pupil, and the distance between the two centers is independent of the location of dS' on the spherical image surface. Therefore, the solid angle subtended by dS' at the exit pupil is simply equal to

dS'/L_i^2. Hence, the flux emerging from a *small* exit pupil and converging on the image element is given by

$$F' = B'dS'\left(S_{ex}/L_i^2\right)\cos\theta'$$

$$= B'dS'\Omega'\left(P_0'\right)\cos\theta' \quad . \tag{2-104}$$

Accordingly, the irradiance of the image element is given by

$$\boxed{E = B\left(n'/n\right)^2\Omega'\left(P_0'\right)\cos\theta' \quad .} \tag{2-105}$$

Thus, *the irradiance of the spherical image formed by a concentric system decreases linearly with the cosine of the angle of the chief ray in the image space.* For an object lying at infinity, we let $L_i = f'$, the focal length of the system.

2.7 PHOTOMETRY

Now we give a brief discussion of *photometry,* the branch of radiometry that is limited to observations with a human eye, which is sensitive only in the visible region of the electromagnetic spectrum called *light.* The theory of photometry in terms of transfer of light from a source to a receiver is the same as discussed earlier, except that the spectral response of the eye must be taken into account to determine the final result of any observation. The names, symbols, and units of photometric quantities are given, along with an equation on how to obtain a photometric quantity from a corresponding radiometric quantity. It is shown that a Lambertian surface appears equally bright at all distances and along all directions of observation. The reason stars can be observed during daytime with the aid of a telescope is also discussed.

2.7.1   Photometric Quantities and Spectral Response of the Human Eye

The units of some of the basic quantities used in photometry are given in Table 2-1, along with their radiometric counterparts. The abbreviation of a unit is indicated in parentheses. To avoid confusion, the corresponding photometric and radiometric terms are distinguished from each other by adding to them the adjectives "luminous" and "radiant," respectively, e.g., luminous flux and radiant flux. It is a common practice to use the term "luminance" in place of "luminous radiance", and "illuminance" in place of "luminous irradiance."

A photometric quantity can be obtained from a corresponding spectral radiometric quantity by weighting it with the spectral response of the eye.[11] The relative spectralresponse of the eye is shown in Figure 2-20 and its numerical values are given in Table 2-2 for both day (photopic) and night (scotopic) vision. The peak values of the two spectral visions lie at 555 nm and 507 nm and correspond to 683 lm/W and 1754 lm/W, respectively. Thus, for example, the absolute daytime response of the eye at 600 nm is given by $0.631 \times 683 = 431$ lm/W. If $V(\lambda)$ represents the relative spectral response of the eye, then the luminous flux Φ_l of a source with a spectral radiant flux $\Phi_r(\lambda)$ is given by

Table 2-1 Photometric and radiometric units of some basic quantities.

Quantity	Photometric Unit	Radiometric Unit
Energy	talbot	joule (J)
Flux	lumen (lm)	watt (W)
Intensity	lumens/steradian (lm/sr) = candela (cd)	W/sr
Radiance (Luminance)	lm/m^2 sr	W/m^2 sr
Irradiance (Illuminance)	lm/m^2 = lux (lx)	W/m^2

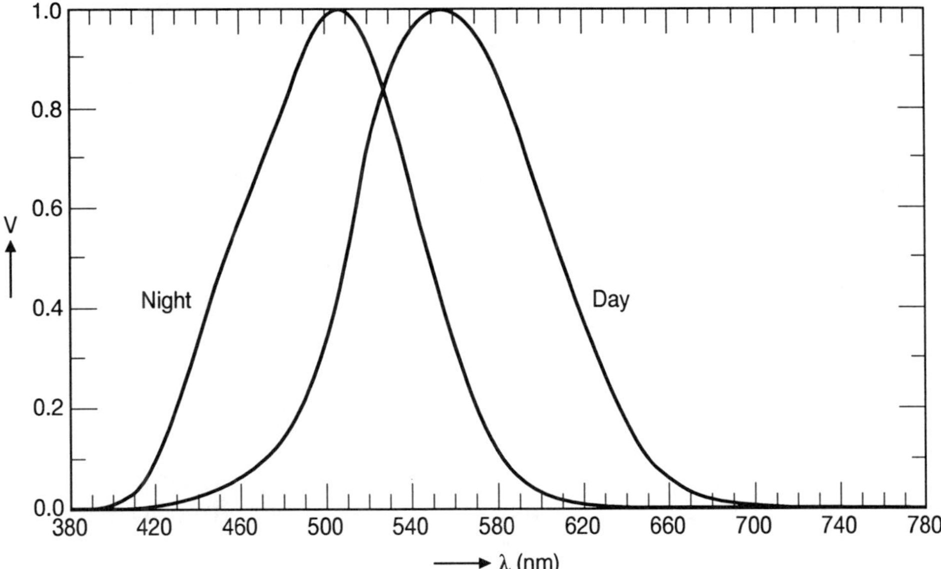

Figure 2-20. Relative spectral response of a human eye for day (photopic) and night (scotopic) visions.

$$\Phi_l = \kappa \int \Phi_r(\lambda) V(\lambda) \, d\lambda \quad , \tag{2-106}$$

where $\kappa = 683$ lm/W or 1754 lm/W, depending upon whether $V(\lambda)$ is for day or night vision.

2.7.2 Imaging by a Human Eye

The human eye consists of an *iris* that acts as the aperture stop and whose diameter increases or decreases, depending on the luminance of an object under observation. It has

Table 2-2. Relative spectral response of human eye for day (photopic) and night (scotopic) visions.

Wavelength λ (nm)	Day (Photopic) V	Night (Scotopic) V
380	0.00004	0.000589
390	0.00012	0.002209
400	0.0004	0.00929
410	0.0012	0.03484
420	0.0040	0.0966
430	0.0116	0.1998
440	0.023	0.3281
450	0.038	0.455
460	0.060	0.567
470	0.091	0.676
480	0.139	0.793
490	0.208	0.904
500	0.323	0.982
507	0.445	**1**
510	0.503	0.997
520	0.710	0.935
530	0.862	0.811
540	0.954	0.650
550	0.995	0.481
555	**1**	0.402
560	0.995	0.3288
570	0.952	0.2076
580	0.870	0.1212
590	0.757	0.0655
600	0.631	0.03315
610	0.503	0.01593
620	0.381	0.00737
630	0.265	0.003335
640	0.175	0.001497
650	0.107	0.000677
660	0.061	0.0003129
670	0.032	0.0001480
680	0.017	0.0000715
690	0.0082	0.00003533
700	0.0041	0.00001780
710	0.0021	0.00000914
720	0.00105	0.00000478
730	0.00052	0.000002546
740	0.00025	0.000001379
750	0.00012	0.000000760
760	0.00006	0.000000425
770	0.00003	0.0000002413
780	0.000015	0.0000001390

a lens that forms images of objects on a light-sensitive screen called the *retina*. It does not have a field stop, but the resolution of the retina decreases rapidly as a function of the distance from its center. In looking at objects, the eye rotates until the image of an object under observation falls on the central portion of the retina. The apparent size of an object is determined by the size of its retinal image.

Consider an object of height h lying at a distance R from the principal point H, as illustrated in Figure 2-21. An image of height h' is formed on the retina at a distance R' from the principal point H'. (See Problems 1.7 and 1.8 for a Gaussian model of the human eye.) The angular sizes β and β' of the object and image as seen from the respective principal points are related to each other according to [see Eq. (1-61)]

$$n\beta = n'\beta' \ , \tag{2-107}$$

where n and n' are the refractive indices of the object and image spaces, respectively. The image height h' is given by

$$\begin{aligned} h' &= R'\beta' \\ &= (n/n')R'\beta \ . \end{aligned} \tag{2-108}$$

As the object distance varies, the eye lens changes its focal length by a process called *accommodation* so that the distance R' remains practically invariant. Consequently, the apparent size of an object is proportional to the angle β it subtends at H independent of the state of accommodation.

2.7.3 Brightness of a Lambertian Surface

Consider a Lambertian surface or a uniform diffuser of area dS_1 and luminance L observed by an eye of pupil area dS_2 lying at a distance R as illustrated in Figure 2-22. The *subjective brightness* of the surface depends on the illuminance on the retina. The total flux entering the eye according to Eq. (2-27) is given by

$$dF = \frac{L\,dS_1\cos\theta_1\,dS_2}{R^2} \ , \tag{2-109}$$

where θ_1 is the angle between the normal to the surface dS_1 and the direction of observation, i.e., the line joining the centers of dS_1 and dS_2. The angle θ_2 is zero since dS_2 is normal to this line.

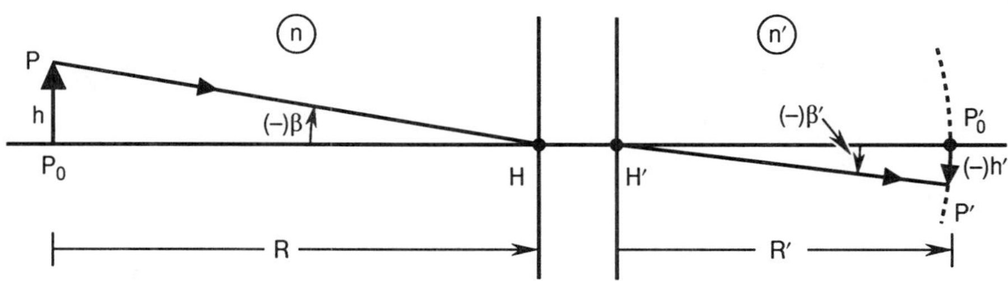

Figure 2-21. Imaging by a human eye.

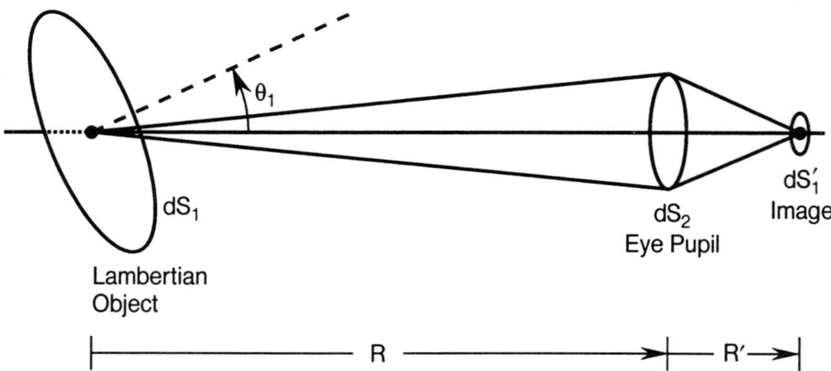

Figure 2-22. Observation of a Lambertian surface.

If η is the transmission factor of the eye, the flux reaching the retina is $\eta\, dF$. This flux is distributed over the retinal image of object dS_1. The projected area of the observed surface normal to the direction of observation is $dS_1 \cos\theta_1$. Therefore, if R' is the image distance, then the area of the image is given by

$$dS_1' = \left(R'/n'R\right)^2 dS_1 \cos\theta_1 \quad, \tag{2-110}$$

where $nR'/n'R$ with $n = 1$ is the (linear) magnification of the image. Hence, the illuminance on the retina is given by

$$E = \eta \frac{dF}{dS_1'}$$
$$= \frac{\eta\, n'^2 L\, dS_2}{R'^2} \quad. \tag{2-111}$$

Since it is independent of θ_1 and R, we note that *a Lambertian surface appears equally bright at all distances along all directions of observation.*

2.7.4   Observing Stars in the Daytime

It is common knowledge that stars are too dim compared with the sky to be observed in the daytime with a naked eye. However, they can be observed with the aid of a telescope. As explained below, this is due to an increase in the star intensity on the retina when a telescope is used.

Consider an observation of a distant object of area S_o and luminance L lying at a distance R. The flux entering an eye with a pupil of area S_e is given by

$$F = L S_o\left(S_e/R^2\right) \quad, \tag{2-112}$$

where S_e/R^2 is the solid angle subtended by the pupil at the object. This flux is spread on the retina over an area S_i given by (with $n = 1$)

$$S_i = \left(\frac{R'}{n'R}\right)^2 S_o \ . \tag{2-113}$$

Hence, the illuminance of the retinal image is given by

$$\boxed{\frac{F}{S_i} = L\, S_e (n'/R')^2 \ .} \tag{2-114}$$

If the same object is observed with the aid of a telescope, then the eye sees the image formed by the telescope. Since the luminance of the telescopic image according to Eq. (2-64) is (at most) equal to the object luminance, it is evident that the retinal illuminance will be the same as in the case of an unaided observation provided the diameter of the exit pupil of the telescope is greater than or equal to the diameter of the eye pupil. If the diameter of the exit pupil is smaller, then S_e is replaced by the area S_{ex} of the exit pupil and the retinal illuminance is reduced by a factor S_{ex}/S_e. Thus, the retinal illuminance of the image of a distant object observed with the aid of a telescope is less than or equal to the corresponding illuminance obtained when the object is observed with the naked eye. However, because of the larger entrance pupil of the telescope compared to its exit pupil or the eye pupil, the flux in the retinal image of a point object increases when it is observed with the aid of a telescope regardless of the size of the exit pupil compared to that of the eye pupil. Hence, star observations with the aid of a telescope are possible during daytime since the sky background appears the same or dimmer but a star appears brighter.

For a more explicit explanation, consider an afocal telescope discussed in Section 1.3.6 and illustrated in Figure 2-23. The aperture stop located at the lens L_1, called the *objective*, has a diameter $D_1 > D_e$, where D_e is the diameter of the eye pupil. The lens L_2, called the *eyepiece,* has a diameter D_2, where $D_2 < D_1$. The lens L_1 is the entrance pupil and its image by lens L_2 is the exit pupil of the telescope. It is easy to show that the diameter of the exit pupil is equal to D_2. Consider a distant extended object of angular radius β. The radii of the images on the retina observed with and without the use of the telescope are given by $R'\beta'/n'$ and $R'\beta/n'$, respectively, where β' is the angular radius of the image formed by the telescope and n' is the refractive index of the eye. Hence, the ratio of the corresponding image areas is given by $(\beta'/\beta)^2$ or M_β^2, where M_β is the angular magnification of the telescope. The ratio of the amounts of light in the two images is given by $(D_1/D_e)^2$, where it is assumed that $D_2 \le D_e$ so that all of the light collected by the telescope enters the eye. The ratio of the retinal image illuminances with and without the use of the telescope is given by

$$\left(D_1/D_e\right)^2 \big/ M_\beta^2 \ = \ \left(D_2/M_t D_e\right)^2 \big/ M_\beta^2$$

$$= \ \left(D_2/D_e\right)^2 \ , \tag{5-215}$$

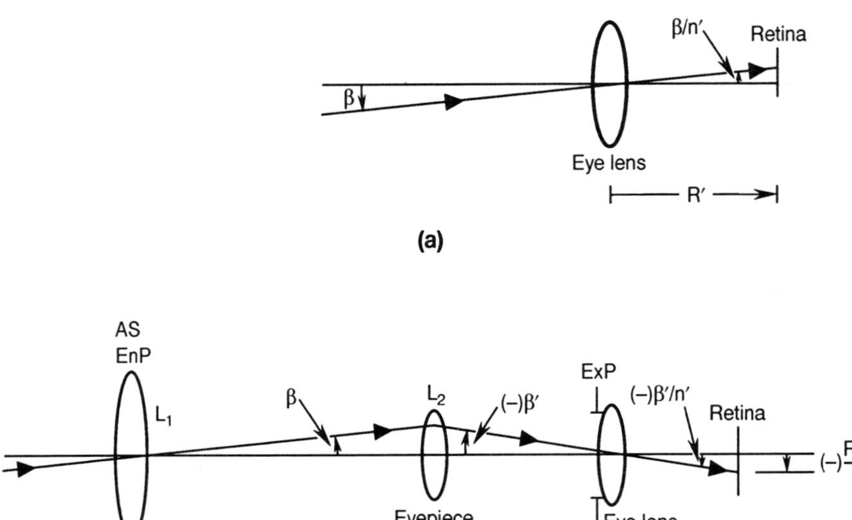

Figure 2-23. Daytime observation of a star against sky background (a) without and (b) with the aid of a telescope. The lenses L_1 and L_2, called the objective and the eyepiece, respectively, with a spacing equal to the sum of their focal lengths constitute the afocal telescope.

where $M_t = D_2/D_1$ is the transverse magnification of the telescope and we have made use of the fact that $M_t M_\beta = 1$ (by Lagrange invariance). If $D_2 = D_e$, then the use of the telescope does not change the illuminance of the retinal image. If $D_2 < D_e$, then the illuminance obtained with the use of a telescope is less than that obtained with the naked eye. In this case, although all of the light collected by the telescope enters the eye, it is spread over a larger retinal image.

If $D_2 > D_e$, then a fraction $\left(D_e/D_2\right)^2$ of the light collected by the telescope lies in the retinal image. In this case, the ratio of the amounts of light in the retinal images with and without the use of the telescope is given by

$$\left(D_1/D_e\right)^2\left(D_e/D_2\right)^2 \;=\; 1/M_t^2 \quad . \tag{5-216}$$

Hence, the ratio of the retinal illuminances with and without the use of a telescope is given by

$$1/M_t^2\, M_\beta^2 \;=\; 1 \quad , \tag{5-217}$$

i.e., the two illuminances are equal. In this case, although all of the light collected by the telescope does not enter the eye, the amount that enters is spread over a correspondingly smaller retinal image. Thus, the retinal illuminance for an extended object observed with a naked eye is the same as that when it is observed through a telescope with $D_2 \geq D_e$, and it is less with a telescope if $D_2 < D_e$.

Now consider the observation of a point object such as a star. The apparent intensity of the star depends on the amount of light in its retinal image. The ratio of the amounts of light in this image with and without a telescope is given by $(D_1/D_e)^2$ provided $D_2 \leq D_e$. If $D_2 > D_e$, then, following the argument given above, the ratio of the amounts of light in the retinal image is given by $1/M_t^2$. In either case, this ratio is greater than 1. Thus, the intensity of the star image on the retina is increased by use of the telescope. Since the illuminance of the sky background on the retina either stays the same or decreases with the use of the telescope and the intensity of the star image on the retina increases with its use, the star visibility or the signal-to-noise ratio increases. Accordingly, it is possible to observe stars in the daytime by the use of a telescope for which $D_1 > D_e$. It should be noted that we have neglected the effects of diffraction, especially in discussing the image of a point object.

APPENDIX: RADIANCE THEOREM

Here we give a general proof of the *radiance theorem* using Hamilton's point characteristic function applicable to imaging systems without any axis of symmetry. Consider two arbitrary points P and P' in the object and image spaces of a system connected by a ray, as illustrated in Figure 2-24, where n and n' are the refractive indices of the two spaces. Let $P(x, y)$ lie in the xy plane of a certain coordinate system x, y, z in the object space. Similarly, let $P'(x', y')$ lie in the $x'y'$ plane of a coordinate system x', y', z' in the image space. Consider small displacements dx and dy of P and small changes dL and dM of the direction cosines of a ray with direction cosines L and M through P. [Note that the direction cosines L and M of a unit vector (L, M, N) predetermine the value of N through $L^2 + M^2 + N^2 = 1$. Hence we need not specify N separately.] Let (dx', dy') and (dL', dM') be the corresponding displacements of P' and changes of the direction cosines of the ray through it.

The direction cosines may be obtained from the Hamilton's point characteristic function $V(P, P')$ or $V(x, y; x', y')$ for the two points P and P'. For example [see Eqs. (1-12)],

$$nL = -\frac{\partial V}{\partial x} \tag{2A-1a}$$

and

$$n'L' = \frac{\partial V}{\partial x'} \quad . \tag{2A-1b}$$

Now

$$\frac{\partial^2 V}{\partial x\, \partial x'}\, \frac{\partial^2 V}{\partial y\, \partial y'} - \frac{\partial^2 V}{\partial y\, \partial x'}\, \frac{\partial^2 V}{\partial x\, \partial y'} = \frac{\partial^2 V}{\partial x'\, \partial x}\, \frac{\partial^2 V}{\partial y'\, \partial y} - \frac{\partial^2 V}{\partial y'\, \partial x}\, \frac{\partial^2 V}{\partial x'\, \partial y} \quad . \tag{2A-2}$$

Or, using Eqs. (2A-1)

$$n'^2 \left(\frac{\partial L'}{\partial x}\, \frac{\partial M'}{\partial y} - \frac{\partial L'}{\partial y}\, \frac{\partial M'}{\partial x} \right) = n^2 \left(\frac{\partial L}{\partial x'}\, \frac{\partial M}{\partial y'} - \frac{\partial L}{\partial y'}\, \frac{\partial M}{\partial x'} \right) , \tag{2A-3a}$$

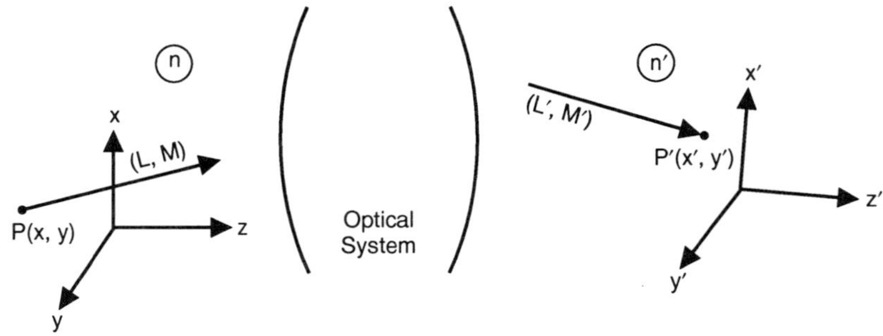

Figure 2-24. Radiance theorem for an arbitrary optical imaging system.

or using Jacobians

$$n'^2 \frac{\partial(L', M')}{\partial(x, y)} = n^2 \frac{\partial(L, M)}{\partial(x', y')} \quad . \tag{2A-3b}$$

Substituting

$$dL\,dM = \frac{\partial(L', M')}{\partial(x, y)} dx\,dy \tag{2A-4a}$$

and

$$dL'\,dM' = \frac{\partial(L, M)}{\partial(x', y')} dx'\,dy' \,, \tag{2A-4b}$$

we obtain

$$\boxed{n'^2 dx'\,dy'\,dL'\,dM' = n^2 dx\,dy\,dL\,dM} \quad . \tag{2A-5}$$

The quantity $n^2 dx\,dy\,dL\,dM$ is called the *generalized Lagrange invariant*.[12] Equation (2A-5) is also an optical analog of the Liouville's theorem in classical mechanics from which Snell's law and the sine condition (discussed in Section 3.7) can be derived.[13]

In spherical polar coordinates, the direction cosines of a ray in the object space are given by (see Figure 2-16)

$$(L, M, N) = (\sin\theta\cos o, \sin\theta\sin\phi, \cos\theta) \quad . \tag{2A-6}$$

Therefore,

$$\begin{aligned}
dL\,dM &= \frac{\partial(L, M)}{\partial(\theta, \phi)} d\theta\,d\phi \\
&= \frac{\partial L}{\partial\theta}\frac{\partial M}{\partial\phi} - \frac{\partial L}{\partial\phi}\frac{\partial M}{\partial\theta} \\
&= \cos\theta\sin\theta\,d\theta\,d\phi \\
&= \cos\theta\,d\Omega \quad ,
\end{aligned} \tag{2A-7}$$

where we have made use of Eq. (2-56) for the solid angle $d\Omega$ of the rays. Thus, $dx\,dy\,dL\,dM$ represents the product of the projected area in the direction of the rays and their solid angle. Similarly, $dx'\,dy'\,dL'\,dM'$ represents the corresponding product in the image space.

Equation (2-64) for the invariance of B/n^2 in optical imaging can be obtained immediately from Eq. (2A-5) as follows. If $B(x, y; L, M)$ is the radiance of the object element in the direction (L, M), then the flux radiated into a cone with an angular width represented by (dL, dM) is given by $B(x, y; L, M)dx\,dy\,dL\,dM$. In the absence of any transmission losses, this flux is received by the image element. If $B'(x', y'; L', M')$ is the radiance of the image element, then the flux radiated by it into a cone with an angular width (dL', dM') is given by $B'(x', y'; L', M')dx'\,dy'\,dL'\,dM'$. Equating the two fluxes and using Eq. (2A-5) yields Eq. (2-64), representing the radiance theorem.

REFERENCES

1.   R. McCluney, *Introduction to Radiometry and Photometry*, Artech, Boston (1994).

2.   P. D. Foote, "Illumination from a radiating disk," *Bull. Bur. Stand.* **12**, 583–586 (1915).

3.   H. F. Gilmore, "The determination of image irradiance in optical systems," *Appl. Opt.* **5**, 1812–1817 (1966).

4.   F. E. Nicodemus, "Radiance," *Am. J. Phys.* **31**, 368–377 (1963).

5.   I. C. Gardner "Validity of the cosine–fourth power law of illumination," *J. Research Nat. Bur. Stand.* **39**, 213–219 (1947).

6.   M. Reiss, "The $\cos^4$ law of illumination," *J. Opt. Soc. Am.* **35**, 283–288 (1945).

7.   M. Reiss, "Notes on the $\cos^4$ law of illumination," *J. Opt. Soc. Am.* **38**, 980–986 (1948).

8   F. Wachendorf, "The condition of equal irradiance and the distribution of light in images formed by optical systems without artificial vignetting," *J. Opt. Soc. Am.* **43**, 1205–1208 (1953).

9.   R. Kingslake, "Illumination in optical images," in *Applied Optics and Optical Engineering*, ed. R. Kingslake, Vol. II, pp. 195–228, Academic Press, New York (1965).

10.   W. H. Steel, "Luminosity, throughput, or etendue," *Appl. Opt.* **13,** 704 (1974); also, "Luminosity, throughput, or etendue ? Further comments," *Appl. Opt.* **14,** 252 (1975).

11.   J. R. Meyer-Arendt, "Radiometry and photometry: units and conversion factors," *Appl. Opt.* **7**, 2081–2084 (1968).

12.   W. T. Welford, *Aberrations of the Symmetrical Optical System*, Section 5.6, Academic Press, New York (1974).

13.   D. Marcuse, *Light Transmission Optics*, Section 3.7, Van Nostrand Reinhold, New York (1989).

PROBLEMS

2.1   A system consisting of two thin lenses of focal lengths 10 cm and 5 cm with apertures of 4 cm are spaced 4 cm apart. A stop 2 cm in diameter is located midway between them. (a) Determine its principal points. (b) Find the position and size of its entrance and exit pupils. (c) Find the position and size of the image of an object placed 10 cm from the first lens. (d) Sketch them on a diagram showing, in addition, the two tangential marginal rays and the chief ray from the top of the object if it is 4 cm high. (e) In the object plane considered, what is the maximum height of a point object for which there is no *vignetting*?

2.2   Consider a system consisting of two thin lenses placed 4 cm apart with a 4 cm aperture placed midway between them. The first lens has a diameter of 4.6 cm and a focal length of 5.8 cm. The second lens has a diameter of 5.8 cm. An object is placed 8 cm from the first lens. (a) Determine the aperture stop of the system. (b) Sketch the *vignetting diagram* for a point object 4 cm from the optical axis.

2.3   An exit pupil with a 3-cm aperture is located 6 cm in front of a convex mirror that has a radius of curvature of 10 cm. An object 1 cm high is centrally located on the axis 12 cm in front of the mirror. (a) Locate the entrance pupil and the image. (b) Find the minimum diameter of the mirror needed to see the entire object from all points of the exit pupil.

2.4   Consider a *Schwarzchild telescope* consisting of two concentric spherical mirrors such that the ratio of their radii of curvature is $\left(3 \pm \sqrt{5}\right)/2$. Its aperture stop is located at the primary mirror. (a) Determine its focal length in terms of the focal lengths of its mirrors. (b) Determine the distance between its focal plane and the mirror close to it. (This distance is often referred to as the *working distance* of a telescope.) (c) Calculate the position and size of its exit pupil. (d) Determine the obscuration ratio of the image-forming beam. (The *obscuration ratio* is the ratio of the inner and outer diameters of the light cone focusing to the image point.) (e) Determine the diameter of its secondary mirror for a field of view of $\pm$ 5 milli-radians. (f) Sketch the system if its focal ratio is 2 and the diameter of its primary mirror is 10 cm.

2.5   Show that the height of a light bulb (assumed to be a point source) from the center of a circular table of radius a for maximum illumination at its edges is given by $2\sqrt{a}$.

2.6   According to the Stefan-Boltzmann law, the exitance (i.e., the power radiated by a unit area) of a blackbody at a temperature T (in Kelvin) is given by σT^4, where $\sigma = 5.67 \times 10^{-8}$ W/m^2 K^4 is the Stefan-Boltzmann constant. Consider the sun to be a blackbody at 6000 K. (a) Determine its radiance. (b) Calculate the solar irradiance on the earth, called the *solar constant* (the solar constant is also expressed as a 2 calories/cm^2 min). (c) Compare it with the irradiance of the solar image formed by a lens with an f-number of 5. (d) Assuming that the moon reradiates 20% of the light incident on it, compare the lunar irradiance on the earth for full moon with solar irradiance in full sunlight. Some of the sizes and distances of interest are as follows: the radius of the sun and its distance from the earth are 6.96×10^8 m and 1.49×10^{11} m, respectively; the radius of the moon and its distance from the earth are 1.77×10^6 m and 3.80×10^8 m, respectively.

2.7　　Consider an optical system imaging a small circular object of radius h centered on its optical axis. Let the circular image be of radius h'. Let β_0 and β_0' be small slope angles of the axial marginal rays in the object and image spaces of the system, as in Figure 2-2. Show by using the Lagrange invariance of Eq. (1-69) that the object and image radiances are related to each other according to Eq. (2-64), where n and n' are the refractive indices of the object and image spaces. The object and image sizes are assumed to be small so that the entrance and exit pupils subtend approximately the same angles at every point on them, respectively.

CHAPTER 3

OPTICAL ABERRATIONS

3.1 Introduction **141**

3.2 Wave and Ray Aberrations...**142**

    3.2.1 Definitions 142

    3.2.2 Relationship Between Wave and Ray Aberrations.................................145

3.3 Defocus Aberration **148**

3.4 Wavefront Tilt **150**

3.5 Aberration Function of a Rotationally Symmetric System **152**

    3.5.1 Rotational Invariants...152

    3.5.2 Power-Series Expansion 155

        3.5.2.1 Explicit Dependence on Object Coordinates 156

        3.5.2.2 No Explicit Dependence on Object Coordinates 159

    3.5.3 Zernike Circle-Polynomial Expansion 163

    3.5.4 Relationships Between Coefficients of Power-Series and
Zernike-Polynomial Expansions..168

3.6 Observation of Aberrations **169**

    3.6.1 Primary Aberrations 172

    3.6.2 Interferograms...173

3.7 Conditions for Perfect Imaging **178**

    3.7.1 Imaging of a 3-D Object...178

    3.7.2 Imaging of a 2-D Transverse Object 181

    3.7.3 Imaging of a 1-D Axial Object 183

    3.7.4 Linear Coma and the Sine Condition 184

    3.7.5 Optical Sine Theorem 186

    3.7.6 Linear Coma and Offense Against the Sine Condition 188

Appendix A: Degree of Approximation in Eq. (3-11) **192**

**Appendix B: Wave and Ray Aberrations: Alternative Definition and
Derivation**..**194**

References **200**

Problems **201**

Chapter 3

Optical Aberrations

3.1 INTRODUCTION

Given the radii of curvature of the surfaces of an imaging system and the refractive indices of the media surrounding them, the position and the size of the Gaussian image of an object can be determined by using the equations given in Chapter 1. By determining the position and the size of its entrance and exit pupils, the irradiance distribution of the image of an object with a certain radiance distribution can be calculated, as discussed in Chapter 2. However, the quality of the image, which depends on the aberrations of the system, was not discussed. In this chapter, the concepts of *wave* and *ray aberrations* are introduced and a relationship between them is derived. The wave aberrations for a certain point object represent the optical deviations of its wavefront at the exit pupil from being spherical. If the wave aberrations are zero, i.e., if the wavefront is spherical, then all the rays converge to its center of curvature and a perfect point image is obtained. The aberrations of a system lead to an imperfect image. The ray aberrations represent the displacement of the rays from the center of curvature in an image plane passing through it.

Although the ray aberrations of a system for a certain point object can be obtained by tracing the rays through the system and up to the image plane, they can also be obtained from the wave aberrations. However, the distribution of rays in an image plane does not represent the true picture of an image, since it does not take into account the diffraction of the wavefront at the exit pupil. Since the wave aberrations play a fundamental role in determining the image quality, their knowledge is essential. It is also noteworthy that the wave aberrations of a multielement system are additive, i.e. , the wave aberration of a ray for the entire system is equal to the sum of its wave aberrations for each of the elements. This is not true of the ray aberrations; the aberration of a ray in the final image plane can not be obtained by adding its values in the image planes for the elements. Of course, the contribution of an element to the ray aberration in the final image plane can be obtained from its contribution, for example, to the primary wave aberrations of the ray according to the equations derived in Chapter 4.

A *defocus wave aberration* is introduced when the image is observed in an image plane other than the one in which the center of curvature lies. It is also introduced if one or more imaging elements of the system are displaced along its optical axis. We derive a relationship between the *longitudinal defocus* of an image and the defocus aberration resulting from it. Similarly, when an imaging element is slightly tilted or displaced perpendicular to the axis, a wavefront tilt is introduced. We show how the *wavefront tilt* is related to the *wavefront tilt aberration*.

Next, the possible aberrations of an imaging system that is *rotationally symmetric* about its optical axis are discussed. The aberration function of the system is expanded in a power series of the object and pupil coordinates, and *primary, secondary,* and *tertiary aberrations* are introduced. It is also expanded in terms of *Zernike circle polynomials,* and relationships between the coefficients of expansions of the two types are given. Interferograms of primary aberrations are discussed briefly to illustrate how such aberrations may be recognized in practice. We end this chapter with a discussion of the conditions under which a system can form a perfect or aberration-free image of an object. In particular, the *sine condition* is derived showing that all orders of coma are zero for off-axis conjugate points satisfying this condition in planes for which axial conjugates are imaged perfectly.

3.2 WAVE AND RAY ABERRATIONS

In this section we define wave and ray aberrations and derive a relationship between the two. We show that the ray aberration of a ray is proportional to the derivative of its wave aberration with respect to its pupil coordinates.

3.2.1   Definitions

Consider an optical system imaging a point object P as illustrated in Figure 3-1. The object radiates a spherical wave. If the image is perfect, the diverging spherical wave incident on the system is converted by it into a spherical wave converging to the Gaussian image point P'. With a few exceptions, the wave exiting from practical systems is only approximately spherical.

We now introduce the concept of wave and ray aberrations associated with an object ray and derive a relationship between the two. The *optical path length* of a ray in a medium of refractive index n is equal to n times its *geometrical path length.* If rays from a point object are traced through the system and up to the exit pupil such that each one travels an optical path length equal to that of the chief ray, the surface passing through their end points is called the system *wavefront* for the point object under consideration. If the wavefront is spherical, with its center of curvature at the Gaussian image point, we say that the image is *perfect.* The rays transmitted by the system have equal optical path lengths in propagating from P to P' and they all pass through P'. If, however, the actual wavefront deviates from this Gaussian spherical wavefront, called the *Gaussian reference sphere,* we say that the image is aberrated. The rays do not have equal optical path lengths and they intersect the Gaussian image plane in the vicinity of P'. The optical deviations (i.e., geometric deviations times the refractive index n_i of the

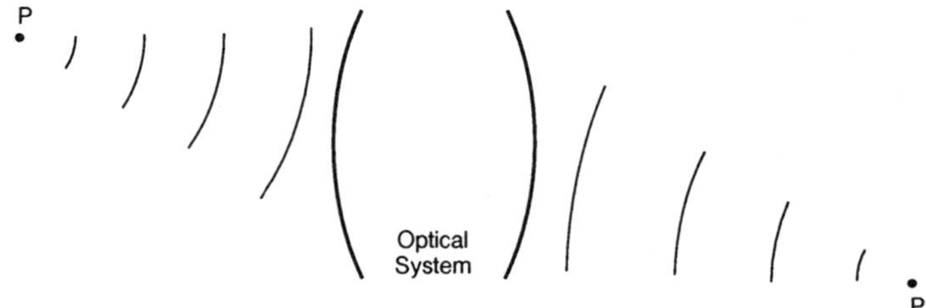

Figure 3-1. Perfect imaging by an optical system. P is the point object and P' is its Gaussian image point.

image space) of the wavefront from a Gaussian reference sphere are called wave aberrations. The *wave aberration* of a ray at a point on the reference sphere where the ray meets it is equal to the optical deviation of the wavefront along that ray from the Gaussian spherical wavefront. It represents the difference between the optical path lengths of the ray under consideration and the chief ray in traveling from the point object to the reference sphere. Accordingly, the wave aberration associated with the chief ray is zero. Since the optical path lengths of the rays from the reference sphere to the Gaussian image point are equal, the wave aberration of a ray is also equal to the difference between its optical path length from the point object P to the Gaussian image point P' and that of the chief ray. This definition will be used in later chapters when we calculate the wave aberrations of actual systems.

The wave aberration of a ray is *positive* if it has to travel an extra optical path length, compared to the chief ray, in order to reach the Gaussian reference sphere. Figures 3-2a and 3-2b illustrate the reference sphere S and the aberrated wavefront W for on-axis and off-axis point objects, respectively. The reference sphere, which is centered at the Gaussian image point P_0' in Figure 3-2a or P' in Figure 3-2b, and the wavefront pass through the center O of the exit pupil. The wave aberration $n_i \overline{Q} Q$ of a general ray GR_0 or GR, where n_i is the refractive index of the image space, as shown in the figures, is numerically positive. The coordinate system is also illustrated in these figures. We choose a right-hand Cartesian coordinate system such that the optical axis lies along the z axis. The object, entrance pupil, exit pupil, and Gaussian image lie in mutually parallel planes that are perpendicular to this axis. Figure 3-3 illustrates the coordinate systems in the object, exit pupil, and image planes. The origin of the coordinate system lies at O and the Gaussian image plane lies at a distance z_g from it along the z axis.

We assume that a point object such as P lies along the x axis. (There is no loss of generality because of this since the system is rotationally symmetric about the optical axis.) The zx plane containing the optical axis and the point object is called the *tangential* or the *meridional plane*. The corresponding Gaussian image point P' lying in the Gaussian image plane along its x axis also lies in the tangential plane. This may be seen by consideration of a tangential object ray and Snell's law, according to which the incident and refracted (or reflected) rays at a surface lie in the same plane. The chief ray always lies in the tangential plane. The plane normal to the tangential plane but containing the chief ray is called the *sagittal plane*. As the chief ray bends when it is refracted or reflected at an optical surface, so does the sagittal plane. It should be evident that only the chief ray lies in both the tangential and sagittal planes, since it lies along the line of intersection of these two planes.

Consider a ray such as GR (general ray) from the object passing through the system and intersecting the Gaussian image plane at $P''(x_i, y_i)$. The displacement $P'P''$ of P'' from the Gaussian image point P' is called the *geometrical* or the *transverse ray aberration*. The distribution of rays in an image plane is called the *ray spot diagram*. The ray aberrations and spot diagrams are discussed in Chapter 4. When the wavefront is

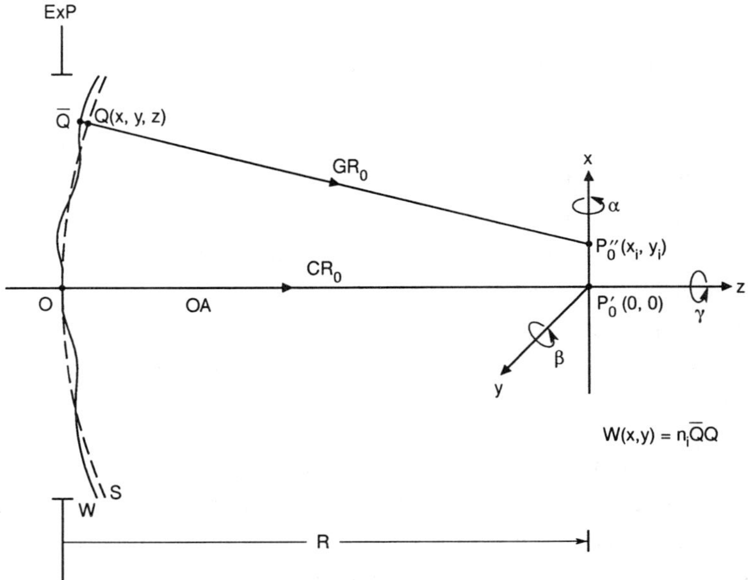

Figure 3-2. (a) Aberrated wavefront for an on-axis point object. The reference sphere S of radius of curvature R is centered at the Gaussian image point P_0'. The wavefront W and reference sphere S pass through the center O of the exit pupil ExP. A right-hand Cartesian coordinate system showing x, y, and z axes is illustrated, where the z axis is along the optical axis of the imaging system. Angular rotations α, β, and γ about the three axes are also indicated. CR_0 is the chief ray and a general ray GR_0 is shown intersecting the Gaussian image plane at P_0''.

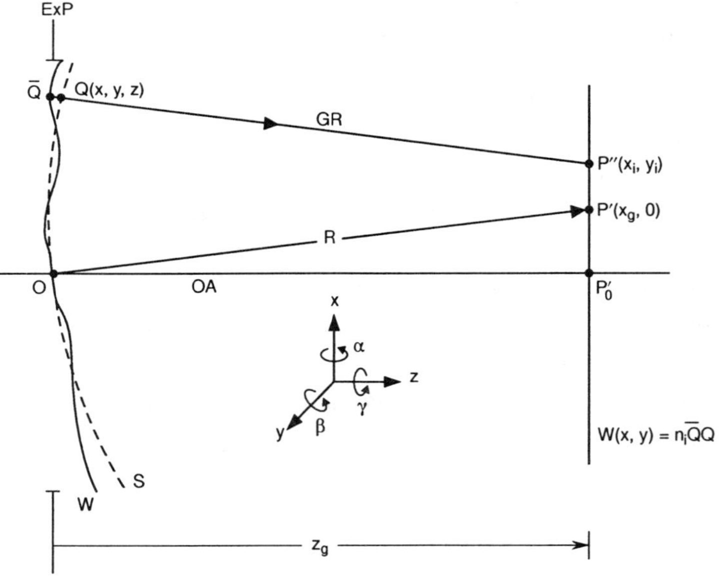

Figure 3-2. (b) Aberrated wavefront for an off-axis point object. The reference sphere S of radius of curvature R is centered at the Gaussian image point P'. The value of R in this figure is slightly larger than its value in Figure 3-2a. GR is a general ray intersecting the Gaussian image plane at the point P''. By definition, the chief ray (not shown) passes through O, but it may or may not pass through P'.

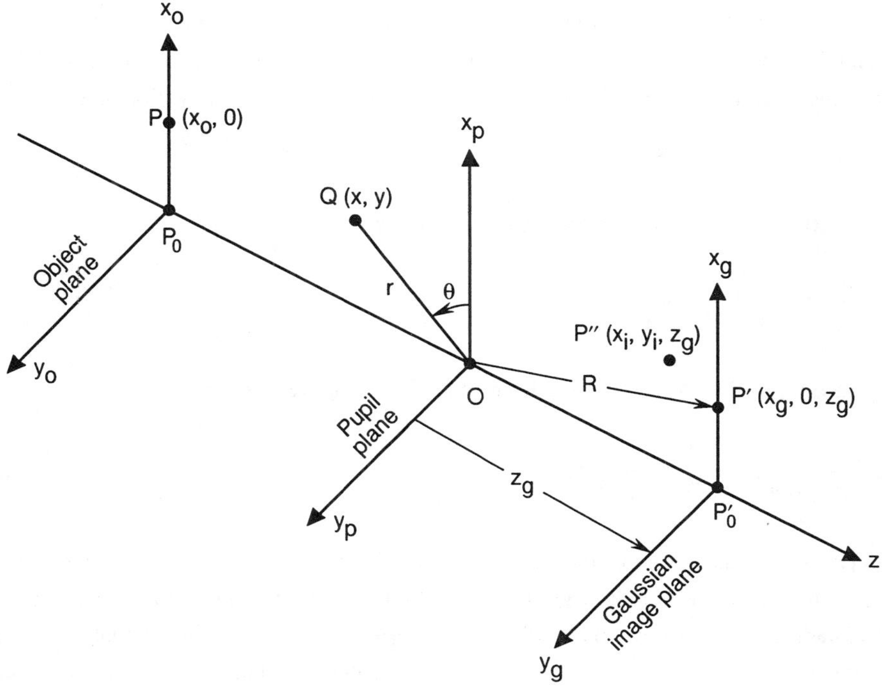

Figure 3-3. Right-hand coordinate system in object, exit pupil, and image planes. The optical axis of the system is along the z axis, and the off-axis point object P is assumed to be along the x axis, thus making zx plane the *tangential plane*.

spherical, with its center of curvature at the Gaussian image point, then the wave and ray aberrations are zero. In that case, all of the object rays transmitted by the system pass through the Gaussian image point, and the image is said to be *perfect*.

3.2.2 Relationship Between Wave and Ray Aberrations

In Figures 3-2, a general ray such as GR_0 or GR, is shown intersecting the wavefront and the reference sphere at points $\overline{Q}$ and Q, respectively. By definition of the wavefront, the optical path length of a ray starting at the point object and ending at $\overline{Q}$ is the same as that of the chief ray ending at O. Hence, $n_i\overline{Q}Q$ gives the *wave aberration* of the ray under consideration, which, as shown in the figures, is numerically positive. Let $W(x, y)$ represent this wave aberration where (x, y, z) are the coordinates of point Q. We need not consider the dependence of W on z explicitly, since z is related to x and y by virtue of Q being on the reference sphere.

Using Hamilton's point characteristic function introduced in Section 1.2.5, we now develop a relationship between the wave and ray aberrations. By its definition, the wave aberration of the ray GR_0 or GR in Figure 3-2 may be written in terms of its characteristic function according to

$$W(x, y) = V(P, Q) - V(P, \overline{Q}) \quad , \tag{3-1}$$

where, for example, $V(P,Q)$ is the Hamilton's point characteristic function (or the optical path length) of a ray from the object point P to the point Q. Since the points $\overline{Q}$ and O lie on the wavefront, therefore, $V(P,\overline{Q}) = V(P,O)$. Hence, Eq. (3-1) may also be written

$$W(x,y) = V(P,Q) - V(P,O) \quad . \tag{3-2}$$

Differentiating Eq. (3-2) with respect to x, we may write

$$\frac{\partial W}{\partial x} = \frac{\partial V}{\partial x} + \frac{\partial V}{\partial z}\frac{\partial z}{\partial x} \quad . \tag{3-3}$$

Applying Eq. (1-12a) to the ray path from Q to P'' in Figure 3-2 we find that

$$\left(\frac{\partial V}{\partial x}, \frac{\partial V}{\partial y}, \frac{\partial V}{\partial z}\right) = \frac{n_i}{R'}\,(x_i - x, y_i - y, z_g - z) \quad , \tag{3-4}$$

where $\left(x_i, y_i, z_g\right)$ are the coordinates of P'' and R' represents the distance QP''. Note that z_g, which is equal to R in Figure 3-2a and approximately equal to R in Figure 3-2b, is the distance between the planes of the exit pupil and the Gaussian image. Since the point object is assumed to lie along the x axis of the object plane, its Gaussian image also lies along the x axis of the image plane. Let $\left(x_g, 0, z_g\right)$ be the coordinates of the Gaussian image point P'. Then, since $OP' = QP' = R$, we may write

$$x_g^2 + z_g^2 = R^2 \tag{3-5}$$

and

$$(x - x_g)^2 + y^2 + (z - z_g)^2 = R^2 \quad . \tag{3-6}$$

Substituting Eq. (3-5) into Eq. (3-6), we obtain

$$x^2 + y^2 + z^2 - 2xx_g - 2zz_g = 0 \quad . \tag{3-7}$$

Differentiating Eq. (3-7) with respect to x, we obtain

$$\frac{\partial z}{\partial x} = -\frac{x - x_g}{z - z_g} \quad . \tag{3-8}$$

Substituting Eqs. (3-4) and (3-8) into Eq. (3-3), we find that

$$\frac{\partial W}{\partial x} = \frac{n_i}{R'}\,(x_i - x_g) \quad ,$$

or

$$\boxed{x_i - x_g = \frac{R'}{n_i}\frac{\partial W}{\partial x}} \quad . \tag{3-9}$$

Similarly, we can show that

$$\boxed{y_i = \frac{R'}{n_i}\frac{\partial W}{\partial y}}\;.$$

(3-10)

Equations (3-9) and (3-10) give an *exact* relationship between the wave aberration $W(x, y)$ at a point (x, y, z) on the reference sphere and the ray aberration $(x_i - x_g, y_i)$ in the Gaussian image plane. However, they involve a distance R' that itself depends on the coordinates (x_i, y_i) of the point P''. Since, in practice, the radius of curvature R of the reference sphere is much larger than the extent of the ray distribution, we may replace R' by R and write

$$\boxed{(x_i, y_i) = \frac{R}{n_i}\left(\frac{\partial W}{\partial x}, \frac{\partial W}{\partial y}\right)}\;,$$

(3-11)

where (x_i, y_i) now represent the coordinates of P'' with respect to those of the Gaussian image point P'. For systems with narrow fields of view, P' lies close to P_0' and we may replace R with z_g. Note that in the case of an axial point object, $R = z_g$. The degree of approximation involved in Eq. (3-11) or in replacing R by z_g is discussed in Appendix A.

Thus, if $W(x, y)$ is the wave aberration of a ray in the exit pupil, the corresponding ray aberration in the image plane is given by its spatial derivative multiplied by the radius of curvature of the Gaussian reference sphere and divided by the refractive index of the image space. Since the rays are normal to a wavefront, the ray aberrations depend on the shape of the wavefront, and, therefore, on its geometrical path length difference from the reference sphere. The division by n_i in Eqs. (3-9) and (3-10) converts optical path length difference into geometrical path length difference. When an image is formed in free space, as is often the case in practice, then $n_i = 1$. An alternative definition of the wave aberration and the derivation of its relationship to the ray aberration is given in Appendix B.

We will refer to the aberration $W(x, y)$ as the wave aberration at a *projected point* (x, y) in the plane of the exit pupil. If (r, θ) are the polar coordinates of this point, as illustrated in Figure 3-3, they are related to its rectangular coordinates (x, y) according to

$$\boxed{(x, y) = r(\cos\theta,\ \sin\theta)}\;.$$

(3-12)

Note that the *tangential rays*, i.e., those lying in the zx plane, lie along the x axis of the exit pupil plane and thus correspond to $\theta = 0$ or π. Similarly, the *sagittal rays*, i.e., those lying in a plane orthogonal to the tangential plane but containing the chief ray lie along the y axis of the exit pupil plane and thus correspond to $\theta = \pi/2$ or $3\pi/2$. If $W(r, \theta)$ represents the aberration in polar coordinates, then the ray aberrations (with respect to the Gaussian image point) are given by

$$\boxed{x_i = \frac{R}{n_i}\left(\cos\theta\frac{\partial W}{\partial r} - \frac{\sin\theta}{r}\frac{\partial W}{\partial \theta}\right)}$$

(3-13a)

and

$$y_i = \frac{R}{n_i}\left(\sin\theta\,\frac{\partial W}{\partial r} + \frac{\cos\theta}{r}\,\frac{\partial W}{\partial\theta}\right) \ . \tag{3-13b}$$

For a radially symmetric aberration $W(r)$, a ray of zone r in the exit pupil plane intersects the Gaussian image plane at a distance r_i from the Gaussian image point given by

$$r_i = \frac{R}{n_i}\,\frac{\partial W}{\partial r} \ . \tag{3-13c}$$

3.3 DEFOCUS ABERRATION

We now discuss defocus wave aberration of a system and relate it to the longitudinal defocus of an image. Consider, as indicated in Figure 3-4, an imaging system for which the expected Gaussian image of a point object is located at P_1. If the system is assembled properly and it is aberration free, a spherical wavefront with its center of curvature at P_1 emerges from its exit pupil, and a perfect image is observed in the Gaussian image plane. However, if one or more of its elements is slightly displaced along its optical axis, then (as discussed in Section 7.2.3) the image is displaced longitudinally to a point, say P_2, so that P_2 lies on the line OP_1 joining the center O of the exit pupil and the Gaussian image point P_1. In that case, the wavefront W for this point object is spherical, with its center of curvature at P_2. The aberration of the wavefront with respect to the Gaussian reference sphere S which is centered at P_1 is the optical deviation between the two along a ray. For a point Q_1 on the reference sphere, this deviation in the figure is given by $n_i Q_2 Q_1$, where n_i is the refractive index of the image space and $Q_2 Q_1$ is approximately equal to the difference in the sags of the reference sphere and the wavefront. (The *sag* of a surface S at

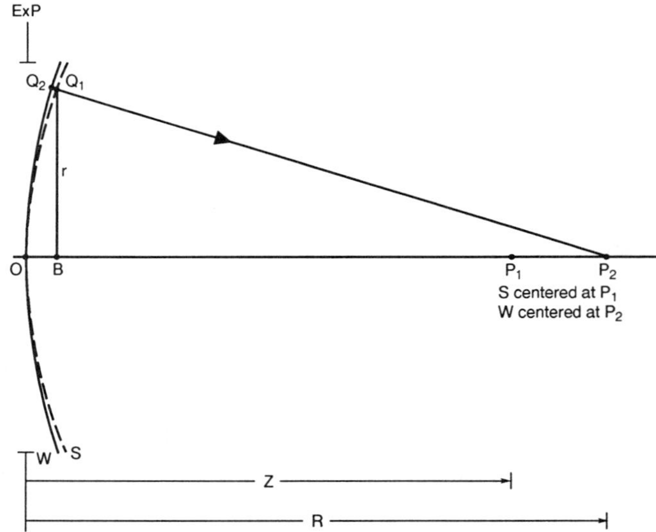

Figure 3-4. Defocused wavefront W is spherical with a radius of curvature R centered at P_2. The reference sphere S with a radius of curvature z is centered at P_1. Both W and S pass through the center O of the exit pupil ExP. The ray $Q_2 P_2$ is normal to the wavefront at Q_2.

a point Q_1, indicated by OB in Figure 3-4, represents its deviation along its axis of symmetry from a plane surface that is tangent to it at its vertex.) It is numerically positive since, compared with the chief ray passing through O, it represents the extra optical path length that a ray passing through Q_1 has to travel in order to reach the reference sphere. Thus, the *defocus wave aberration* at the point Q_1 is given by

$$\boxed{W(r) = \frac{n_i}{2}\left(\frac{1}{z} - \frac{1}{R}\right)r^2} \quad , \tag{3-14}$$

where z and R are the radii of curvature of the reference sphere S and the spherical wavefront W centered at P_1 and P_2, respectively, passing through the center O of the exit pupil, and r is the distance of Q_1 from the optical axis. We note that the defocus wave aberration is proportional to r^2. If $z \simeq R$, then Eq. (3-14) may be written

$$\boxed{W(r) \simeq -\frac{n_i}{2}\frac{\Delta R}{R^2}r^2} \quad , \tag{3-15a}$$

where

$$\boxed{\Delta R = z - R} \tag{3-15b}$$

is called the *longitudinal defocus*. We note that the defocus wave aberration and the longitudianl defocus have numerically opposite signs.

A defocus aberration is also introduced if the system is assembled properly, but the image is observed in a plane other than the Gaussian image plane. Consider, for example, an imaging system forming an aberration-free image at the Gaussian image point P_2. (Note that the Gaussian image is now located at P_2 in Figure 3-4.) Thus, the wavefront at the exit pupil is spherical, passing through its center O with its center of curvature at P_2. Let the image be observed in a defocused plane passing through a point P_1 which lies on the line joining O and P_2. For the observed image at P_1 to be aberration free, the wavefront at the exit pupil must be spherical, with its center of curvature at P_1. Such a wavefront forms the reference sphere with respect to which the aberration of the actual wavefront must be defined. Once again, the aberration of the wavefront at a point Q_1 on the reference sphere is given by Eq. (3-14).

For a system with a *circular exit pupil* of radius a, Eq. (3-15a) may be written

$$W(r) = \frac{n_i}{2}\left(\frac{1}{z} - \frac{1}{R}\right)a^2\rho^2 \tag{3-16a}$$

$$= B_d\rho^2 \quad , \tag{3-16b}$$

where

$$\boxed{\rho = r/a} \tag{3-17}$$

is the normalized distance of a point in the pupil plane from its center,

$$\boxed{\begin{aligned} B_d &= \frac{n_i}{2}\left(\frac{1}{z} - \frac{1}{R}\right) \\ &\simeq -n_i \Delta R / 8F^2 \end{aligned}}$$

$$(3\text{-}18\text{a})$$
$$(3\text{-}18\text{b})$$

is the *peak value* of the defocus aberration. The quantity F in Eq. (3-18b) is the *focal ratio* of the image-forming light cone. It is given by

$$\boxed{F = R/2a} \;.$$

$$(3\text{-}19)$$

We note that a positive value of B_d implies a negative value of the longitudinal defocus ΔR, or $z < R$. Thus, an imaging system having a positive value of defocus aberration B_d can be made defocus free if the image is observed in a plane lying farther from the exit pupil, compared with the defocused image plane, by a distance $8B_d F^2 / n_i$. Similarly, a positive defocus aberration of $B_d = -n_i \Delta R / 8F^2$ is introduced into the system if the image is observed in a plane lying closer to the exit pupil, compared with the defocus-free image plane, by a (numerically negative) distance ΔR.

3.4 WAVEFRONT TILT

Next, we consider a wavefront tilt angle and the corresponding *wavefront tilt aberration*. We consider a system that has one or more of its optical elements inadvertently tilted and/or decentered slightly, resulting in a transverse displacement of the image of a point object from its Gaussian image at P_1 to P_2, as indicated in Figure 3-5. (See Section 7.2 for a discussion of the effects of a tilt and or a decenter of an optical element on the image.) Thus, a spherical wavefront with its center of curvature at P_2 emerges from the exit pupil of the system. The Gaussian reference sphere is, of course centered at P_1 . The aberration of the wavefront at a point Q_1 on the reference sphere is its optical deviation $n_i Q_2 Q_1$ from the reference sphere along the ray passing through Q_1. It is evident that for small values of the ray aberration $P_1 P_2$, the wavefront and the reference sphere are tilted with respect to each other by a small angle β. The ray and the wave aberrations can be written

$$x_i = R\beta$$

$$(3\text{-}20)$$

and

$$W(r,\theta) = n_i \beta r \cos\theta \;,$$

$$(3\text{-}21)$$

respectively, where (r, θ) are the polar coordinates of the point Q_1 projected onto the plane of the exit pupil. Both the wave and ray aberrations are numerically positive in Figures 3-5.

Once again, for a system with a circular exit pupil of radius a, Eq. (3-21) may be written

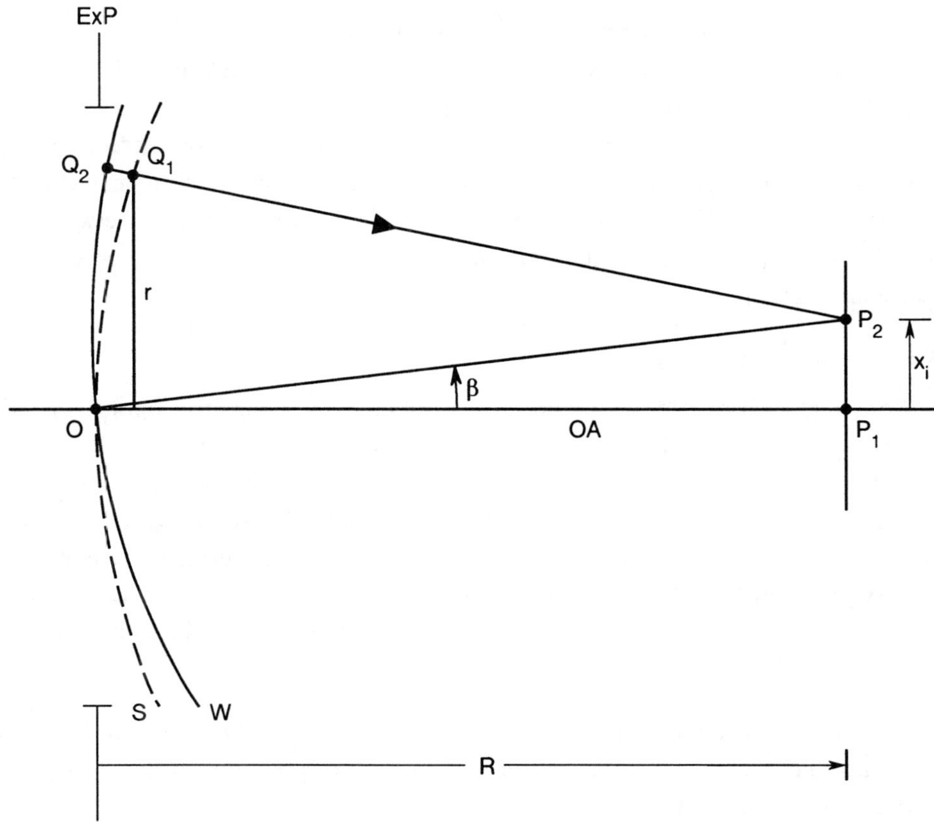

Figure 3-5. Wavefront tilt. The spherical wavefront W is centered at P_2 while the reference sphere S is centered at P_1. Thus, for small values of P_1P_2, the two spherical surfaces are tilted with respect to each other by a small angle $\beta = P_1P_2/R$, where R is their radius of curvature. The ray Q_2P_2 is normal to the wavefront at Q_2.

$$W(\rho,\theta) \;=\; n_i a\beta\rho\cos\theta \quad,\tag{3-22a}$$

or

$$W(\rho,\theta) \;=\; B_t\,\rho\cos\theta \quad,\tag{3-22b}$$

where

$$\boxed{B_t \;=\; n_i a\beta}\tag{3-23}$$

is the *peak value* of the tilt aberration. Note that a positive value of B_t implies that the wavefront tilt angle β is also positive. Thus, if an aberration-free wavefront is centered at P_2, then an observation with respect to P_1 as the origin implies that we have introduced a tilt aberration of $B_t\,\rho\cos\theta$.

3.5 ABERRATION FUNCTION OF A ROTATIONALLY SYMMETRIC SYSTEM

We now consider the aberrations of a rotationally symmetric optical system imaging a point object. We show that these aberrations depend on the object height h (or image height h') and pupil coordinates (r, θ) through three *rotational invariants* h^2, r^2, and $hr\cos\theta$. A *power-series expansion* of the aberration function is considered, thereby introducing *primary, secondary,* and *tertiary aberrations.* The aberration function is also expanded in terms of *Zernike circle polynomials,* and the relationships between the coefficients of a power-series expansion and the coefficients of a Zernike-polynomial expansion are given.

3.5.1 Rotational Invariants

Consider, as illustrated in Figure 3-6, a rotationally symmetric optical system imaging a point object P. The axis of rotational symmetry, namely, the optical axis, lies along the z axis. Let the position vector of the object point be $\vec{h}$ with rectangular coordinates (x_o, y_o) in a plane orthogonal to the optical axis. Similarly, let $\vec{r}$ be the position vector of a point with rectangular coordinates (x, y) in the plane of the exit pupil of the system, which is also orthogonal to the optical axis. The origins of (x_o, y_o) and (x, y) lie on the optical axis and we assume, for example, that the x_o and x axes are coplanar.

In its most general form, the aberration function $W\left(\vec{h}; \vec{r}\right)$ of the system for the point object under consideration can be written as a power series in terms of the rectangular coordinates of the object and pupil points in the form

$$W(x_o, y_o; x, y) = \sum_{j=0}^{\infty} x_o^j \sum_{k=0}^{\infty} y_o^k \sum_{l=0}^{\infty} x^l \sum_{m=0}^{\infty} y^m a_{jklm} \quad , \tag{3-24}$$

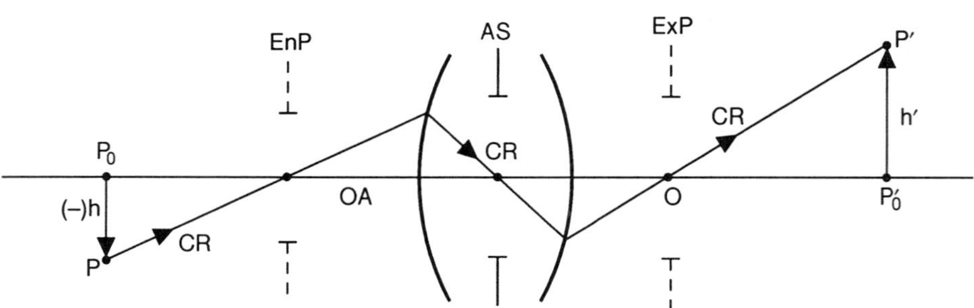

Figure 3-6. Schematic of an optical imaging system. A point object P at a height h from the optical axis OA is imaged at P' at a height h'. AS is the aperture stop, EnP is the entrance pupil, ExP is the exit pupil, and CR is the chief ray. The plane containing the optical axis and the point object (and, therefore, its Gaussian image) is the tangential plane.

where a_{jklm} are the expansion coefficients. The coefficients depend on the construction parameters of the system, such as the radii of curvature of its surfaces, the refractive indices of the spaces between them, and their spacings. Note that the series consists of terms with nonnegative integral powers (including zero) of the four rectangular coordinates x_o, y_o, x, and y.

Since the pupils of optical systems are generally circular, it is convenient to use polar coordinates. Let (h, θ_o) and (r, θ) be the polar coordinates corresponding to the rectangular coordinates (x_o, y_o), and (x, y) of the object and pupil points respectively, so that

$$(x_o, y_o) = h(\cos\theta_o, \sin\theta_o) \tag{3-25a}$$

and

$$(x, y) = (r\cos\theta, \sin\theta) \quad . \tag{3-25b}$$

Now, quantities that are invariant under rotation of the optical system about its axis of symmetry are the three scalars $\left|\vec{h}\right|$, $\left|\vec{r}\right|$, and $\vec{h} \cdot \vec{r}$, where

$$\left|\vec{h}\right| = h = \left(x_o^2 + y_o^2\right)^{1/2} \quad , \tag{3-26a}$$

$$\left|\vec{r}\right| = r = \left(x^2 + y^2\right)^{1/2} \quad , \tag{3-26b}$$

and

$$\vec{h} \cdot \vec{r} = hr\cos(\theta - \theta_o) \tag{3-26c}$$

$$= x_o x + y_o y \quad . \tag{3-26d}$$

In order that the aberration function consist of terms with nonnegative integral powers of the four rectangular coordinates, it must depend on the first two through h^2 and r^2. If we rotate the system about the optical axis by a certain angle, the aberration function must not change. We note that this is indeed the case. As the system rotates, so do the x and y axes in each plane. Both θ and θ_o change by the angle of rotation, but h, r, and $\theta - \theta_o$ do not change. Thus, because of rotational symmetry, the aberration function depends on the four variables (x_o, y_o) and (x, y) only through the three combinations h^2, r^2, and $hr\cos(\theta - \theta_o)$. These combinations are called the *rotational invariants* of the aberration function of an optical imaging system with an axis of rotational symmetry. When written in terms of these invariants, the aberration function will consist of three summations instead of the four in Eq. (3-24). This is discussed further in the next section.

The dependence of the aberration function on the four rectangular coordinates only through the three rotational invariants can also be obtained as follows. For simplicity (but without loss of generality), we assume that the angle θ is measured from the tangential plane (which passes through the optical axis and the object point) so that θ replaces

$\theta - \theta_o$. Indeed we choose the x_o axis to pass through the point object so that $x_o = h$, $y_o = 0$, and the tangential plane is zx.

For an axial point object (i.e., for $h = 0$), the rotational symmetry of the optical system about its optical axis implies that the aberration function must be radially symmetric. Hence, those terms of the aberration function $W(h; x, y)$ that do not depend on h must vary as $x^2 + y^2$ or as its integral powers. In polar coordinates, such terms must vary as r^2 or as its integral powers. The terms for which $h \neq 0$, referring to Figure 3-7, symmetry about the tangential plane zx implies that

$$W(h; Q_1) = W(h; Q_2)$$

i.e.,

$$W(h; x, y) = W(h; x, -y) \tag{3-27a}$$

or

$$W(h; r, \theta) = W(h; r, -\theta) \ . \tag{3-27b}$$

Therefore, their dependence on y must be with even powers. Or, alternatively, their dependence on θ must be a function of $\cos\theta$.

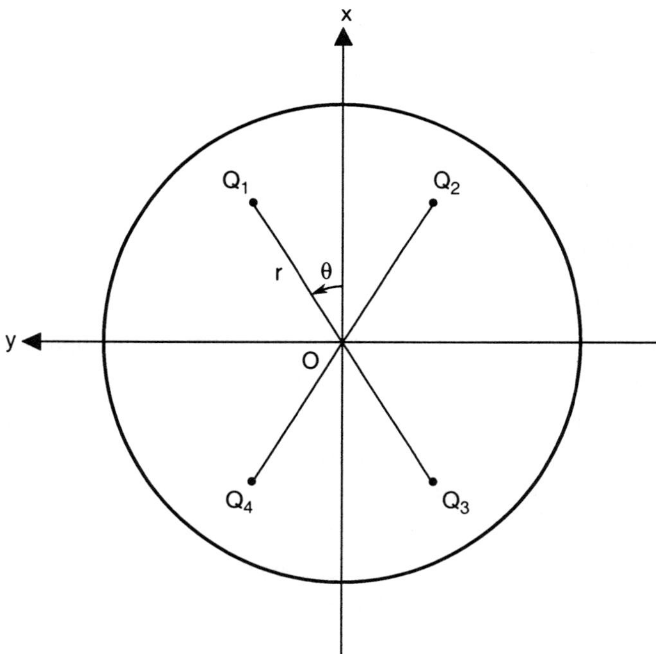

Figure 3-7. The exit pupil as seen from the Gaussian image plane. (r, θ) represent the polar coordinates of a point Q_1 in the plane of the pupil. The point Q_3 is diagonally opposite to Q_1 from the origin O. The point Q_2 is symmetric to point Q_1 about the tangential plane zx. Similarly, the point Q_4 is symmetric to point Q_3 about the tangential plane.

Because of the rotational symmetry, the aberration $W(h; Q_1)$ corresponding to a point object at a height h above the axis must equal the aberration $W(-h; Q_3)$ for a point object at a height h below the axis, where Q_3 is diagonally opposite to Q_1 from the origin O, the point of intersection of the chief ray (or its extension) with the exit pupil. Thus,

$$W(h; x, y) = W(-h; -x, -y) \quad , \tag{3-28a}$$

or

$$W(h; r, \theta) = W(-h; r, \pi + \theta) \quad . \tag{3-28b}$$

Hence, those terms of the aberration function $W(h; r, \theta)$ that do not depend on θ must be functions of $h^2 r^2$ or its integral powers.

Because of the symmetry about the tangential plane, we may also write

$$W(-h; Q_3) = W(-h; Q_4) \quad , \tag{3-29a}$$

or

$$W(-h; -x, -y) = W(-h, -x, y) \quad , \tag{3-29b}$$

or

$$W(-h; r, \pi + \theta) = W(-h; r, \pi - \theta) \quad . \tag{3-29c}$$

From Eqs. (3-28) and (3-29), we obtain

$$W(h; x, y) = W(-h; -x, y) \quad , \tag{3-30a}$$

or

$$W(h; r,) = W(-h; r, \Pi -) \quad . \tag{3-30b}$$

Hence, those terms that depend on θ must be functions of hx or $hr\cos\theta$. Combining this with the θ-independent terms, we find that the aberration function consists of terms containing h^2, r^2, and $hr\cos\theta$ factors. Generally, we will consider the aberration function in terms of the image height h' instead of the object height h.

3.5.2 Power-Series Expansion

Now we consider a power-series expansion of the aberration function in terms of the three rotational invariants. In particular, we discuss *primary*, *secondary*, and *tertiary aberrations*. A power-series expansion in which explicit dependence on the image height is suppressed is also considered. This is done by combining aberration terms having different dependence on image height but the same dependence on pupil coordinates.

3.5.2.1   Explicit Dependence on Object Coordinates

If $\vec{h}$ and $\vec{r}$ represent the position vectors of object and pupil points, then because of the rotational symmetry, the corresponding aberration function will consist of terms containing one or more of the three rotational invariants h^2, r^2 and $\vec{h} \cdot \vec{r}$. A *power-series expansion* of the aberration function in terms of these invariants may be written in the form

$$W\left(\vec{h};\vec{r}\right) = \sum_{l=0}^{\infty} \sum_{p=0}^{\infty} \sum_{m=0}^{\infty} C_{lpm}\left(h^2\right)^l \left(r^2\right)^p \left[hr\cos(\theta - \theta_o)\right]^m \tag{3-31a}$$

$$= \sum_{l=0}^{\infty} \sum_{p=0}^{\infty} \sum_{m=0}^{\infty} C_{lpm} h^{2l+m} r^{2p+m} \cos^m\left(\theta - \theta_o\right) \ , \tag{3-31b}$$

where we have used polar coordinates according to $\vec{h} = \left(h, \theta_o\right)$, $\vec{r} = \left(r, \theta\right)$, C_{lpm} are the expansion coefficients; and l, p, and m are positive integers, including zero. As before, the aberration function is defined with respect to the Gaussian reference sphere of radius of curvature R that passes through the center of the exit pupil and whose center of curvature lies at the Gaussian image point at a height h' from the optical axis. The magnification of the image is $M = h'/h$.

It is evident that the *degree* of each term of the series in the object and pupil coordinates is even and given by $2(l + p + m)$. Any terms for which $2p + m = 0$, i.e., those terms that do not depend on r, must add up to zero since the aberration associated with the chief ray (for which $r = 0$) is zero. Thus, the zero-degree term C_{000} and terms such as $C_{100} h^2$, $C_{200} h^4$, etc., do not appear in Eq. (3-31). There is also no term of second degree. The term $C_{010} r^2$ represents a defocus aberration that is independent of h. It is eliminated if the image is observed in a slightly different plane, i.e., by a longitudinal shift of the image plane. However, that would imply that the Gaussian image point with respect to which the aberration function is defined must be incorrect. Hence, this term must be zero. Similarly, the term $C_{001} hr\cos\theta$ represents a wavefront tilt aberration that depends on h. It can be corrected by a transverse shift of the image by an amount $C_{001} Rh$, implying an image magnification of $M\left(1 + C_{001} R\right)$. It must be zero since it contradicts the fact that the image magnification is M. Hence, a power series expansion of the aberration function consists of terms of degree 4, 6, 8, etc. The corresponding aberrations are referred to as the *primary, secondary, tertiary aberrations*, etc.

For simplicity, we let the object point be along the x_o axis as in Figure 3-3 so that $\theta_0 = 0$. Accordingly, Eq. (3-31b) may be written

$$\boxed{W(h';r,\theta) = \sum_{l=0}^{\infty} \sum_{n=1}^{\infty} \sum_{m=0}^{n} {}_{2l+m}a_{nm}\, h'^{2l + m} r^n \cos^m\theta \ ,} \tag{3-31c}$$

where

$$n = 2p + m \tag{3-31d}$$

is a positive integer *not* including zero, h' is the height of the Gaussian image point, and $_{2l+m}a_{nm}$ are the expansion coefficients. From Eq. (3-31d), we note that $n - m = 2p \geq 0$ and even. The *order i* of an aberration term, which is equal to its degree in the object and pupil coordinates, is given by

$$\boxed{i \;=\; 2l + m + n \;\;.}$$

(3-32)

The *order i is always even*, as may be seen by substituting Eq. (3-31d) into Eq. (3-32). The number of terms N_i of a certain order i, i.e., the number of integer sets satisfying Eq. (3-32) with $n - m \geq 0$ and even, is given by

$$\boxed{N_i \;=\; (i+2)(i+4)/8 \;\;.}$$

(3-33)

This number includes a term with $n = 0 = m$, called *piston aberration*, although such a term does not constitute an aberration (since it corresponds to the chief ray which has a zero aberration associated with it).

For *primary* (or *Seidel*) *aberrations*, i.e., for $i = 4$, the values of the indices and the corresponding aberration terms are listed in Table 3-1. Although a term with $l = 2, n = m = 0$ corresponds to $i = 4$, it does not constitute an aberration. Hence, it is not listed in the table. Thus, there are five aberration terms of fourth order, and the primary aberration function may be written

$$\boxed{\begin{aligned} W(h';r,\theta) \;=\;\; & _{0}a_{40}r^4 + {}_{1}a_{31}h'r^3 \cos\theta + {}_{2}a_{22}h'^2 r^2 \cos^2\theta \\ & + {}_{2}a_{20}h'^2 r^2 + {}_{3}a_{11}h'^3 r \cos\theta \;\;, \end{aligned}}$$

(3-34)

corresponding to the *five primary aberrations*. The coefficients $_{0}a_{40}, {}_{1}a_{31}, {}_{2}a_{22}, {}_{2}a_{20}$, and $_{3}a_{11}$ represent the coefficients of *spherical aberration, coma, astigmatism, field curvature,* and *distortion*.

We note that the dependence of the field curvature term on the pupil coordinates is just like the defocus aberration discussed in Section 3.3. Hence, this term is a defocus

Table 3-1. Primary aberrations; $i = 2l + m + n = 4$.

l	n	m	$2l + m$	Aberration Term $_{2l+m}a_{nm}h'^{2l+m}r^n \cos^m\theta$	Aberration Name*
0	4	0	0	$_{0}a_{40}\,r^4$	Spherical
0	3	1	1	$_{1}a_{31}h'r^3 \cos\theta$	Coma
0	2	2	2	$_{2}a_{22}h'^2 r^2 \cos^2\theta$	Astigmatism
1	2	0	2	$_{2}a_{20}h'^2 r^2$	Field curvature
1	1	1	3	$_{3}a_{11}h'^3 r \cos\theta$	Distortion

*The word "primary" is to be associated with these names, e.g., *primary spherical.*

whose coefficient varies with the height of the point object. It can be eliminated by observing the image of a planar object on a curved surface (typically spherical as discussed in Section 4.3.3); hence, the name *field curvature.*

Similarly, the dependence of the distortion term on the pupil coordinates is just like the wavefront tilt aberration discussed in Section 3.4. Hence, this term is a wavefront tilt aberration whose coefficient varies with the height of the point object. Accordingly, the image of a point object in the presence of distortion is perfect, but it is transversally displaced from the Gaussian image point; the amount of the displacement depends on the height of the point object. The reason for the name distortion becomes clear when the image of an extended object is considered. (For an example, see Section 4.3.3 where the distorted image of a square grid is considered.)

For *secondary* (or *Schwarzschild*) *aberrations*, i.e., for $i = 6$, the values of the indices and the corresponding aberration terms are listed in Table 3-2. A term with $l = 3$, $n = m = 0$ corresponding to $i = 6$ does not constitute an aberration and is not listed in the table. There are nine aberration terms of sixth order. Four of these correspond to $l = 0$. The remaining five corresponding to $l \neq 0$ and called *lateral aberrations* are similar to the corresponding primary aberrations except for their dependence on the image height h'. The *lateral spherical aberration* $_2a_{40}h'^2r^4$ is also called the *oblique spherical aberration.*

Table 3-2. Secondary aberrations; $i = 2l + m + n = 6$.

l	n	m	$2l + m$	Aberration Term $_{2l+m}a_{nm}h'^{2l+m}r^n \cos^m \theta$	Aberration Name
0	6	0	0	$_0a_{60}r^6$	Spherical*
0	5	1	1	$_1a_{51}h'r^5 \cos\theta$	Coma*
0	4	2	2	$_2a_{42}h'^2r^4 \cos^2\theta$	Astigmatism* (wings or Flügelfehler)
0	3	3	3	$_3a_{33}h'^3r^3 \cos^3\theta$	Arrows or Pfeilfehler
1	4	0	2	$_2a_{40}h'^2r^4$	Lateral spherical
1	3	1	3	$_3a_{31}h'^3r^3 \cos\theta$	Lateral coma
1	2	2	4	$_4a_{22}h'^4r^2 \cos^2\theta$	Lateral astigmatism
2	2	0	4	$_4a_{20}h'^4r^2$	(Lateral) field curvature*
2	1	1	5	$_5a_{11}h'^5r \cos\theta$	(Lateral) distortion*

*The word "secondary" is to be associated with these aberrations, e.g., *secondary spherical.*

Next we consider *tertiary aberrations*, i.e., those with $i = 8$. The values of the indices l, m, and n giving $i = 8$ and the corresponding aberrations are listed in Table 3-3. A term with $l = 4$, $n = m = 0$ does not constitute an aberration and is, therefore, not listed the table. We note that there are fourteen aberration terms of eighth order. Only five of these have dependencies on pupil coordinates that are different from those of the secondary or primary aberrations. Four have dependence on these coordinates as for the secondary aberrations, and the remaining five have the same dependence as the primary aberrations. Their difference lies in their dependence on the image height.

3.5.2.2 No Explicit Dependence on Object Coordinates

For a system imaging a given point object, aberration terms of the power-series expansion may be written so that their explicit dependence on the image height h' is suppressed. We may also let

$$\rho = r/a \ , \tag{3-35}$$

Table 3-3. Tertiary aberrations; $i = 2l + m + n = 8$.

l	n	m	$2l + m$	Aberration Term $_{2l+m}a_{nm}h'^{2l+m}r^{n}\cos^{m}\theta$	Aberration Name*
0	8	0	0	$_{0}a_{80}r^{8}$	Spherical
0	7	1	1	$_{1}a_{71}h'r^{7}\cos\theta$	Coma
0	6	2	2	$_{2}a_{62}h'^{2}r^{6}\cos^{2}\theta$	Astigmatism
0	5	3	3	$_{3}a_{53}h'^{3}r^{5}\cos^{3}\theta$	
0	4	4	4	$_{4}a_{44}h'^{4}r^{4}\cos^{4}\theta$	
1	6	0	2	$_{2}a_{60}h'^{2}r^{6}$	
1	5	1	3	$_{3}a_{51}h'^{3}r^{5}\cos^{3}\theta$	
1	4	2	4	$_{4}a_{42}h'^{4}r^{4}\cos^{2}\theta$	
1	3	3	5	$_{5}a_{11}h'^{5}r^{3}\cos^{3}\theta$	
2	4	0	4	$_{4}a_{40}h'^{4}r^{4}$	
2	3	1	5	$_{5}a_{31}h'^{5}r^{3}\cos\theta$	
2	2	2	6	$_{6}a_{22}h'^{6}r^{2}\cos^{2}\theta$	
3	2	0	6	$_{6}a_{20}h'^{6}r^{2}$	Field curvature
3	1	1	7	$_{7}a_{11}h'^{7}r\cos\theta$	Distortion

*The word "tertiary" is to be associated with these names, e.g., *tertiary spherical.*

where a is the radius of the exit pupil of the system. Combining the aberration terms having different dependencies on the object coordinates but the same dependence on pupil coordinates so that there is only one term for each pair of (n, m) values, Eq. (3-31c) for the power-series expansion of the aberration function may be written

$$W(\rho, \theta) = \sum_{n=1}^{\infty} \sum_{m=0}^{n} a_{nm} \rho^n \cos^m \theta \quad , \tag{3-36}$$

where

$$a_{nm} = a^n \sum_{l=0}^{\infty} {}_{2l+m} a_{nm} h'^{2l+m} \quad . \tag{3-37}$$

As stated earlier, n and m are positive integers, including zero, and $n - m \geq 0$ and even [see Eq. (3-31d)]. Each aberration coefficient a_{nm} depends on the image height h', and, since $0 \leq \rho \leq 1$ and $|\cos\theta| \leq 1$, it represents the *peak value* or half of the *peak-to-valley value* of the corresponding aberration term, depending on whether m is even or odd, respectively. The indices n and m represent the powers of ρ and $\cos\theta$, respectively. The index m also represents the minimum power of h' dependence of a coefficient (with the exception of tilt and defocus terms corresponding to $n - m \geq 0$ and 2, respectively). The maximum power of h' dependence is given by $i - n$. Moreover, the powers of h' dependence are even or odd according to whether n and m are even or odd, respectively. The number of terms through a certain order i in the *reduced power-series expansion* of the aberration function given by Eq. (3-36) is also given by Eq. (3-33). This number includes a nonaberration piston term corresponding to $n = 0 = m$. The terms of Eq. (3-30) through a certain order i correspond to those terms of Eq. (3-36) for which $n + m \leq i$.

The primary aberrations written in this simplified form are listed in Table 3-4, along with the values of the indices n and m of the new aberration coefficients a_{nm}. They correspond to $n+m \leq 4$. The primary aberration function of Eq. (3-34) may be written in terms of these coefficients in the form

$$W_P(\rho, \theta) = a_{11}\rho\cos\theta + a_{20}\rho^2 + a_{22}\rho^2\cos^2\theta + a_{31}\rho^3\cos\theta + a_{40}\rho^4 \quad , \tag{3-38}$$

where

$$a_{11} = {}_3a_{11}h'^3 a = a_t h'^3 a = A_t \quad , \tag{3-39a}$$

$$a_{20} = {}_2a_{20}h'^2 a^2 = a_d h'^2 a^2 = A_d \quad , \tag{3-39b}$$

$$a_{22} = {}_2a_{22}h'^2 a^2 = a_a h'^2 a^2 = A_a \quad , \tag{3-39c}$$

$$a_{31} = {}_1a_{31}h' a^3 = a_c h' a^3 = A_c \quad , \tag{3-39d}$$

and

$$a_{40} = {}_0a_{40}a^4 = a_s a^4 = A_s \quad , \tag{3-39e}$$

Table 3-4. Primary aberrations in a simplified form. $i = 4$, $n+m \leq 4$.

n	m	Aberration Term $a_{nm}\rho^n \cos^m \theta$	Aberration Name
1	1	$a_{11}\rho\cos\theta$	Distortion
2	0	$a_{20}\rho^2$	Field curvature
2	2	$a_{22}\rho^2 \cos^2\theta$	Astigmatism
3	1	$a_{31}\rho^3 \cos\theta$	Coma
4	0	$a_{40}\rho^4$	Spherical

and we have introduced aberration coefficients a_i and A_i with abbreviated notation which will be used later.

Comparing the distortion term given in Table 3-4 with the wavefront tilt aberration given by Eq. (3-22b), we note that while the two are similar in their dependence on the pupil coordinates, their coefficients depend on the image height differently. The distortion coefficient a_{11} (or A_t) varies with h' as h'^3, but the tilt coefficient B_t is independent of h'. Similarly, comparing the field curvature term with the defocus wave aberration given by Eq. (3-16b), we note that their dependence on the pupil coordinates is the same. However, whereas the field curvature coefficient a_{20} (or A_d) varies with h' as h'^2, the defocus coefficient B_d is independent of h'.

The aberration function through the sixth order i.e., for $i \leq 6$ or $n+m \leq 6$, may be written

$$\boxed{\begin{aligned} W_S(\rho,\theta) &= a_{11}\rho\cos\theta + a_{20}\rho^2 + a_{22}\rho^2\cos^2\theta + a_{31}\rho^3\cos\theta + a_{33}\rho^3\cos^3\theta \\ &\quad + a_{40}\rho^4 + a_{42}\rho^4\cos^2\theta + a_{51}\rho^5\cos\theta + a_{60}\rho^6 \quad, \end{aligned}}$$

$$(3\text{-}40)$$

where

$$a_{11} = \left(_3a_{11}h'^3 + {_5}a_{11}h'^5\right)a \quad, \tag{3-41a}$$

$$a_{20} = \left(_2a_{20}h'^2 + {_4}a_{20}h'^4\right)a^2 \quad, \tag{3-41b}$$

$$a_{22} = \left(_2a_{22}h'^2 + {_4}a_{22}h'^4\right)a^2 \quad, \tag{3-41c}$$

$$a_{31} = \left(_1a_{31}h' + {_3}a_{31}h'^3\right)a^3 \quad, \tag{3-41d}$$

$$a_{33} = {_3}a_{33}h'^3a^3 \quad, \tag{3-41e}$$

$$a_{40} = \left(_0a_{40} + {_2}a_{40}h'^2\right)a^4 \quad, \tag{3-41f}$$

$$a_{42} = {}_2a_{42}h'^2a^4 \ , \tag{3-41g}$$

$$a_{51} = {}_1a_{51}h'a^5 \ , \tag{3-41h}$$

and

$$a_{60} = {}_0a_{60}a^6 \ . \tag{3-41i}$$

Written in this form, the aberration function has nine aberration terms through the sixth order. For convenience, the values of the indices n and m and the combined aberration terms along with their names are listed in Table 3-5. Since the dependence of an aberration term on the image height h' is contained in the aberration coefficient a_{nm}, it should be noted that the primary aberrations (including distortion and field curvature terms) are not the same as those discussed earlier since they contain aberration components not only of the fourth degree, but the sixth degree as well.

The aberration function through the eighth order may be written

$$W_T(\rho, \theta)) = a_{11}\rho\cos\theta + a_{20}\rho^2 + a_{22}\rho^2\cos^2\theta + a_{31}\rho^3\cos\theta + a_{33}\rho^3\cos^3\theta$$

$$+ a_{40}\rho^4 + a_{42}\rho^4\cos^2\theta + a_{44}\rho^4\cos^4\theta + a_{51}\rho^5\cos\theta$$

$$+ a_{60}\rho^6 + a_{62}\rho^6\cos^2\theta + a_{71}\rho^7\cos\theta + a_{80}\rho^8 \ , \tag{3-42}$$

where the aberration coefficients a_{nm} are given by

Table 3-5. Combined primary and secondary aberrations. $i \le 6$, $n+m \le 6$.

n	m	Aberration Term $a_{nm}\rho^n\cos^m\theta$	Aberration Name
1	1	$a_{11}\rho\cos\theta$	Distortion
2	0	$a_{20}\rho^2$	Field curvature
2	2	$a_{22}\rho^2\cos^2\theta$	Primary astigmatism
3	1	$a_{31}\rho^3\cos\theta$	Primary coma
3	3	$a_{33}\rho^3\cos^3\theta$	Elliptical coma (arrows)
4	0	$a_{40}\rho^4$	Primary spherical
4	2	$a_{42}\rho^4\cos^2\theta$	Secondary astigmatism
5	1	$a_{51}\rho^5\cos\theta$	Secondary coma
6	0	$a_{60}\rho^6$	Secondary spherical

$$a_{11} = \left({}_3 a_{11} h'^3 + {}_5 a_{11} h'^5 + {}_7 a_{11} h'^7 \right) a \quad , \tag{3-43a}$$

$$a_{20} = \left({}_2 a_{20} h'^2 + {}_4 a_{20} h'^4 + {}_6 a_{20} h'^6 \right) a^2 \quad , \tag{3-43b}$$

$$a_{22} = \left({}_2 a_{22} h'^2 + {}_4 a_{22} h'^4 + {}_6 a_{22} h'^6 \right) a^2 \quad , \tag{3-43c}$$

$$a_{31} = \left({}_1 a_{31} h' + {}_3 a_{31} h'^3 + {}_5 a_{31} h'^5 \right) a^3 \quad , \tag{3-43d}$$

$$a_{33} = \left({}_3 a_{33} h'^3 + {}_5 a_{33} h'^5 \right) a^3 \quad , \tag{3-43e}$$

$$a_{40} = \left({}_0 a_{40} + {}_2 a_{40} h'^2 + {}_4 a_{40} h'^4 \right) a^4 \quad , \tag{3-43f}$$

$$a_{42} = \left({}_2 a_{42} h'^2 + {}_4 a_{42} h'^4 \right) a^4 \quad , \tag{3-43g}$$

$$a_{44} = {}_4 a_{44} h'^4 a^4 \quad , \tag{3-43h}$$

$$a_{51} = \left({}_1 a_{51} h' + {}_3 a_{51} h'^3 \right) a^5 \quad , \tag{3-43i}$$

$$a_{53} = {}_3 a_{53} h'^3 a^5 \quad , \tag{3-43j}$$

$$a_{60} = \left({}_0 a_{60} + {}_2 a_{60} h'^2 \right) a^6 \quad , \tag{3-43k}$$

$$a_{62} = {}_2 a_{62} h'^2 a^6 \quad , \tag{3-43l}$$

$$a_{71} = {}_1 a_{71} h' a^7 \quad , \tag{3-43m}$$

and

$$a_{80} = {}_0 a_{80} a^6 \quad . \tag{3-43n}$$

The values of the indices n and m and the combined aberrations are listed in Table 3-6. Once again it should be noted that a primary or a secondary aberration listed in this table is not the same as the corresponding aberration in Table 3-1 or Table 3-2, respectively. For example, the coefficient a_{20} of the field curvature term depends on the image height h' in a complex manner according to Eq. (3-43b) and consists of terms of the fourth, sixth, and eighth degrees. Similarly, secondary spherical aberration consists of terms of the sixth and eighth degrees, as may be seen from Eq. (3-43k). It is convenient to refer to the aberration terms of a power-series expansion as the *classical aberrations*, e.g., a term in ρ^4 may be referred to as the *classical primary spherical aberration*.

3.5.3 Zernike Circle-Polynomial Expansion

For a given point object, the aberration function of a rotationally symmetric optical system can also be expanded in terms of a complete set of *Zernike circle polynomials* $R_n^m(\rho)\cos m\theta$ that are orthogonal over a unit circle in the form

Table 3-6. Combined primary, secondary, and tertiary aberrations. $i \le 8, n + m \le 8$.

n	m	Aberration Term $a_{nm}\rho^n \cos^m \theta$	Aberration Name
1	1	$a_{11}\rho \cos \theta$	Distortion
2	0	$a_{20}\rho^2$	Field curvature
2	2	$a_{22}\rho^2 \cos^2 \theta$	Primary astigmatism
3	1	$a_{31}\rho^3 \cos \theta$	Primary spherical
3	3	$a_{33}\rho^3 \cos^3 \theta$	Elliptical coma (arrows)
4	0	$a_{40}\rho^4$	Primary spherical
4	2	$a_{42}\rho^4 \cos^2 \theta$	Secondary astigmatism
5	1	$a_{51}\rho^5 \cos \theta$	Secondary coma
4	4	$a_{44}\rho^4 \cos^4 \theta$	
5	3	$a_{53}\rho^5 \cos^3 \theta$	
6	0	$a_{60}\rho^6$	Secondary spherical
6	2	$a_{62}\rho^6 \cos^2 \theta$	Tertiary astigmatism
7	1	$a_{71}\rho^7 \cos \theta$	Tertiary coma
8	0	$a_{80}\rho^8$	Tertiary spherical

$$W(\rho,\theta) \;=\; \sum_{n=0}^{\infty} \sum_{m=0}^{n} c_{nm}\left[2(n+1)/(1+\delta_{m0})\right]^{1/2} R_n^m(\rho)\cos m\theta \;, \tag{3-44}$$

where c_{nm} are the expansion coefficients which depend on the image height h', and n and m are positive integers including zero, $n - m \ge 0$ and even, δ_{ij} is a Kronecker delta, and

$$R_n^m(\rho) \;=\; \sum_{s=0}^{(n-m)/2} \frac{(-1)^s (n-s)!}{s!\left(\dfrac{n+m}{2}-s\right)!\left(\dfrac{n-m}{2}-s\right)!}\,\rho^{n-2s} \tag{3-45}$$

is a polynomial of degree n in ρ containing terms in ρ^n, ρ^{n-2}, ..., and ρ^m. The radial polynomials $R_n^m(\rho)$ are even or odd in ρ depending on whether n (or m) is even or odd. Note that

$$R_n^n(1) \;=\; 1 \tag{3-46}$$

and

$$R_n^n(\rho) = \rho^n \quad .$$

(3-47)

Although a complete set of Zernike polynomials would imply inclusion of terms varying as $\sin m\theta$ on the right-hand side of Eq. (3-44), they do not appear because their coefficients are zero owing to the rotational symmetry of the aberrated system about its optical axis. (In the case of random aberrations introduced by *atmospheric turbulence*, terms varying as $\sin m\theta$ also appear.)

The *orthogonalities* of the radial polynomials and the angular functions are given by

$$\int_0^1 R_n^m(\rho)R_{n'}^m(\rho)\rho\,d\rho = \frac{1}{2(n+1)}\delta_{nn'}$$

(3-48)

and

$$\int_0^{2\pi}\cos m\theta \cos m'\theta\,d\theta = \pi\left(1+\delta_{m0}\right)\delta_{mm'} \quad .$$

(3-49)

The Zernike expansion coefficients are given by

$$c_{nm} = (1/\pi)\left[2(n+1)(1+\delta_{m0})\right]^{1/2}\int_0^1\int_0^{2\pi} W(\rho,\theta)R_n^m(\rho)\cos m\theta\,\rho\,d\rho\,d\theta \quad ,$$

(3-50)

as may be seen by substituting Eq. (3-44) into Eq. (3-50). We note that the angular dependence of an aberration term consists of the cosine of the integral multiple of angle θ rather than the integral power of the cosine of the angle as in the power-series expansion of Eq. (3-31) or Eq. (3-36). Because of their orthogonality, we will refer to the aberration terms of a Zernike-polynomial expansion as the *orthogonal aberrations*. The *orthonormal Zernike aberrations* and the names associated with some of them are also listed in Table 3-7 for $n \leq 8$. The number of aberration terms in the expansion of the aberration function through a certain order n is given by

$$\boxed{N_n = (n+2)(n+4)/8} \quad ,$$

(3-51)

which is similar to Eq. (3-33).

We note that each Zernike aberration is made up of one or more classical aberrations. For example, the Zernike primary spherical aberration is made up of classical primary spherical aberration and defocus. Similarly, the Zernike secondary spherical aberration is made up of classical secondary and primary spherical aberrations and defocus. Inclusion of the piston aberration makes the mean value of these orthogonal aberrations zero. The relative amounts of the classical aberrations in a certain Zernike aberration are such that it is orthogonal to the other Zernike aberrations. Since unity $\left[R_0^0(\rho)\right]$ is one of the Zernike aberrations, the orthogonality of a Zernike aberration to others also implies that its mean value is zero. Similarly, for example, the Zernike primary coma is made up of classical primary coma and tilt.

The Zernike polynomials are unique in that they are the only polynomials in two coordinate variables ρ and θ that (a) are orthogonal over a unit circle, (b) are invariant in form with respect to rotation of the axes about the origin, and (c) include a polynomial for each permissible pair of n and m. The reason for expressing the aberration function in terms of Zernike polynomials is that each polynomial represents a combination of power-series terms that is optimally balanced to give minimum variance across the pupil. For example, Zernike polynomial $R_4^0(\rho)$ represents a balanced spherical aberration in that spherical aberration $(\rho^4 \text{ term})$ is combined with defocus $(\rho^2 \text{ term})$ so that the variance of the aberration across the circular exit pupil of a system is minimum.

An advantage of the orthogonal-polynomial expansion of the aberration function in the form of Eq. (3-44) is that each aberration coefficient c_{nm} represents the standard deviation of the corresponding aberration term across the exit pupil, and, therefore, it is very easy to determine the standard deviation of the aberration function once the expansion coefficients are known. We note that the *mean* and *mean square values* of the aberration function are given by

$$<W(\rho,\theta)> = \int_0^1 \int_0^{2\pi} W(\rho,\theta)\rho\, d\rho\, d\theta \Big/ \int_0^1 \int_0^{2\pi} \rho\, d\rho\, d\theta$$

$$= c_{00} \tag{3-52}$$

(since $\displaystyle\int_0^{2\pi} \cos m\theta\, d\theta = 2\pi\delta_{m0}$), and

$$<W^2(\rho,\theta)> = \int_0^1 \int_0^{2\pi} W^2(\rho,\theta)\rho\, d\rho\, d\theta \Big/ \int_0^1 \int_0^{2\pi} \rho\, d\rho\, d\theta$$

$$= \sum_{n=0}^{\infty} \sum_{m=0}^{n} c_{nm}^2 \tag{3-53}$$

as may be seen by substituting Eq. (3-44) and using the orthogonality Eqs. (3-48) and (3-49). Hence, the *variance* of the aberration function is given by

$$\sigma_w^2 = <W^2(\rho,\theta)> - <W(\rho,\theta)>^2$$

$$= \sum_{n=1}^{\infty} \sum_{m=0}^{n} c_{nm}^2 \quad , \tag{3-54}$$

Table 3-7. Orthonormal Zernike (circle polynomial) aberrations.

n	m	Orthonormal Zernike Polynomial $$Z_n^m(\rho, \theta) = \left[\frac{2(n+1)}{1+\delta_{m0}}\right]^{1/2} R_n^m(\rho)\cos m\theta$$	Aberration Name*
0	0	1	Piston
1	1	$\rho\cos\theta$	Distortion (tilt)
2	0	$\sqrt{3}\left(2\rho^2 - 1\right)$	Field curvature (defocus)
2	2	$\sqrt{6}\,\rho^2\cos 2\theta$	Primary astigmatism
3	1	$\sqrt{8}\left(3\rho^3 - 2\rho\right)\cos\theta$	Primary coma
3	3	$\sqrt{8}\,\rho^3\cos 3\theta$	
4	0	$\sqrt{5}\left(6\rho^4 - 6\rho^2 + 1\right)$	Primary spherical
4	2	$\sqrt{10}\left(4\rho^4 - 3\rho^2\right)\cos 2\theta$	Secondary astigmatism
4	4	$\sqrt{10}\,\rho^4\cos 4\theta$	
5	1	$\sqrt{12}\left(10\rho^5 - 12\rho^3 + 3\rho\right)\cos\theta$	Secondary coma
5	3	$\sqrt{12}\left(5\rho^5 - 4\rho^3\right)\cos 3\theta$	
5	5	$\sqrt{12}\,\rho^5\cos 5\theta$	
6	0	$\sqrt{7}\left(20\rho^6 - 30\rho^4 + 12\rho^2 - 1\right)$	Secondary spherical
6	2	$\sqrt{14}\left(15\rho^6 - 20\rho^4 + 6\rho^2\right)\cos 2\theta$	Tertiary astigmatism
6	4	$\sqrt{14}\left(6\rho^6 - 5\rho^4\right)\cos 4\theta$	
6	6	$\sqrt{14}\,\rho^6\cos 6\theta$	
7	1	$4\left(35\rho^7 - 60\rho^5 + 30\rho^3 - 4\rho\right)\cos\theta$	Tertiary coma
7	3	$4\left(21\rho^7 - 30\rho^5 + 10\rho^3\right)\cos 3\theta$	
7	5	$4\left(7\rho^7 - 6\rho^5\right)\cos 5\theta$	
7	7	$4\rho^7\cos 7\theta$	
8	0	$3\left(70\rho^8 - 140\rho^6 + 90\rho^4 - 20\rho^2 + 1\right)$	Tertiary spherical

*The words "orthonormal Zernike" are to be associated with these names, e.g., *orthonormal Zernike primary astigmatism.*

where σ_w is its *standard deviation*. Note that $\sigma_w \neq W_{rms}$, where $W_{rms} = <W^2>^{1/2}$ is the *root-mean-square* (rms) value of the aberration, unless the mean value of the aberration $<W> = 0$. The Zernike circle polynomials are suitable for systems with circular pupils. The polynomoals that are suitable for systems with annular pupils, such as the astronomical telescopes discussed in Section 6.8, are the Zernike annular polynomials.[2]

3.5.4 Relationships Between Coefficients of Power-Series and Zernike-Polynomial Expansions

To relate the *power-series coefficients* a_{nm} of Eq. (3-36) to the *Zernike coefficients* c_{nm} of Eq. (3-44), we first change the indices n and m of coefficients a_{nm} to k and l (not to be confused with the l used earlier), respectively, and write the power-series expansion of the aberration function in the form

$$W(\rho,\theta) = \sum_{k=0}^{\infty} \sum_{l=0}^{k} a_{kl}\rho^k\cos^l\theta \quad , \tag{3-55}$$

where k and l are positive integers including zero and $k - l \geq 0$ and even. The coefficients a_{kl} and c_{nm} can be related to each other by comparing Eqs. (3-44) and (3-55) and using the identities

$$\cos m\theta = \left(1 + \delta_{m0}\right)2^{m-1}\cos^m\theta + m\sum_{q=1}^{m-1} \frac{(-1)^q(m-q-1)!\,2^{m-2q-1}}{q(q-1)!(m-2q)!}(\cos\theta)^{m-2q} \tag{3-56}$$

and

$$\cos^l\theta = \frac{1}{2l}\sum_{q=0}^{l} \frac{l!}{q!(l-q)!}\cos(l-2q)\theta \quad . \tag{3-57}$$

If the Zernike coefficients c_{nm} are known, the power-series coefficients a_{kl} can be obtained from them as follows. Substituting Eqs. (3-45) and (3-56) into Eq. (3-44) and equating the coefficient of the term containing the factor $\rho^k\cos^l\theta$ in the equation thus obtained to the corresponding coefficient in Eq. (3-55), we find that

$$a_{00} = c_{00} \tag{3-58}$$

and

$$a_{kl} = \sum_{n=1}^{\infty} \sum_{m=0}^{n} b_{klnm}c_{nm} \quad , \tag{3-59}$$

where

$$
b_{klnm} = \begin{cases} \left[2(n+1)(1+\delta_{m0})\right]^{1/2} \dfrac{(-1)^{\frac{n-k}{2}}\left(\dfrac{n+k}{2}\right)!\,2^{m-1}}{\left(\dfrac{n-k}{2}\right)!\left(\dfrac{k+m}{2}\right)!\left(\dfrac{k-m}{2}\right)!} \quad , \ m = l & \text{(3-60a)} \\[3em] \left[\dfrac{2(n+1)}{1+\delta_{m0}}\right]^{1/2} \dfrac{(-1)^{\frac{n-k}{2}}\left(\dfrac{n+k}{2}\right)!\,m}{\left(\dfrac{n-k}{2}\right)!\left(\dfrac{k+m}{2}\right)!\left(\dfrac{k-m}{2}\right)!}\,\dfrac{2^{l}(-1)^{\frac{m-l}{2}}\left(\dfrac{m+l}{2}-1\right)!}{l!(m-l)\left(\dfrac{m-l}{2}-1\right)!} \quad , \ m \neq l \ . \end{cases}
$$

$$\text{(3-60b)}$$

The values of b_{klnm} for $k \le 8$ and $n \le 8$ are given in Table 3-8. Note that only those coefficients exist for which $k - l$, $n - m$, $n - k$, $k - m$, and $m - l$ are all positive even integers including zero.

If the power-series coefficients a_{kl} are known, the Zernike coefficients c_{nm} can be obtained from them as follows. Substituting Eq. (3-57) into Eq. (3-55) and then substituting the equation thus obtained and Eq. (3-45) into Eq. (3-50), we obtain

$$c_{nm} = \sum_{k=0}^{\infty} \sum_{l=0}^{\infty} d_{nmkl} a_{kl} \quad , \tag{3-61}$$

where

$$d_{nmkl} = \left[\frac{2(n+1)}{1+\delta_{m0}}\right]^{1/2} \frac{l!}{2^{l}\left(\dfrac{l-m}{2}\right)!\left(\dfrac{l+m}{2}\right)!}$$

$$\times \sum_{s=0}^{(n-m)/2} \frac{(-1)^{s}(n-s)!}{s!\left(\dfrac{n+m}{2}-s\right)!\left(\dfrac{n-m}{2}-s\right)!(n-2s+k+2)} \quad . \tag{3-62}$$

The values of d_{nmkl} for $n \le 8$ and $k \le 8$ are given in Table 3-9. Note that only those coefficients exist for which $n - m$, $k - l$, and $l - m$ are all positive even integers including zero.

3.6 OBSERVATION OF ABERRATIONS

Now we describe briefly how the aberrations of an optical system may be observed. The emphasis of our discussion is on how to *recognize a primary aberration and not on how to measure it precisely*. Since the optical frequencies are very high ($10^{14} - 10^{15}$ Hz), optical wavefronts, aberrated or not, cannot be observed directly; optical detectors simply do not respond at these frequencies. The image of a monochromatic point object formed by an aberrated system is characteristically different for a different aberration. Another and more powerful way to recognize an aberration is to form an *interferogram* by combining two parts of a light beam, one of which has been transmitted through the

Table 3-8. Values of b_{klnm} used for obtaining power-series aberration coefficients a_{kl} from Zernike aberration coefficients c_{nm}.

n	m (k→)	0	1	2	2	3	3	4	4	4	5	5	5	6	6	6	6	7	7	7	7	8
(l→)		0	1	0	2	1	3	0	2	4	1	3	5	0	2	4	6	1	3	5	7	0
0	0	$\sqrt{2}$																				
1	1		2																			
2	0	$\sqrt{3}$		$2\sqrt{3}$																		
2	2			$-\sqrt{6}$	$2\sqrt{6}$																	
3	1		$-4\sqrt{2}$			$6\sqrt{2}$																
3	3					$-6\sqrt{2}$	$8\sqrt{2}$	$6\sqrt{5}$														
4	0	$\sqrt{5}$		$-6\sqrt{5}$																		
4	2			$3\sqrt{10}$	$-6\sqrt{10}$			$-4\sqrt{10}$	$8\sqrt{10}$													
4	4							$\sqrt{10}$	$-8\sqrt{10}$	$8\sqrt{10}$												
5	1		$6\sqrt{3}$			$-24\sqrt{3}$					$20\sqrt{3}$											
5	3					$24\sqrt{3}$	$-32\sqrt{3}$				$-30\sqrt{3}$	$40\sqrt{3}$										
5	5										$10\sqrt{3}$	$-40\sqrt{3}$	$32\sqrt{3}$									
6	0	$\sqrt{7}$		$6\sqrt{7}$				$-30\sqrt{7}$						$20\sqrt{7}$								
6	2			$-6\sqrt{14}$	$12\sqrt{14}$			$20\sqrt{14}$	$-40\sqrt{14}$					$-15\sqrt{14}$	$30\sqrt{14}$							
6	4							$-5\sqrt{14}$	$40\sqrt{14}$	$-40\sqrt{14}$				$6\sqrt{14}$	$-48\sqrt{14}$	$48\sqrt{14}$						
6	6													$-\sqrt{14}$	$18\sqrt{14}$	$-48\sqrt{14}$	$32\sqrt{14}$					
7	1		-16			120					-240							140				
7	3					-120	160				360	-480						-252	336			
7	5										-120	480	-384					140	-560	448		
7	7																	-28	224	-448	256	
8	0	3		-60				270						-420								210

Table 3-9. Values of d_{nmkl} for obtaining Zernike aberration coefficients c_{nm} from the power-series aberration coefficients a_{kl}.

n	m	$k=0$	1	2	2	3	3	4	4	4	5	5	5	6	6	6	6	7	7	7	7	8
	$l=$	0	1	0	2	1	3	0	2	4	1	3	5	0	2	4	6	1	3	5	7	0
0	0	1		$\frac{1}{2}$	$\frac{1}{4}$			$\frac{1}{3}$	$\frac{1}{6}$	$\frac{1}{8}$				$\frac{1}{4}$	$\frac{1}{8}$	$\frac{3}{32}$	$\frac{5}{64}$					$\frac{1}{5}$
1	1		$\frac{1}{2}$			$\frac{1}{3}$	$\frac{1}{4}$				$\frac{1}{4}$	$\frac{3}{16}$	$\frac{5}{32}$					$\frac{1}{5}$	$\frac{3}{20}$	$\frac{1}{8}$	$\frac{7}{64}$	
2	0			$\frac{1}{2\sqrt{3}}$	$\frac{1}{4\sqrt{3}}$			$\frac{1}{2\sqrt{3}}$	$\frac{1}{4\sqrt{3}}$	$\frac{\sqrt{3}}{16}$				$\frac{3\sqrt{3}}{20}$	$\frac{3\sqrt{3}}{40}$	$\frac{9\sqrt{3}}{160}$	$\frac{3\sqrt{3}}{64}$					$\frac{2}{5\sqrt{3}}$
2	2				$\frac{1}{2\sqrt{6}}$				$\frac{1}{8}\sqrt{\frac{3}{2}}$	$\frac{1}{8}\sqrt{\frac{3}{2}}$					$\frac{1}{10}\sqrt{\frac{3}{2}}$	$\frac{1}{10}\sqrt{\frac{3}{2}}$	$\frac{3}{32}\sqrt{\frac{3}{2}}$					
3	1					$\frac{1}{6\sqrt{2}}$	$\frac{1}{8\sqrt{2}}$				$\frac{1}{5\sqrt{2}}$	$\frac{3}{20\sqrt{2}}$	$\frac{1}{8\sqrt{2}}$					$\frac{1}{5\sqrt{2}}$	$\frac{3}{20\sqrt{2}}$	$\frac{1}{8\sqrt{2}}$	$\frac{7}{64\sqrt{2}}$	
3	3						$\frac{1}{8\sqrt{2}}$					$\frac{1}{10\sqrt{2}}$	$\frac{1}{8\sqrt{2}}$						$\frac{1}{12\sqrt{2}}$	$\frac{5}{48\sqrt{2}}$	$\frac{7}{64\sqrt{2}}$	
4	0							$\frac{1}{6\sqrt{5}}$	$\frac{1}{12\sqrt{5}}$	$\frac{1}{16\sqrt{5}}$				$\frac{1}{4\sqrt{5}}$	$\frac{1}{8\sqrt{5}}$	$\frac{3}{32\sqrt{5}}$	$\frac{\sqrt{5}}{64}$					$\frac{2}{7\sqrt{5}}$
4	2								$\frac{1}{8\sqrt{10}}$	$\frac{1}{8\sqrt{10}}$					$\frac{1}{6\sqrt{10}}$	$\frac{1}{6\sqrt{10}}$	$\frac{1}{32}\sqrt{\frac{5}{2}}$					
4	4									$\frac{1}{8\sqrt{5}}$					$\frac{\sqrt{5}}{48}$	$\frac{\sqrt{5}}{32}$						
5	1										$\frac{1}{20\sqrt{3}}$	$\frac{\sqrt{3}}{80}$	$\frac{1}{32\sqrt{3}}$					$\frac{\sqrt{5}}{35}$	$\frac{3\sqrt{3}}{140}$	$\frac{\sqrt{3}}{56}$	$\frac{\sqrt{3}}{64}$	
5	3											$\frac{1}{40\sqrt{3}}$	$\frac{1}{32\sqrt{3}}$						$\frac{5}{140\sqrt{3}}$	$\frac{5}{112\sqrt{3}}$	$\frac{\sqrt{3}}{64}$	
5	5												$\frac{1}{32\sqrt{3}}$							$\frac{\sqrt{3}}{112}$	$\frac{\sqrt{3}}{64}$	
6	0													$\frac{1}{20\sqrt{7}}$	$\frac{1}{40\sqrt{7}}$	$\frac{3}{160\sqrt{7}}$	$\frac{1}{64\sqrt{7}}$					$\frac{1}{10\sqrt{7}}$
6	2														$\frac{1}{30\sqrt{14}}$	$\frac{1}{30\sqrt{14}}$	$\frac{1}{32\sqrt{14}}$					
6	4															$\frac{1}{48\sqrt{14}}$	$\frac{1}{32\sqrt{14}}$					
6	6																$\frac{1}{32\sqrt{14}}$					
7	1																	$\frac{1}{140}$	$\frac{3}{560}$	$\frac{1}{224}$	$\frac{1}{256}$	
7	3																		$\frac{1}{336}$	$\frac{5}{1344}$	$\frac{1}{256}$	
7	5																			$\frac{1}{448}$	$\frac{1}{256}$	
7	7																				$\frac{1}{256}$	
8	0																					$\frac{1}{210}$

system. An aberration in the system yields an *interference pattern* that is characteristically different for a different aberration. Here, we briefly discuss the interference patterns for primary aberrations.[1]

3.6.1 Primary Aberrations

Considering an optical system with a circular exit pupil of radius a and letting (r, θ) be the polar coordinates of a point in the plane of its exit pupil, the functional form of the *primary phase aberrations* may be written

$$\Phi(\rho,\theta) = \begin{cases} A_s\rho^4 + B_d\rho^2, & \text{Spherical combined with defocus} & (3\text{-}63) \\ A_c\rho^3\cos\theta + B_t\rho\cos\theta, & \text{Coma combined with tilt} & (3\text{-}64) \\ A_a\rho^2\cos^2\theta + B_d\rho^2, & \text{Astigmatism combined with defocus} & (3\text{-}65) \\ A_d\rho^2, & \text{Field curvature} & (3\text{-}66) \\ A_t\rho\cos\theta, & \text{Distortion}, & (3\text{-}67) \end{cases}$$

where A_i or B_i is a peak aberration coefficient representing the maximum value of the corresponding aberration across the pupil and $\rho = r/a$ is a normalized radial variable. The phase and the wave aberrations are related to each other according to $\Phi = (2\pi/\lambda)W$, where λ is the wavelength of the optical radiation. When $\Phi(\rho,\theta) = 0$, the wavefront passing through the center of the exit pupil, for a point object, is spherical centered at the Gaussian image point. Let its radius of curvature be R. For an aberrated system, $\Phi(\rho,\theta)$ represents the optical deviation of the wavefront from being spherical at a point (ρ,θ).

In Eq. (3-63), when $B_d = 0$, the aberration is *spherical*. A nonzero value of B_d implies that the aberration is combined with defocus, i.e., the aberration is not defined with respect to a reference sphere centered at the Gaussian image point but with respect to another sphere centered at a distance z from the plane of the exit pupil according to Eq. (3-14). As discussed in Section 4.3.1, the reference sphere is centered at the marginal image point, the center of the circle of least confusion, and the point midway between the marginal and Gaussian image points when $B_d/A_s = -2$, -1.5, and -1, respectively. The midway point corresponds to minimum variance of the aberration as, may be seen by comparing the aberration thus obtained with the Zernike polynomial $Z_4^0(\rho)$.

In Eq. (3-64), when $B_t = 0$, the aberration is *coma*. A nonzero value of B_t implies that the aberration is combined with tilt, or that it is defined with respect to a reference sphere centered at a point $(2FB_t, 0)$ in the image plane, where F is the focal ratio or the f-number of the image-forming light cone. The variance of the aberration is minimum when $B_t/A_c = -2/3$, as may be seen by comparing the aberration thus obtained with the Zernike polynomial $Z_3^1(\rho, \theta)$.

In Eq. (3-65), when $B_d = 0$, the aberration is *astigmatism*. A nonzero value of B_d implies that it is combined with defocus. The variance of the aberration is minimum when $B_d/A_a = -1/2$. When $B_d/A_a = 0$ or -1, we obtain the so-called *tangential* and *sagittal images* of a point object discussed in Section 4.3.3. Equations (3-66) and (3-67) represent field curvature and distortion aberrations, respectively. The coefficients A_d and A_t of these aberrations vary with the image height as h'^2 and h'^3, respectively. However, for a given image height, these aberrations are equivalent to defocus and tilt aberrations, respectively. Figure 3-8 shows a 3-D plot of the various aberrations.

3.6.2 Interferograms

There are a variety of interferometers that are used for detecting and measuring aberrations of optical systems.[1] Figure 3-9 schematically illustrates a *Twyman-Green interferometer* in which a collimated laser beam is divided into two parts by a beam splitter *BS*. One part, called the *test beam*, is incident on the system under test, indicated by the lens *L*, and the other, called the *reference beam*, is incident on a plane mirror M_1. The focus *F* of the lens system lies at the center of curvature *C* of a spherical mirror M_2. As the angle of the incident light is changed to study the off-axis aberrations of the system, the mirror is tilted so that its center of curvature lies at the current focus of the beam. In this arrangement the mirror does not introduce any aberration since it is forming the image of an object lying at its center of curvature (see Section 6.4). The two reflected beams interfere in the region of their overlap. The lens *L'* is used to observe the

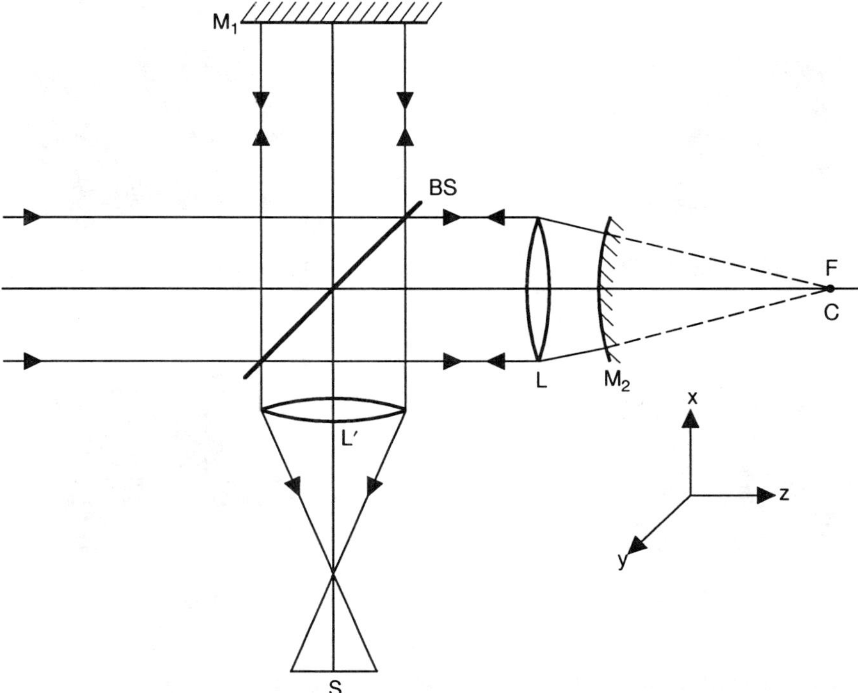

Figure 3-9. Twyman-Green interferometer for testing a lens system *L*. *F* is the image-space focal point of *L* and *C* is the center of curvature of a spherical mirror M_2. The interfering beams are focused by a lens *L'* and the interference pattern is observed on a screen *S*.

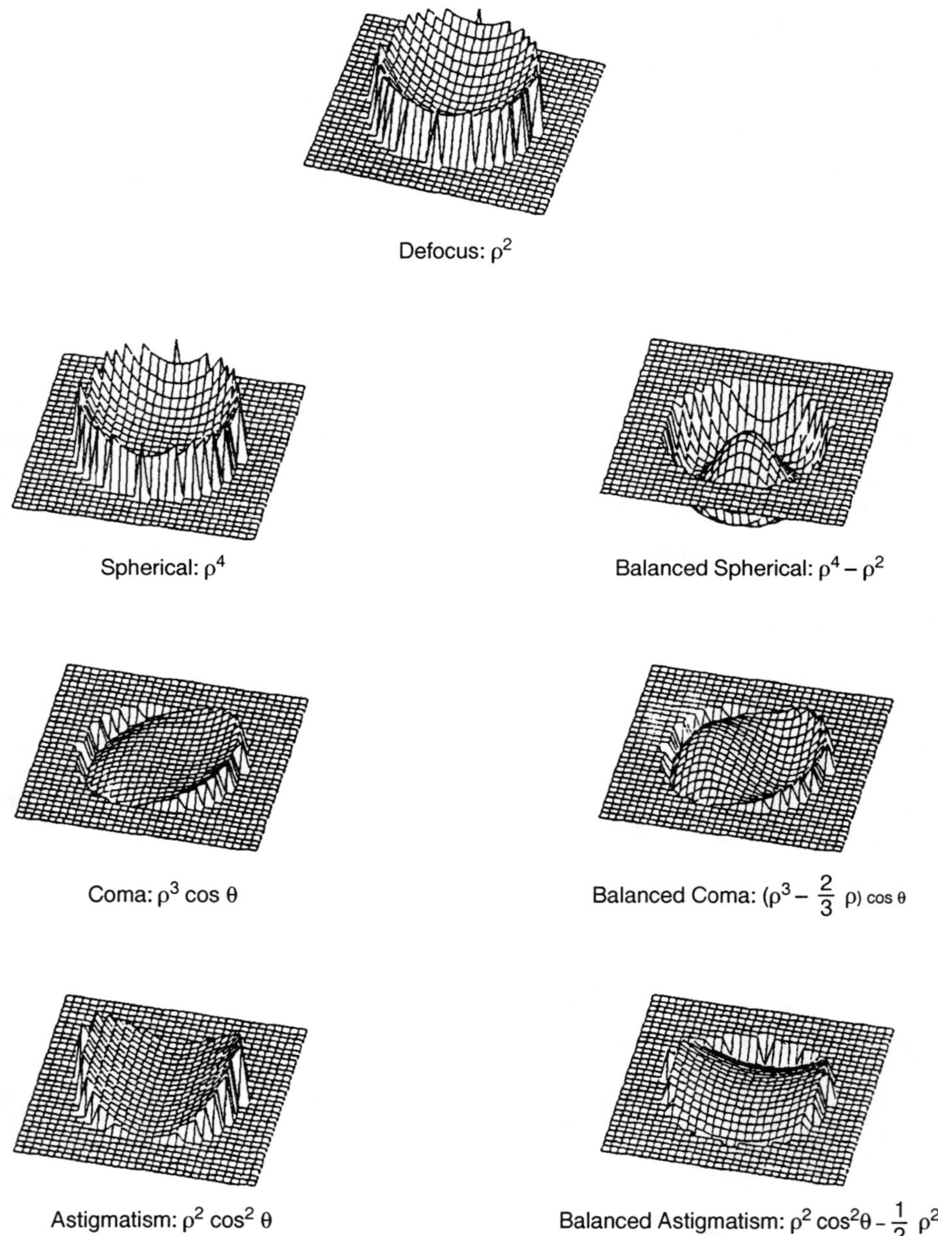

Figure 3-8. Shape of primary aberrations representing the difference between an ideal wavefront (typically, spherical) and an actual wavefront.

interference pattern on a screen S. A record of the interference pattern is called an *interferogram*. Note that since the test beam goes through the lens system L twice, its aberration is twice that of the system.

If the reference beam has uniform phase and the test beam has a phase distribution $\Phi(x, y)$, and if their amplitudes are equal to each other, the irradiance distribution of their interference pattern is given by

$$I(x,y) = I_0\left|1 + \exp\left[i\Phi(x,y)\right]\right|^2$$

$$= 2I_0\left\{1 + \cos\left[\Phi(x,y)\right]\right\} \quad , \tag{3-68}$$

where I_0 is the irradiance when only one beam is present. The irradiance has a maximum value equal to $4I_0$ at those points for which

$$\Phi(x,y) = 2\pi n \tag{3-69a}$$

and a minimum value equal to zero wherever

$$\Phi(x,y) = 2\pi(n+1/2) \quad , \tag{3-69b}$$

where n is a positive or negative integer, including zero. Each *fringe* in the interference pattern represents a certain value of n, which in turn corresponds to the locus of (x, y) points with the phase aberration given by Eq. (3-69a) for a bright fringe and Eq. (3-69b) for a dark fringe. If the test beam is aberration free $\left[\Phi(x, y) = 0\right]$, then the interference pattern has a uniform irradiance of $2I_0$.

Figure 3-10 shows interferograms when the lens system L under test suffers from 3λ of a primary aberration, corresponding to 6λ of an aberration of the interfering test beam. In our discussion, we give the value of an aberration coefficient in wavelength units, rather than in radians, as is customary in optics. For defocus and spherical aberration, the interference pattern consists of concentric circular interference fringes. The fringe spacing depends on the type of aberration. Figure 3-10a shows the interferogram obtained when the system is aberration free but is misfocused, i.e., when its focus F lies to the left or right of the center of curvature C of the spherical mirror M_2 by an amount corresponding to 3λ of the defocus aberration. [See Eqs. (3-19) for the relationship between the longitudinal defocus, i.e., the axial spacing between F and C, and the peak defocus aberration B_d, which is 3λ in our example.]

Figure 3-10b shows the interferograms obtained when the system has 3λ of spherical aberration (i.e., $A_s = 3\lambda$) and a certain amount of defocus. The case $B_d = 0$ (i.e., F and C coincident) represents such a system with an image of a certain object being observed in its Gaussian or paraxial image plane. Similarly, the interferogram obtained for $B_d/A_s = -2$ represents the system when the image is observed in its marginal image plane. For a system with a positive spherical aberration, its marginal focus lies farther from it than its paraxial focus (see Figure 4-3). Hence, this interferogram is obtained when points F and C are separated from each other axially, according to Eq. (3-19b), by $48\lambda F^2$, i.e., when F lies to the left of C by $48\lambda F^2$. The other two interferograms, $B_d = -A_s$ and $B_d = -1.5\,A_s$, represent the system when the image is observed in the minimum-aberration-variance plane and the circle-of-least-confusion plane, respectively.

Figure 3-10c shows the interferograms obtained when light is incident at a certain angle from the axis of the system so that it suffers from 3λ of coma. The fringes in this case are cubic curves. The case $B_t = 0$ corresponds to two parallel interfering beams (F and C are coincident in this case). The case $B_t = -2A_c/3$ represents the system corresponding to a minimum aberration variance. A tilt aberration with a peak value of B_t may be obtained by transversally displacing C from F by $\left(-2FB_t, 0\right)$. It may also be obtained by tilting the plane mirror M_1 by an angle B_t/a, where a is the radius of the test beam [see Eq. (3-23) and note the factors of 2 resulting from the reflection of the reference beam by mirror M_1 and the doubling of the system aberration in the test beam].

Figure 3-10d shows the interferograms obtained when the system suffers from 3λ of astigmatism. When $B_d = 0$ or $-A_a$, representing the system with an image being observed in a plane containing one or the other astigmatic focal line, respectively, we obtain an interferogram with straight-line fringes, since the aberration then depends on either x or y (but not both). However, the fringe spacing is not uniform. When $B_d = -A_a/2$, the fringe pattern consists of rectangular hyperbolas. If the system under test is aberration free but the two interfering beams are tilted with respect to each other, representing a wavefront tilt error, we obtain straight-line fringes that are uniformly spaced. The fringe spacing is inversely proportional to the tilt angle.

So far we have discussed interferograms of primary aberrations when only one of them is present. These interferograms are relatively simple and the aberration type may be recognized from the shape of the fringes. It should be evident that a general aberration consisting of a mixture of these aberrations and/or others will yield a much more complex interferogram. As an example of a general aberration, Figure 3-11a shows a possible aberration introduced by atmospheric turbulence as in ground-based astronomical observations. It corresponds to $D/r_0 = 10$, where D is the diameter of the aperture stop and r_0 is the *atmospheric coherence length*. On the average, the standard deviation of the instantaneous aberration introduced is given by $[0.134\,(D/r_0)^{5/3}]^{1/2}$ which, for $D/r_0 = 10$, is 2.494 radians or 0.397λ. The interferogram for this aberration is shown in Figure 3-11b. When 25λ of tilt are added to the aberration, the interferogram appears as in Figure 3-11c. Doubling of the aberration as in a Twyman-Green interferometer is not considered in Figure 3-11. If the aberration were zero, then Figure 3-11a would appear as a plane, Figure 3-11b as uniformly bright, and Figure 3-11c as uniformly spaced straight lines.

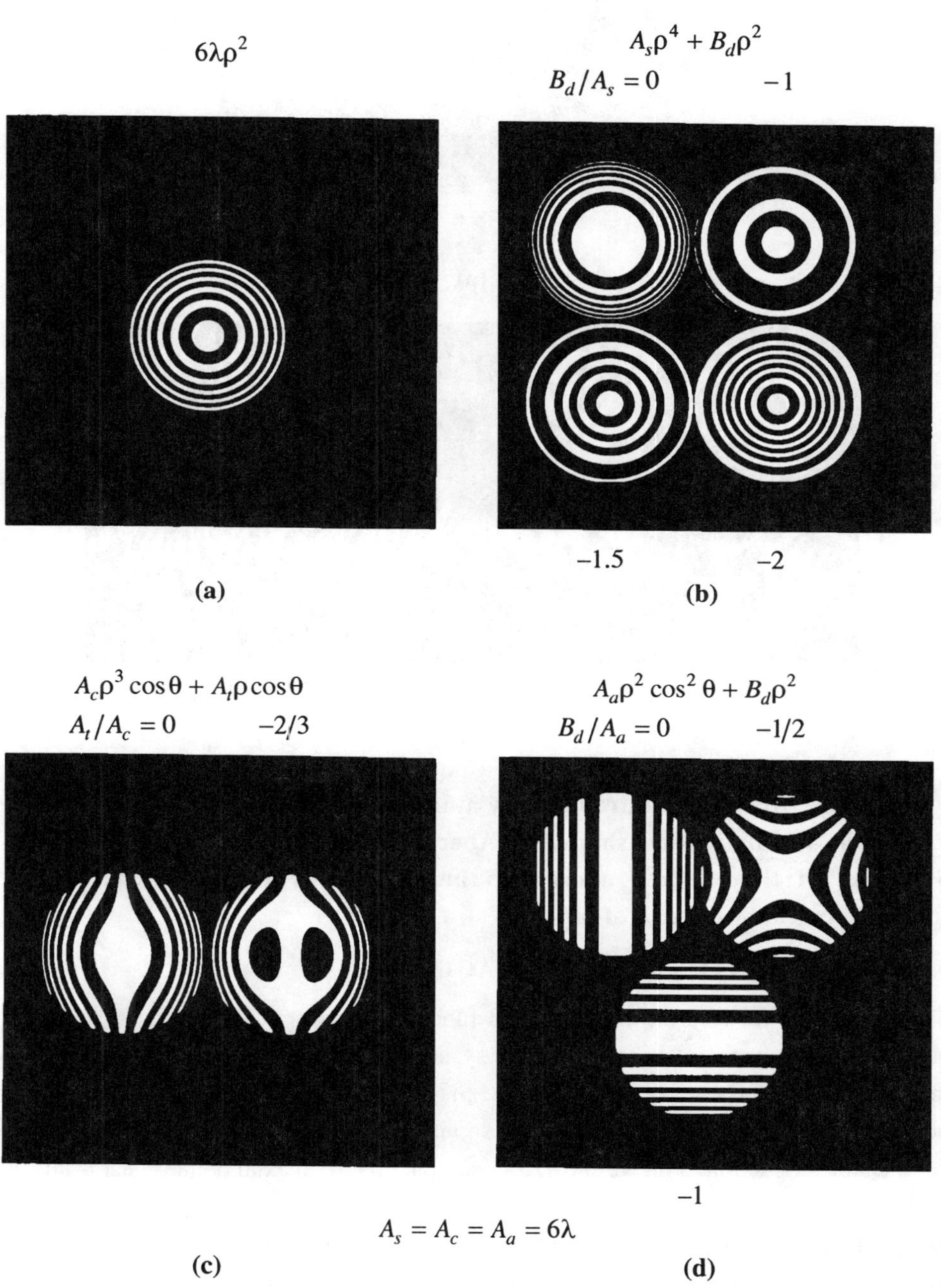

Figure 3-10. Interferograms of primary aberrations: (a) defocus, (b) spherical combined with defocus, (c) coma combined with tilt, (d) astigmatism combined with defocus. The aberrations in the interferograms are twice their corresponding values in the system under test because the test beam goes through the system twice.

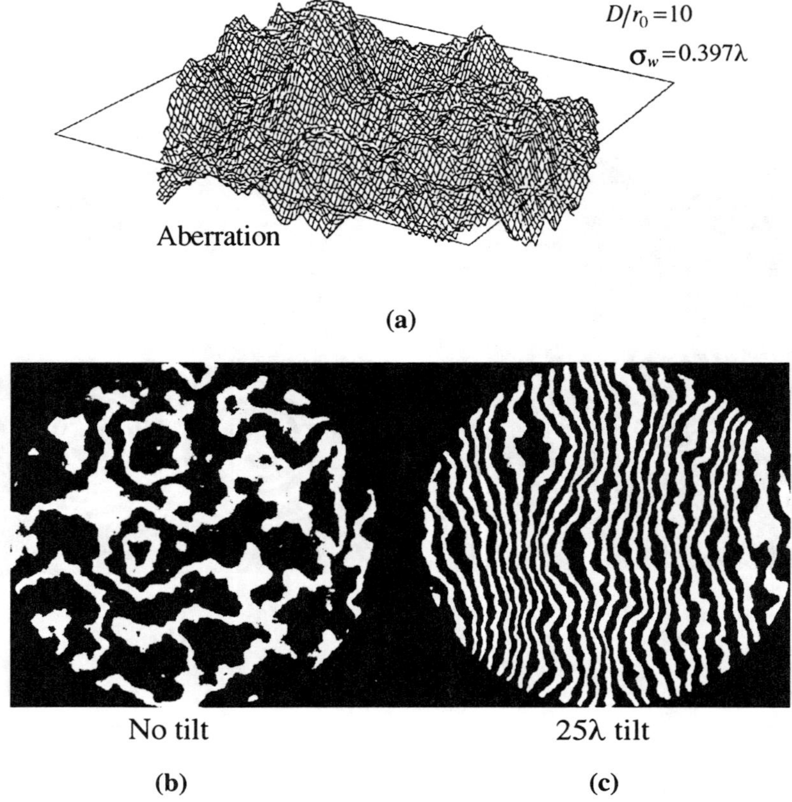

Figure 3-11. Aberration introduced by atmospheric turbulence corresponding to $D/r_0 = 10$. (a) Aberration shape. (b) Aberration interferogram. The standard deviation of the tilt-free aberration introduced by turbulence is 0.396λ. (c) Interferogram with 25λ of tilt.

3.7 CONDITIONS FOR PERFECT IMAGING

So far we have focused on the aberration function of an optical imaging system with emphasis on the types of aberrations that systems with an axis of rotational symmetry may suffer from. In this final section of the chapter, we inquire if a system can image an object aberration free or perfectly, and, if so, under what conditions.[3] Using Hamilton's point characteristic function, we show, for example, that if an axial point object is imaged perfectly by a system, then a small object lying in its neighborhood in a plane containing this point and perpendicular to the optical axis is also imaged perfectly by it, provided it satisfies what is called the sine condition. The linear coma of a system that does not satisfy this condition is defined and an expression for its magnitude is obtained. A quantity called offense against the sine condition is defined that gives a dimensionless measure of linear coma.

3.7.1   Imaging of a 3-D Object

Consider a point object P imaged perfectly onto a point P' by an optical system as illustrated in Figure 3-12. Thus, all the rays emanating from P and transmitted by the

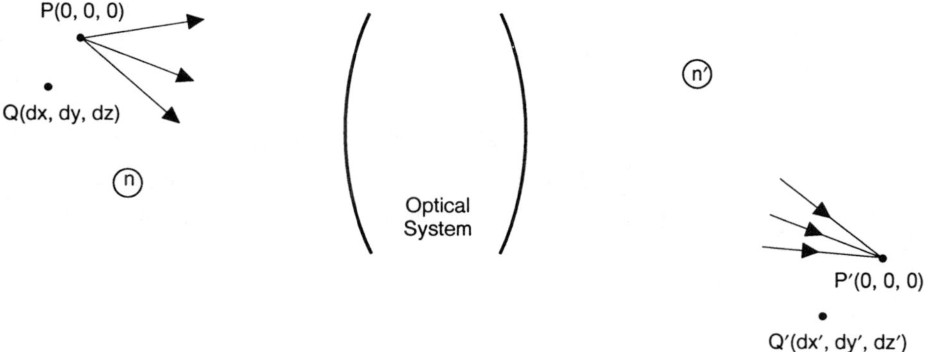

Figure 3-12. Perfect imaging of a small 3-D object.

system pass through P' and travel equal optical path lengths from P to P'. Let n and n' be the refractive indices of the object and image spaces, respectively. We now determine the conditions under which a small 3-D region in the neighborhood of P is imaged perfectly into a small 3-D region in the neighborhood of P'. Let Q and Q' be the object and image points in these regions with coordinates (dx, dy, dz) and (dx', dy', dz') in rectangular coordinate systems (x, y, z) and (x', y', z') with their origins at P and P', respectively. The coordinates of the points Q and Q' are related to each other according to

$$\begin{pmatrix} dx' \\ dy' \\ dz' \end{pmatrix} = \begin{pmatrix} a_{xx} & a_{xy} & a_{xz} \\ a_{yx} & a_{yy} & a_{yz} \\ a_{zx} & a_{zy} & a_{zz} \end{pmatrix} \begin{pmatrix} dx \\ dy \\ dz \end{pmatrix} \quad . \tag{3-70}$$

where

$$a_{xy} = \frac{dx'}{dy} \quad , \tag{3-71}$$

etc., are the derivatives of the image point coordinates with respect to the object point coordinates.

Consider a certain ray starting at P with direction cosines (L, M, N) and ending at P' with direction cosines (L', M', N'). From Section 1.2.5, the difference in the optical path lengths of the rays from Q to Q' and from P to P' may be written

$$dV = [QQ'] - [PP']$$

$$= n'(L'dx' + M'dy' + N'dz') - n(Ldx + Mdy + Ndz) \quad . \tag{3-72}$$

If Q' is the perfect image of Q, then the optical path length of a ray from Q to Q' is fully determined by the position of Q alone and, of course, it is independent of which ray from P to P' is considered to calculate dV. Thus, we may write

$$
\begin{aligned}
dV &= dF(x,y,z) \\
&= \frac{\partial F}{\partial x}dx + \frac{\partial F}{\partial y}dy + \frac{\partial F}{\partial z}dz \quad ,
\end{aligned}
\tag{3-73}
$$

where F is a function of (x,y,z) such that the derivatives $\partial F/\partial x$, etc., are independent of the direction cosines (L, M, N) and (L', M', N'). Substituting for dx', etc., from Eq. (3-70) into Eq. (3-72) and comparing the results obtained with Eq. (3-73), we obtain

$$
\begin{aligned}
&\left[n'\left(a_{xx}L' + a_{yx}M' + a_{zx}N' \right) - nL \right]dx + \left[n'\left(a_{xy}L' + a_{yy}M' + a_{zy}N' \right) - nM \right]dy \\
&\quad + \left[n'\left(a_{xz}L' + a_{yz}M' + a_{zz}N' \right) - nN \right]dz \\
&\quad\quad = \frac{\partial F}{\partial x}dx + \frac{\partial F}{\partial y}dy + \frac{\partial F}{\partial z}dz \quad .
\end{aligned}
\tag{3-74}
$$

Since the derivatives $\partial F/\partial x$, etc., are independent of the direction cosines, the expressions inside the square brackets on the left-hand side of Eq. (3-74) must also be independent of them. Thus, each of these expressions must be equal to zero, showing that the unit vectors (L, M, N) and (L', M', N') are related to each other according to

$$
\begin{pmatrix} L \\ M \\ N \end{pmatrix} = \frac{n'}{n} \begin{pmatrix} a_{xx} & a_{xy} & a_{xz} \\ a_{yx} & a_{yy} & a_{yz} \\ a_{zx} & a_{zy} & a_{zz} \end{pmatrix} \begin{pmatrix} L' \\ M' \\ N' \end{pmatrix} ,
\tag{3-75}
$$

or that the matrix

$$
\frac{n'}{n} \begin{pmatrix} a_{xx} & a_{xy} & a_{xz} \\ a_{yx} & a_{yy} & a_{yz} \\ a_{zx} & a_{zy} & a_{zz} \end{pmatrix}
\tag{3-76}
$$

is unitary. Consequently the angle between two rays emanating from P is equal to the angle between them as they converge on P'. Moreover, it follows from Eq. (3-70) that the ratio $P'Q'/PQ$ of the lengths of the image and object line elements $P'Q'$ and PQ, respectively, is n/n'. Thus, the image $P'Q'$ of the object PQ is perfect when the transverse magnification of the image is n/n', in which case the angle between two rays in the object space is equal to the angle between them in the image space [see Eqs. (3-87) and (3-88) also]. Other than a plane mirror, there is no imaging system that satisfies this requirement. However, the above analysis does not show how close a practical system can be to forming a perfect image, i.e., "good quality" imaging is not excluded.

Next we consider two special cases. Instead of a 3-D object, we consider a small 2-D object in a plane perpendicular to the optical axis of an imaging system, and a small 1-D or linear object along the axis.

3.7.2 Imaging of a 2-D Transverse Object

Consider a line element P_0P of a small object lying in a plane perpendicular to the optical axis, which we assume to be the z axis, as shown in Figure 3-13a. Because of the rotational symmetry of the optical system about its axis, there is no loss of generality if we let P_0P be along the x axis. We assume that the axial point object P_0 is imaged perfectly at its Gaussian conjugate P_0' and determine the condition under which a neighboring off-axis point object P is imaged perfectly at its conjugate P'. Equation (3-72) for the difference in optical path lengths of rays from P to P' and from P_0 to P_0' for perfect imaging of P_0 and P at P_0' and P', respectively, reduces to

$$dV = [PP'] - [P_0P_0']$$
$$= n'L'dx' - nLdx \quad ,$$

(3-77)

and must be independent of which ray from P_0 to P_0' is used to calculate the difference, i.e., dV must be independent of L and L'. It is indeed zero, as may be seen by considering the axial ray $P_0VV'P_0'$ for which $L = 0 = L'$. Hence, the transverse magnification of the image may be written

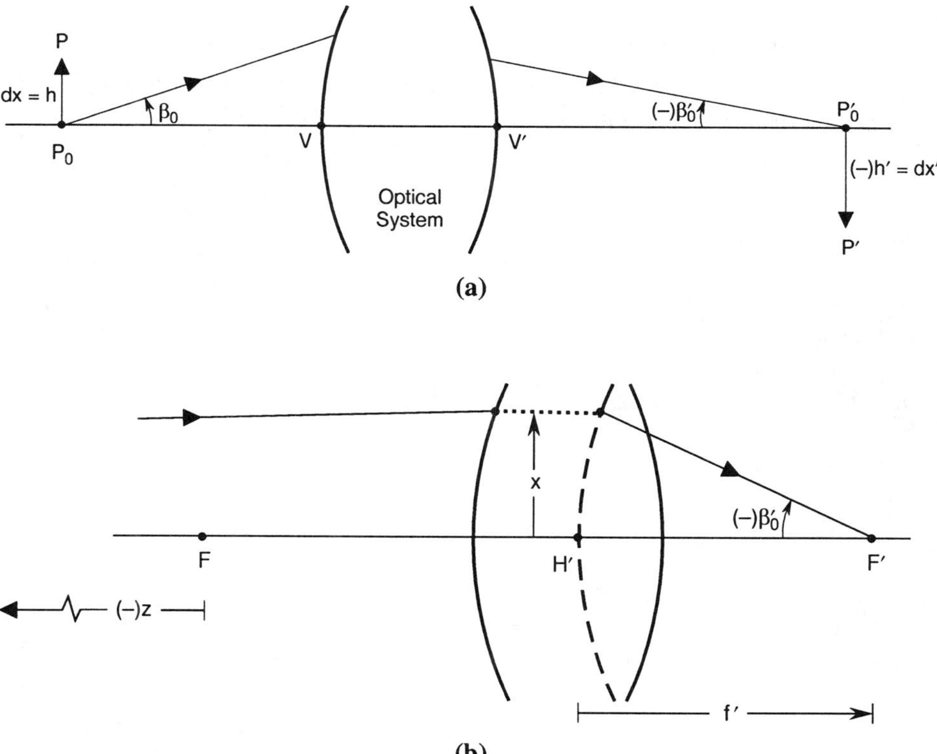

Figure 3-13. Perfect imaging of a small 2-D object. (a) Nearby object. (b) Faraway object.

$$M_t = \frac{dx'}{dx}$$

$$= \frac{nL}{n'L'}$$

$$= \frac{n \sin\beta_0}{n' \sin\beta_0'} \quad , \tag{3-78}$$

where β_0 and β_0' are the angles a ray starting at P_0 and ending at P_0' makes with the optical axis. If we let $dx = h$ and $dx' = h'$, where h and h' are the heights of the object and image points P and P', respectively, Eq. (3-78) may be written

$$\boxed{n'h' \sin\beta_0' = nh \sin\beta_0} \quad . \tag{3-79}$$

Note that although h and h' are assumed to be small (so that the aberrations depending nonlinearly on them are negligible), there is no limitation on the values of the angles β_0 and β_0'. Equation (3-79) is called the *sine condition* under which a line element P_0P perpendicular to the optical axis is imaged perfectly as $P_0'P'$, given that the point P_0 on the axis is imaged perfectly at P_0'. Axial points that are perfect images of each other and which, in addition, have the property that conjugate rays passing through them satisfy the sine condition, are called *aplanatic conjugates*.

If an object lies at a large (numerically negative) distance z from the object-space focal point F, as in Figure 3-13b, then a ray from an axial point object making an angle β_0 with the axis intersects the first surface at a height x given by

$$x \simeq -z \sin\beta_0 \quad , \tag{3-80}$$

or, keeping x constant,

$$\frac{x}{z \sin\beta_0} \to -1 \text{ as } z \to -\infty \quad . \tag{3-81}$$

Hence, the sine condition of Eq. (3-79) may be written

$$n'h' \sin\beta_0' = -\frac{nhx}{z} \quad ,$$

or

$$\frac{x}{\sin\beta_0'} = -\frac{n'}{n}\frac{h'}{h}z$$

$$= \frac{n'}{n}f$$

$$= -f' \quad , \tag{3-82}$$

where f' is the image-space focal length of the system and we have made use of Eqs. (1-63) and (1-77). As illustrated in Figure 3-13b, Eq. (3-82) implies that a ray incident parallel to the axis intersects its conjugate ray on a sphere of radius f' centered at the image-space focal point F'.

3.7.3 Imaging of a 1-D Axial Object

Now we determine the condition under which an axial point object Q_0 in the neighborhood of an axial point object P_0 (see Figure 3-14) is imaged perfectly at its Gaussian conjugate Q_0', given that P_0 is imaged perfectly at its conjugate point P_0'. In this case, Eq. (3-72) for the difference in optical path lengths of the rays from Q_0 to Q_0' and from P_0 to P_0' for perfect imaging of P_0 and Q_0 at P_0' and Q_0', respectively, reduces to

$$dV = [Q_0 Q_0'] - [P_0 P_0']$$

$$= n'N'dz' - nNdz \quad , \tag{3-83}$$

and must be independent of which ray from P_0 to P_0' is used to calculate the difference, i.e., dV must be independent of N and N'. Along the axis, both N and N' are equal to unity. Hence, we must have

$$n'N'dz' - nNdz = n'dz' - ndz \quad . \tag{3-84}$$

Substituting $N = \cos\beta_0$ and $N' = \cos\beta_0'$ into Eq. (3-84), we obtain

$$M_l = \frac{dz'}{dz}$$

$$= \frac{n\sin^2(\beta_0/2)}{n'\sin^2(\beta_0'/2)} \quad , \tag{3-85}$$

where M_l is the longitudinal magnification. Equation (3-85) gives the condition under which a line element along the optical axis of a system is imaged perfectly by it, given that a point on it is imaged perfectly. It is called the *Herschel condition*. Substituting Eqs. (1-64) and (1-71) into Eq. (3-85), we obtain a different form of the Herschel condition, namely,

$$n'h'\sin(\beta_0'/2) = nh\sin(\beta_0/2) \quad . \tag{3-86}$$

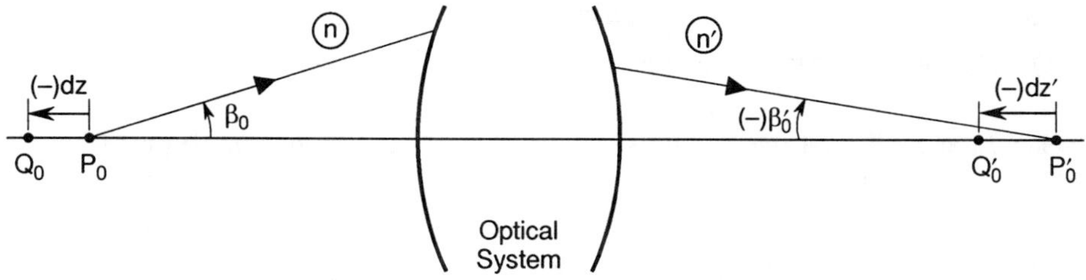

Figure 3-14. Perfect imaging of a small axial object $P_0 Q_0$.

We note from Eqs. (3-79) and (3-86) that the sine and Herschel conditions can be satisfied simultaneously only when $\beta_0' = \pm\beta_0$, i.e., when the angular magnification

$$M_\beta = \frac{\beta_0'}{\beta_0}$$

$$= \pm 1 \ . \tag{3-87}$$

In that case the transverse and longitudinal magnifications are both equal in magnitude to the ratio of the refractive indices of the object and image spaces, i.e.,

$$M_l = n/n' = \pm M_t \ , \tag{3-88}$$

where the positive and negative signs hold for a virtual and a real image, respectively. Note that β_0' and h' are numerically negative quantities in Figure 3-13, where the image is real. Equations (3-87) and (3-88) are the results discussed earlier for 3-D imaging.

3.7.4   Linear Coma and the Sine Condition

From the form of the aberration terms, namely, $_{2l+m}a_{nm}h'^{2l+m}r^n\cos m\theta$, terms that depend linearly on h' are those for which $l=0$, $m=1$, and n is an odd integer. These terms such as $_1a_{31}h'r^3\cos\theta$ called primary (or Seidel) coma, $_1a_{51}h'r^5\cos\theta$ called secondary coma, and $_1a_{71}h'r^7\cos\theta$ called tertiary coma, are together called *linear coma*. Individually, each such term represents linear coma of a certain order, e.g., primary linear coma. We now show explicitly that when an optical system images an axial point object perfectly, so that all orders of spherical aberration are zero, and it satisfies the sine condition for the conjugate planes passing through the perfect axial conjugates, then the system is free from all orders of linear coma for the conjugate points in those planes.

Consider an optical system imaging an axial point object P_0 perfectly at the axial point P_0' as shown in Figure 3-15. Let W and W' be the spherical wavefronts at the entrance and exit pupils EnP and ExP, respectively, for the axial conjugates; thus W is centered at P_0 and passes through the center O of the entrance pupil, and W' is centered at P_0' and passes through the center O' of the exit pupil. Since P_0 and P_0' are perfect conjugates, a ray such as P_0A incident on the system emerges from it as a ray $A'P_0'$ passing through P_0'. Let a point P' at a height h' be the Gaussian conjugate of an off-axis point object P at a height h. We assume that h and h' are small enough that the aberration terms depending on their squares and higher powers can be neglected. The chief ray PO incident on the system emerges from it as the ray $O'P'$.

Since P_0 and P_0' are perfect conjugates, the optical path lengths of all rays from P_0 to P_0' are equal to each other. Thus

$$\left[P_0AA'P_0'\right] = \left[P_0OO'P_0'\right] \ . \tag{3-89}$$

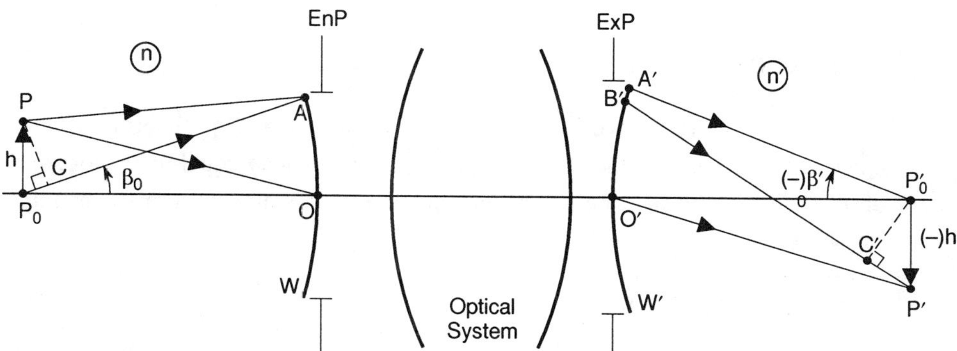

Figure 3-15. Schematic showing perfect imaging of an off-axis point object.

Now consider a ray such as PA incident on the system. In order that P' be a perfect image of P, it must emerge as a ray $B'P'$ passing through P', where B' is a point in the neighborhood of A'. In that case, the optical path length of the ray $PAB'P'$ is equal to that of the chief ray $POO'P'$. If their optical path lengths are not equal, then it emerges as a ray, say, $B'P''$, where P'' is a point in the image plane in the neighborhood of P'. The wave aberration W of this ray is equal to the difference in optical path lengths $[PAB'P']$ and $[POO'P']$, i.e.,

$$W = [PAB'P'] - [POO'P'] \ . \tag{3-90}$$

We note from the right-angle triangles P_0PO and $P_0'P'O'$ that

$$[POO'P'] = [P_0OO'P_0'] + O(h'^2) \ , \tag{3-91}$$

where we have used the fact that

$$O(h^2) = O(h'^2) \tag{3-92}$$

since h and h' are related to each other linearly by the transverse magnification. Substituting Eqs. (3-89) and (3-91) into Eq. (3-90), we may write

$$W = [PAB'P'] - [P_0AA'P_0'] + O(h'^2) \ ,$$

or

$$W = [PAB'C'] + [C'P'] - [P_0C] - [CAA'P_0'] + O(h'^2) \ , \tag{3-93}$$

where PC and $P_0'C'$ are perpendicular to P_0A and $B'P'$, respectively. Since $PAB'P'$ and $P_0AA'P_0'$ are *neighboring rays* (see Section 1.2.3),

$$[PAB'C'] = [CAA'P_0'] + O(h'^2) \ . \tag{3-94}$$

$$W = [C'P'] - [P_0 C] + O\left(h'^2\right) \ ,\tag{3-95}$$

or

$$W = n'h'\sin\beta_0' - nh\sin\beta_0 + O\left(h'^2\right) \ ,\tag{3-96}$$

where n and n' are the refractive indices of the object and image spaces, respectively, and we have used the fact that the angle $C'P_0'P'$ is approximately equal to β_0' for small values of h'. It is evident that the aberration function W contains no terms depending linearly on h' if

$$\boxed{n'h'\sin\beta_0' = nh\sin\beta_0 \ ,}\tag{3-97}$$

which is indeed the *sine condition*. Thus, all orders of coma are zero for off-axis conjugate points satisfying this condition in planes for which the axial conjugates are imaged perfectly.

3.7.5   Optical Sine Theorem

We now establish what is called the *optical sine theorem*. It gives a relationship between the slope angles of two conjugate rays for an axial point object and the object and sagittal image heights in the corresponding conjugate planes.

Consider a spherical refracting surface of radius of curvature R separating media of refractive indices n and n' imaging an object $P_0 P$ of height h as shown in Figure 3-16. Let the height of its Gaussian image $P_0'P'$ be h'. Consider a ray $P_0 Q$ incident on the surface from the point object P_0. Let the refracted ray be QM_0', where M_0' is the point of its intersection with the optical axis. The distance $P_0'M_0'$ is called the *longitudinal spherical aberration*, discussed in Section 4.3.1, and it is numerically negative in the figure. From the triangle $P_0 CQ$, where C is the center of curvature of the refracting

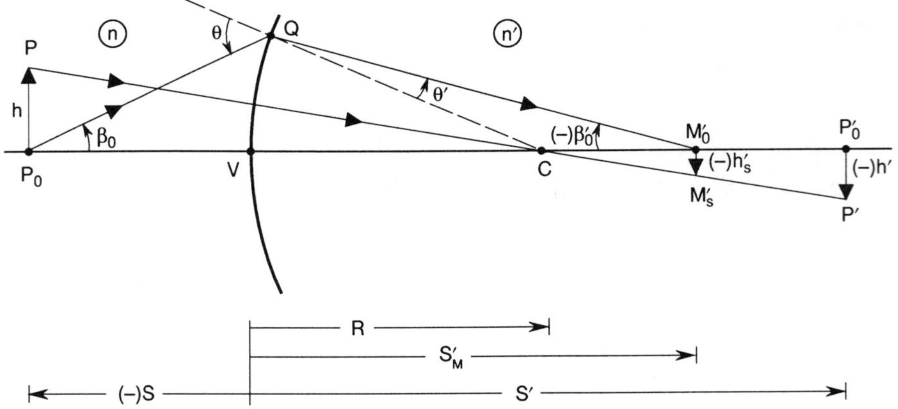

Figure 3-16. Optical sine theorem for a spherical refracting surface. M_0' lies to the left of P_0' as expected for a negative spherical aberration (see Sections 5.2 and 4.3.1). The conjugate points P and P' lie in mutually parallel planes that are perpendicular to the plane containing the points $P_0, Q,$ and P_0'.

surface, we note that

$$\frac{\sin\beta_0}{QC} = \frac{\sin(\pi - \theta)}{P_0C} \quad,$$

or

$$\frac{\sin\beta_0}{R} = \frac{\sin\theta}{R - S} \quad, \tag{3-98}$$

where β_0 is the slope angle of the incident ray P_0Q with the optical axis and θ is its angle of incidence on the surface. Similarly, from the triangle $CM_0'Q$, we note that

$$-\frac{\sin\beta_0'}{QC} = \frac{\sin\theta'}{CM_0'} \quad,$$

or

$$\frac{\sin\beta_0'}{R} = -\frac{\sin\theta'}{S_M' - R} \quad, \tag{3-99}$$

where β_0' is the slope angle of the refracted ray QM_0' and θ' is its angle of refraction. The negative sign in Eq. (3-99) is due to β_0' being numerically negative. The angles θ and θ' are related to each other according to Snell's law, i.e.,

$$n'\sin\theta' = n\sin\theta \quad. \tag{3-100}$$

From Eqs. (3-98) and (3-99), we obtain

$$\frac{S_M' - R}{S + R} = -\frac{\sin\theta'}{\sin\theta}\frac{\sin\beta_0}{\sin\beta_0'}$$

$$= -\frac{n}{n'}\frac{\sin\beta_0}{\sin\beta_0'} \quad, \tag{3-101}$$

where in the last step we have used Eq. (3-100).

A ray from the point object P incident in the direction of C is refracted undeviated by the surface and passes through its Gaussian conjugate point P'. Because of the symmetry of the sagittal rays about the tangential plane P_0PC, the two sagittal rays of the same zone as that of the point Q intersect the *auxiliary axis* PC at the same point after refraction. For a small object height h [so that field curvature can be neglected, resulting in a planar sagittal surface, as may be seen from Eq. (4-45)], this point is the same as M_s' at a height h_s' from M_0'.

From similar triangles P_0PC and $CM_0'M_s'$, we find that

$$\frac{M_0'M_s'}{P_0P} = \frac{CM_0'}{P_0C} \quad,$$

or

$$\frac{h'_s}{h} = -\frac{S'_M - R}{S + R} \quad . \tag{3-102}$$

The negative sign on the right-hand side of Eq. (3-102) accounts for the fact that the image point M'_s lies below the axis and, therefore, h'_s is numerically negative. From Eqs. (3-101) and (3-102), we obtain the final result that

$$\boxed{n'h'_s \sin \beta'_0 = nh \sin \beta_0 \quad ,} \tag{3-103}$$

which is called the *optical sine theorem*. It relates the slope angles of two conjugate rays for an axial point object and the object and sagittal image heights in the corresponding conjugate planes. The theorem holds for a multisurface system also, as may be seen by applying Eq. (3-103) successively to its surfaces. For such a system, the quantities on the left-hand side of Eq. (3-103) belong to its final image space. The theorem is used in the next section to obtain a generalized sine condition for the case when spherical aberration is also present. For small angles β_0 and β'_0, Eq. (3-103) reduces to the Lagrange invariance Eq. (1-69) of Gaussian optics.

3.7.6 Linear Coma and Offense Against the Sine Condition

We have shown in Section 3.7.4 that a system with zero spherical aberration (so that axial Gaussian conjugates are perfect images of each other) will have zero linear coma if it satisfies the sine condition. Now, we consider a system with nonzero spherical aberration and show how to determine its linear coma for a certain zone from the ray-trace data for an axial point object for the same zone. We also determine the condition under which linear coma is zero. This new condition is similar to the sine condition and reduces to it if we let the spherical aberration be zero. A quantity called offense against the sine condition is defined, which gives a dimensionless measure of linear coma.

From the form of the aberration terms, linear coma as an aberration may be written

$$W_{lc}(x,y;h') = h' \sum_{j=1}^{\infty} c_j x \left(x^2 + y^2\right)^j \quad , \tag{3-104}$$

where c_j's are the coefficients of the terms. They are related to the coefficients in Eq. (3-31c) according to

$$c_j \equiv {}_1 a_{2j+1,1} \quad . \tag{3-105}$$

The terms with $j = 1, 2, 3$, etc., represent primary, secondary, tertiary, etc., coma, respectively. Using the normalized radial variable ρ, Eq. (3-104) may be written in polar coordinates in the form

$$W_{lc}(\rho,\theta;h') = h' \cos\theta \sum_{j=1}^{\infty} c_j a^{2j+1} \rho^{2j+1} \quad . \tag{3-106}$$

The x component of the corresponding ray aberrations is given by

$$x_i = \frac{R}{n'}\frac{\partial W}{\partial x}$$

$$= \frac{Rh'}{n'}\sum_{j=1}^{\infty}c_j\left[(2j+1)x^2+y^2\right]\left(x^2+y^2\right)^{j-1} \quad . \tag{3-107}$$

Note that in Eq. (3-107), $R = OP'$ is the radius of curvature of the reference sphere (see Figure 3-17). For small values of h', $R \simeq OP_0'$. The value of x_i for the marginal sagittal rays, i.e., for

$$(x,y) = (0,\pm a) \quad , \tag{3-108}$$

is given by

$$x_{is} = \frac{Rh'}{n'}\sum_j c_j\,a^{2j}$$

$$= \frac{R}{n'a}W_{lc}(a,0;h') \quad , \tag{3-109}$$

where $W_{lc}(a,0;h')$ is the *peak value of linear coma* for marginal rays according to Eq. (3-104). Next, we show that this value can be obtained from the ray-trace data for the axial point object.

Let P_0' and P' be the Gaussian image points corresponding to on- and off-axis object points P_0 and P, respectively, as shown in Figure 3-17. Because of spherical aberration alone, let M_0' and M' be the corresponding marginal images, i.e., M_0' is the point where the marginal rays intersect the optical axis and M' is the point where the chief ray intersects the marginal image plane. In the presence of linear coma, the marginal sagittal rays intersect the marginal image plane at a point M_s' called the *sagittal image point* at a distance x_{is} from M'. The quantity $x_{is} = M'M_s'$ is called *sagittal coma*, which is discussed further in Section 4.3.2. The chief ray is shown to be passing through the Gaussian image point P', i.e., it is assumed to coincide with the auxiliary axis in image space. In other words, distortion, which represents the distance between P' and the point of intersection of the chief ray with the Gaussian image plane, is assumed to be zero. Let h_M' and h_s' be the heights of M' and M_s'. We note from similar triangles $OM_0'M'$ and $OP_0'P'$ that

$$\frac{h_M'}{h'} = \frac{R+\Delta}{R} \quad , \tag{3-110}$$

where $\Delta = P_0'M_0'$ is the *longitudinal spherical aberration* discussed in Chapter 4 (see Section 4.3.1). Hence, we may write

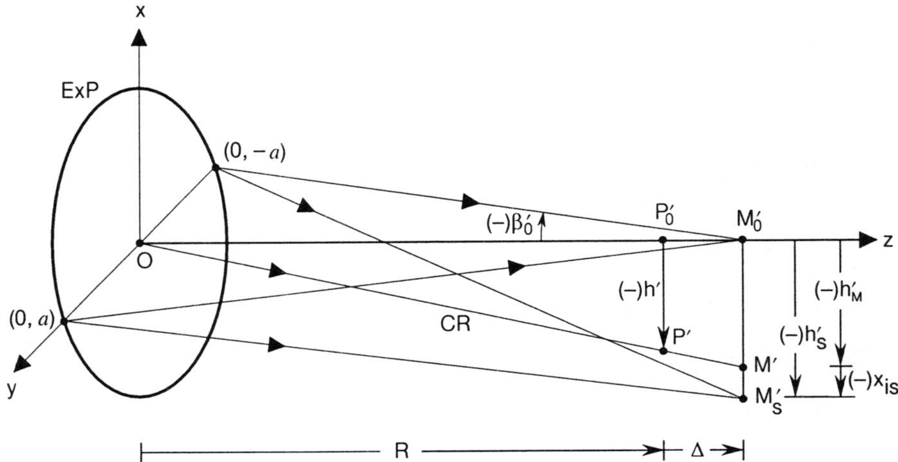

Figure 3-17. Schematic showing the effect of spherical aberration and linear coma of an imaging system. P_0' **and** P' **are the Gaussian images of on- and off-axis point objects** P_0 **and** P**, respectively. The spherical aberration is assumed to be positive and, therefore, the marginal axial image** M_0' **lies to the right-hand side of** P_0'**.** M_s' **is the sagittal marginal image of the point object** P**. It coincides with** M' **if linear coma is zero.**

$$x_{is} = M'M_s'$$

$$= h_s' - h_M'$$

$$= h_s' - \frac{R+\Delta}{R} h' \quad .$$

(3-111)

Note that if spherical aberration is zero, then the point M_0' coincides with the point P_0', M' coincides with P', and M_s' (which may be called P_s') lies below P' at a distance $h_s' - h'$ from it.

From the optical sine theorem for rays with conjugates P_0 and M_0' and slope angles β_0 and β_0', we may write

$$n'h_s' \sin\beta_0' = nh \sin\beta_0$$

or

$$h_s' = \frac{nh \sin\beta_0}{n' \sin\beta_0'} \quad .$$

(3-112)

From Eq. (3-109), the peak value of marginal linear coma may be written

$$W_{lc}(a, 0; h') = \frac{n'a}{R} x_{is}$$

$$= -n' \sin\beta_0' x_{is} \quad ,$$

(3-113a)

where we have let

$$a/R \; \simeq \; -\sin\beta_0' \quad . \tag{3-113b}$$

Again, the negative sign in Eqs. (3-113) is due to β_0' being numerically negative. Substituting Eq. (3-112) into Eq. (3-111) and substituting the result obtained into Eq. (3-113a), we obtain an expression for the peak value of linear coma in terms of the axial ray-trace data:

$$W_{lc}(a,0;h') \; = \; \frac{R+\Delta}{R}\,n'h'\sin\beta_0' \; - \; nh\sin\beta_0 \quad , \tag{3-114}$$

We note that the peak value of marginal linear coma depends on the location of the pupil [through the value of R in Eq. (3-114)] only when spherical aberration is present. (The dependence of coma on the location of the pupil of a system in the presence of spherical aberration is discussed further in Section 5.9. As an example, the dependence of coma of a mirror on the location of its aperture stop is discussed in Section 6.4.) Moreover, it is zero when

$$\frac{R+\Delta}{R}\,n'h'\sin\beta_0' \; = \; nh\sin\beta_0 \quad . \tag{3-115}$$

Or, if spherical aberration Δ is zero, then

$$\boxed{n'h'\sin\beta_0' \; = \; nh\sin\beta_0} \tag{3-116}$$

is the condition for zero linear coma. Equation (3-116) is the *sine condition* discussed in Sections 3.7.2 and 3.7.4. The dimensionless quantity x_{is}/h_M' is a measure of linear coma and it is called the *offense against the sine condition* (OSC). Using Eqs. (3-111) and (3-112), it may be written

$$OSC \; = \; x_{is}/h_M'$$

$$= \; \frac{x_{is}}{h_s' - x_{is}}$$

$$= \; \frac{R}{R+\Delta}\,\frac{nh\sin\beta_0}{n'h'\sin\beta_0'} - 1 \quad . \tag{3-117}$$

Although Eq. (3-114) was obtained for the marginal sagittal rays, a similar equation is obtained for a sagittal ray of any zone of the exit pupil. All that is required is that the slope angles β_0 and β_0' correspond to an axial ray for that zone. Similarly, Eqs. (3-115) and (3-116) represent the conditions for zero linear coma at the zone corresponding to those slope angles. However, if we assume that, for example, primary coma represented by the coefficient c_1 in Eq. (3-104) dominates and other coma terms are negligible, then if Eq. (3-115) or Eq. (3-116) holds, c_1 must be zero. Accordingly, coma is zero for any zone. Alternatively, linear coma is zero if Eq. (3-115) or Eq. (3-116) holds for all zones of the pupil.

APPENDIX A: DEGREE OF APPROXIMATION IN EQ. (3-11)

To understand the degree of approximation[4] involved in Eq. (3-11), we introduce a small quantity of first order $\mu = a/R \simeq a/z_g$. Then the rays exiting from the exit pupil may be assumed to make small angles $O(\mu)$ in radians with the optical axis, where the symbol $O(\mu)$ means "not exceeding a moderate multiple" of μ. To determine the error involved in replacing R' by R or z_g in Eqs. (3-9) and (3-10), we proceed as follows. The distance $R' = QP''$ is given by

$$
\begin{aligned}
R'^2 &= (x - x_i)^2 + (y - y_i)^2 + (z - z_g)^2 \\
&= R^2 + x_i^2 - x_g^2 + y_i^2 - 2x(x_i - x_g) - 2yy_i \quad ,
\end{aligned}
\tag{3A-1}
$$

where we have made use of Eq. (3-6). It is shown in Section 3.5.2.1 that the aberration function of a rotationally symmetric system is a polynomial of even degree in pupil and object (or image) coordinates and the lowest degeree of a term is four. Hence

$$
W(x, y) = O(R\mu^4) \quad ,
\tag{3A-2}
$$

where $\mu = x/R$ or $\mu = y/R$ and $O(\mu) = x_g/R$. Accordingly,

$$
\frac{\partial W}{\partial x} = O(\mu^3)
\tag{3A-3a}
$$

and

$$
R\frac{\partial W}{\partial x} = O(R\mu^3) \quad .
\tag{3A-3b}
$$

Therefore, Eq. (3-9) yields

$$
x_i = x_g + O(R\mu^3)
\tag{3A-4a}
$$

or

$$
x_i^2 = x_g^2 + RO(R\mu^4) \quad .
\tag{3A-4b}
$$

Similarly

$$
y_i = O(R\mu^3)
\tag{3A-5a}
$$

and

$$
y_i^2 = RO(R\mu^4) \quad .
\tag{3A-5b}
$$

Substituting Eqs. (3A-4) and (3A-5) into Eq. (3A-1), we obtain

$$R'^2 = R^2 + RO\left(R\mu^4\right)$$

or

$$R' = R + O\left(R\mu^4\right) \quad . \tag{3A-6}$$

Substituting Eqs. (3A-3a) and (3A-6) into Eqs. (3-9) and (3-10), we may write

$$\boxed{\left(x_i, y_i\right) = \frac{R}{n}\left(\frac{\partial W}{\partial x}, \frac{\partial W}{\partial y}\right) + O\left(R\mu^7\right) \quad ,} \tag{3A-7}$$

where, as in Eq. (3-11), $\left(x_i, y_i\right)$ are the coordinates of P'' with respect to P'. Comparing Eq. (3A-7) with Eq. (3-11), we find that the error associated in replacing R' by R as in Eq. (3-11) is $O\left(R\mu^7\right)$. From Eq. (3-5)

$$R = z_g\left(1 + x_g^2 / z_g^2\right)^{1/2}$$

$$= z_g + O\left(z_g\,\mu^2\right) \quad . \tag{3A-8}$$

Substituting Eq. (3A-8) into Eq. (3A-7), we obtain

$$\boxed{\left(x_i, y_i\right) = \frac{z_g}{n}\left(\frac{\partial W}{\partial x}, \frac{\partial W}{\partial y}\right) + O\left(z_g\,\mu^5\right) \quad ,} \tag{3A-9}$$

i.e., the error term in replacing R' by z_g is $O\left(z_g\,\mu^5\right)$. It should be noted that the degree of error term in Eq. (3A-7) does not change if we replace R by $z_g + \left(x_g^2 / 2\,z_g\right)$, although it is easy enough to calculate R from Eq. (3-5).

APPENDIX B: WAVE AND RAY ABERRATIONS: ALTERNATIVE DEFINITION AND DERIVATION

The wave aberrations of an optical imaging system for a certain point object P are determined by tracing rays from it through the system and up to the Gaussian reference sphere S, which is a spherical surface passing through the center of O of the exit pupil and centered at the Gaussian image point P'. In Section 3.2, we defined the wave aberration of a ray at the point Q on the reference sphere where the ray intersects it as the difference between its optical path length and that of the chief ray. It represents the optical deviation $n_i\overline{Q}Q$ of the wavefront W, which is a surface obtained by tracing rays of optical path lengths equal to that of the chief ray, from the reference sphere along the ray, $\overline{Q}$ being the point where the ray intersects the wavefront. It is positive when the ray travels an extra optical path length $n_i\overline{Q}Q$ in reaching the point Q on the reference sphere, as in Figure 3-18, compared to the optical path length $[P\cdot\!\cdot O]$ of the chief ray. The ray intersects the Gaussian image plane at a point P'' whose coordinates (x_i, y_i) are given by Eqs. (3-10), where (x, y) are the coordinates of the point Q on the reference sphere at which the ray intersects it, $W(x, y)$ is the wave aberration at the point Q, R' is the distance between Q and P'', and n_i is the refractive index of the image space. In Figure 3-2b, $(x_g, 0)$ are the coordinates of the Gaussian image point P'. The displacement $(x_i - x_g, y_i)$ of the point P'' from P' is the ray aberration. We now give an alternative definition of the wave aberration and derive its relationships to the ray aberration.

An alternative[5] but equivalent definition of the wave aberration at a point $\overline{Q}$ on the wavefront, rather than at Q on the reference sphere, is the optical deviation $n_i\overline{Q}A$ (see Figure 3-18) of the wavefront from the reference sphere along the radial line $\overline{Q}AP'$. Here, A is the point of intersection of the line $\overline{Q}P'$ with the reference sphere. Thus, the wave aberration at $\overline{Q}$ with coordinates $(\overline{x}, \overline{y}, \overline{z})$ may be written

$$W(\overline{x}, \overline{y}) = [\overline{Q}P'] - [AP']$$
$$= n_i(\overline{Q}P' - R) \quad ,$$

(3B-1)

where $R = AP'$ is the radius of curvature of the reference sphere. As in Figure 3-2, we assume a coordinate system with its origin at the center O of the exit pupil, z axis along the optical axis, and x axis lying in the tangential plane (which contains the Gaussian image point P'). The Gaussian image plane lies at a distance z_g from the pupil plane. Thus, the coordinates of P' are $(x_g, 0, z_g)$. The wavefront may be specified by a surface equation

$$F(\overline{x}, \overline{y}, \overline{z}) = 0$$

(3B-2a)

or

$$\overline{z} = f(\overline{x}, \overline{y}) \quad .$$

(3B-2b)

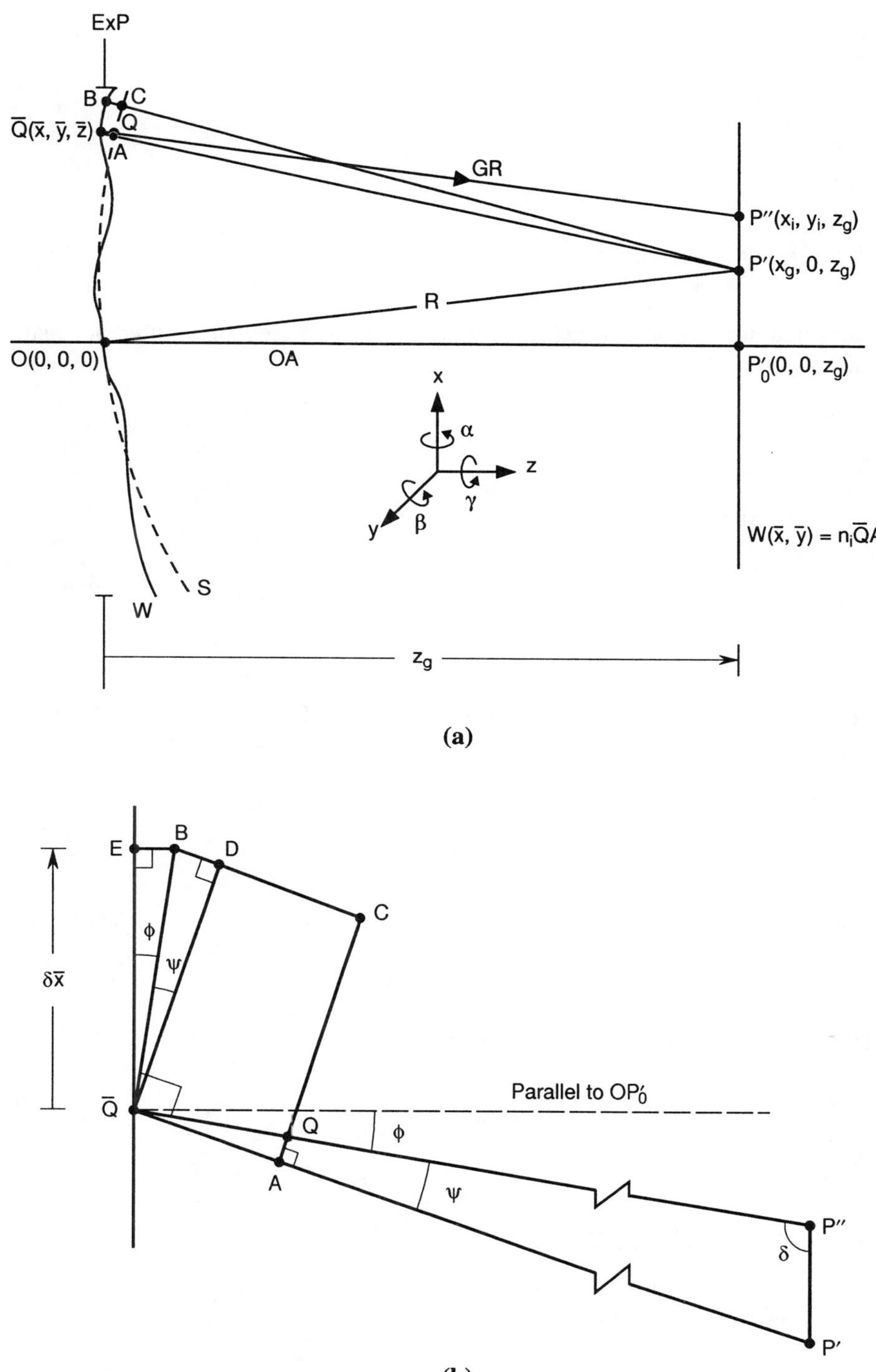

Figure 3-18. Wave and ray aberrations. (a) The wave aberration at a point $\overline{Q}$ on the wavefront is its optical deviation $n_i\overline{Q}A$ from the reference sphere along a radius $\overline{Q}P'$ of the sphere passing through the point. (b) Enlargement of the region in the vicinity of the point $\overline{Q}$.

If we let

$$s = \overline{Q}P' \quad , \tag{3B-3}$$

we may write

$$\left(x_g - \bar{x}\right)^2 + \bar{y}^2 + \left(z_g - \bar{z}\right)^2 - s^2 = 0 \quad . \tag{3B-4a}$$

If we substitute for $\bar{z}$ from Eq. (3B-2b) into Eq. (3B-4a), we obtain a different equation of the wavefront:

$$s = g(\bar{x}, \bar{y}) \quad . \tag{3B-2c}$$

It may also be written in an alternative form

$$G(\bar{x}, \bar{y}, \bar{z}; s) \equiv \left(x_g - \bar{x}\right)^2 + \bar{y}^2 + \left(z_g - \bar{z}\right)^2 - s^2 \tag{3B-4b}$$

$$= 0 \quad . \tag{3B-4c}$$

The total differentials of the explicit functions f and g of the wavefront are

$$d\bar{z} = \frac{\partial \bar{z}}{\partial \bar{x}} d\bar{x} + \frac{\partial \bar{z}}{\partial \bar{y}} d\bar{y} \tag{3B-5}$$

and

$$ds = \frac{\partial s}{\partial \bar{x}} d\bar{x} + \frac{\partial s}{\partial \bar{y}} d\bar{y} \quad , \tag{3B-6}$$

respectively. Differentiating the implicit functions F and G, we obtain

$$F_{\bar{x}} d\bar{x} + F_{\bar{y}} d\bar{y} + F_{\bar{z}} d\bar{z} = 0 \tag{3B-7}$$

and

$$G_{\bar{x}} d\bar{x} + G_{\bar{y}} d\bar{y} + G_{\bar{z}} d\bar{z} + G_s ds = 0 \quad , \tag{3B-8}$$

respectively, where, for example

$$F_{\bar{x}} = \frac{\partial F}{\partial \bar{x}} \quad . \tag{3B-9}$$

Substituting for $d\bar{z}$ from Eq. (3B-5) into Eq. (3B-7), we obtain

$$\left(F_{\bar{x}} + F_{\bar{y}} \frac{\partial \bar{z}}{\partial \bar{x}}\right) d\bar{x} + \left(F_{\bar{y}} + F_{\bar{z}} \frac{\partial \bar{z}}{\partial \bar{y}}\right) d\bar{y} = 0 \quad . \tag{3B-10}$$

Similarly, substituting for $d\bar{z}$ and ds from Eqs. (3B-5) and (3B-6), respectively, into Eq. (3B-8), we obtain

$$\left(G_{\bar{x}} + G_{\bar{z}}\frac{\partial\bar{z}}{\partial\bar{x}} + G_s\frac{\partial s}{\partial\bar{x}}\right)d\bar{x} + \left(G_{\bar{y}} + G_{\bar{z}}\frac{\partial\bar{z}}{\partial\bar{y}} + G_s\frac{\partial s}{\partial\bar{y}}\right)d\bar{y} = 0 \quad . \tag{3B-11}$$

Since $d\bar{x}$ and $d\bar{y}$ are independent infinitesimal variables, their coefficients in Eqs. (3B-10) and (3B-11) must be individually equal to zero. Thus,

$$F_{\bar{x}} + F_{\bar{z}}\frac{\partial\bar{z}}{\partial\bar{x}} = 0 \quad , \tag{3B-12a}$$

$$F_{\bar{y}} + F_{\bar{z}}\frac{\partial\bar{z}}{\partial\bar{y}} = 0 \quad , \tag{3B-12b}$$

$$G_{\bar{x}} + G_{\bar{z}}\frac{\partial\bar{z}}{\partial\bar{x}} + G_s\frac{\partial s}{\partial\bar{x}} = 0 \quad , \tag{3B-13a}$$

and

$$G_{\bar{y}} + G_{\bar{z}}\frac{\partial\bar{z}}{\partial\bar{y}} + G_s\frac{\partial s}{\partial\bar{y}} = 0 \quad . \tag{3B-13b}$$

Substituting for $\partial\bar{z}/\partial\bar{x}$ from Eq. (3B-12a) into (3B-13a), we obtain

$$\frac{\partial s}{\partial\bar{x}} = \frac{1}{G_s}\left(G_{\bar{z}}\frac{F_{\bar{x}}}{F_{\bar{z}}} - G_{\bar{x}}\right) \quad . \tag{3B-14a}$$

Similarly, substituting for $\partial\bar{z}/\partial\bar{y}$ from Eq. (3B-12b) into Eq. (3B-13b), we obtain

$$\frac{\partial s}{\partial\bar{y}} = \frac{1}{G_s}\left(G_{\bar{z}}\frac{F_{\bar{y}}}{F_{\bar{z}}} - G_{\bar{y}}\right) \quad . \tag{3B-14b}$$

From Eq. (3B-4b), the partial derivatives of G are given by

$$G_{\bar{x}} = -2\left(x_g - \bar{x}\right),\ G_{\bar{y}} = 2\bar{y},\ G_{\bar{z}} = -2\left(z_g - \bar{z}\right),\ G_s = -2s \quad . \tag{3B-15}$$

Since a ray is perpendicular to the wavefront, the partial derivatives of F are proportional to its direction cosines according to

$$\left(F_{\bar{x}},\ F_{\bar{y}},\ F_{\bar{z}}\right) = \left(\cos\alpha,\ \cos\beta,\ \cos\gamma\right)\Big/\left(F_{\bar{x}}^2 + F_{\bar{y}}^2 + F_{\bar{z}}^2\right)^{1/2} \quad , \tag{3B-16}$$

where $(\alpha,\ \beta,\ \gamma)$ are the angles the ray $\overline{Q}P''$ makes with the x, y, z axes. The direction cosines are given by

$$(\cos\alpha,\ \cos\beta,\ \cos\gamma) = \left(x_i - \bar{x},\ y_i - \bar{y},\ z_g - \bar{z}\right)\Big/\overline{Q}P'' \quad . \tag{3B-17}$$

From Eqs. (3B-16) and (3B-17), we find that

$$\frac{F_{\bar{x}}}{F_{\bar{z}}} = \frac{x_i - \bar{x}}{z_g - \bar{z}}$$

(3B-18a)

and

$$\frac{F_{\bar{y}}}{F_{\bar{z}}} = \frac{y_i - \bar{y}}{z_g - \bar{z}} \quad .$$

(3B-18b)

Substituting Eqs. (3B-15) and (3B-18) into Eqs. (3B-14), we obtain

$$\frac{\partial s}{\partial \bar{x}} = \frac{x_i - x_g}{s}$$

(3B-19a)

and

$$\frac{\partial s}{\partial y} = \frac{y_i}{s} \quad .$$

(3B-19b)

From Eqs. (3B-1) and (3B-3), we find that

$$s = R + \frac{1}{n_i} W(\bar{x}, \bar{y}) \quad .$$

(3B-20)

Substituting Eq. (3B-20) into Eqs. (3B-19), we finally obtain

$$\left(x_i - x_g, y_i \right) = \frac{1}{n_i} \left[R + \frac{1}{n_i} W(\bar{x}, \bar{y}) \right] \left(\frac{\partial W}{\partial \bar{x}}, \frac{\partial W}{\partial \bar{y}} \right) \quad .$$

(3B-21)

In practice, since W is no more than a few wavelengths, it is much smaller than R. Hence, Eq. (3B-21) may be written

$$\boxed{\left(x_i - x_g, y_i \right) = \frac{R}{n_i} \left(\frac{\partial W}{\partial \bar{x}}, \frac{\partial W}{\partial \bar{y}} \right) \quad .}$$

(3B-22)

Except for the definition of $W(\bar{x}, \bar{y})$, Eq. (3B-22) is similar to Eq. (3-11).

To gain some physical insight into the above derivation, we consider a point B on the wavefront in the vicinity of the point $\bar{Q}$. The aberration at this point is $n_i BC$, where C is the point of intersection of the line BP' with the reference sphere. The segments AC of the sphere and $\bar{Q}B$ of the wavefront are so small that they may be assumed to be straight lines as illustrated in Figure 3-18b. Let $\bar{Q}D$ be parallel to AC. Then the difference δW in the wave aberration at the points and $\bar{Q}$ and B may be written

$$\delta W = n_i \left(BC - \bar{Q}A \right)$$
$$= n_i BD \quad .$$

(3B-23)

The ray $\overline{Q}P''$ passing through the point $\overline{Q}$ is perpendicular to the wavefront at the point $\overline{Q}$. It intersects the chief ray OP' at M' (which is not shown in Figure 3-18) and the Gaussian image plane at the point P''. The distance $P'P''$ along the x axis is the ray aberration of the ray corresponding to the wave aberration $\overline{Q}A$. The angle $P'\overline{Q}P''$, or ψ, is called the *angular aberration*. Since $\overline{Q}D$ is parallel to AC, which is perpendicular to $\overline{Q}P'$, and $\overline{Q}B$ is perpendicular to $\overline{Q}P''$, the angle $D\overline{Q}B$ is also equal to ψ. Therefore, from the right-angle triangle $\overline{Q}DB$,

$$BD = \overline{Q}B \sin\psi \quad . \tag{3B-24}$$

From the triangle $\overline{Q}P'P''$

$$\frac{\sin\psi}{P'P''} = \frac{\sin\delta}{\overline{Q}P'} \quad ,$$

or

$$\sin\psi = \frac{x_i - x_g}{\overline{Q}P'}\cos\phi \quad , \tag{3B-25}$$

where we have used the fact that $\delta = \phi + \pi/2$. If $\bar{x}$ and $\bar{x} + \delta\bar{x}$ are the x-coordinates of the points $\overline{Q}$ and B, we note from the right-angle triangle $\overline{Q}EB$ that

$$\cos\phi = \frac{\delta\bar{x}}{\overline{Q}B} \quad . \tag{3B-26}$$

Substituting Eq. (3B-26) into Eq. (3B-25), and substituting the result obtained into Eq. (3B-24), we obtain

$$BD = \frac{x_i - x_g}{\overline{Q}P'}\delta\bar{x} \quad . \tag{3B-27}$$

Substituting Eq. (3B-27) into Eq. (3B-23), we finally obtain

$$\delta W = \frac{n_i}{\overline{Q}P'}\left(x_i - x_g\right)\delta\bar{x} \quad ,$$

or

$$\boxed{x_i - x_g = \frac{\overline{Q}P'}{n_i}\frac{\partial W}{\partial\bar{x}}} \quad . \tag{3B-28}$$

Similarly, we can show that the y_i coordinate of P'' with respect to that of P' (which is assumed to be zero) is given by

$$\boxed{y_i = \frac{\overline{Q}P'}{n_i}\frac{\partial W}{\partial\bar{y}}} \quad . \tag{3B-29}$$

Equations (3B-28) and (3B-29) are the same as Eq. (3B-21), where $s = \overline{Q}P'$.

REFERENCES

1. For a detailed discussion of different methods of aberration measurement, see D. Malacara, ed., *Optical Shop Testing*, Wiley, New York (1977).

2. V. N. Mahajan, "Zernike annualr polynomials for imaging systems with annular pupils," *J. Opt. Soc. Am.* **71**, 75-78, 1408 (1981), and **A1**, 685 (1984); also "Zernike annular polynomials and optical aberrations of systems with annular pupils," *Appl. Opt.* **33**, 8125-8127 (1994).

3. W. Brouwer and A. Walther, "Geometric Optics," Chapter 16, Sections 1.6 and 1.7 in *Advanced Optical Techniques*, ed. A. C. S. Van Heel, North Holland (1967). Also, A. Walther, *The Ray and Wave Theory of Lenses,* Chapter 8, Cambridge University Press, New York (1995).

4. E. Wolf, "On a new aberration function of optical instruments," *J. Opt. Soc. Am.* **42**, 547–552 (1952). Also, M. Born and E. Wolf, *Principles of Optics*, Section 5.1, Pergamon, New York (1985).

5. J. L. Rayces, "Exact relations between wave aberration and ray aberration," *Optica Acta* **11**, 85–88 (1964).

PROBLEMS

3.1 Show that the *defocus wave aberration* introduced by a lens of image-space focal length f' is given by $W(r) = -\left(n_i / 2f'\right)r^2$, where n_i is the refractive index of the image space.

3.2 The *field curvature* aberration of an imaging system may be written $W(r) = a_d h'^2 r^2$, where a_d is the aberration coefficient and h' is the height of a Gaussian image point. Show that the effect of the aberration is eliminated if the image is observed on a spherical surface of radius of curvature $1/4\, a_d R^2$ passing through a corresponding axial Gaussian image point, where R is the radius of curvature of the reference sphere with respect to which the aberration is defined. The refractive index of the image space is assumed to be unity.

3.3 If the Gaussian image of an object is formed at infinity by an imaging system at an angle β from its optical axis, but the system suffers from *field curvature* according to $W(r) = b_d \beta^2 r^2$, what is the distance at which the image rays come to focus?

3.4 Consider an imaging system suffering from *distortion aberration* given by $W(r,\theta) = a_t h'^3 r \cos\theta$, where a_t is the aberration coefficient and h' is the height of a Gaussian image point. Determine the height of the actual image point.

3.5 Consider a primary aberration function given by Eq. (3-38). Write it in terms of Zernike circle polynomials and determine its standard deviation.

CHAPTER 4

GEOMETRICAL POINT-SPREAD FUNCTION

4.1   **Introduction** **205**

4.2   **Theory** **205**

4.3   **Application to Primary Aberrations** **209**

     4.3.1   Spherical Aberration 210

     4.3.2   Coma 217

     4.3.3   Astigmatism and Field Curvature 224

     4.3.4   Distortion 233

4.4   **Balanced Aberrations for Minimum RMS Spot Radius** **235**

4.5   **Spot Diagrams** **236**

4.6   **Summary of Results** **239**

     4.6.1   Spherical Aberration 240

     4.6.2   Coma 240

     4.6.3   Astigmatism and Field Curvature 241

     4.6.4   Distortion 242

     4.6.5   Aberration Tolerance 242

References **243**

Problems **244**

Chapter 4

Geometrical Point-Spread Function

4.1 INTRODUCTION

In this chapter, we discuss the distribution of rays in the image of a point object formed by an optical system for a given primary aberration. Such a distribution is referred to as the *spot diagram* and its extent is called the *spot size*. The distribution of the density of rays is called the *geometrical point-spread function*. We define its *centroid* and *root mean square radius* and calculate them for primary aberrations. In the case of spherical aberration and astigmatism, ray distribution and spot size are considered in image planes other than the Gaussian as well, thereby introducing the concept of *aberration balancing*. In the early stages of the design of an optical imaging system, one often considers its transverse ray aberrations in an image plane for a set of rays lying along a certain line in the plane of the exit pupil and passing through its center. Such a set of rays is called a *ray fan* and, often, rays along the x and y axes are used for investigating the ray aberrations and thereby the quality of the system, where the point object is assumed to be along the x axis of the object plane. The set of rays along the x axis of the exit pupil plane is called the *tangential ray fan*, and the one along its y axis is called the *sagittal ray fan*. The wave and ray aberrations for the two types of ray fans are discussed for each primary aberration. Also discussed are the balanced aberrations for the *minimum root-mean-square radius* in terms of Zernike circle polynomials. The characteristics of the ray spots and *tolerance* for primary aberrations are summarized in the last section of this chapter.

4.2 THEORY

Consider an optical system consisting of a series of rotationally symmetric coaxial refracting and/or reflecting surfaces imaging a point object P lying at a height h from the optical axis. As in earlier chapters (see Figure 3-3), we assume that the point object lies along the x axis, and the z axis is along the optical axis of the system. In Chapter 3, we showed [see Eq. (3-38)] that the *primary aberration function* at its exit pupil may be written

$$W(r,\theta;h') = a_s r^4 + a_c h' r^3 \cos\theta + a_a h'^2 r^2 \cos^2\theta + a_d h'^2 r^2 + a_t h'^3 r \cos\theta \quad , \tag{4-1}$$

where (r, θ) are the polar coordinates of a point in the plane xy of the exit pupil of the system, h' is the height of the Gaussian image point P', and a_s, a_c, a_a, a_d, and a_t represent the coefficients of *spherical aberration, coma, astigmatism, field curvature,* and *distortion*, respectively. The angle θ is equal to zero or π for points lying in the *tangential* or *meridional plane* (i.e., the zx plane containing the optical axis and the point object and, therefore, its Gaussian image). The *chief ray*, which by definition passes through the center of the exit pupil, always lies in this plane. The plane normal to the tangential plane but containing the chief ray is called the *sagittal plane*. The angle θ is equal to $\pi/2$ or $3\pi/2$ for points lying in the sagittal plane. As the chief ray bends when it is refracted or reflected by a surface, so does the sagittal plane. The rays lying in the tangential plane are referred to as the *tangential ray fan* and those lying in the sagittal plane are referred to as the *sagittal ray fan*.

For an optical system with a *circular exit pupil*, say of radius a, it is convenient to use normalized coordinates (ρ, θ) where $\rho = r/a$, $0 \le \rho \le 1$, $0 \le \theta < 2\pi$, suppress the explicit dependence on h', and write the aberration function in the form

$$\boxed{W(\rho, \theta) = A_s \rho^4 + A_c \rho^3 \cos\theta + A_a \rho^2 \cos^2\theta + A_d \rho^2 + A_t \rho \cos\theta \quad,} \tag{4-2}$$

where the new aberration coefficients A_i are related to the a_i used in Eq. (4-1) according to

$$\boxed{A_s = a_s a^4, \quad A_c = a_c h' a^3, \quad A_a = a_a h'^2 a^2, \quad A_d = a_d h'^2 a^2, \quad A_t = a_t h'^3 a \quad.} \tag{4-3}$$

If (x, y) represent the rectangular coordinates of a pupil point, the corresponding normalized coordinates (ξ, η) are given by

$$(\xi, \eta) = \frac{1}{a}(x, y) \tag{4-4a}$$

$$= \rho(\cos\theta, \sin\theta) \quad, \tag{4-4b}$$

where $-1 \le \xi \le 1$, $-1 \le \eta \le 1$, and $\xi^2 + \eta^2 = \rho^2 \le 1$. The aberration function defined in the form of Eq. (4-2) has the advantage that an aberration coefficient A_i has the dimensions of length (i.e., dimensions of the wave aberration), and represents the peak or the maximum value of the corresponding primary aberration. For example, if $A_s = 1\lambda$, where λ is the wavelength of the object radiation, we speak of one wave of spherical aberration.

In rectangular coordinates, Eq. (4-2) for the primary aberration function may be written

$$\boxed{W(\xi, \eta) = A_s\left(\xi^2 + \eta^2\right)^2 + A_c \xi\left(\xi^2 + \eta^2\right) + A_a \xi^2 + A_d\left(\xi^2 + \eta^2\right) + A_t \xi \quad.} \tag{4-5}$$

An aberration term is *even* in pupil coordinates if $W(-\xi, -\eta) = W(\xi, \eta)$; it is *odd* if $W(-\xi, -\eta) = -W(\xi, \eta)$. Among the five primary aberrations in Eq. (4-5), only coma and distortion are odd aberrations; the other three are even. Of course, spherical aberration and field curvature are radially symmetric.

The distribution of rays in an image plane is called the *ray spot diagram*. The distribution of their density (i.e., the number of rays per unit area) is called the *geometrical point-spread function (PSF)*. If the system is aberration free, then the wavefront at the exit pupil is spherical and all the object rays transmitted by the system converge to the Gaussian image point. When the wavefront is aberrated, a ray passing through a point (ξ, η) or (ρ, θ) in the plane of the exit pupil intersects the Gaussian image plane at a point (x_i, y_i) which, following Eq. (3-11), may be written

$$(x_i, y_i) = 2F\left(\frac{\partial W}{\partial \xi}, \frac{\partial W}{\partial \eta}\right) \tag{4-6a}$$

$$= 2F\left(\cos\theta\,\frac{\partial W}{\partial \rho} - \frac{\sin\theta}{\rho}\frac{\partial W}{\partial \theta}, \ \sin\theta\,\frac{\partial W}{\partial \rho} + \frac{\cos\theta}{\rho}\frac{\partial W}{\partial \theta}\right), \tag{4-6b}$$

where $F = R/2a$ is the focal ratio of the image-forming light cone. Here, R is the radius of curvature of the Gaussian reference sphere with respect to which the aberration $W(\rho, \theta)$ is defined, and (x_i, y_i) are the coordinates of the point of intersection of the ray in the Gaussian image plane with respect to the Gaussian image point and represent its ray aberrations. The reference sphere is centered at the Gaussian image point and, like the aberrated wavefront, passes through the center of the exit pupil. In Eqs. (4-6), we have assumed that the refractive index n_i of the the image space is unity since it is often the case in practice.

For a radially symmetric aberration, i.e., one for which $W(\rho, \theta) = W(\rho)$, we note from Eq. (4-6b) that the PSF is also radially symmetric. The radial distance r_i of a ray from the Gaussian image point in that case is given by

$$r_i = \left(x_i^2 + y_i^2\right)^{1/2}$$

$$= 2F\left|\frac{\partial W(\rho)}{\partial \rho}\right|, \tag{4-7}$$

where the vertical bars ensure that r_i is a numerically positive quantity.

The geometrical PSF of an aberrated system can be obtained by noting that an element of area $dS_p = dx\,dy$ centered at a point (x, y) in the plane of the exit pupil is mapped into an element of area $dS_i = dx_i\,dy_i$ centered at the point (x_i, y_i) in the image plane according to

$$dS_i = J\,dS_p, \tag{4-8a}$$

where

$$J = \frac{\partial(x_i, y_i)}{\partial(x, y)} \tag{4-8b}$$

is the Jacobian of transformation between the corresponding area elements in the two planes. If $I_p(x, y)$ represents the irradiance or the density of rays at the point (x, y) in the pupil plane, then the irradiance or the density of rays $I_g(x_i, y_i)$ at a point (x_i, y_i) in the image plane is given by

$$I_g(x_i, y_i)\,dS_i = I_p(x, y)\,dS_p, \tag{4-9}$$

or

$$I_g(x_i, y_i) = I_p(x, y)\frac{dS_p}{dS_i}$$

$$= I_p(x, y)|J|^{-1}$$

$$= I_p(x, y)\begin{vmatrix} \dfrac{\partial x_i}{\partial x} & \dfrac{\partial x_i}{\partial y} \\[2mm] \dfrac{\partial y_i}{\partial x} & \dfrac{\partial y_i}{\partial y} \end{vmatrix}^{-1} . \qquad (4\text{-}10)$$

Integrating both sides of Eq. (4-9), we obtain the fact that the total power or the number of rays in the image plane is the same as in the pupil plane. Substituting Eq. (4-6a) into Eq. (4-10) we obtain

$$I_g(x_i, y_i) = \frac{I_p(x, y)a^4}{R^2}\begin{vmatrix} \dfrac{\partial^2 W}{\partial \xi^2} & \dfrac{\partial^2 W}{\partial \xi \partial \eta} \\[2mm] \dfrac{\partial^2 W}{\partial \xi \partial \eta} & \dfrac{\partial^2 W}{\partial \eta^2} \end{vmatrix}^{-1} ,$$

or

$$\boxed{I_g(x_i, y_i) = \frac{I_p(x, y)a^4}{R^2}\left| \frac{\partial^2 W}{\partial \xi^2}\frac{\partial^2 W}{\partial \eta^2} - \left(\frac{\partial^2 W}{\partial \xi \partial \eta}\right)^2 \right|^{-1} .} \qquad (4\text{-}11)$$

For a radially symmetric pupil with illumination $I_p(\rho)$ and an aberration function $W(\rho)$, consider the rays lying in an annular region of radius ρ and width $d\rho$ in the pupil plane. The rays lie in a corrresponding annular region of radius r_i and width dr_i in the image plane. Therefore, the area elements in Eq. (4-9) are given by

$$dS_p = 2\pi a^2 \rho\, d\rho$$

and

$$dS_i = 2\pi r_i\, dr_i .$$

Hence, Eq. (4-9) may be written

$$I_g(r_i) = I_p(\rho)\frac{\rho a^2}{r_i}\left|\frac{\partial \rho}{\partial r_i}\right| . \qquad (4\text{-}12a)$$

Or, using Eq. (4-7), we may write

$$\boxed{I_g(r_i) = I_p(\rho)(a^2/R)^2 \rho \left|\frac{\partial W}{\partial \rho}\frac{\partial^2 W}{\partial \rho^2}\right|^{-1} .} \qquad (4\text{-}12b)$$

The right-hand side of Eq. (4-11), which is in terms of ξ and η, can be written in terms of x_i and y_i for a particular aberation by use of Eq. (4-6a). Similarly, the right-hand side of Eq. (4-12b) can be written in terms of r_i by use of Eq. (4-7).

By definition, the *centroid* (or the center of gravity) of a PSF is given by

$$
\begin{aligned}
\left(x_c, y_c\right) &= \left\langle x_i, y_i\right\rangle \\
&= \frac{\iint \left(x_i, y_i\right) I_g\left(x_i, y_i\right) dx_i\, dy_i}{\iint I_g\left(x_i, y_i\right) dx_i\, dy_i} \quad,
\end{aligned}
\tag{4-13}
$$

where the angular brackets indicate a mean value. However, it can be obtained in a simple manner by substituting Eqs. (4-6a) and (4-9) into Eq. (4-13). Thus, for a *uniformly illuminated pupil*, i.e., for constant $I_p(x, y)$, say I_p, we may write

$$
\begin{aligned}
\left(x_c, y_c\right) &= \frac{2F \iint \left(\dfrac{\partial W}{\partial \xi}, \dfrac{\partial W}{\partial \eta}\right) d\xi\, d\eta}{\iint d\xi\, d\eta} \\[2mm]
&= (2F/\pi) \iint \left(\frac{\partial W}{\partial \xi}, \frac{\partial W}{\partial \eta}\right) d\xi\, d\eta \quad.
\end{aligned}
\tag{4-14}
$$

The *root-mean-square* (rms) *radius* (or the radius of gyration) of the image distribution for a uniformly illuminated pupil is given by

$$
r_{irms} = \left\langle x_i^2 + y_i^2\right\rangle^{1/2}
\tag{4-15a}
$$

$$
= 2F \left\{ \frac{1}{\pi} \iint \left[\left(\frac{\partial W}{\partial \xi}\right)^2 + \left(\frac{\partial W}{\partial \eta}\right)^2 \right] d\xi\, d\eta \right\}^{1/2}
\tag{4-15b}
$$

$$
= \left[\frac{1}{\pi} \int_0^1 \int_0^{2\pi} r_i^2\, \rho\, d\rho\, d\theta \right]^{1/2} \quad.
\tag{4-15c}
$$

Next, we discuss the characteristics of an image aberrated by a primary aberration. To be definite, we assume that each of the aberration coefficients A_i is positive, unless stated otherwise. If two or more of these aberrations are present simultaneously, the image coordinates $\left(x_i, y_i\right)$ of a ray are given by the sum of the coordinates for each aberration.

4.3 APPLICATION TO PRIMARY ABERRATIONS

In this section we discuss the geometrical PSFs, including their shapes and sizes, for primary aberrations and uniform pupil illumination I_p using the equations given in the above section. The concept of aberration balancing is introduced whereby a given aberration is mixed with another to reduce the spot size. The wave and ray aberrations for tangential and sagittal ray fans are also considered.

4.3.1 Spherical Aberration

Consider a wavefront aberrated by a spherical aberration

$$W(\rho) = A_s \rho^4$$ (4-16)

with respect to a reference sphere centered at the Gaussian image point P_0' of an axial point object P_0. Substituting Eq. (4-16) into Eq. (4-7), we find that a ray of zone ρ in the plane of the exit pupil intersects the Gaussian image plane at a distance

$$r_i = 8FA_s \rho^3$$ (4-17)

from P_0'. Thus, the rays lying on a circle of radius ρ in the exit pupil lie on a circle of radius r_i given by Eq. (4-17) in the Gaussian image plane. The maximum value of r_i is $8FA_s$ and corresponds to rays with $\rho = 1$; i.e., it corresponds to the marginal rays. We will refer to the maximum value of r_i as the *radius of the image spot*. For an off-axis point object, since A_s is independent of the height h of the point object from the optical axis, the ray distribution owing to spherical aberration alone is also independent of h.

Let us consider the ray distribution in a slightly defocused image plane by introducing a defocus aberration B_d. The aberration with respect to a new reference sphere centered at a defocused point lying at a distance z from the plane of the exit pupil may be written

$$W(\rho) = A_s \rho^4 + B_d \rho^2 \quad ,$$ (4-18)

where for $\Delta R = z - R$ and $z \simeq R$, the defocus coefficient B_d, following Eq. (3-18), is given by

$$B_d = \frac{1}{2}\left(\frac{1}{z} - \frac{1}{R}\right)a^2$$ (4-19a)

$$\simeq -\frac{\Delta R}{8F^2} \quad .$$ (4-19b)

Note that B_d is numerically negative for $z > R$, i.e., if the defocused image plane lies farther from the exit pupil than the Gaussian image plane, or the longitudinal defocus ΔR is positive. Figure 4-1 shows how the wave aberration given by Eq. (4-18) varies across the exit pupil for values of B_d corresponding to paraxial $(B_d = 0)$, marginal $(B_d = -2A_s)$, midway $(B_d = -A_s)$, and least-confusion $(B_d = -1.5A_s)$ image planes. The names of the image planes given here will become clear from what follows. We note that for a negative value of B_d, the aberration is negative everywhere except at the center and the edge of the pupil where it is zero.

The rays of zone ρ now lie in the defocused image plane on a circle of radius

$$r_i = 8FA_s \left| \rho^3 + (B_d/2A_s)\rho \right| \quad .$$ (4-20)

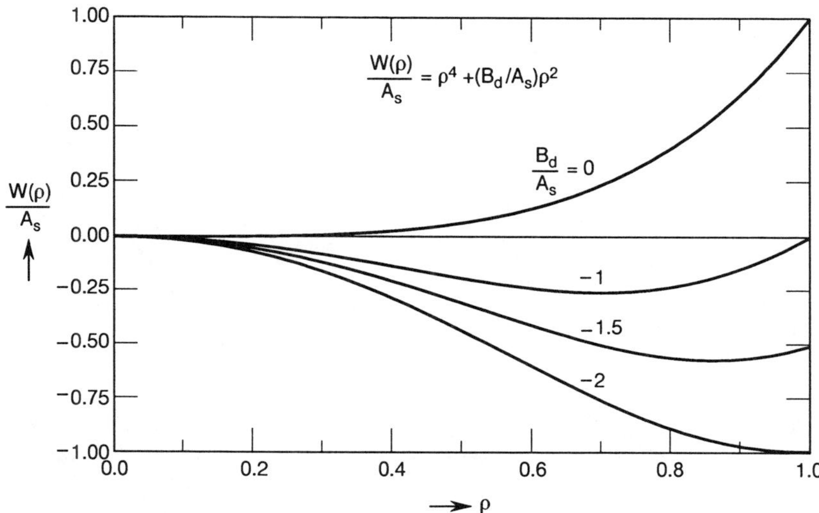

Figure 4-1. Variation of spherical aberration across the exit pupil in units of A_s combined with different amounts of defocus B_d.

The circle in the image plane is traced out in the same sense as in the pupil plane as θ varies from 0 to 2π to complete a circle of rays. In a given image plane, i.e., for a given value of B_d, the maximum value of r_i as ρ varies from 0 to 1 is the spot radius in that plane. It occurs either at the stationary value of ρ obtained by letting $\partial r_i/\partial \rho = 0$ or at the end value $\rho = 1$. We note that $\rho = 0$ at the other end point $r_i = 0$, implying that the chief ray passes through the center of the image. When B_d is negative, $r_i = 0$ also for rays with $\rho = \sqrt{-B_d/2A_s}$.

How r_i varies with ρ is shown in Figure 4-2 for the values of B_d considered above. We note that only when $B_d = 0$, a given value of r_i corresponds to a certain value of ρ. When $B_d = -2A_s$, there are two different values of ρ lying between zero and one that correspond to a given value of r_i; i.e., rays lying on two different circles in the pupil plane lie on the same circle in the image plane. When $B_d = -A_s$, or $B_d = -3A_s/2$, there are three different values of ρ lying between zero and one that correspond to a given value of r_i for $0 < r_i < 1/3\sqrt{6}$ or $0 < r_i < 1/4$, respectively; i.e., rays lying on three different circles in the pupil plane lie on the same circle in the image plane. A circle of rays with a larger value of r_i up to $r_i = 1/2$ corresponds to only one circle of rays in the pupil plane when $B_d = -A_s$. There are two circles of rays in the pupil plane with $\rho = 1/2$ and 1 that correspond to $r_i = 1/4$ when $B_d = -3A_s/2$.

For the marginal rays, i.e., for $\rho = 1$, $r_i \to 0$ if $B_d = -2A_s$. From Eq. (4-19), we find that the marginal rays intersect the axis at a distance

$$\Delta R = -8F^2 B_d \qquad\qquad (4\text{-}21a)$$

$$= 16F^2 A_s \qquad\qquad (4\text{-}21b)$$

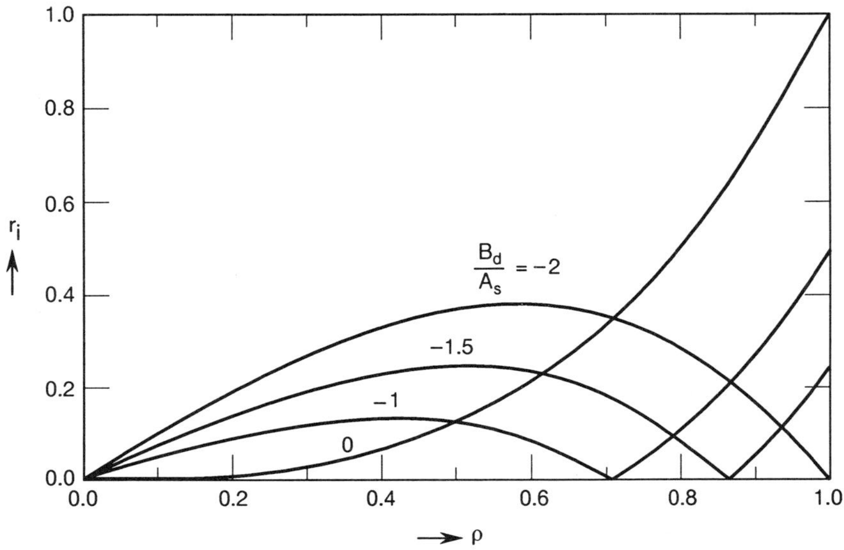

Figure 4-2. Radius r_i of a circle of rays in units of $8FA_s$ in various image planes characterized by the value of B_d as a function of corresponding radius ρ in the pupil plane.

from P_0'. A positive value of ΔR implies that, compared with the old reference sphere, the new reference sphere is centered at a point that is farther from the center of the exit pupil, or that the defocused image plane lies farther from the exit pupil than the Gaussian image plane. Hence, the point of intersection M of the marginal rays lies to the right of P_0', as shown in Figure 4-3. This is to be expected since, as may be seen from Figure 4-3, the wavefront W is less curved than the reference sphere S for positive values of A_s. The points P_0' and M are called the *Gaussian* or *paraxial* (meaning for very small values of ρ) and the *marginal image points*, respectively. Substituting $B_d = -2A_s$ into Eq. (4-20), we find that the maximum value of r_i in the marginal image plane occurs for rays of zone $\rho = 1/\sqrt{3}$. This maximum value, i.e., the spot radius, is $2/3\sqrt{3}$ (or 0.385) times the corresponding value in the Gaussian image plane. Thus, the marginal spot radius is considerably smaller than the paraxial spot radius. The quantity ΔR given by Eq. (4-21b) is called the *longitudinal spherical aberration*. It represents the distance of the marginal image point from the Gaussian image point. If we consider the variation of longitudinal spherical aberration with ρ, i.e., if we determine the distance of the point where the rays of a zone ρ intersect the optical axis from P_0', we find from Eqs. (4-19) and (4-20) that it varies quadratically with ρ according to

$$\boxed{\Delta R = 16F^2A_s\rho^2} \tag{4-21c}$$

The image plane MW lying midway between the Gaussian and marginal planes corresponds to $B_d = -A_s$. The spot radius in this plane is half of that in the Gaussian image plane G and corresponds to marginal rays. The image plane that has the smallest

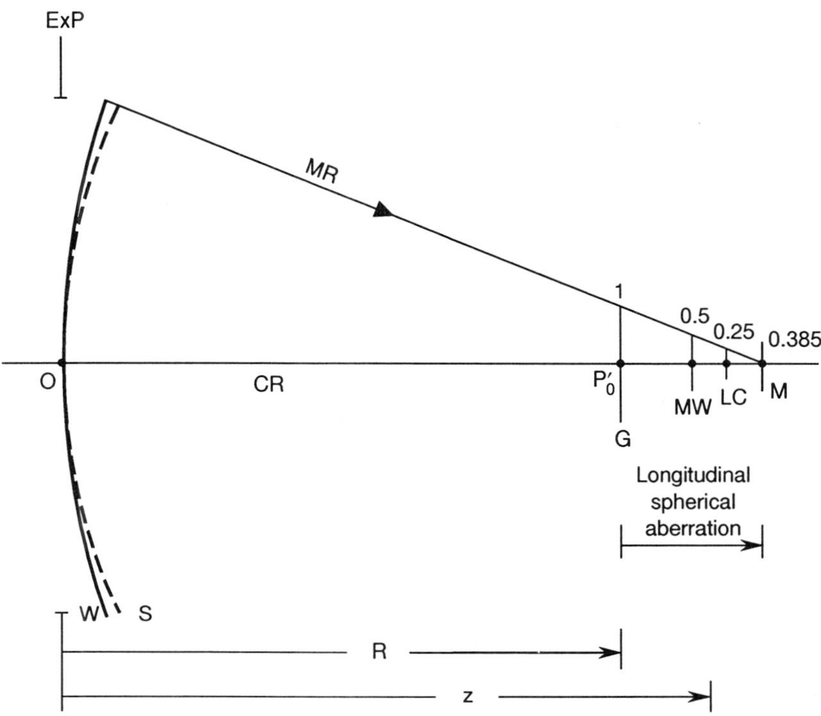

Figure 4-3. Ray spot radii in various image planes for a wavefront W aberrated by spherical aberration. G – Gaussian or paraxial, M – marginal, MW – midway, LC – least confusion. The reference sphere S is centered at a Gaussian image point P_0'.

spot radius corresponds to that value of B_d which minimizes the maximum value of r_i as ρ varies from 0 to 1 in Eq. (4-20). This optimization problem is similar to the one of determining the shape of a Schmidt plate that introduces minimum spherochromatism discussed in Section 6.6.2. It is evident from Eq. (4-20) that B_d must be negative; a positive value of B_d can only increase the value of r_i for any value of ρ. The value of ρ corresponding to the spot radius is either $\rho_1 = \sqrt{c/6}$ obtained by letting $\partial r_i/\partial \rho = 0$, where $c = -B_d / A_s$, or $\rho_2 = 1$. In units of $8FA_s$, the corresponding values of the spot radius are $r_1 = c^{3/2}/3\sqrt{6}$ and $r_2 = |1 - c/2|$, respectively. Figure 4-4 shows that r_1 increases monotonically as c increases, but r_2 first decreases, approaches zero as $c \rightarrow 2$, and then increases monotonically. The value of c that gives the minimum spot radius is the one obtained by letting $r_1 = r_2$. This equality yields a cubic equation in c with solutions $c = 6, 6$, and $3/2$. The value $3/2$ yields the minimum spot radius. Hence, the spot radius is minimum in a plane LC (for least confusion) corresponding to $B_d = -3A_s/2$, i.e., a plane that is 3/4 of the way from the Gaussian image plane to the marginal image plane. The spot radius in this case is 1/4 of the Gaussian spot radius and corresponds to the rays of zone $\rho = 1/2$ and 1. This spot is called the *circle of least confusion*. The spot radii in the various image planes considered here are listed in Table 4-1. Note that they increase linearly with F and A_s. odd (or from odd to even as in the

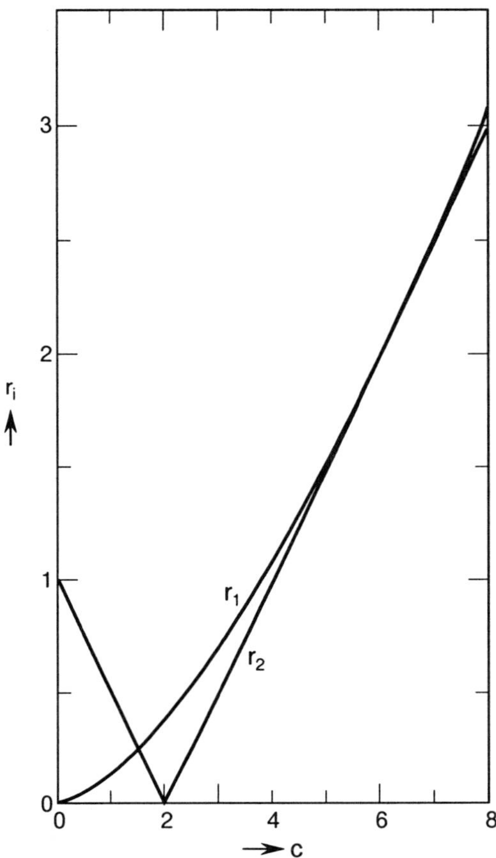

Figure 4-4. Variation of image spot radius with $c = -B_d/A_s$.

Table 4-1. Ray spot radius in units of $8\,FA_s$, **for peak spherical aberration** A_s.

Image Plane	Balancing Defocus B_d/A_s	Spot Radius r_{imax}	RMS Radius r_{irms}
Gaussian	0	1	0.5
Marginal	-2	0.385	0.289
Midway	-1	0.5	0.204
Minimum rms radius	$-4/3$	1/3	0.167
Least confusion	$-3/2$	0.25	0.177

case of coma discussed below) is simply a consequence of the relation expressed by Eq. (4-6) between the wave and ray aberrations.

Because of the radial symmetry of spherical aberration, the wave and ray aberrations of any ray fan can be written immediately from Eqs. (4-18) and (4-20), respectively. For example, for the tangential ray fan, i.e., for the $\eta = 0$ rays, we may write

$$W(\xi,0) = A_s\left[\xi^4 + (B_d/A_s)\xi^2\right] \tag{4-22a}$$

and

$$(x_i, y_i) = 8FA_s\left[\xi^3 + (B_d/2A_s)\xi, 0\right] . \tag{4-22b}$$

Figure 4-5 shows how the wave and ray aberrations vary with ξ for defocus values listed in Table 4-1. (Not considered in the figure but listed in the table is the defocus value for minimum rms radius discussed in Section 4.4.) Note that the wave aberration is even in pupil coordinates, but the ray aberration is odd. The change in symmetry from even to odd (or from odd to even as in the case of coma discussed below) is simply a consequence of the relation expressed by Eq. (4-6) between the wave and ray aberrations.

Since the wave aberration given by Eq. (4-18) is radially symmetric, the PSF is also radially symmetric. Substituting Eq. (4-18) into Eq. (4-12b), or Eq. (4-20) into Eq. (4-12a), we obtain

$$I_g(r_i) = I_p\left(a^2/2A_sR\right)^2 \sum \left|12\rho^4 + 8(B_d/A_s)\rho^2 + (B_d/A_s)^2\right|^{-1} , \tag{4-23}$$

where r_i is obtained from ρ by use of Eq. (4-20). The summation sign on the right-hand side of Eq. (4-23) represents the rays with different values of ρ but the same value of r_i. Note that generally rays with different values of ρ but the same value of r_i give different values of $I_g(r_i)$. How $I_g(r_i)$ varies with r_i is shown in Figure 4-6 for the values of B_d considered in Table 4-1. In the Gaussian image plane $B_d = 0$, it is given by

$$I_g(r_i) = I_p\frac{a^2 r_i^{-4/3}}{12(FA_s)^{2/3}} . \tag{4-24}$$

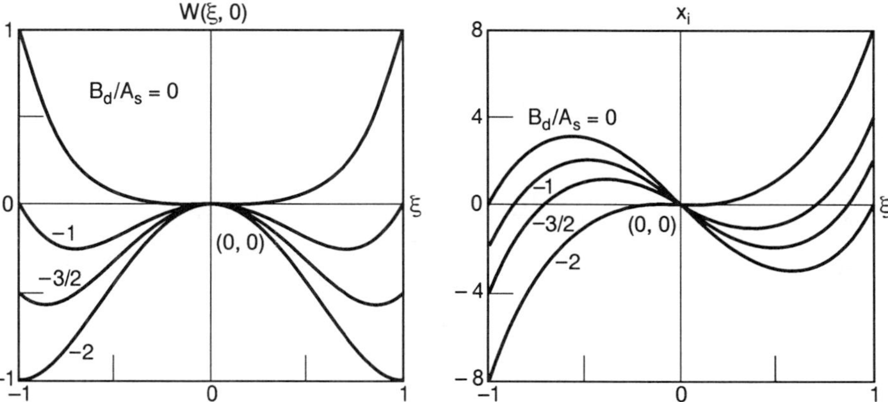

Figure 4-5. Wave and ray aberrations for a ray fan for spherical aberration corresponding to various image planes. The wave aberration is in units of A_s and the ray aberration is in units of FA_s.

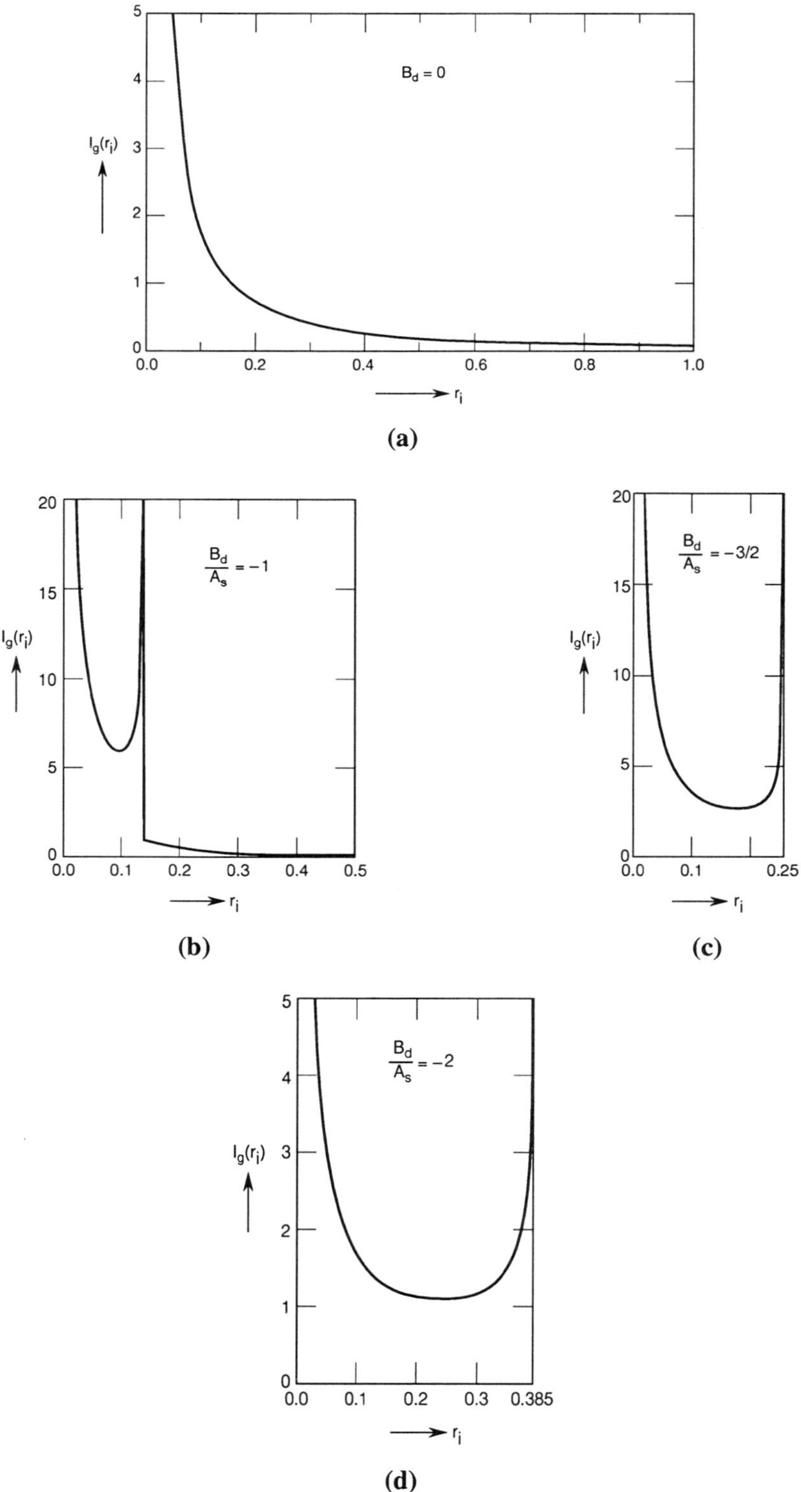

Figure 4-6. Geometrical PSF for spherical aberration A_s in various image planes characterized by the value of B_d. (a) Gaussian, (b) midway, (c) least confusion, and (d) marginal. The irradiance is in units of $I_p\left(a^2/2A_sR\right)^2$ and r_i is in units of $8FA_s$.

We note that I_g approaches infinity for rays with ρ values given by

$$\boxed{12\rho^4 + \left(8\,B_d/A_s\right)\rho^2 + \left(B_d/A_s\right)^2 = 0}\qquad\text{,}$$

(4-25)

or for $\rho^2 = -B_d/2A_s$ and $\rho^2 = -B_d/6A_s$, which in turn correspond to $r_i = 0$ regardless of the value of B_d, and $r_i = 1/3\sqrt{6}$, $1/4$, and $2/3\sqrt{3}$ in units of $8FA_s$ in the midway, least-confusion, and marginal image planes, respectively. These values also correspond to $\partial r_i/\partial\rho = 0$, i.e., r_i is maximum at these values for $0 < \rho < 1$, as may be seen from Figure 4-2. Infinite irradiances, which also occur at the Gaussian image point for the aberration-free systems, correspond to Dirac delta functions so that although the function is infinite at some points, its integral over the image plane is finite and equal to the total power exiting from the exit pupil.

Because of its radial symmetry, the centroid of the PSF lies at the Gaussian image point $(0,0)$. Substituting Eq. (4-20) into Eq. (4-15c), we obtain the *rms radius* of the image spot

$$\boxed{r_{irms} = 8FA_s\left\{\frac{1}{4} + \frac{B_d}{3A_s} + \frac{1}{2}\left(\frac{B_d}{2A_s}\right)^2\right\}^{1/2}}\ .$$

(4-26a)

Letting

$$\frac{\partial r_{irms}}{\partial A_d} = 0\ ,$$

(4-26b)

we find that the rms radius is minimum when $B_d = -(4/3)A_s$. Its value is equal to $4FA_s/3$ compared with its value of $4FA_s$ in the Gaussian image plane. We note that the rms radius is minimum in a plane that is different from the least-confusion plane in which the spot radius is minimum. The values of r_{irms} in various image planes are listed in Table 4-1. The variation of r_{irms} with defocus is shown in Figure 4-7.

The deliberate mixing of one aberration with one or more other aberrations is called *aberration balancing*. Here, we have balanced spherical aberration with defocus in order to minimize the spot radius or its rms value. The amount of defocus that gives the smallest ray spot or its rms value may be called the *optimum defocus* based on geometrical optics. The balanced aberration giving the smallest ray spot is $A_s[\rho^4 - (3/2)\rho^2]$. Similarly, the balanced aberration that gives the smallest rms radius is $A_s[\rho^4 - (4/3)\rho^2]$. Based on *diffraction*, the optimum amount of defocus corresponds to the midway plane, since in that case it is used to reduce the *variance* of the aberration across the exit pupil, i.e., the balanced aberration giving minimum variance is $A_s\left(\rho^4 - \rho^2\right)$, similar to the Zernike polynomial $Z_4^0(\rho)$ (see Table 3-7).

4.3.2 Coma

The coma wave aberration is given by

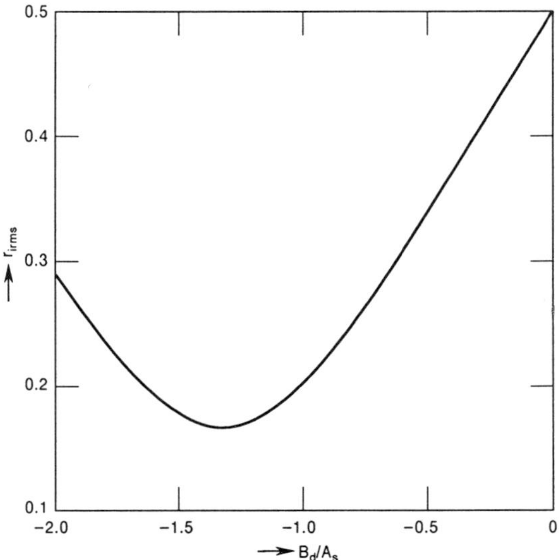

Figure 4-7. Variation of r_{irms} in units of $8FA_s$ for spherical aberration with defocus.

$$\boxed{W(\rho,\theta) \; = \; A_c\rho^3\cos\theta \;,}$$

(4-27a)

or

$$\boxed{W(\xi,\eta) \; = \; A_c\xi\!\left(\xi^2+\eta^2\right) \;.}$$

(4-27b)

Substituting Eq. (4-27) into Eq. (4-6), we obtain the corresponding ray aberrations in the Gaussian image plane with respect to the Gaussian image point

$$\left(x_i,y_i\right) \; = \; 2FA_c\rho^2\left(2+\cos2\theta,\ \sin2\theta\right)$$

(4-28a)

$$= \; 2FA_c\left(\rho^2+2\xi^2,\ 2\xi\eta\right) \;.$$

(4-28b)

For a given value of ρ, the locus of the points of intersection of the rays in the Gaussian image plane is given by

$$\boxed{\left(x_i-4FA_c\rho^2\right)^2+y_i^2 \; = \; \left(2FA_c\rho^2\right)^2 \;.}$$

(4-29)

Thus, the rays coming from a circle of radius ρ in the exit pupil lie on a circle of radius $2FA_c\rho^2$ centered at $\left(4FA_c\rho^2,0\right)$ in the image plane. The circle in the image plane is traced out twice in the same sense as in the pupil plane as θ varies from 0 to 2π to complete a circle of rays. As illustrated in Figure 4-8, since $CB/CP' = 1/2$, all of the rays in the image plane are contained in a cone of semiangle of 30° bounded by a circle of radius $2FA_c$ centered at $\left(4FA_c,0\right)$ corresponding to the marginal rays. Here C is the center of the circle formed by the marginal rays and $P'A$ and $P'B$ are tangents to the circle. The vertex of the cone, of course, coincides with the Gaussian image point P'.

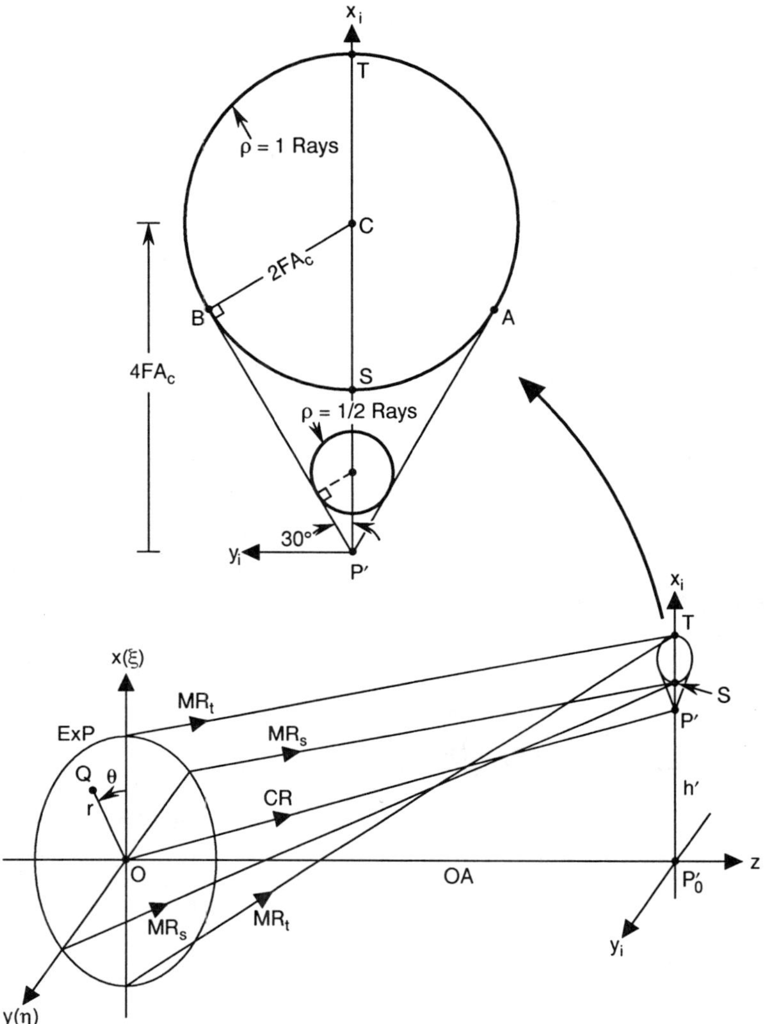

Figure 4-8. Ray spot diagram for coma. The tangential marginal rays MR_t are focused at the point T and the sagittal marginal rays MR_s are focused at the point S. All rays in the image plane lie in a cone of semiangle 30° with its vertex at the Gaussian image point $P′$ bounded by the upper arc of a circle of radius $2FA_c$ centered at $(4FA_c, 0)$. The cone angle is 30° because $CB/CP′ = 1/2$.

Only the chief ray passes through $P′$. Rays in the image plane corresponding to a zone of $\rho = 1/2$ are also shown in the figure. They lie on a circle of radius $FA_c/2$ centered at $(FA_c, 0)$ in the image plane. Since the spot diagram has the shape of a comet, the aberration is appropriately called *coma*. Note that the tangential marginal rays $MR_t\ (\rho = 1,\ \theta = 0,\ \pi)$ intersect this plane at a point T at a distance $6FA_c$ from $P′$ along the x_i axis, and the sagittal marginal rays $MR_s\ (\rho = 1,\ \theta = \pi/2,\ 3\pi/2)$ intersect the image plane at a point S at a distance $2FA_c$ from $P′$. Accordingly, the length $6FA_c$ and half-width $2FA_c$ of the coma pattern are called *tangential* and *sagittal coma*, respectively.

According to Eq. (4-27b), the wave aberration for the tangential ray fan is given by

$$W_t(\xi, 0) = A_c \xi^3 \ . \tag{4-30}$$

It is zero for the sagittal ray fan. The ray aberrations given by Eq. (4-28b) may be written for the two types of rays in the form

$$(x_i, y_i)_t = 6FA_c(\xi^2, 0) \tag{4-31a}$$

and

$$(x_i, y_i)_s = 2FA_c(\eta^2, 0) \ . \tag{4-31b}$$

We note that even though the wave aberration of the rays in the sagittal fan is zero, their ray aberration is not; the rays are displaced along the x (or ξ) axis in the image plane. Figure 4-9 shows the variation of wave and ray aberrations with pupil coordinates. We note that the wave aberration is odd and the ray aberration is even in pupil coordinates. Of course, this is also evident from Eqs. (4-27b) and (4-28b).

Substituting Eq. (4-27b) into Eq. (4-11) we obtain the PSF for coma

$$\boxed{I_g(x_i, y_i) = I_p\left(a^2/2RA_c\right)^2 \Sigma \left|3\xi^2 - \eta^2\right|^{-1}} \ . \tag{4-32}$$

The summation sign on the right-hand side represents rays with different values of (ξ, η) but the same value of (x_i, y_i). There are four rays with coordinates $(\pm\xi, \pm\eta)$ and $\left(\pm\eta/\sqrt{3}, \pm\sqrt{3}\xi\right)$ in the pupil plane satisfying

$$3\xi^2 + \eta^2/3 \le 1 \tag{4-33}$$

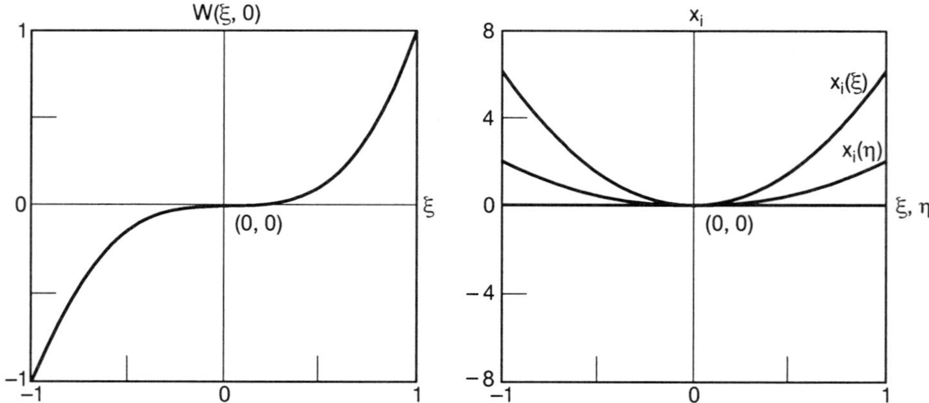

Figure 4-9. Wave and ray aberrations for tangential and sagittal ray fans for coma. The wave aberration is in units of A_c and the ray aberration is in units of FA_c. The wave aberration is zero for the sagittal ray fan.

that have the same coordinates (x_i, y_i) in the image plane, as may be seen from Eq. (4-28b). Each of the four rays yields the same value of $I_g(x_i, y_i)$, as may be seen from Eq. (4-32). There are two rays with coordinates $(\pm\xi, \pm\eta)$ satisfying

$$3\xi^2 + \eta^2/3 > 1 \tag{4-34a}$$

but

$$\xi^2 + \eta^2 \leq 1 \tag{4-34b}$$

that have the same coordinates (x_i, y_i). Both of these rays give the same value of $I_g(x_i, y_i)$.

From Eq. (4-28b), we find that

$$\xi^2 = \frac{1}{12FA_c}\left[x_i \pm \left(x_i^2 - 3y_i^2\right)^{1/2}\right] \tag{4-35a}$$

and

$$\eta^2 = \frac{1}{4FA_c}\left[x_i \mp \left(x_i^2 - 3y_i^2\right)^{1/2}\right] \ . \tag{4-35b}$$

It is evident from Eqs. (4-35) that (x_i, y_i) must satisfy $|y_i| \leq x_i/\sqrt{3}$ corresponding to a cone of semiangle of $30°$. Substituting Eqs. (4-35) into Eq.(4-33), we find that, as illustrated in Figure 4-10a, the rays satisfying Eq. (4-33) in the pupil plane lie in the image plane in a cone of semiangle $30°$ with its vertex at the Gaussian image point $(0, 0)$ and lower arc BSA of a circle of radius $2FA_c$ centered at $(4FA_c, 0)$ and encompassed by the cone. The end points B and A of the lower arc lie at $(3, \pm\sqrt{3})FA_c$ where the cone lines $|y_i| = x_i/3$ are tangent to the ray circle in the image plane corresponding to the marginal rays. Every point in this region in the image plane corresponds to four points in the pupil plane. Similarly, substituting Eqs. (4-35) into Eqs. (4-34), we find that every point in the image plane within the above circle and on its upper arc BTA corresponds to two rays in the pupil plane. Hence, substituting Eqs. (4-35) into Eq. (4-32) and considering the multiplicity of the rays in the image plane, we obtain

$$\boxed{I_g(x_i, y_i) = I_p \frac{a^3}{R A_c}\left(x_i^2 - 3y_i^2\right)^{-1/2}\begin{cases} 1, \text{Region I} \\[2ex] 1/2, \text{Region II} \end{cases}} \tag{4-36a}$$

where, as indicated in Figure 4-10a, region I is the conical region from its vertex P' to and including the lower arc BSA, and region II is the circular region, including its upper arc BTA. From Eqs. (4-32) and (4-36a), we note that the PSF approaches infinity for those rays for which $3\xi^2 = \eta^2$ in the pupil plane, i.e., at points (x_i, y_i) in the image plane for which $x_i^2 = 3y_i^2$. Thus, the PSF along the tangent lines $P'A$ and $P'B$ is infinity. Moreover, since $y_i = 0$ for tangential $(\eta = 0)$ and sagittal $(\xi = 0)$ rays, the PSF value for these rays is given by

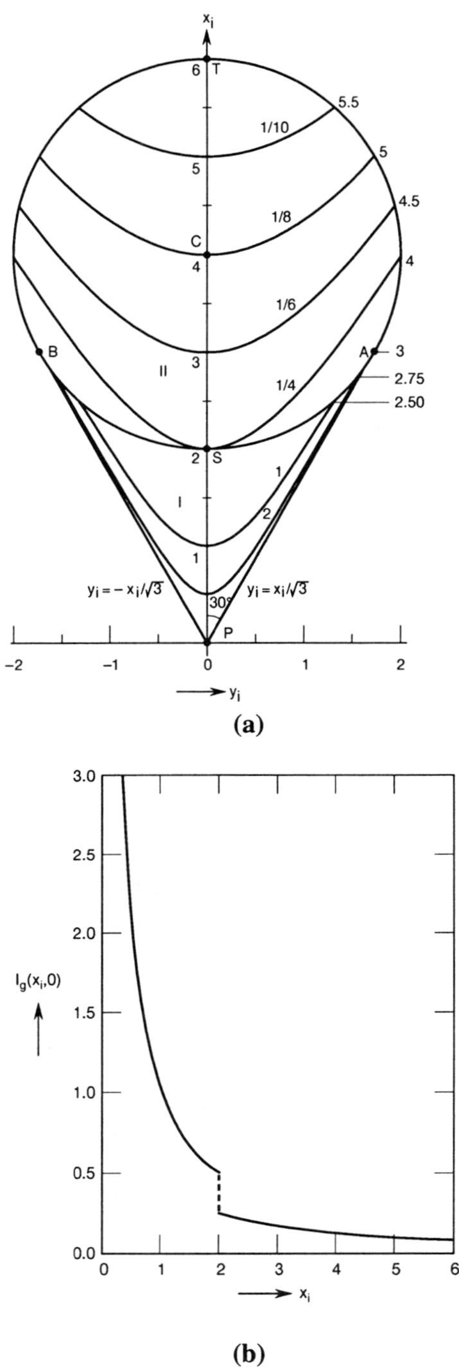

Figure 4-10. (a) Geometrical PSF for coma showing irradiance contours in units of $I_p a^3/RA_c$. As in Figure 4-8, C is the center of the circle formed by the marginal rays. x_i and y_i are in units of FA_c. Each contour ends on the marginal ray circle at a point whose x_i value is indicated on its extreme right. For example, the contour with a PSF value of $1/8$ ends on the circle at a point for which $x_i = 5$. (b) PSF along its symmetry axis x_i.

$$I_g(x_i, 0) = I_p \frac{a^3}{RA_c x_i} \begin{cases} 1, \text{Region I} \\ \\ 1/2, \text{Region II} \end{cases}, \tag{4-36b}$$

i.e., it gives the PSF along the x_i axis. Figure 4-10a shows several contours of the PSF, which, of course, is highly asymmetric about the Gaussian image point P'. The contour values are shown on the right-hand side of the x_i axis and the x_i values of the points where they meet the circle are shown on the extreme right. For example, the contour with a PSF value of 1/8 meets the circle at a point for which $x_i = 5$. How the PSF varies along its symmetry axis x_i is shown in Figure 4-10b.

Since the PSF is highly asymmetric about the Gaussian image point P', its centroid does not lie at it. Substituting Eq. (4-28b) into Eq. (4-14), we obtain the coordinates of the centroid

$$(x_c, y_c) = (2FA_c, 0) . \tag{4-37}$$

Thus, the centroid lies at the point S in Figure 4-8 where the sagittal marginal rays intersect the image plane. Measuring the ray coordinates in the image plane with respect to a point other than the Gaussian image point is equivalent to introducing a wavefront tilt aberration in the aberration function. Thus, we consider an aberration function in the form

$$W(\rho, \theta) = A_c \rho^3 \cos\theta + B_t \rho \cos\theta , \tag{4-38}$$

where B_t is the peak value of the balancing tilt aberration and corresponds to measuring the wave aberration with respect to a reference sphere centered at a point in the image plane with coordinates $(-2FB_t, 0)$, or equivalently, measuring the ray coordinates with respect to this point. Thus, a ray with coordinates (ρ, θ) in the pupil plane has its coordinates in the image plane given by

$$(x_i, y_i) = 2FA_c \left[\rho^2 (2 + \cos 2\theta) + B_t/A_c , \rho^2 \sin 2\theta \right] . \tag{4-39}$$

Substituting Eq. (4-39) into Eq. (4-15a), we obtain the rms radius of the image spot:

$$
\begin{aligned}
r_{i\,rms} &= 2FA_c \left\langle \left[\rho^2 (2 + \cos 2\theta) + B_t/A_c \right]^2 + \rho^4 \sin^2 2\theta \right\rangle^{1/2} \\
&= 2FA_c \left[\frac{5}{3} + 2\frac{B_t}{A_c} + \left(\frac{B_t}{A_c} \right)^2 \right]^{1/2} .
\end{aligned}
\tag{4-40a}
$$

The variation of $r_{i\,rms}$ with tilt is shown in Figure 4-11. Letting

$$\frac{\partial r_{i\,rms}}{\partial A_t} = 0 , \tag{4-40b}$$

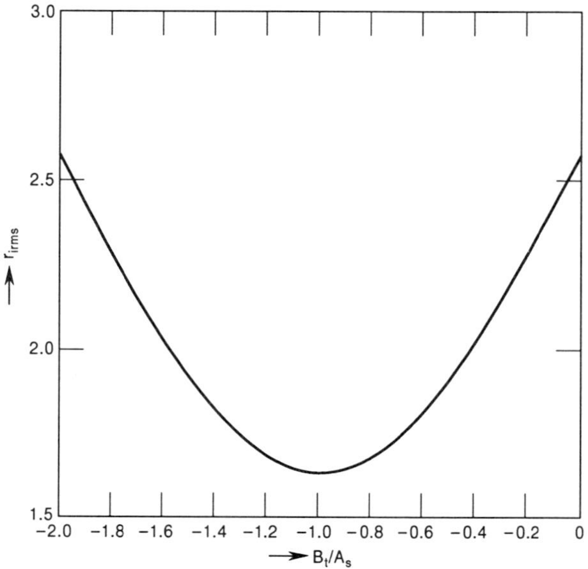

Figure 4-11. Variation of r_{irms} in units of FA_c for coma with tilt.

we find that the rms radius is minimum and equal to $2\sqrt{2/3}\,FA_c$ when $B_t = -A_c$, i.e., when it is measured with respect to the centroid of the PSF. The corresponding balanced aberration is given by $A_c\left(\rho^3 - \rho\right)\cos\theta$. The variance of the aberration is minimum when $B_t = -(2/3)A_c$, i.e., if the balanced aberration is $A_c\left[\rho^3 - (2/3)\rho\right]\cos\theta$, similar to the Zernike polynomial $Z_3^1(\rho,\theta)$.

4.3.3 Astigmatism and Field Curvature

Next, we consider PSFs aberrated by astigmatism and field curvature. If the image of a point object is observed in a defocused plane, the aberration function may be written

$$W(\rho,\theta) \;=\; A_a\rho^2\cos^2\theta + A_d\rho^2 + B_d\rho^2 \tag{4-41a}$$

or

$$W(\xi,\eta) \;=\; \left(A_a + A_d + B_d\right)\xi^2 + \left(A_d + B_d\right)\eta^2 \quad, \tag{4-41b}$$

where A_a and A_d are both proportional to h'^2 and the balancing defocus coefficient B_d is related to the longitudinal defocus ΔR according to Eq. (4-21a). The corresponding ray aberrations are given by

$$(x_i,y_i) \;=\; 4F\rho\left[\left(A_a + A_d + B_d\right)\cos\theta,\left(A_d + B_d\right)\sin\theta\right] \tag{4-42a}$$

$$=\; 4F\left[\left(A_a + A_d + B_d\right)\xi,\left(A_d + B_d\right)\eta\right] \quad. \tag{4-42b}$$

For a given value of ρ, the locus of the points of intersection of the rays in the defocused image plane is given by

$$\left(\frac{x_i}{A}\right)^2 + \left(\frac{y_i}{B}\right)^2 \;=\; 1 \quad, \tag{4-43}$$

where

$$A = 4F(A_a + A_d + B_d)\rho \tag{4-44a}$$

and

$$B = 4F(A_d + B_d)\rho \ . \tag{4-44b}$$

Thus, the rays lying on a circle of radius ρ in the exit pupil, in general, lie in a defocused image plane on an ellipse whose semiaxes are given by A and B, respectively. The largest ellipse is obtained for the marginal rays.

The Gaussian image $(B_d = 0)$ is an elliptical spot with semiaxes $4F(A_a + A_d)$ and $4FA_d$, as illustrated in Figure 4-12. We note that if $B_d = -A_d$, corresponding to $\Delta R_s = 8F^2 A_d$, the ellipse reduces to a line S of full length $8FA_a$ parallel to the x_i axis. The line image is called the *sagittal* (or *radial*) *image* because the sagittal rays converge to a point at its center. It lies in the tangential (or meridional) plane zx, containing the point object (which lies along the x_i axis in the object plane) and the optical axis. If, however, $B_d = -(A_a + A_d)$, corresponding to $\Delta R_t = 8F^2(A_a + A_d)$, then the ellipse reduces to a line T parallel to the y_i axis. The full length of this line image is the same as that of the line image S. This line image is called the *tangential image* because

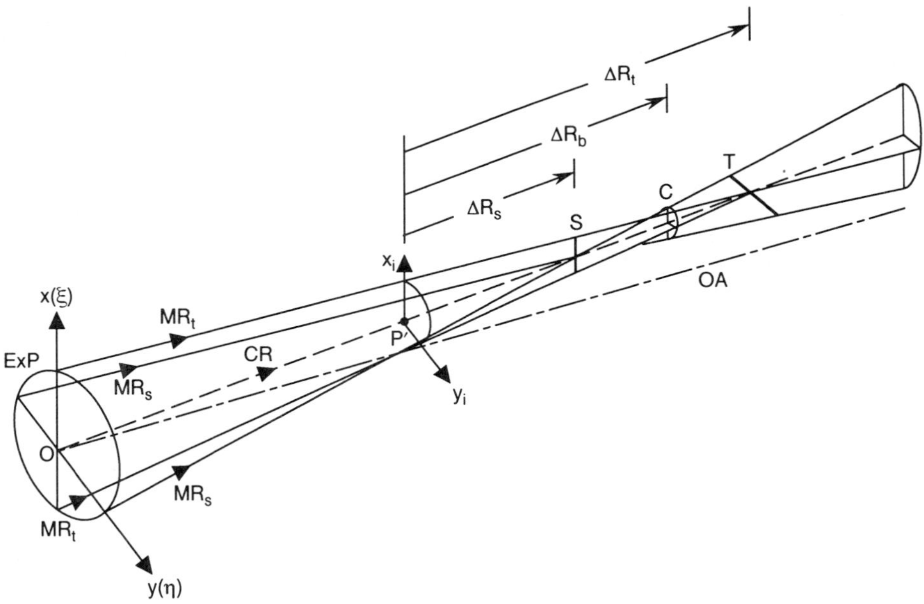

Figure 4-12. Spot diagrams for astigmatism and field curvature showing elliptical image spots and astigmatic focal lines. The sagittal marginal rays MR_s are shown converging on the sagittal line image S and the tangential marginal rays MR_t are shown converging on the tangential line image T. The line images S and T and the circle of least confusion C are special cases of the elliptical spots.

the tangential rays converge to a point at its center, and it lies in the sagittal plane. The distance $8F^2A_a$ between the two line images is called *longitudinal astigmatism*. It should be evident that it is independent of the zone value ρ of the rays. The two line images are called the *astigmatic focal lines*. (The terms "radial" and "tangential" images also become evident by consideration of Figure 4-18, where these images are shown for a point object P as well as straight and circular line objects.)

If $B_d = -\left(A_a + 2A_d\right)/2$, corresponding to $\Delta R_b = 4F^2\left(A_a + 2A_d\right)$, the ellipse reduces to a circle C of diameter $4FA_a$, which is half the full length of the two line images. Since this circle is the smallest of all the possible images, Gaussian or defocused, it is called the *circle of least (astigmatic) confusion*. The circle in the image plane is traced out once in the opposite sense of that in the pupil plane as θ varies from 0 to 2π to complete a circle of rays, as may be seen from Eq. (4-42a). Substituting $B_d = -\left(A_a + 2A_d\right)/2$ into Eq. (4-41a), we obtain the balanced aberration $\left(A_a/2\right)\rho^2\cos 2\theta$, similar to the Zernike polynomial $Z_2^2(\rho,\theta)$. Astigmatism balanced in this manner not only gives the smallest spot but also yields minimum variance of the aberration. We will refer to the image thus obtained as the *best image*.

Since both A_a and $A_d \sim h'^2$, the length of the sagittal and tangential line images of a point object increases quadratically with the height h' of the Gaussian image point. Similarly, ΔR_s, ΔR_t, ΔR_b, and longitudinal astigmatism increase as h'^2. For a line object, equating ΔR to the sag of a curved line image, we find that the sagittal, tangential, and best images are parabolic with the vertex radii of curvature given by

$$R_s = h'^2/16F^2A_d \tag{4-45a}$$

$$= 1/4R^2a_d \quad, \tag{4-45b}$$

$$R_t = h'^2/16F^2\left(A_a + A_d\right) \tag{4-46a}$$

$$= 1/4R^2\left(a_a + a_d\right) \quad, \tag{4-46b}$$

and

$$R_b = \frac{h'^2}{8F^2\left(A_a + 2A_d\right)} \tag{4-47a}$$

$$= \frac{1}{2R^2\left(a_a + 2a_d\right)} \quad, \tag{4-47b}$$

respectively. Note that a positive value of R_s, for example, corresponds to positive values of A_d and ΔR_s. The images of a planar object centered on the optical axis are the corresponding paraboloids symmetric about the optical axis.

From Eqs. (4-45b) and (4-46b) we note that

$$\frac{3}{R_s} - \frac{1}{R_t} = 4R^2\left(2a_d - a_a\right) \ . \tag{4-48}$$

Following Eq. (5-117), Eq. (4-48) may be written

$$\boxed{\frac{3}{R_s} - \frac{1}{R_t} = \frac{2}{R_p}} \ , \tag{4-49}$$

where R_p is the radius of curvature of the Petzval image surface given by Eq. (5-98). Since the sag of a surface is inversely proportional to its (vertex) radius of curvature, Eq. (4-49) has the consequence that, as illustrated in Figure 4-13, *the Petzval surface is three times as far from the tangential surface as it is from the sagittal surface.* Moreover, *the sagittal surface always lies between the tangential and the Petzval surfaces.* When astigmatism is zero, the sagittal and the tangential surfaces reduce to the Petzval surface. We also note from Eqs. (4-45) through (4-47) that

$$\boxed{\frac{1}{R_b} = \frac{1}{2}\left(\frac{1}{R_s} + \frac{1}{R_t}\right)} \ , \tag{4-50}$$

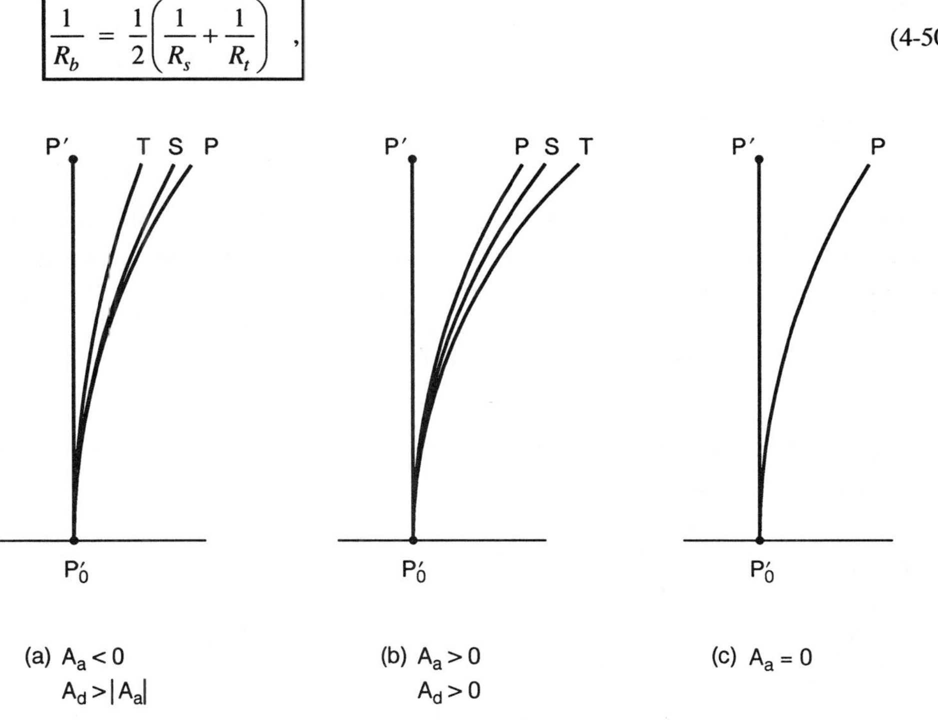

Figure 4-13. Parabolic image surfaces. S – sagittal, T – tangential, and P – Petzval. The sagittal and tangential surfaces correspond to astigmatism, and the Petzval surface corresponds to field curvature. The sagittal surface lies between the tangential and Petzval surfaces as in (a) and (b) when astigmatism is nonzero. The Petzval surface is three times as far from the tangential surface as it is from the sagittal surface. The sagittal and tangential surfaces coincide with the Petzval surface as in (c) when astigmatism is zero. $P_0' P'$ is the Gaussian image of a planar object.

i.e., the vertex curvature of the best-image surface is equal to the mean value of the vertex curvatures of the sagittal and tangential surfaces. The best-image surface is planar when $a_a = -2a_d$. In that case, $R_s = -R_t$, i.e., the sagittal and tangential image surfaces have equal but opposite vertex curvatures.

The wave and ray aberrations of a tangential ray fan are given by Eqs. (4-41b) and (4-42b) according to

$$W_t(\xi,0) = (A_a + A_d + B_d)\xi^2 \tag{4-51}$$

and

$$(x_i, y_i)_t = 4F(A_a + A_d + B_d)(\xi, 0) \quad , \tag{4-52}$$

respectively. Similarly, for the sagittal ray fan, they are given by

$$W_s(0,\eta) = (A_d + B_d)\eta^2 \tag{4-53}$$

and

$$(x_i, y_i)_s = 4F(A_d + B_d)(\eta, 0) \quad . \tag{4-54}$$

The wave and ray aberrations for $A_d + B_d = 0$, $-A_a/2$, and $-A_a$ are illustrated in Figure 4-14. It is evident that the wave aberration varies quadratically with a pupil coordinate and the ray aberration varies linearly with it.

Substituting Eq. (4-41b) into Eq. (4-11), we obtain the PSF

$$I_g(x_i, y_i) = \frac{I_p a^4}{4R^2 \left| (A_d + B_d)(A_a + A_d + B_d) \right|} \quad . \tag{4-55}$$

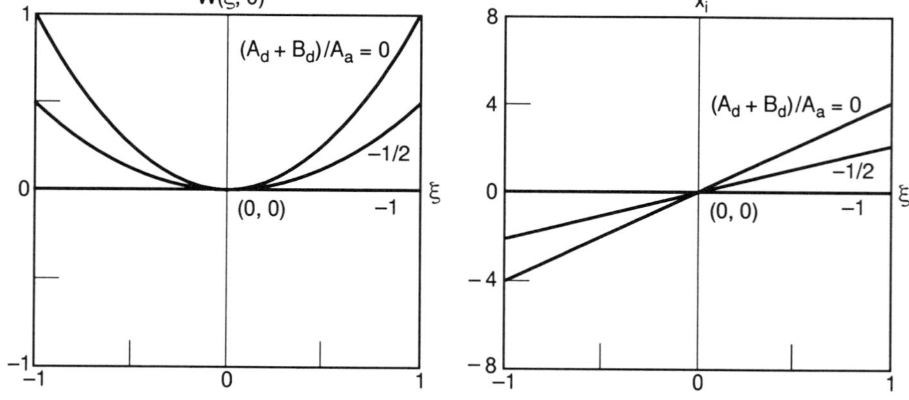

Figure 4-14. Wave and ray aberrations for tangential and sagittal ray fans for astigmatism corresponding to various image planes. The wave aberration is in units of A_a and the ray aberration is in units of FA_a.

Thus, within the range of (x_i, y_i) values given by Eqs. (4-42), the PSF is uniform regardless of the value of B_d; the value of the PSF, of course, depends on the value of B_d. The line images corresponding to $A_d + B_d = 0$ and $A_d + B_d = -A_a$ have infinite irradiance. The circle of least confusion, which corresponds to $A_d + B_d = -A_a/2$, has a uniform irradiance of $I_p\left(a^2/A_a R\right)^2$. The centroid of the PSF lies at the Gaussian image point $(0,0)$ since it is symmetric about both the x_i and y_i axes. The rms radius of the image spot may be obtained by substituting Eq. (4-42a) into Eq. (4-15c). Thus,

$$r_{irms} = 2FA_a\left[1 + 2\frac{A_d + B_d}{A_a} + 2\left(\frac{A_d + B_d}{A_a}\right)^2\right]^{1/2} . \tag{4-56}$$

The variation of r_{irms} with $A_d + B_d$ is shown in Figure 4-15. Letting

$$\frac{\partial r_{irms}}{\partial B_d} = 0 \quad , \tag{4-57}$$

we find that the rms radius is minimum and equal to $\sqrt{2}\, FA_a$ when $A_d + B_d = -A_a/2$, i.e., in the plane of the circle of least confusion, as expected for uniform irradiance. The spot shape and size, including its rms radius, in an image plane defined by the balancing defocus are summarized in Table 4-2.

If astigmatism is the only aberration present, i.e., if the field curvature coefficient $A_d = 0$ in Eqs. (4-41), then all of the object rays transmitted by the exit pupil intersect the Gaussian image plane on a line S of full length $8FA_a$ along the x_i axis centered at the Gaussian image point P', as illustrated in Figure 4-16. This is the sagittal image of a

Table 4-2. Ray spot shape, size, and rms radius for astigmatism A_a and field curvature A_d in various image planes defined by defocus B_d.

Image Plane	Balancing Defocus B_d	Spot Shape and Size*	RMS Radius $r_{irms}/2FA_a$
General	B_d	Elliptical, $4F(A_a + A_d + B_d)$ $\times\, 4F(A_d + B_d)$	$\left[1 + 2\dfrac{A_d + B_d}{A_a} + 2\left(\dfrac{A_d + B_d}{A_a}\right)^2\right]^{1/2}$
Gaussian	0	Elliptical, $4F(A_d + A_d)$ $\times\, 4FA_d$	$\left[1 + 2\dfrac{A_d}{A_a} + 2\left(\dfrac{A_d}{A_a}\right)^2\right]^{1/2}$
Sagittal	$-A_d$	Line along x_i axis, $8FA_a$	1
Tangential	$-(A_d + A_a)$	Line along y_i axis, $8FA_a$	1
Best	$-(A_d + A_a/2)$	Circular, $4FA_a$	$1/\sqrt{2}$

*Spot sizes are full major and minor axes of an elliptical image, full length of a line image, and diameter of a circular image.

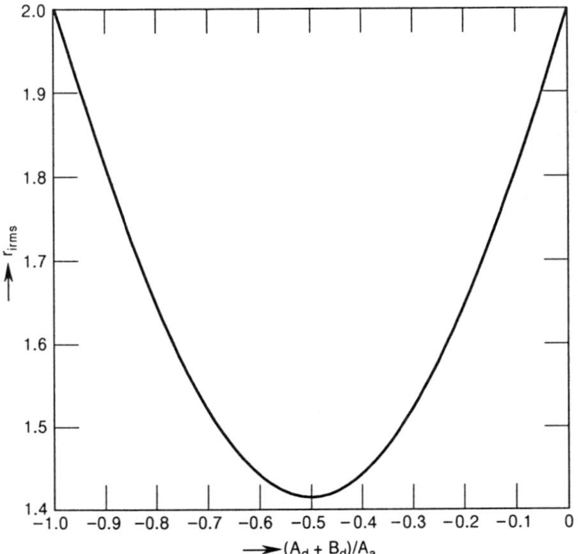

Figure 4-15. Variation of r_{irms} in units of FA_a for astigmatism with $A_d + B_d$.

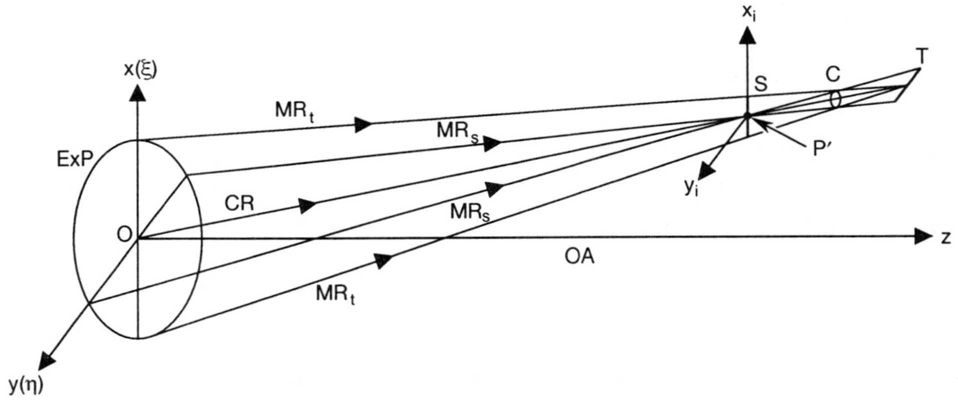

Figure 4-16. Astigmatic focal lines when only astigmatism is present. The tangential marginal rays MR_t are focused at a point on the tangential focal line T. Similarly, the sagittal marginal rays MR_s are focused at the Gaussian image point P' on the sagittal focal line S. The focal lines S and T lie in the tangential and sagittal planes, respectively. The circle of least confusion C lies in a plane midway between the planes of line images S and T.

point object. The sagittal rays converge on the Gaussian image point. Similarly, a tangential line image T of the same full length as the sagittal line image is obtained in a defocused image plane corresponding to $B_d = - A_d$, The tangential rays converge to a point at its center.The sagittal image of a line object is also a line that is slightly longer (by an amount $8FA_a$) than but coincident with its Gaussian line image. However, its tangential image is parabolic with a vertex radius of curvature of $h'^2/16F^2A_a$ or $1/4R^2a_a$. Note that the longitudinal astigmatism in this case represents the sag of the tangential image surface. Similarly, the sagittal image of a planar object will be planar, but its tangential image will be paraboloidal.

We now consider the case when field curvature is the only aberration present, i.e., when the wave aberration is given by

$$W(\rho) = A_d\rho^2 \quad .$$

$$(4\text{-}58)$$

Since the wave aberration is radially symmetric, the distribution of rays in the Gaussian image plane is also radially symmetric. For rays lying on a circle of radius ρ in the exit pupil, the radius of the corresponding circle of rays in the image plane, following Eq. (4-7), is given by

$$\boxed{r_i \;=\; 4FA_d\rho \quad .}$$

$$(4\text{-}59)$$

Its maximum value is $4FA_d$ and corresponds to the marginal rays. The circle in the image plane is traced out in the same sense as in the pupil as θ varies from 0 to 2π.

From the discussion in Section 3.3, we note that a defocus aberration represented by Eq. (4-58) implies that the wavefront is spherical, but it is not centered at the Gaussian image point. Instead, it is centered at a distance

$$\boxed{\Delta R \;=\; 8F^2A_d}$$

$$(4\text{-}60)$$

from the Gaussian image point along the optical axis (strictly speaking, it is centered on the line joining the center of the exit pupil and the Gaussian image point). Since the aberration coefficient $A_d \sim h'^2$, ΔR also increases as h'^2. Hence, the sagittal image of a line object will be parabolic with a vertex radius of curvature of $h'^2/16F^2A_d$, or $1/4R^2a_d$. Similarly, the image of a planar object will be paraboloidal. The paraboloidal surface for a system with zero astigmatism is called the *Petzval image surface*.

As in the case of spherical aberration, because of the radial symmetry of field curvature, the wave and ray aberrations of any ray fan can be written immediately from Eqs. (4-58) and (4-59), respectively. For example, for the tangential ray fan, we may write

$$W_t(\xi,0) \;=\; A_d\xi^2$$

$$(4\text{-}61a)$$

and

$$(x_i,y_i)_t \;=\; (4FA_d\xi,0) \quad .$$

$$(4\text{-}61b)$$

Figure 4-17 shows how the wave and ray aberrations vary with ξ. The PSF in this case has a uniform irradiance of $I_p\left(a^2/2R\,A_d\right)^2$ across a circle of radius $4FA_d$, as may be seen from Eq. (4-53) by letting $A_a = 0 = B_d$, or directly from Eq. (4-12b).

Figure 4-18 illustrates the effect of astigmatism and field curvature on the image of a spoked wheel where the images formed on the sagittal and tangential surfaces are shown. A magnification of -1 is assumed in the figure. As discussed earlier, a point object P is imaged as a sagittal or radial line P_s' on the sagittal surface and as a tangential line P_t' on the tangential surface. Each point on the object is imaged in this manner, so that the

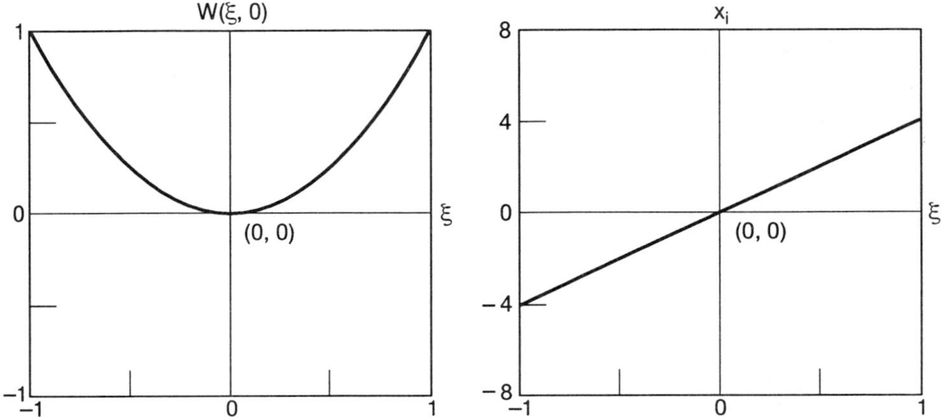

Figure 4-17. Wave and ray aberrations of a ray fan for field curvature. The wave aberration is in units of A_d and the ray aberration is in units of FA_d.

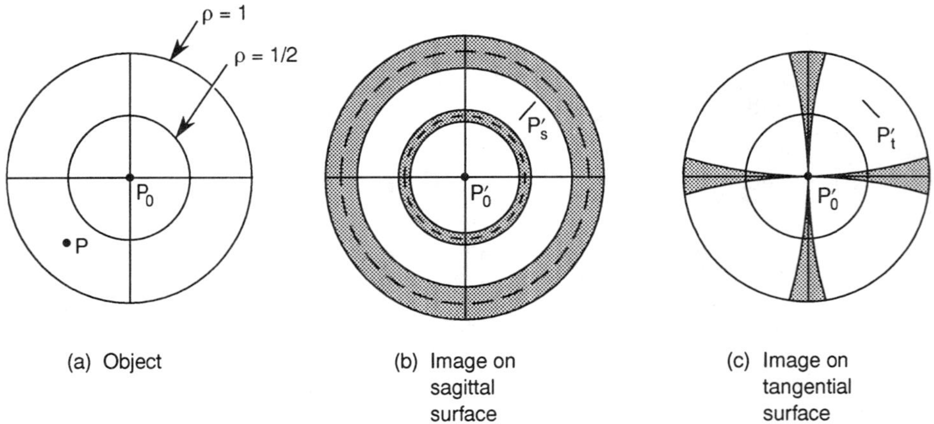

(a) Object

(b) Image on sagittal surface

(c) Image on tangential surface

Figure 4-18. Astigmatic images of a spoked wheel. Gaussian magnification of the image is assumed to be -1. The sagittal and tangential images P_s' and P_t' of a point object P are shown very much exaggerated. The dashed circles in (b) are the Gaussian images of the object circles.

sagittal image consists of sharp radial lines and diffuse circles while the tangential image consists of sharp circles and diffuse radial lines. If the object contains lines that are neither radial nor tangential, they will not be sharply imaged on any surface.

It should be understood that the astigmatism discussed here is for a system that is rotationally symmetric about its optical axis, and its value reduces to zero for an axial point object. It is different from the *astigmatism of the eye* which is caused by one or more of its refracting surfaces, usually the cornea, that is curved more in one plane than another. The refracting surface that is normally spherical acquires a small cylindrical component, i.e., it becomes toric. Such a surface forms a line image of a point object even when it lies on its axis. Hence, a person afflicted with astigmatism sees points as lines. If the object consists of vertical and horizontal lines as in the wires of a window screen, such a person can focus (by accommodation) only on the vertical or the horizontal lines at a time. This is analogous to the spoked wheel example where the rim is in focus in one observation plane and the spokes are in focus in another.

4.3.4 Distortion

The distortion wave aberration is given by

$$\boxed{W(\rho,\theta) = A_t\rho\cos\theta} \tag{4-62a}$$

or

$$\boxed{W(\xi,\eta) = A_t\xi \ ,} \tag{4-62b}$$

where the aberration coefficient A_t is proportional to h'^3. The corresponding ray aberrations are given by

$$(x_i, y_i) = (2FA_t, 0) \tag{4-63a}$$

$$= (Ra_t h'^3, 0) \ . \tag{4-63b}$$

Since the ray aberrations are independent of the coordinates (ρ, θ) of a ray in the exit pupil, all the rays converge at the image point $(2FA_t, 0)$, which lies along the x_i axis at a distance $2FA_t$ from the Gaussian image point. Thus, a wavefront aberrated by distortion is tilted with respect to the Gaussian reference sphere by an angleM

$$\beta = A_t/a \ . \tag{4-64}$$

This angle is proportional to h'^3. Similarly, the distance $2FA_t$ of the perfect image point from the Gaussian image point is proportional to h'^3. Distortion is often measured as a fraction of the image height. Thus, for example, percent distortion is $100(Ra_t h'^2)$.

It should be noted that although the ray aberration for distortion is independent of the ray coordinates in the pupil plane, all the rays converge at the point $(2FA_t, 0)$ if distortion is the only wave aberration present. However, if other wave aberrations are present, then different rays will intersect the Gaussian image plane at different points. But, the chief ray will still intersect the Gaussian image plane at the point $(2FA_t, 0)$ since its ray aberration due to the other wave aberrations is zero. Hence, the ray distortion aberration is the distance of the point where the chief ray intersects the Gaussian image plane from the Gaussian image point, i.e., it represents the distance between the points of intersection of the actual (within the approximation of primary aberration) and the paraxial chief rays in the Gaussian image plane.

If we consider a line object $L_1 L_2$ as illustrated in Figure 4-19 at a distance h_1 from the optical axis, its Gaussian image is also a line parallel to it at a distance h_1' from the optical axis, where h_1 and h_1' are related to each other by the Gaussian magnification of the system (just as h and h' are related to each other). A magnification of -1.5 is assumed in the figure.

Because of distortion, the image of any point object is displaced from its Gaussian image point by an amount $2FA_t$ along a line joining the axial image point and the Gaussian image point under consideration. We consider imaging of point objects P_1 and P_2 which are at distances h_1 and h_2, respectively from the axial point object P_0. Their Gaussian images P_1' and P_2' are located at distances h_1' and h_2', respectively, from the Gaussian image P_0' of the axial object P_0. Because of distortion, the images are displaced to positions P_1'' and P_2'' so that the displacements $P_1'P_1''$ and $P_2'P_2''$ are proportional to $h_1'^3$, and $h_2'^3$, respectively.

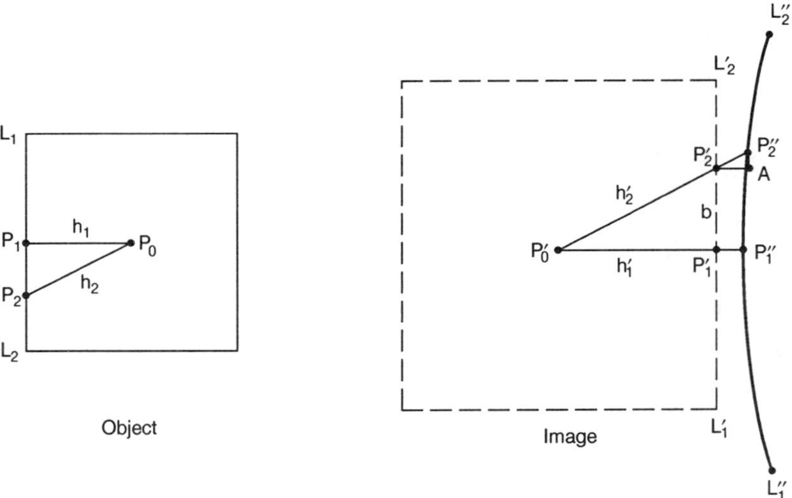

Figure 4-19. Image of a square in the presence of distortion. The dashed square is the Gaussian image. $L_1'L_2'$ and $L_1''L_2''$ are the Gaussian and distorted images of the line object L_1L_2, respectively. A magnification of -1.5 is assumed in the figure.

We note from similar triangles $P_0' \, P_1' \, P_2'$ and $P_2' \, A \, P_2''$ in Figure 4-19 that

$$\frac{P_2' A}{h_1'} = \frac{P_2' \, P_2''}{h_2'} = \frac{A P_2''}{b} \quad , \tag{4-65}$$

where $b = P_1' \, P_2'$. Therefore

$$P_2' A = \left(h_1'/h_2'\right) P_2' P_2''$$

$$= Ra_t h_1' h_2'^2$$

$$= Ra_t h_1' \left(h_1'^2 + b^2\right) \quad . \tag{4-66}$$

Since $P_1' P_1'' = Ra_t h_1'^3$, therefore,

$$P_2' A - P_1' P_1'' = Ra_t h_1' b^2 \quad , \tag{4-67}$$

which represents the sag of P_2'' from a line parallel to the Gaussian line image $L_1' L_2'$ but passing through P_1''. Now from Eq. (4-65)

$$A P_2'' = \left(b/h_2'\right) P_2' P_2''$$

$$= Ra_t b h_2'^2 \quad . \tag{4-68}$$

For small values of a_t, $A P_2''$ is also small; therefore, $P_1'' P_2'' \simeq P_1' P_2' = b$. From Eq. (4-67) we note then that the sag of P_2'' is proportional to the square of its distance b from P_1''. Hence, the locus of P_2'' represents a parabola with a vertex at P_1'' and a vertex radius of curvature of $1/2Ra_t h_1'$. If a_t is positive, the parabolic image is curved away from the Gaussian image line as shown in Figure 4-19. If it is negative, the parabolic image will be curved toward the Gaussian image line. We note from Eq. (4-67) that if the line object intersects the optical axis so that h_1' is zero, then the sag of P_2'' is also zero. Accordingly, the image P_2'' of a point object P_2 is simply displaced along the image line. Thus, the image of a line object intersecting the optical axis is also a line differing from the Gaussian image line only in that it is slightly longer. This discussion can be easily extended to obtain the distorted images of a square grid shown in Figure 4-20. It should be evident that when A_t is positive, we speak of a *pincushion distortion*. Similarly, when A_t is negative, we speak of a *barrel distortion*.

4.4 BALANCED ABERRATIONS FOR MINIMUM RMS SPOT RADIUS

We note that to obtain the smallest spot radius we have combined spherical aberration and astigmatism with defocus. In the case of coma, since the centroid does not lie at the origin, measuring the spot radius with respect to the centroid is equivalent to adding a certain amount of wavefront tilt. For a higher-order aberration, the balancing lower-order aberrations can be obtained in a similar manner. A balanced aberration, giving the smallest rms radius, in terms of *Zernike circle polynomials* $R_n^m(\rho)\cos m\theta$, is given by $B_n^m(\rho)\cos m\theta$, where[2]

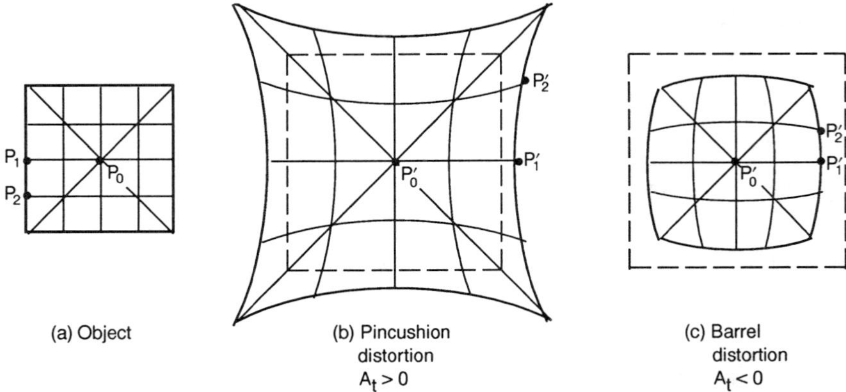

Figure 4-20. Images of a square grid in the presence of distortion. When the distortion aberration coefficient A_t is positive, we obtain pincushion distortion as in (b). When A_t is negative, we obtain barrel distribution as in (c). The dashed squares represent the Gaussian image of the square object with a magnification of -1.5.

$$B_n^m(\rho) = R_n^m(\rho) - R_{n-2}^m(\rho) \quad . \tag{4-69}$$

These polynomials are listed in Table 4-3 and may be obtained from the Zernike polynomials given in Table 3-7. If the aberration function is written in terms of these polynomials, e.g.,

$$W(\rho,\theta) = \sum_{n=0}^{\infty} \sum_{m=0}^{n} b_n^m B_n^m(\rho)\cos m\theta \quad , \tag{4-70}$$

then the rms radius of the image spot, obtained by substituting Eq. (4-70) into Eqs. (4-15c), is given by[2]

$$r_{irms} = 2F\left\{ \sum_{n/2=1}^{\infty} 4n\left(b_n^0\right)^2 + \sum_{m=1}^{\infty}\left[m\left(b_m^m\right)^2 + \sum_{i=1}^{\infty} 2(2i+m)\left(b_{2i+m}^m\right)^2 \right] \right\}^{1/2} \quad . \tag{4-71}$$

We note that the polynomials $B_4^0(\rho)$, $B_3^1(\rho)\cos\theta$, and $B_2^2(\rho)\cos 2\theta$ represent balanced spherical aberration, coma, and astigmatism giving a minimum rms spot radius, which is in agreement with the results obtained earlier in Sections 4.3.1, 4.3.2, and 4.3.3.

4.5 SPOT DIAGRAMS

If an optical system is aberration free, the wavefront at its exit pupil corresponding to a certain point object is spherical and all the object rays lying in the pupil plane converge to the Gaussian image point. For an aberrated system, the wavefront is nonspherical and the rays are distributed in a finite region of an image plane. This distribution of rays is called a *spot diagram*. We now illustrate the distribution of rays in an image plane for a system aberrated with a primary aberration. For each aberration, we consider the distribution for rays from four zones of the exit pupil, $\rho = 1/4$, $1/2$, $3/4$, and 1. In Figure 4-21, the rays from these zones are indicated by different symbols.

Table 4-3. Balanced polynomials for minimum rms spot radius.

n	m	$B_n^m(\rho)\cos m\theta$	Balanced Aberration
0	0	1	Piston
1	1	$\rho\cos\theta$	Tilt
2	0	$2(\rho^2-1)$	Defocus
2	2	$\rho^2\cos 2\theta$	Primary astigmatism
3	1	$3(\rho^3-\rho)\cos\theta$	Primary coma
3	3	$\rho^3\cos 3\theta$	
4	0	$2(3\rho^4-4\rho^2+1)$	Primary spherical
4	2	$4(\rho^4-\rho^2)\cos 2\theta$	Secondary astigmatism
4	4	$\rho^4\cos 4\theta$	
5	1	$5(2\rho^5-3\rho^3+\rho)\cos\theta$	Secondary coma
5	3	$5(\rho^5-\rho^3)\cos 3\theta$	
5	5	$\rho^5\cos 5\theta$	
6	0	$2(10\rho^6-18\rho^4+9\rho^2-1)$	Secondary spherical
6	2	$3(5\rho^6-8\rho^4+3\rho^2)\cos 2\theta$	Tertiary astigmatism
6	4	$6(\rho^6-\rho^4)\cos 4\theta$	
6	6	$\rho^6\cos 6\theta$	
7	1	$7(5\rho^7-10\rho^5+6\rho^3-\rho)\cos\theta$	Tertiary coma
7	3	$7(3\rho^7-5\rho^5+2\rho^3)\cos 3\theta$	
7	5	$7(\rho^7-\rho^5)\cos 5\theta$	
7	7	$\rho^7\cos 7\theta$	
8	0	$2(35\rho^8-80\rho^6+60\rho^4-16\rho^2+1)$	Tertiary spherical

Figure 4-22 illustrates the distribution of rays for spherical aberration in the Gaussian or paraxial $(B_d=0)$, midway $(B_d=-A_s)$, least-confusion $(B_d=-3/2\,A_s)$, and marginal $(B_d=-2A_s)$ planes. We note that in the plane of least confusion, rays from zones $\rho=1/2$ and 1 arrive on the same circle. By definition, the marginal rays $(\rho=1)$ intersect the optical axis at the marginal image point. The spot radius in the marginal image plane corresponds to rays of zone $\rho=1/\sqrt{3}=0.577$ and they are indicated by Δ in the figure.

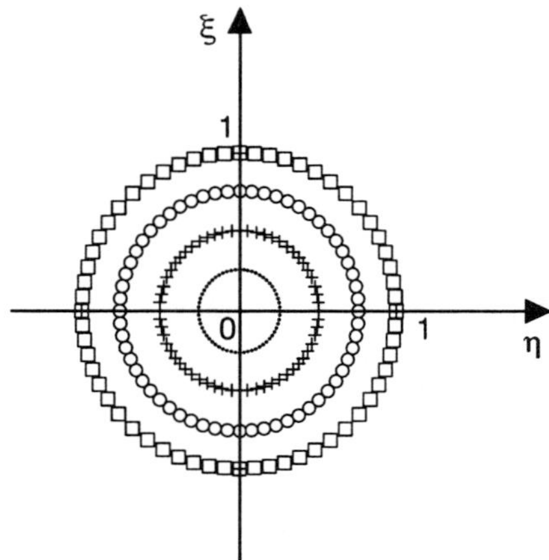

Figure 4-21. Zonal rays in the pupil plane corresponding to four zones, $\rho = 1/4$, 1/2, 3/4, and 1.

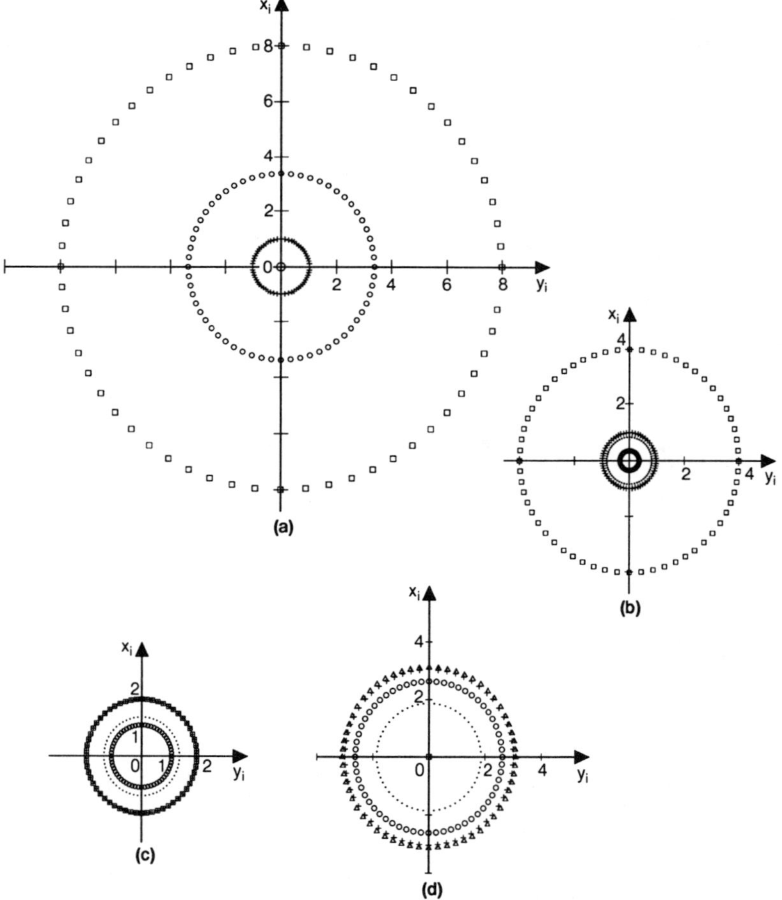

Figure 4-22. Ray distribution for spherical aberration in (a) Gaussian, (b) midway, (c) least-confusion, and (d) marginal image planes. The units of x_i and y_i are FA_s.

Figure 4-23 illustrates the distribution of rays for coma in the Gaussian image plane. As in Figure 4-6, all rays lie in a cone of semiangle 30° bounded by a circle of marginal rays of radius $2FA_c$ entered at $\left(4FA_c, 0\right)$

Figure 4-24 illustrates the ray distribution of various images for astigmatism. The images shown are (a) sagittal line, (b) least-confusion circle, (c) tangential line, and (d) elliptical that is symmetrically opposite to the least-confusion circle. The value of B_d for these images is given by $\left(A_d + B_d\right)/A_a = 0,\ -1/2,\ -1,$ and $1/2$, respectively.

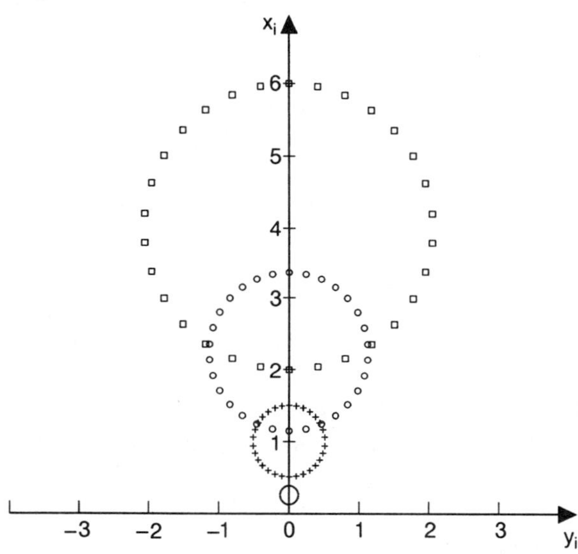

Figure 4-23 Ray distribution for coma in the paraxial image plane. The units of x_i and y_i are FA_c.

The ray distribution for field curvature alone in the Gaussian image plane is identical to the one for astigmatism in the plane of least confusion if $B_d = A_a/2$. Comparing Figures 4-22a , 4-23, and 4-24b, we note that rays of a given zone ρ lie on a circle whose radius is proportional to ρ^3 in the case of spherical aberration, ρ^2 in the case of coma, and ρ in the case of field curvature. In the case of astigmatism also they lie on a circle whose radius is proportional to ρ in the least-confusion image plane. Note, however, that the circles are not concentric in the case of coma; they are centered at points along its symmetry axis at distances from the Gaussian image point that vary as ρ^2.

4.6 SUMMARY OF RESULTS

The size of a geometrical ray image spot corresponding to a Gaussian image at a height h' from the axis of an optical system aberrated with a primary aberration of peak value A_i is given below. The refractive index of the medium in which the ray spot is formed is assumed to be unity and the f-number of the image-forming light cone is F.

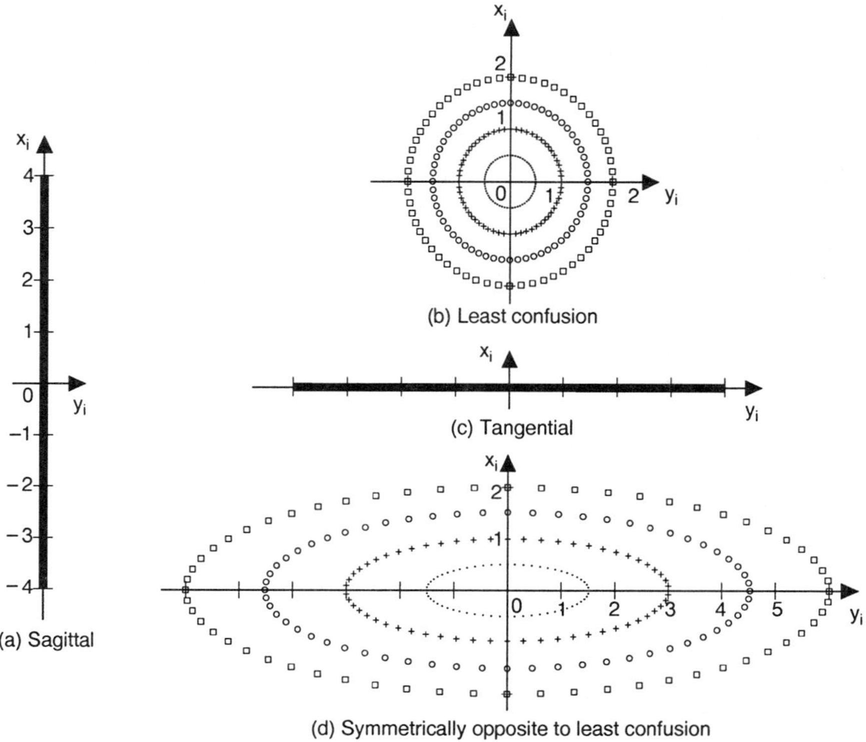

Figure 4-24 Ray distribution of various images for astigmatism. (a) sagittal, (b) least confusion, (c) tangential, and (d) symmetrically opposite to least confusion. The units of x_i and y_i are FA_a.

4.6.1 Spherical Aberration $\left(A_s\rho^4\right)$

$$\text{Longitudinal spherical aberration} = 16F^2A_s \quad . \tag{4-72}$$

A positive value of longitudinal spherical aberration implies that the marginal image lies farther from the exit pupil than the Gaussian image corresponding to $B_d = -2A_s$.

$$\text{Radius of circle of least confusion} = 2FA_s \quad . \tag{4-73}$$

The circle of least confusion lies in a plane that is 3/4 of the way from the paraxial to the marginal image plane.

$$\textit{PSF} \text{ centroid}, \left(x_c, y_c\right) = \left(0, 0\right) \quad . \tag{4-74}$$

$$\textit{RMS} \text{ radius of circle of least confusion} = \sqrt{2}FA_s \quad . \tag{4-75}$$

4.6.2 Coma $\left(A_c\,\rho^3\cos\theta\right)$

$$\text{Sagittal coma} = 2FA_c \quad . \tag{4-76a}$$

$$\text{Tangential coma} = 6FA_c \quad . \tag{4-76b}$$

$$PSF \text{ centroid}, \left(x_c, y_c\right) = \left(2FA_c, 0\right) \quad . \tag{4-77}$$

$$RMS \text{ radius of the image spot with respect to its centroid } = 2\sqrt{2/3}\, FA_c \quad . \tag{4-78}$$

4.6.3 Astigmatism and Field Curvature $\left(A_a \rho^2 \cos^2\theta + A_d \rho^2\right)$

$$\text{Full length of sagittal focal line } = 8FA_a \quad . \tag{4-79}$$

This line is centered on the chief ray at a distance at a distance $\Delta R_s = 8F^2 A_d$ from the Gaussian image point and lies along the x_i axis.

$$\text{Full length of tangential focal line } = 8FA_a \quad . \tag{4-80}$$

This line is centered on the chief ray at a distance $\Delta R_t = 8F^2\left(A_a + A_d\right)$ from the Gaussian image point and lies along the y_i axis. The distance $8F^2 A_a$ between the two line images is the longitudinal astigmatism.

$$\text{Diameter of circle of least confusion } = 4FA_a \quad . \tag{4-81}$$

This circle is centered on the chief ray and lies in a plane that is midway between the sagittal and tangential focal line images. It is referred to as the best image.

$$PSF \text{ centroid}, \left(x_c, y_c\right) = \left(0, 0\right) \quad . \tag{4-82}$$

$$RMS \text{ radius of circle of least confusion } = \sqrt{2}FA_a \quad . \tag{4-83}$$

The radii of curvature of the sagittal, tangential, Petzval, and best-image surfaces are given by

$$R_s = \frac{h'^2}{16F^2 A_d} \tag{4-84a}$$

$$= 1/4\, R^2 a_d \quad , \tag{4-84b}$$

$$R_t = \frac{h'^2}{16F^2\left(A_d + A_a\right)} \tag{6-85a}$$

$$= \frac{1}{4R^2\left(a_a + a_d\right)} \quad , \tag{4-85b}$$

$$\frac{2}{R_p} = \frac{3}{R_s} - \frac{1}{R_t} \tag{4-86a}$$

$$= \frac{16F^2}{h'^2}\left(2A_d - A_a\right) \tag{4-86b}$$

$$= 4R^2\left(2a_d - a_a\right) \quad , \tag{4-86c}$$

and

$$R_b = \frac{h'^2}{8F^2(A_a + 2A_d)} \tag{4-87a}$$

$$= \frac{1}{2R^2(a_a + 2a_d)} \ . \tag{4-87b}$$

Moreover,

$$\frac{1}{R_b} = \frac{1}{2}\left(\frac{1}{R_s} + \frac{1}{R_t}\right) \ . \tag{4-88}$$

In the absence of astigmatism, an image of radius $4FA_d$ is obtained if it is observed in an image plane at a distance $\Delta R = 8F^2 A_d$ from the gaussian image palne.

4.6.4   Distortion $\left(A_t \rho \cos\theta\right)$

A distortion wave aberration of $A_t\, \rho \cos\theta$ corresponds to a wavefront tilt of

$$\beta = A_t / a \ , \tag{4-89}$$

where a is the radius of the exit pupil. The image point lies at $\left(2FA_t, 0\right)$ relative to the Gaussian image point.

4.6.5   Aberration Tolerance

It should be understood that the spot diagrams give only a qualitative description of the actual light distribution in the image of a point object. Nevertheless, they are often used to assess the quality of a system in the early stages of its design. Aberration tolerances based on the spot size can be obtained from the equations discussed in this chapter. For example, the *depth of focus* (giving the tolerance on the location of the plane for observing the image) can be determined from Eq. (4-60), where A_d is the aberration value based on a certain tolerance for the spot radius $4FA_d$. Alternatively, the *depth of field* (giving the tolerance on the object location for a fixed observation plane) can be determined from the depth of focus using Eq. (1-71) for the longitudinal magnification. Similarly, distortion tolerance for a certain amount of *line-of-sight error* can be obtained from Eq. (4-89).

As the design improves and the aberrations reduce, a quantitative assessment of its quality is made on the basis of the diffraction image of a point object. For example, for small aberrations, maximum irradiance is obtained at the Gaussian image point when the variance of the aberration across the exit pupil is minimum. Thus, tolerance for an aberration can be specified in terms of its variance or standard deviation for a certain loss of irradiance at the Gaussian image point.

REFERENCES

1. K. Miyamoto, "On a comparison between wave optics and geometrical optics by using Fourier analysis. I. General Theory," *J. Opt. Soc. Am.* **48**, 57–63 (1958); "II. Astigmatism, coma, spherical aberration" **48**, 567–575 (1958); "III. Image evaluation by spot diagram," **49**, 35–40 (1959); also, "Wave optics and geometrical optics in optical design," *Progress in Optics*, Vol. **1**, 31–65 (1960).

2. J. Braat, "Polynomial expansion of severely aberrated wavefronts," *J. Opt. Soc. Am.* **A4**, 643–650 (1987).

PROBLEMS

4.1   Sketch the geometrical PSF of a system with a uniformly illuminated circular exit pupil aberrated by *spherical aberration* $W(\rho) = A_s\,\rho^4$ in the Gaussian, marginal, least confusion, and midway image planes for $A_s = 1\,\lambda$, $\lambda = 0.5\,\mu m$, and $F = 10$, and total image power of $1W$. Give the location of these image planes with respect to the Gaussian image plane. Calculate the size and the rms radius of the image spot in these planes.

4.2   Consider the imaging system of Problem 4.1 except that it is aberrated by *astigmatism* $W(\rho,\theta) = A_a\rho^2\cos^2\theta$, where $A_a = \lambda/4$. Calculate the size, location, and irradiance of the tangential, sagittal, and least confusion images of a point object.

4.3   Consider an imaging system forming the image of a point object at a distance of 15 cm from the plane of its exit pupil at a height of 0.2 cm from its optical axis. Let the image be aberrated by $\lambda/4$ each of *astigmatism and field curvature*. If the radius of the exit pupil is 1 cm, determine and sketch the *tangential, sagittal,* and *Petzval image surfaces* for $\lambda = 0.5\,\mu m$.

4.4   Sketch the pattern of the image of the point object considered in Problem 4.3 if it is aberrated by *coma* given by $W(\rho,\theta) = A_c\rho^3\cos\theta$, where $A_c = \lambda/4$. Illustrate the tangential and sagittal coma on this sketch. Show how the irradiance of the image varies along the symmetry axis x_i. Determine the rms radius and centroid of the image spot.

4.5   Sketch the pattern of the image of a point object aberrated by *secondary coma* $A_5\rho^5\cos\theta$, where A_5 is the peak value of the aberration. Illustrate the tangential and sagittal coma on the sketch for $F = 4$. and $A_5 = 1.5\,\lambda$, where $\lambda = 3\,\mu m$. Also determine the centroid of the image and its rms radius.

CHAPTER 5

CALCULATION OF PRIMARY ABERRATIONS: REFRACTING SYSTEMS

5.1 Introduction **247**

5.2 Spherical Refracting Surface with Aperture Stop at the Surface......................**249**

    5.2.1   On-Axis Point Object 249

    5.2.2   Off-Axis Point Object..252

            5.2.2.1   Aberrations with Respect to Petzval Image Point.....................253

            5.2.2.2   Aberrations with Respect to Gaussian Image Point 259

5.3 Spherical Refracting Surface with Aperture Stop Not at the Surface **261**

    5.3.1   On-Axis Point Object 262

    5.3.2   Off-Axis Point Object..264

5.4 Aplanatic Points of a Spherical Refracting Surface **266**

5.5 Conic Refracting Surface...**271**

    5.5.1   Sag of a Conic Surface 271

    5.5.2   On-Axis Point Object 275

    5.5.3   Off-Axis Point Object..278

5.6 General Aspherical Refracting Surface **281**

5.7 Series of Coaxial Refracting (and Reflecting) Surfaces **281**

    5.7.1   General Imaging System...282

    5.7.2   Petzval Curvature and Corresponding Field Curvature Wave
            Aberration 282

    5.7.3   Relationship Among Petzval Curvature, Field Curvature, and
            Astigmatism Wave Aberration Coefficients................................287

5.8 Aberration Function in Terms of Seidel Sums or Seidel Coefficients **287**

5.9 Effect of Change in Aperture Stop Position on the Aberration Function.........**290**

    5.9.1   Change of Peak Aberration Coefficients 291

    5.9.2   Illustration of the Effect of Aperture-Stop Shift on Coma and
            Distortion 295

    5.9.3   Aberrations of a Spherical Refracting Surface with Aperture Stop
            Not at the Surface Obtained from Those with Stop at the Surface.........297

5.10 Thin Lens **299**

    5.10.1   Imaging Relations 300

    5.10.2   Thin Lens with Spherical Surfaces and Aperture Stop at the Lens.........301

    5.10.3   Petzval Surface 306

    5.10.4   Spherical Aberration and Coma 307

    5.10.5   Aplanatic Lens...310

    5.10.6   Thin Lens with Conic Surfaces..312

    5.10.7   Thin Lens with Aperture Stop Not at the Lens.............................313

5.11 Field Flattener **314**

    5.11.1 Imaging Relations 315

    5.11.2 Aberration Function...316

5.12 Plane-Parallel Plate..**318**

    5.12.1 Introduction..318

    5.12.2 Imaging Relations 318

    5.12.3 Aberration Function...321

5.13 Chromatic Aberrations...**323**

    5.13.1 Introduction..323

    5.13.2 Single Refracting Surface 323

    5.13.3 Thin Lens...327

    5.13.4 General System: Surface-by-Surface Approach 331

    5.13.5 General System: Use of Principal and Focal Points 336

    5.13.6 Chromatic Aberrations as Wave Aberrations 347

5.14 Symmetrical Principal **348**

5.15 Pupil Aberrations and Conjugate-Shift Equations...............**349**

    5.15.1 Introduction..349

    5.15.2 Pupil Aberrations 350

    5.15.3 Conjugate – Shift Equations 354

    5.15.4 Invariance of Image Aberrations 357

    5.15.5 Simultaneous Correction of Aberrations for Two or More Object Positions 359

References..**360**

Problems **361**

Chapter 5

Calculation of Primary Aberrations: Refracting Systems

5.1 INTRODUCTION

In Chapter 1, we discussed how to determine the Gaussian image of an object formed by an imaging system. In Chapter 2, we defined the entrance and exit pupils of a system, which determine the light cone diverging from a point object that enters the system and the light cone that exits from it converging to the Gaussian image point, respectively. Although we defined wave and ray aberrations of a system in Chapter 3 and determined the image spot shapes and sizes in Chapter 4, we did not discuss the *quality* of the images formed by it. The quality of an image formed by a system depends on its wave aberrations for the point object under consideration. Thus, before we can discuss the quality of an image, we must first determine the aberrations of the system corresponding to this image. This is done by considering the wave aberration of a ray as the difference between its optical path length from the object point to the Gaussian image point and that of the chief ray. It is possible to determine the shape of a refracting surface for which the optical path length of all the rays from a given point object to its Gaussian image point is the same. (See Section 5.4 and Problems 5.1 and 5.2 for an example.) For such a surface, called a *Cartesian surface*, the rays from the given point object all pass through the image point after refraction by it, and the image is said to be *perfect* or aberration free. However, this is not true for any other point object; the images of other point objects are aberrated. Accordingly, such surfaces are not very practical for imaging of extended objects.

In this chapter, we give a step-by-step derivation of the *monochromatic primary* (or *Seidel) aberrations* of systems with an axis of rotational symmetry and express them in the plane of the exit pupil in terms of the pupil coordinates and the image height. The term "monochromatic" implies that the refractive indices used in the derivation are for a certain optical wavelength of the object radiation. We start with a derivation of the primary aberrations of a spherical refracting surface with the aperture stop located at the surface so that the exit pupil is also located there. An axial point object is considered first and because of the rotational symmetry of the problem, only spherical aberration is obtained. Next, the aberrations are determined for an off-axis point object with respect to its Petzval image point. Corrections are then determined and applied to obtain the aberrations with respect to the Gaussian image point. Next, the aberrations are determined at the exit pupil when the aperture stop is not located at the surface. These expressions are used to determine the aplanatic points of a spherical refracting surface.

The aberrations introduced by a (aspheric) conic surface are determined by considering the additional optical path lengths of the rays due to the difference in its shape from a spherical surface whose radius of curvature is equal to the vertex radius of curvature of the conic surface. This is then generalized to an arbitrary rotationally symmetric surface. This completes the derivation of the primary aberrations of a single refracting surface. Next, a procedure is outlined and illustrated by an example for the determination of the aberrations of a multisurface imaging system. In particular, it is shown how the aberration contributions of the surfaces described at their respective exit pupils are added to determine the aberration of the system at its exit pupil.

The primary aberration function of a system can be expressed in a number of different but equivalent ways. Two of these, the so-called *Seidel sums* and *Seidel coefficients*, are considered in the next section. Relationships between the various forms of the coefficients of a given primary aberration are also given. Next we derive the *stop-shift equations* showing how the primary aberrations of a system change with a change in the position of its aperture stop. It is shown, for example, that the coma of a system with nonzero spherical aberration can be made zero by selecting an appropriate position of its aperture stop. It also shows why a certain aberration may not change when the position of the aperture stop is changed. This is followed by determination of the primary aberrations of a few simple systems. A thin lens with an aperture stop at the lens is considered first. It is shown that the spherical aberration of a thin lens cannot be zero for real conjugates. An aplanatic thin lens is also discussed. Aberrations of a thin lens with conic surfaces are considered next and they are then generalized for an arbitrary location of its aperture stop. The aberrations of a field flattener (i.e., a field-flattening thin lens) are considered next, and it is shown that since it is placed at an image formed by a certain system, it introduces only Petzval field curvature and distortion. Finally, the aberrations of a plane-parallel plate with an aperture stop located at its front surface are discussed.

Since the refractive index of a transparent substance varies with the optical wavelength, the angle of refraction of a ray also varies with it. Hence, even the Gaussian image of a multiwavelength point object formed by a refracting system is generally not a point. The distance and height of the image vary with the wavelength. The axial and transverse extents of the image are called *longitudinal* and *transverse chromatic aberrations* or *axial* and *lateral colors*, respectively. They are discussed next for a single refracting surface, a thin lens, an achromatic doublet, and, finally, a general refracting system. The stop-shift equations for chromatic aberrations are also considered. The monochromatic aberrations of a refracting system also vary with the wavelength, but such a variation is small for a small change in the wavelength and is usually negligible. Since the chromatic aberrations represent the variation of image distance or height, they are also expressed as wavefront defocus or tilt aberrations, respectively. A *symmetrical principle* is discussed showing that a system that is symmetrical about its aperture stop images objects with a magnification of -1 that are free of aberrations odd in field angle or image height, e.g., coma, distortion, and lateral color.

Finally, considering the entrance pupil of a system as an object, the aberrations of its exit pupil are obtained by interchanging the roles of the object and the entrance pupil. The expressions for pupil aberrations are used to obtain the *conjugate-shift equations* that relate the aberrations of the image of one object to those of another. From these equations, conditions are obtained under which one or more aberrations of a system can be made zero and invariant with object position. Although all primary aberrations can not be made zero and invariant, one or more of them can be corrected for more than one object position.

Our discussion in this chapter and elsewhere is limited to primary aberrations for simplicity and because they are often the dominant aberrations in a system. Moreover, only Gaussian parameters of a system are needed to determine them. If, however, they do not adequately describe a system, i.e., if the higher-order aberrations are not negligible, the aberrations may be determined by ray tracing the system using a computer. Indeed this is what is often done in practice. A preliminary design of an optical imaging system is carried out based on Gaussian optics, giving a layout of the system. The primary aberrations are determined giving an approximate image quality, and the final design is obtained with the aid of a standard computerized ray-trace program. For an analytical approach to determining the secondary aberrations, the reader may refer to Buchdahl.[1]

5.2 SPHERICAL REFRACTING SURFACE WITH APERTURE STOP AT THE SURFACE

In this section, we derive expressions for the primary aberrations of a spherical refracting surface imaging an axial or a nonaxial point object. The aperture stop is located at the surface so that the entrance and exit pupils are also located there. For an on-axis point object, an object ray incident on the surface has radial symmetry about the optical axis. Hence, the only aberration obtained is spherical aberration. For an off-axis point object, an auxiliary axis is defined and the primary aberrations are first obtained with respect to the Petzval image point. Corrections are then applied to obtain the aberrations with respect to the Gaussian image point.

5.2.1   On-Axis Point Object

As indicated in Figure 5-1, consider a spherical refracting surface of radius of curvature R separating media of refractive indices n and n' corresponding to some wavelength of the object radiation, where $n' > n$. The line joining its vertex V_0 and its center of curvature C is called the *optical axis*. Consider an axial point object P_0 at a (numerically negative) distance S from the vertex. Let P_0' be its Gaussian image at a distance S' from the vertex. We assume that a circular aperture stop is located at the surface. Accordingly, the entrance and exit pupil planes, which are images of the stop by the surface, are also located there.

We now determine the aberration $W_0(r)$ of a ray P_0QP_0' from the point object P_0 passing through a point Q on the surface at a radial distrance r from the optical axis with respect to the chief ray $P_0V_0P_0'$ passing through the center V_0 of the exit pupil. According to Gaussian optics, a ray such as P_0Q from the object P_0 incident on the surface at a certain point Q is refracted as a ray QP_0' passing through the Gaussian image point P_0'. In reality, however, the refracted ray passes through P_0' only if the wave aberration of the ray P_0QP_0' and, therefore, its ray aberration with respect to the chief ray $P_0V_0P_0'$ is zero. A nonzero wave aberration implies that the ray under consideration will intersect the image plane at a point other than P_0' according to Eq. (3-11).

The wave aberration is given by the difference in the optical path lengths of the ray P_0QP_0' and the chief ray $P_0V_0P_0'$ in traveling from P_0 to P_0'; i.e.,

$$W_0(r) = [P_0QP_0'] - [P_0V_0P_0']$$

$$= (nP_0Q + n'QP_0') - (n'S' - nS) \ , \tag{5-1}$$

where the square brackets indicate the optical path length of a ray. {In Figure 5-1 where $n' > n$, both the object and the image are real. For $n' < n$, the refracted ray will bend upward and intersect the optical axis (when extended backwards) at a virtual image point

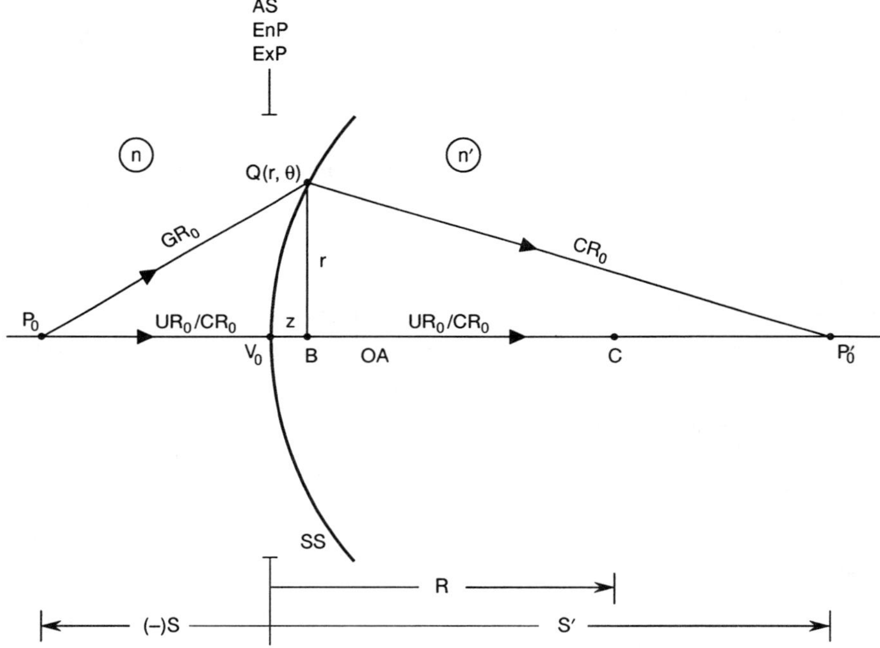

Figure 5-1. On-axis imaging by spherical refracting surface SS of radius of curvature R centered at C separating media of refractive indices n and n', where $n' > n$. P_0 is an on-axis point object at a (numerically negative) distance S from its vertex V_0 whose Gaussian image lies at P_0' at a distance S' from V. The undeviated ray UR_0 passing through the center of curvature C is also the chief ray CR_0 since it also passes through the center of the aperture stop AS. Since the aperture stop is located at the refracting surface, the exit pupil ExP and the entrance pupil EnP are also located there. GR_0 is a general ray from the axial point object passing through a point Q on the refracting surface. The point Q lies at a distance r from the optical axis V_0C.

P_0' to the left of V_0. The optical path length $[P_0 Q P_0']$ in that case would be equal to $n P_0 Q - n' Q P_0'$, i.e., the virtual optical path segment $[Q P_0']$ will be treated as a numerically negative quantity.} Since both Q and V_0 are at a distance R from C, then letting $z = V_0 B$ be the sag of Q, we have

$$R^2 = (R - z)^2 + r^2 \; ,$$

or

$$z^2 - 2Rz + r^2 = 0 \; . \tag{5-2a}$$

Solving the quadratic equation in z and neglecting terms of an order higher than four in r, we may write

$$z \simeq \frac{r^2}{2R} + \frac{r^4}{8R^3} \; . \tag{5-2b}$$

Now

$$P_0 Q = \left[(z - S)^2 + r^2 \right]^{1/2}$$

$$= -S + \frac{1}{2}\left(\frac{1}{R} - \frac{1}{S} \right) r^2 + \left[\frac{1}{8R^2}\left(\frac{1}{R} - \frac{1}{S} \right) + \frac{1}{8S}\left(\frac{1}{R} - \frac{1}{S} \right)^2 \right] r^4 \; . \tag{5-3a}$$

where we have substituted Eq. (5-2) and neglected terms in r of an order higher than four. Similarly,

$$Q P_0' = \left[(S' - z)^2 + r^2 \right]^{1/2}$$

$$= S' + \frac{1}{2}\left(\frac{1}{S'} - \frac{1}{R} \right) r^2 + \left[\frac{1}{8R^2}\left(\frac{1}{S'} - \frac{1}{R} \right) - \frac{1}{8S'}\left(\frac{1}{S'} - \frac{1}{R} \right)^2 \right] r^4 \; . \tag{5-3b}$$

Substituting Eqs. (5-3a) and (5-3b) into Eq. (5-1), we obtain

$$W_0(r) = \left(\frac{n'}{S'} - \frac{n}{S} - \frac{n' - n}{R} \right)\left(\frac{r^2}{2} + \frac{r^4}{8R^2} \right)$$

$$- \frac{1}{8}\left[\frac{n'}{S'}\left(\frac{1}{R} - \frac{1}{S'} \right)^2 - \frac{n}{S}\left(\frac{1}{R} - \frac{1}{S} \right)^2 \right] r^4 \; . \tag{5-4}$$

The first term on the right-hand side of Eq. (5-4) is zero if

$$\frac{n'}{S'} - \frac{n}{S} = \frac{n' - n}{R} \; , \tag{5-5}$$

i.e., if P_0' is the Gaussian image of P_0, as may be seen by comparing Eq. (5-5) with Eq. (1-19). That being the case, Eq. (5-4) reduces to

$$W_0(r) = a_s r^4 \quad , \tag{5-6}$$

where

$$a_s = -\frac{1}{8}\left[\frac{n'}{S'}\left(\frac{1}{R} - \frac{1}{S'}\right)^2 - \frac{n}{S}\left(\frac{1}{R} - \frac{1}{S}\right)^2\right] \quad . \tag{5-7a}$$

Using Eq. (5-5) to express S in terms of S', we can write Eq. (5-7) in the form

$$a_s = -\frac{n'(n'-n)}{8n^2}\left(\frac{1}{R} - \frac{1}{S'}\right)^2\left(\frac{n'}{R} - \frac{n+n'}{S'}\right) \tag{5-7b}$$

$$= -\frac{n'^2}{8}\left(\frac{1}{R} - \frac{1}{S'}\right)^2\left(\frac{1}{n'S'} - \frac{1}{nS}\right) \quad . \tag{5-7c}$$

From our discussion in Section 3.6.2, it should be clear that $W_0(r)$ represents the *fourth-order spherical wave aberration*, and a_s represents its *coefficient*. Moreover, we note from Eq. (5-7a) that when both the object and the image are real (so that $S < 0$ and $S' > 0$), a_s is numerically negative, indicating that the optical path length $[P_0 Q P_0']$ of a ray passing through a point Q on the surface is shorter than that of the chief ray. In other words, the optical path length $[P_0 V_0 P_0']$ of the chief ray is the longest among the path lengths of all the image-forming rays. A negative spherical aberration also implies that the wavefront for the axial point object passing through V_0 is more curved than the Gaussian reference sphere. Thus, the true refracted ray in Figure 5-1 intersects the optical axis at a point to the left of P_0'.

It can be shown (see Problem 5.3) that if we choose r to be the chord $V_0 Q$ instead of being the distance of Q from the optical axis as assumed above, and neglected terms in r of an order higher than four, we would obtain Eq. (5-6) for the aberration $W_0(r)$. This may also be seen from the right-angle triangle $V_0 B Q$ according to which

$$V_0 Q^2 = r^2 + z^2 \quad . \tag{5-8a}$$

Substituting for z from Eq. (5-2b), we find that

$$V_0 Q^4 = r^4 + O(r^6) \quad . \tag{5-8b}$$

Thus, up to the fourth order in r, Eq. (5-6) is independent of the precise definition of r, and we may write the spherical aberration interchangeably in terms of $V_0 Q$ and r.

5.2.2 Off-Axis Point Object

We now consider an off-axis point object and determine the aberrations associated with its Gaussian image. We first determine the aberrations with respect to its Petzval image point, thereby introducing the concept of a spherical image surface called the Petzval image or surface, and then modify them for the Gaussian image. The difference between the aberrations of the two images lies in the field curvature and distortion terms.

5.2.2.1 Aberrations with Respect to Petzval Image Point

We now consider an off-axis point object P located along the x axis at a height h from the optical axis, as indicated in Figure 5-2. Its Gaussian image P' is also located along the x axis at a height h'. It is seen from similar triangles P_0CP and $CP_0'P'$ that the *transverse magnification* of the image is given by

$$M_t = h'/h \tag{5-9a}$$

$$= \frac{S' - R}{S - R} \tag{5-9b}$$

$$= \frac{n\,S'}{n'\,S} \;, \tag{5-9c}$$

where, in the last step, we have made use of the Gaussian imaging Eq. (5-5). Except for notation, this equation is the same as Eq. (1-25a).

The aberration $W(Q)$ of a ray passing through a point Q on the refracting surface is given by the difference in optical path lengths of the ray PQP' and the chief ray PV_0P' passing through the center V_0 of the aperture stop i.e.,

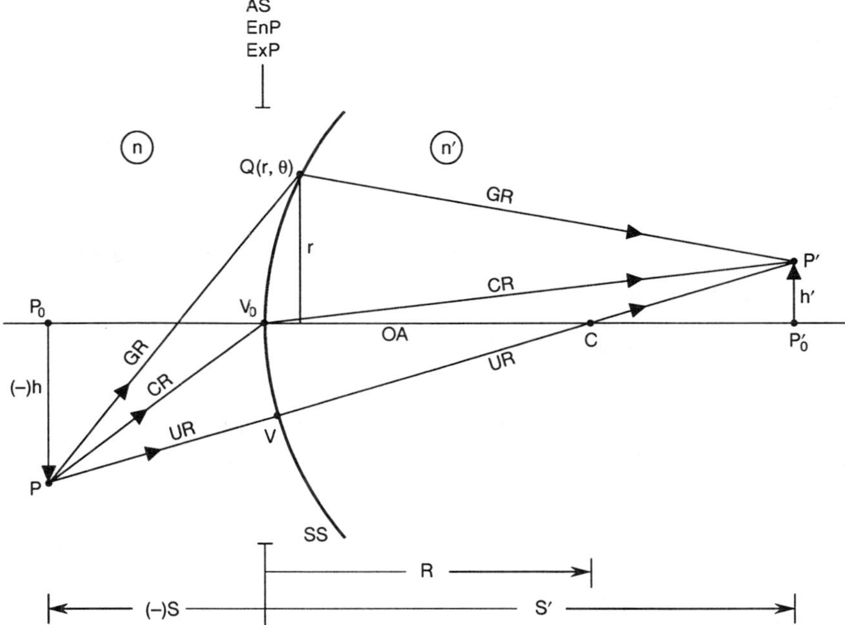

Figure 5-2. Off-axis imaging by a sperical refracting surface. P **is an off-axis object point at a height** h **from the optical axis** V_0C, **and** P' **is its Gaussian image at a height** h'. **The aperture stop** AS, **entrance pupil** EnP, **and exit pupil** ExP **are located at the refracting surface.** CR **– chief ray passing through the center of the exit pupil,** UR **– undeviated ray passing through the center of curvature** C, GR **– general ray passing through a point** Q **on the surface at a distance** r **from the optical axis.**

$$W(Q) = [PQP'] - [PV_0P'] \ . \tag{5-10a}$$

The two optical path lengths in Eq. (5-10a) can be easily determined with respect to the optical path length $[PVP']$, since PVP', passing through the center of curvature C, now forms a reference axis, called the *auxiliary axis*, just as $P_0V_0P_0'$ did in the case of an axial point object P_0. Thus, we write

$$\begin{aligned}
W(Q) &= \{[PQP'] - [PVP']\} - \{[PV_0P'] - [PVP']\} \\
&= a_s\left(VQ^4 - VV_0^4\right)
\end{aligned} \tag{5-10b}$$

We note that the point V, where the undeviated ray PCP' intersects the refracting surface, lies in the tangential plane. Hence, its projection in the plane of the exit pupil lies along the x axis, as shown in Figure 5-3, where we have used the same symbol V for the projected point. From the approximately similar triangles VV_0C and $CP_0'P'$ in Figure 5-2 we note that

$$\frac{VV_0}{P_0'P'} \simeq \frac{V_0C}{CP_0'}$$

or

$$VV_0 \simeq bh' \ , \tag{5-11a}$$

where

$$b = \frac{R}{S' - R} \ . \tag{5-11b}$$

Let (r, θ) be the polar coordinates of the point representing the projection of Q in the plane of the exit pupil, with V_0 as the origin. We note from Figure 5-3 that

$$VQ^2 = r^2 + VV_0^2 + 2rVV_0 \cos\theta \ .$$

Squaring both sides, substituting in Eq. (5-10b), and using Eq. (5-11a), we obtain the *primary aberration function*

$$W(r,\theta;h') = a_s\left(r^4 + 4bh'r^3 \cos\theta + 4b^2h'^2r^2 \cos^2\theta \right.$$

$$\left. + 2b^2h'^2r^2 + 4b^3h'^3r \cos\theta\right) \ . \tag{5-12a}$$

A slightly different approach to obtaining Eq. (5-12a) from Eq. (5-6) may be described as follows. Considering PVP' as the axis of the system, P is an axial point object. Accordingly, following Eq. (5-6), the aberration of the ray PQP' with respect to the ray PVP' is given by a_sVQ^4. It is evident that the aberration function thus obtained has a value of zero at the point V, a point with respect to which the pupil is eccentric. In order that the aberration function be defined with respect to the center V_0 of the pupil and

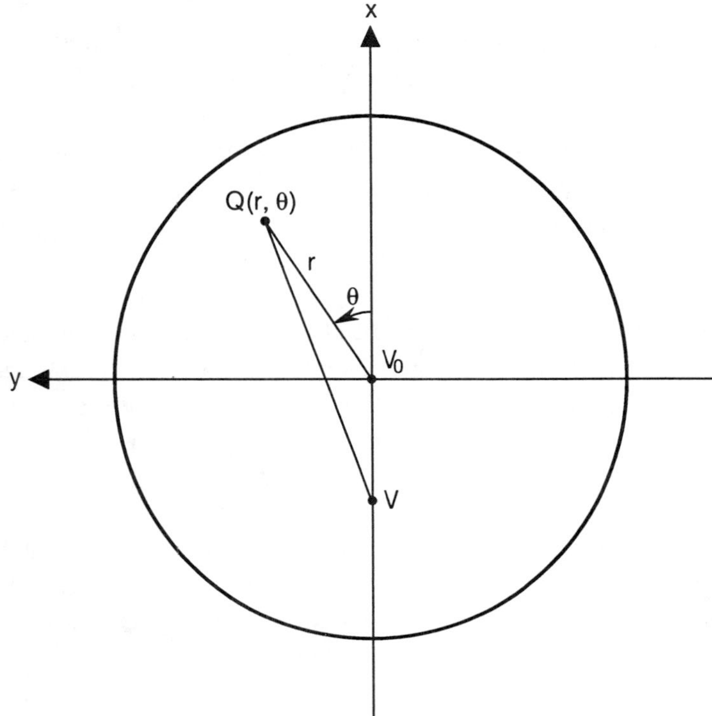

Figure 5-3. Exit pupil showing coordinates (r, θ) of a pupil point Q. V_0 lies at the center of the exit pupil. Although the points Q and V lie on the refracting surface in Figure 5-2, here they are used in the sense of their projections on the exit pupil plane. The x and y axes in this figure (and in Figures 5-7 and 5-12) are shown as observed from the image plane.

have a value of zero at that point, we shift the origin of the coordinate system from V to V_0 and subtract the aberration at point V_0 from the aberration function thus obtained. Letting V_0 be the origin, the rectangular coordinates of point V are $(-bh', 0)$. Accordingly, if (x, y) are the coordinates of point Q, with V_0 as the origin, then $VQ^2 = (x + bh')^2 + y^2$ and the aberration function becomes

$$W(x, y; h') = a_s\left\{\left[(x + bh')^2 + y^2\right]^2 - (bh')^4\right\} \ . \tag{5-12b}$$

Using polar coordinates (r, θ) for the point Q, where $(x, y) = r(\cos\theta, \sin\theta)$, this equation reduces to Eq. (5-12b).

We note that the primary aberration function given by Eq. (5-12a) consists of five terms, as discussed in Section 3.5. The first term is independent of the angle θ of a pupil point Q. It is called *spherical aberration*. The second term, which depends on the pupil coordinates as $r^3 \cos\theta$ and linearly on h', is called *coma*. The third term, which depends on pupil coordinates as $r^2 \cos^2\theta$ and quadratically on h', is called *astigmatism*. The fourth term, which varies as $h'^2 r^2$, is called *field curvature* while the fifth term, which varies as $h'^3 r \cos\theta$, is called *distortion*. We also note that the degree of each term in the coordinates of the object (h) [or image (h') since h and h' are linearly related to each other according to Eq. (5-9)] and the exit pupil (r) is 4. Accordingly, these aberrations are called *fourth-order wave aberrations*, or *primary wave aberrations*. They are also called the *Seidel aberrations*, after Seidel, who first investigated them. Since the ray aberrations are related to the wave aberrations by spatial derivatives [see Eq. (3-11)], their degree is lower by one. Accordingly, primary aberrations are also called *third-order ray aberrations*.

Since we have used axial distances S and S' for the off-axis point object P and its Gaussian image P' respectively, the aberration function of Eq. (5-12a) is with respect to a reference sphere centered at the *Petzval image point* P'_p of P and not its Gaussian image point P'. The Petzval image is obtained from Eq. (5-5) by letting $S = VP$ and determining $S' = VP'_p$, where P'_p lies on the auxiliary axis PCP'.

It should be evident from the symmetry of the spherical refracting surface that the image of a concentric spherical object surface $P_0 P_1$ will also be a concentric spherical surface $P'_0 P'_1$, as indicated in Figure 5-4. Thus, P_1 and P'_1 are Gaussian conjugate points on the auxiliary axis PCP' just as P_0 and P'_0 are on the optical axis $P_0 CP'_0$. We now show that the image of a planar object $P_0 P$ is also a spherical surface, and not a plane surface as assumed in Figure 5-2.

For generality, we first consider a spherical object surface $P_0 P_2$ of radius of curvature R_0 with its center of curvature lying on the optical axis. Let $P'_0 P'_2$ be the corresponding Petzval image surface, where P'_2 is the Gaussian conjugate of P_2 on the auxiliary axis. The object and the image surfaces intersect the optical axis at P_0 and P'_0 at distances S and S', respectively, from the vertex V_0. By differentiating Eq. (5-5), we find that if the object distance changes by a small amount ΔS, then the correspsonding change $\Delta S'$ in the image distance is given by

$$M_l \equiv \frac{\Delta S'}{\Delta S} \tag{5-13a}$$

$$= \frac{n\,S'^2}{n'\,S^2} \tag{5-13b}$$

$$= \frac{n'}{n}\left(\frac{h'}{h}\right)^2 \tag{5-13c}$$

$$= \frac{n'}{n}M_t^2 \;, \tag{5-13d}$$

where we have made use of Eqs. (5-9a) and (5-9c). The quantity M_l representing the ratio $\Delta S'/\Delta S$ is the longitudinal magnification of the image and we note that it is proportional to the square of its transverse magnification. In Figure 5-4, $P'_1 P'_2$ gives the

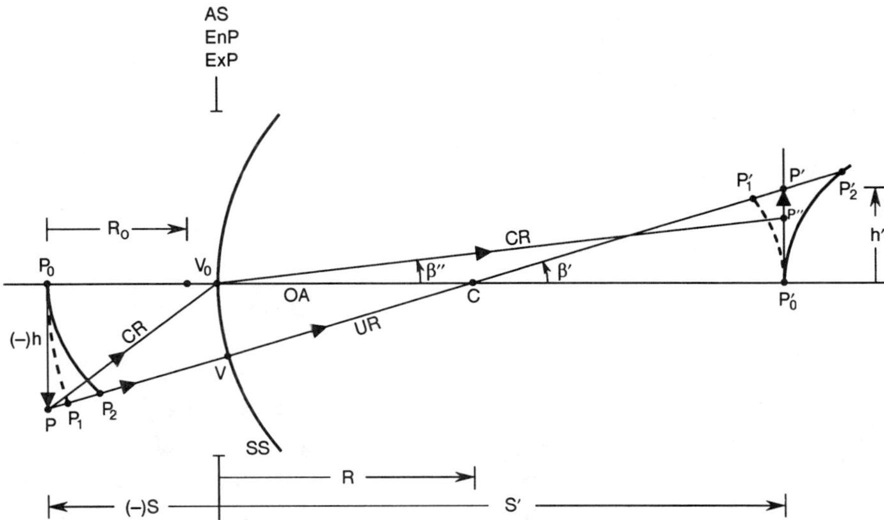

Figure 5-4. Petzval curvature of images. $P_0'P'$ **is the Gaussian image of object**
P_0P**.** $P_0'P_1'$ **is the** *Petzval image* **of object** P_0P_1**; both the object and the image are**
concentric with the refracting surface. $P_0'P_2'$ **is the spherical Petzval image of a**
spherical object P_0P_2**. Note that** $VP_1 = S$ **and** $VP_1' = S'$**.**

increase in image distance $VP_1' = S'$ corresponding to an increase of P_1P_2 in the
(numerically negative) object distance $VP_1 = S$ of conjugates P_1 and P_1'. Now, P_1P_2 is
approximately equal to the difference in the sags of points P_2 and P_1. Since the heights of
P_1 and P_2 from the optical axis are approximately equal to h, referring to Eq. (5-2b), we
may write

$$\Delta S = VP_2 - VP_1$$
$$= P_1P_2$$
$$\approx -\frac{h^2}{2}\left(\frac{1}{R_0} - \frac{1}{R-S}\right) \; , \tag{5-14a}$$

where we have neglected the sag terms of an order higher than the quadratic in h. Note
that since P_2 lies to the right of P_1 in the figure, the center of curvature of the object lies
between P_0 and C and, therefore, $R_0 < R - S$. Similarly, $\Delta S'$ is equal to the difference in
the sags of points P_2' and P_1', i.e.,

$$\Delta S' = VP_2' - VP_1'$$
$$= P_1'P_2'$$
$$\approx \frac{h'^2}{2}\left(\frac{1}{R_i} - \frac{1}{R-S'}\right) \; , \tag{5-14b}$$

where R_i is the radius of curvature of the image surface $P_0'P_2'$. Substituting Eqs. (5-14)
into Eq. (5-13c) and using Eq. (5-5), we obtain (after some manipulations)

$$\boxed{\frac{1}{n'R_i} - \frac{1}{nR_0} = \frac{1}{R}\left(\frac{1}{n'} - \frac{1}{n}\right)} \; . \qquad (5\text{-}15)$$

By letting $R_0 \to \infty$, we obtain the radius of curvature of the spherical image surface for a planar object P_0P:

$$\boxed{\frac{1}{R_i} = \frac{n - n'}{nR}} \; , \qquad (5\text{-}16)$$

which is numerically negative for $n' > n$. This image surface, shown in Figure 5-5 as $P_p'P_p''$, is called the *Petzval surface*, and its radius of curvature R_i is called the *Petzval radius of curvature*. Note that R_i does not depend on the object distance S or the image distance S'; it depends only on the radius of curvature of the refracting surface and the refractive indices of the media that this surface separates. The location of the Petzval image P_p' of an off-axis point object P is the point of intersection of the auxiliary axis PCP' and a spherical surface of radius of curvature R_i centered on the optical axis and passing through the axial image point P_0'.

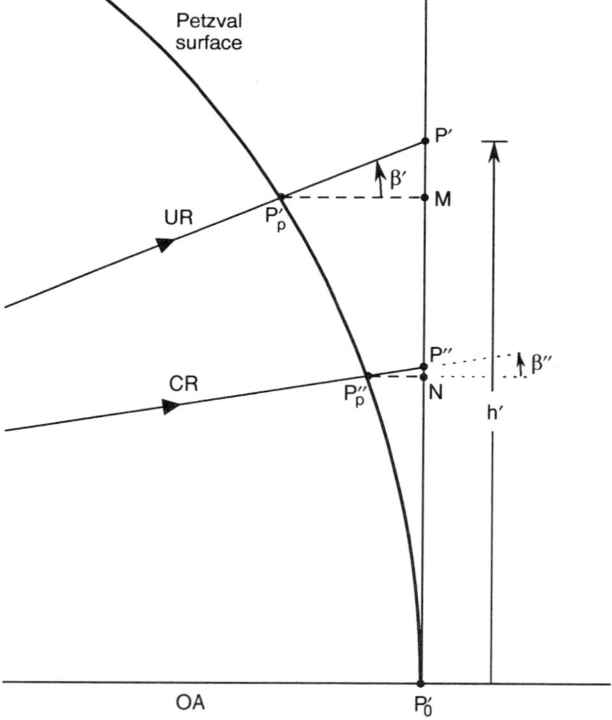

Figure 5-5. Ray distortion $P_p'P_p'' \simeq MN$ on the Petzval surface and $P'P''$ in the Gaussian image plane. $P'P''$ is numerically negative since P'' lies below P', while h' is positive.

5.2.2.2 Aberrations with Respect to Gaussian Image Point

It should be clear from the foregoing discussion that the aberration function for a point object P obtained in Eq. (5-12a) is with respect to a reference sphere centered at its Petzval image point P'_p and not centered at its Gaussian image point P'. To determine the aberration with respect to the Gaussian image point P', we must take into account the effect of (numerically positive) longitudinal defocus error $P'_p P'$. This introduces not only a field curvature aberration term but also a distortion term, as discussed below.

Following Eq. (3-15) and noting that the longitudinal defocus $P'_p P'$ is approximately equal in magnitude but opposite in sign to the (numerically negative) sag $MP'_p = h'^2 / 2R_i$ of the Petzval image point P'_p , we may write the field curvature aberration corresponding to a longitudinal defocus of $P'_p P'$,

$$\Delta W_d(r; h') = -\frac{n'}{2}\left(\frac{r}{S'}\right)^2 P'_p P' \tag{5-17a}$$

$$= \frac{n'}{4R_i}\left(\frac{h'}{S'}\right)^2 r^2 \tag{5-17b}$$

$$= -\frac{n(n'-n)}{4nR}\left(\frac{h'}{S'}\right)^2 r^2 \quad . \tag{5-17c}$$

As indicated in Figure 5-5, the presence of a distortion term in Eq. (5-12a) implies that if it were the only aberration, all the rays, including the chief ray, do not intersect the Petzval surface at the point P'_p; instead, they intersect at a point, say, P''_p. Then $P'_p P''_p \simeq MN$ represents the ray aberration on the Petzval surface corresponding to the distortion wave aberration in Eq. (5-12a). The ray aberration in the Gaussian image plane is, however, given by $P'P''$, where P' and P'' are the points of intersection of the undeviated ray passing through C and the chief ray passing through V_0 with the Gaussian image plane, respectively. We note from Figure 5-5 that

$$P'P'' = MN + \left(P'M - P''N\right)$$

$$\simeq P'_p P''_p + \left(P'M - P''N\right) \quad ,$$

Now

$$P'M = MP'_p \tan\beta'$$

$$\simeq \frac{h'^3}{2R_i\left(S' - R\right)}$$

and

$$P''N = NP''_p \tan\beta''$$

$$\simeq \frac{h'^3}{2R_1 S'} \quad ,$$

where from Figure 5-4,

$$\tan\beta' = \frac{h'}{S'-R}$$

and

$$\tan\beta'' = \frac{h'}{S'} \quad .$$

Thus, following Eqs. (3-20) and (3-21), the additional distortion wave aberration term is given by

$$\Delta W_t(r,\theta;h') = \frac{n'}{S'}\left(P'M - P''N\right)r\cos\theta \tag{5-18a}$$

$$= \frac{n'}{2S'R_i}\left(\frac{1}{S'-R} - \frac{1}{S'}\right)h'^3 r\cos\theta \tag{5-18b}$$

$$= -\frac{n'(n'-n)b}{2nRS'^2}h'^3 r\cos\theta \quad , \tag{5-18c}$$

where in the last step we used Eqs. (5-11b) and (5-16).

Combining the additional field curvature and distortion aberration terms given by Eqs. (5-17) and (5-18c), respectively, with the aberrations of a spherical refracting surface given by Eq. (5-12a) with respect to the Petzval image point P'_p, we obtain its aberrations with respect to a Gaussian image point P'. Hence, the aberration function with respect to the Gaussian image point may be written

$$\boxed{\begin{aligned} W(r,\theta;h') &= a_s r^4 + a_c h' r^3 \cos\theta + a_a h'^2 r^2 \cos^2\theta \\ &\quad + a_d h'^2 r^2 + a_t h'^3 r\cos\theta \quad , \end{aligned}} \tag{5-19}$$

where

$$a_c = 4ba_s \tag{5-20a}$$

$$= -\frac{n'(n'-n)}{2n^2 S'}\left(\frac{1}{R} - \frac{1}{S'}\right)\left(\frac{n'}{R} - \frac{n+n'}{S'}\right) \tag{5-20b}$$

$$= -\frac{n'^2}{2S'}\left(\frac{1}{R} - \frac{1}{S'}\right)\left(\frac{1}{n'S'} - \frac{1}{nS}\right) \quad , \tag{5-20c}$$

$$a_a = 4b^2 a_s \tag{5-21a}$$

$$= -\frac{n'(n'-n)}{2n^2 S'^2}\left(\frac{n'}{R} - \frac{n+n'}{S'}\right) \tag{5-21b}$$

$$= -\frac{n'^2}{2S'^2}\left(\frac{1}{n'S'} - \frac{1}{nS}\right) \;, \tag{5-21c}$$

$$a_d = 2b^2 a_s - \frac{n'(n'-n)}{4nRS'^2} \tag{5-22a}$$

$$= \frac{1}{2}\left[a_a - \frac{n'(n'-n)}{2nRS'^2}\right] \tag{5-22b}$$

$$= \frac{n'^3}{4S'^2}\left(\frac{1}{R} - \frac{1}{S'}\right)\left(\frac{1}{n'^2} - \frac{1}{n^2}\right) \;, \tag{5-22c}$$

and

$$a_t = 4b^3 a_s - \frac{n'(n'-n)b}{2nRS'^2} \tag{5-23a}$$

$$= \frac{n'^3}{2S'^3}\left(\frac{1}{n'^2} - \frac{1}{n^2}\right) \;. \tag{5-23b}$$

The second term on the right-hand side of Eq. (5-22a) is equal to $a_d - a_a/2$ and represents the *field curvature contribution* in going from the *Petzval* to the Gaussian image point. We will refer to it as the *coefficient of Petzval curvature*. For $n' > n$, a_t according to Eq. (5-23b) is numerically negative. It is consistent with Figure 5-5, where the ray distortion $P'P''$ is also numerically negative since P'' lies below P'.

It should be evident from Eq. (5-10a) that if the roles of P and P' are interchanged, the primary aberration of the ray PQP' does not change. This can be shown explicitly by considering P' as the object point and P as its Gaussian image point (see Problem 5.5). It should also be evident that since QP' is generally not the actual refracted ray, the higher-order aberrations for P' as the image point would be different from those for P as the image point.

The results of this section are used in Section 5.10 to obtain the aberrations of a thin lens with an aperture stop located at the lens.

5.3 SPHERICAL REFRACTING SURFACE WITH APERTURE STOP NOT AT THE SURFACE

In Section 5.2, we obtained expressions for the primary aberration functions for a spherical refracting surface imaging axial and non axial point objects. We assumed that the aperture stop was located at the surface, so that the exit pupil was also located there. In this section, we let the aperture stop be located at some arbitrary position, so that the exit pupil is no longer located at the surface, and determine how the primary aberration function changes from that given in Section 5.2. It should be noted that the optical path length of a ray, or its optical path length difference with respect to another, does not change with a change in the position of the aperture stop. However, its radial coordinate in the plane of the exit pupil is different from that in the plane of the surface. Hence, the wave aberration of a ray from an axial point object appears to be scaled. For an off-axis point object, the chief ray also changes and the new aberration function merely describes the wave aberration of a ray with respect to the new chief ray.

5.3.1   On-Axis Point Object

As in Figure 5-1, we consider the imaging at P_0' of an axial point object P_0 by a spherical refracting surface of radius of curvature R, vertex V_0 , and center of curvature C. The object and image lie in media of refractive indices n and n' at distances S and S' from the vertex, respectively. However, whereas in Figure 5-1, the aperture stop was located at the refracting surface, we now consider it at a position such that the image P_0' lies at a distance L from it, as indicated in Figure 5-6a. It is evident from the geometry of the figure that the aperture stop is also the exit pupil of the imaging system. L is numerically positive in this figure since the image lies to the right of the exit pupil.

The aberration of a ray P_0AP_0' from P_0 passing through a point A on the refracting surface with respect to the axial chief ray $P_0V_0P_0'$ passing through the center O of the exit pupil is given by

$$W_0(A) = \left[P_0AP_0'\right] - \left[P_0V_0P_0'\right]$$

$$= a_sV_0A^4 \quad , \tag{5-24}$$

where a_s is given by Eq. (5-8). It represents the optical deviation, along the ray, of a wavefront passing through a certain point such as V_0 or O from a Gaussian reference sphere also passing through the same point. Thus, the aberration of a ray at a point of its intersection with the plane of an exit pupil is independent of the location of the pupil. The ray under consideration passes through a point Q in the plane of the exit pupil. We note from approximately similar triangles V_0AP_0' and OQP_0' in Figure 5-6a that

$$\frac{V_0A}{OQ} \simeq \frac{S'}{L} \quad . \tag{5-25}$$

Hence, the aberration of the ray in the exit pupil may be written

$$W_0(Q) = a_s(S'/L)^4 OQ^4 \quad , \tag{5-26}$$

or

$$W_0(r) = a_s(S'/L)^4 r^4 \quad , \tag{5-27}$$

where $r = OQ$ is the radial distance of Q from the center O of the exit pupil. Thus, comparing Eqs. (5-6) and (5-27), we note that owing to the exit pupil not being at the refracting surface, the aberration is scaled by the factor $(S'/L)^4$. Of course, according to Eq. (5-25), the radial coordinate r in Eq. (5-27) is smaller by a factor of (S'/L) compared with the coordinate r in Eq. (5-26). Hence, the numerical value of the aberration of the ray has not changed.

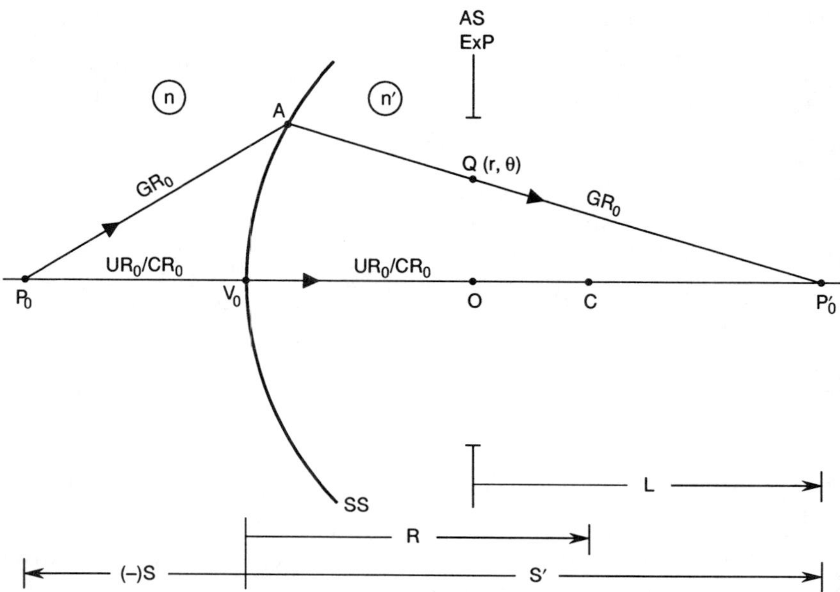

Figure 5-6a. Imaging of an on-axis point object P_0 by a spherical refracting surface of radius of curvature R and center of curvature C when the aperture stop and, therefore, the exit pupil are not located at the surface. The Gaussian image is located at P_0'. L is numerically positive in the figure since the image lies to the right of the exit pupil.

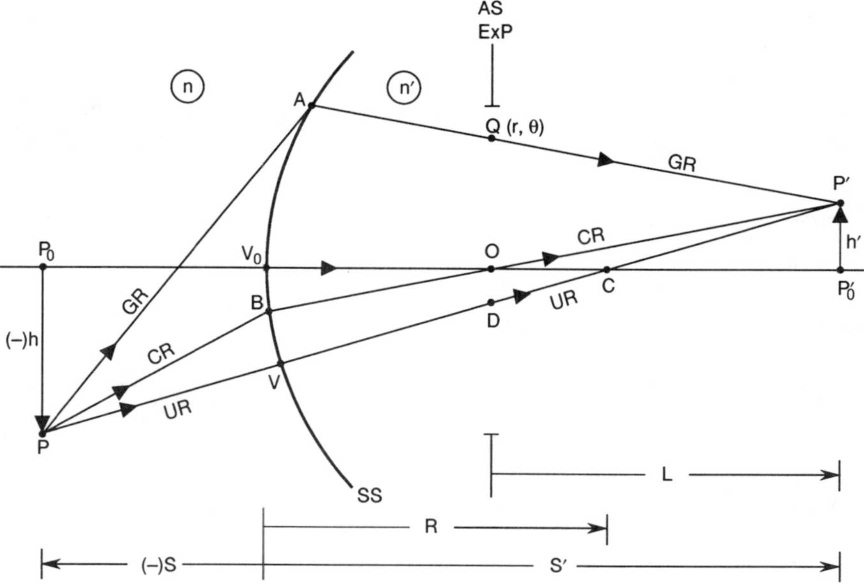

Figure 5-6b. Imaging of an off-axis point object P by a spherical refracting surface of radius of curvature R, center of curvature C when the aperture stop and, therefore, the exit pupil are not located at the surface. The Gaussian image is located at P'.

5.3.2   Off-Axis Point Object

In Figure 5-6b, the aberration of a ray PAP' from an off-axis point object P passing through a point A on the refracting surface with respect to the off-axis chief ray PBP' passing through the center O of the exit pupil is given by

$$W(A) = [PAP'] - [PBP']$$

$$= \{[PAP'] - [PVP']\} - \{[PBP'] - [PVP']\}$$

$$= a_s(VA^4 - VB^4) \quad,$$

or

$$W(Q) = a_s(S'/L^4)(DQ^4 - DO^4) \quad, \tag{5-28}$$

where Q is the point at which the refracted ray AP' intersects the plane of the exit pupil. From similar triangles ODC and $CP_0'P'$ we note that

$$\frac{DO}{P_0'P'} = \frac{OC}{CP_0'} \quad,$$

or

$$DO = dh' \quad, \tag{5-29a}$$

where

$$d = \frac{R - S' + L}{S' - R} \quad. \tag{5-29b}$$

Letting r and θ be the polar coordinates of Q and O as the origin, we note from Figure 5-7 that

$$DQ^2 = r^2 + DO^2 + 2rDO\cos\theta \quad. \tag{5-30}$$

Substituting Eq. (5-30) into Eq. (5-28) and using Eq. (5-29a), we obtain

$$W(r,\theta;h') = a_s(S'/L)^4(r^4 + 4dh'r^3\cos\theta + 4d^2h'^2r^2\cos^2\theta$$

$$+\ 2d^2h'^2r^2 + 4d^3h'^3r\cos\theta) \quad. \tag{5-31}$$

Comparing Eqs. (5-12a) and (5-31), we find that the effect of the exit pupil not being at the refracting surface is to scale the aberration by a factor of $(S'/L)^4$ and to replace the quantity b by the quantity d. Of course, as the exit pupil moves to the refracting surface (i.e., as $L \to S'$), Eq. (5-31) reduces to Eq. (5-12a).

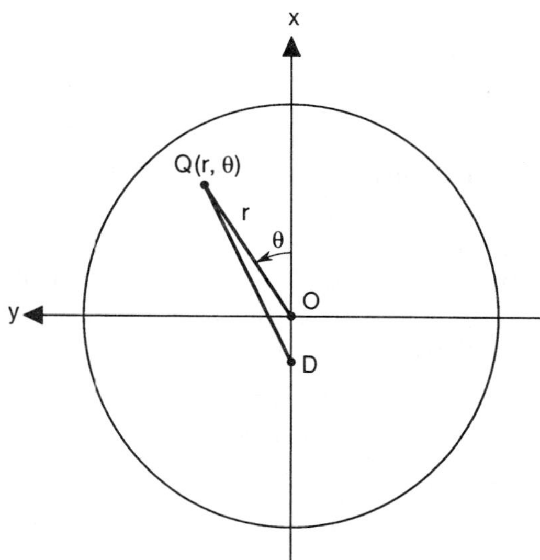

Figure 5-7. Coordinates (r, θ) **of a pupil point** Q**. As in Figure 3-6b,** D **is the point of intersection of the undeviated ray with the plane of the exit pupil.**

Once again we note that the aberration function obtained in Eq. (5-31) represents the aberration of a ray with respect to the image point on a Petzval surface. The aberration function with respect to the Gaussian image point P' is obtained by adding the defocus and distortion terms similar to those given by Eqs. (5-17) and (5-18). However, since we are considering aberration in the plane of the exit pupil, S' in Eqs. (5-17) and (5-18a) must be replaced by L. Moreover, since the chief ray now passes through the center of the exit pupil, $\tan \beta'' = h'/L$ and, therefore, $P''N$ is equal to $h'^3/2R_iL$. Thus, the primary aberration function of a spherical surface with respect to the Gaussian image point P' may be written

$$
\boxed{
\begin{aligned}
W_s(r,\theta;h') \;=\; & a_{ss}r^4 + a_{cs}h'r^3\cos\theta + a_{as}h'^2r^2\cos^2\theta \\
& + a_{ds}h'^2r^2 + a_{ts}h'^3r\cos\theta \quad,
\end{aligned}
}
\tag{5-32}
$$

where

$$
a_{ss} \;=\; (S'/L)^4 a_s \quad,
\tag{5-33}
$$

$$
a_{cs} \;=\; 4d a_{ss} \quad,
\tag{5-34}
$$

$$
a_{as} \;=\; 4d^2 a_{ss} \quad,
\tag{5-35}
$$

$$
a_{ds} \;=\; 2d^2 a_{ss} - \frac{n'(n'-n)}{4n\,RL^2}
\tag{5-36a}
$$

$$
\;=\; \frac{1}{2}\left[a_{as} - \frac{n'(n'-n)}{2n\,RL^2} \right] \quad,
\tag{5-36b}
$$

and

$$a_{ts} = 4d^3 a_{ss} - \frac{n'(n'-n)d}{2nRL^2} \quad .$$

(5-37)

The second term on the right-hand side of Eqs. (5-36) is equal to $a_{ds} - a_{as}/2$ and represents the *coefficient of Petzval curvature.*

Since the aberrations depend on the location of the exit pupil through the value of L, it is possible to select its location so that one or more aberrations are zero. For example, if the exit pupil is located at the center of curvature of the refracting surface, then $L = S' - R$ and $d = 0$. Accordingly, coma, astigmatism, and distortion vanish. This is the principle of *Schmidt* and *Bouwers-Maksutov cameras*, which utilize a spherical mirror with an aperture stop located at its center of curvature, as we will see in Section 6.6. The fact that distortion is zero implies that the chief ray CR passes through the Gaussian image point P'. Indeed, the undeviated ray UR is the chief ray in this case. As stated at the end of Section 5.3.1, the numerical value of the spherical aberration of a ray does not change with a change in the position of the aperture stop. A general discussion of the dependence of the aberrations of a system on the location of its exit pupil is given in Section 5.9.

5.4 APLANATIC POINTS OF A SPHERICAL REFRACTING SURFACE

An optical imaging system that is free of spherical aberration and coma is called an *aplanatic system.* Conjugate points that are free of these aberrations are called *aplanatic points*. From Eqs. (5-33) and (5-34), we now determine the values of the image distance S' for which both a_{ss} and a_{cs} are zero. For these values of S', we then determine the other aberration coefficients according to Eqs. (5-35) through (5-37).

Substituting Eq. (5-7b) into Eq. (5-33), we may write

$$a_{ss} = -\frac{n'(n'-n)}{8n^2} \frac{S'(S'-R)^2\left[n'S' - (n+n')R\right]}{R^3 L^4}$$

(5-38)

$$= 0 \begin{cases} \text{for } S' = 0, \text{ or} & \text{(5-39a)} \\[2ex] S' = R, \text{ or} & \text{(5-39b)} \\[2ex] S' = \dfrac{n+n'}{n'} R \quad . & \text{(5-39c)} \end{cases}$$

Substituting Eqs. (5-29b) and (5-38) into Eq. (5-34), we find that

$$a_{cs} = -\frac{n'(n'-n')}{2n^2} \frac{S'(R-S'+L)(S'-R)\left[n'S' - (n+n')R\right]}{R^3 L^4}$$

(5-40)

$$= 0 \begin{cases} \text{for } S' = 0, \text{ or} & \text{(5-41a)} \\[2ex] S' = R, \text{ or} & \text{(5-41b)} \\[2ex] S' = \dfrac{n+n'}{n'}\, R \quad . & \text{(5-41c)} \end{cases}$$

It is clear from Eqs. (5-39) and (5-41) that there are three values of S', namely, $0, R$, and $(n+n')\,R/n'$ for which both spherical aberration and coma are zero. The corresponding values of the object distances as given by Eq. (5-5) are $0, R$, and $(n+n')\,R/n$, respectively, as illustrated in Figure 5-8. The corresponding image magnifications given by Eq. (5-9c) are 1, n/n', and $(n/n')^2$, respectively. [The unity magnification for zero object distance may be obtained from Eq. (5-5) as follows: For very small object distances S, the right-hand side of Eq. (5-5), which is constant, becomes negligible compared with the terms on the left-hand side, and the image distance is given approximately by $S' = n'\,S/n$. This leads to unity magnification when substituted into Eq. (5-9c). As the object distance approaches zero, so does the image distance.] In each of the three cases, either the object or the image is virtual. The conjugate points in any of the three conjugate planes are aplanatic, and conjugate planes may be called *aplanatic planes*.

Substituting Eqs. (5-29b) and (5-38) into Eq. (5-35), we find that the astigmatism coefficient is given by

$$a_{as} = -\frac{n'(n'-n)}{2n^2}\,\frac{S'(R-S'+L)^2\left[n'S'-(n+n')R\right]}{R^3 L^4} \tag{5-42}$$

$$= \begin{cases} 0 \text{ for } S' = 0 \quad , & \text{(5-43a)} \\[2ex] \dfrac{n'(n'-n)}{2nRL^2} \text{ for } S' = R \quad , & \text{(5-43b)} \\[2ex] 0 \text{ for } S' = (n+n')\,R/n \quad . & \text{(5-43c)} \end{cases}$$

Similarly, the field curvature coefficient given by Eq. (5-36) reduces to

$$a_{ds} = -\frac{n'(n'-n)}{4nRL^2}\left\{\frac{S'(R-S'+L)^2\left[n'S'-(n+n')R\right]}{nR^2L^2}+1\right\} \tag{5-44}$$

$$= \begin{cases} -\dfrac{n'(n'-n)}{4nRL^2} \quad \text{for } S' = 0 \quad, & (5\text{-}45a) \\[2em] 0 \quad \text{for } S' = R \quad, & (5\text{-}45b) \\[2em] -\dfrac{n'(n'-n)}{4nRL^2} \quad \text{for } S' = (n+n')\,R/n' \quad. & (5\text{-}45c) \end{cases}$$

From Eqs. (5-43) and (5-45), we find that as expected, the coefficient of Petzval curvature given by $a_{ds} - a_{as}/2$ is equal to $-n'(n'-n)/4nRL^2$ in each case.

Finally, the distortion coefficient given by Eq. (5-37) may be written

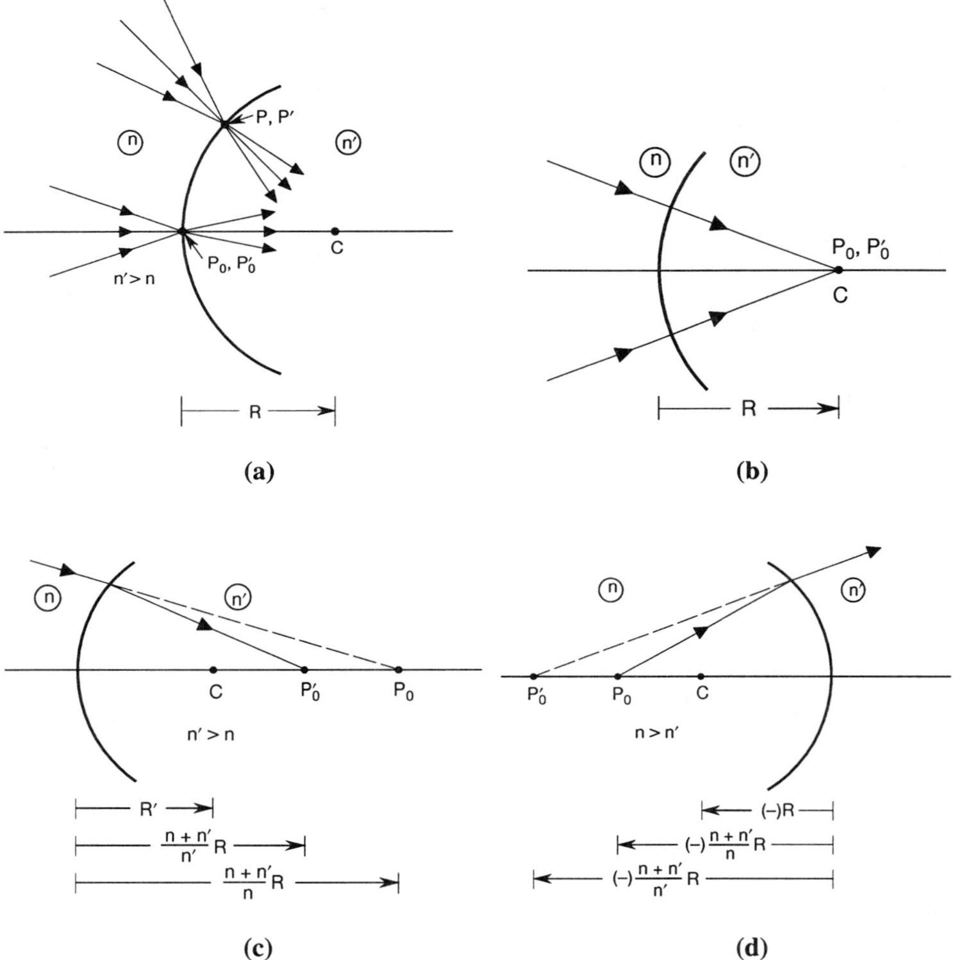

Figure 5-8. Aplanatic points of a spherical refracting surface. (a) Object on the surface. (b) Virtual object at the center of curvature. (c) Virtual object at an aplanatic point P_0. (d) Real object at an aplanatic point P_0.

$$a_{ts} = -\frac{n'(n'-n)}{2nRL^2}\frac{R-S'+L}{S'-R}\left\{\frac{S'(R-S'+L)^2}{nR^2L^2}\left[n'(S'-R)-nR\right]+1\right\} \tag{5-46}$$

$$= -\frac{n'(n'-n)}{2nR^2L^4}(R-S'+L)\left\{\frac{n'S'(R-S'+L)^2}{nR}-\left[L^2-2LS'-S'(R-S')\right]\right\} \tag{5-47}$$

$$= \begin{cases} \dfrac{n'(n'-n)}{2nR^2L^2}(R+L) & \text{for } S' = 0 \ , & \text{(5-48a)} \\[2em] -\dfrac{n'(n'-n)}{nRL}\left(\dfrac{n'-n}{2nR}+\dfrac{1}{L}\right) & \text{for } S' = R \ , & \text{(5-48b)} \\[2em] -\dfrac{n'(n'-n)}{2nRL^2}\left(\dfrac{n'}{n}\dfrac{L}{R}-1\right) & \text{for } S' = (n+n')R/n' \ . & \text{(5-48c)} \end{cases}$$

It is evident that although $d \to \infty$ for $S' = R$, each aberration coefficient is finite.

Substituting the values of the aberration coefficients from Eqs. (5-39), (5-41), (5-43), (5-45), and (5-48) into Eq. (5-32), we obtain the aberration function for each of the three pairs of aplanatic points:

$$W(r,\theta;h') = \begin{cases} -\dfrac{(n'-n)}{4nRL^2}h'^2r^2 + \dfrac{n'(n'-n)(R+L)}{2nR^2L^2}h'^3r\cos\theta & \text{for } S' = 0 \ , & \text{(5-49a)} \\[2em] \dfrac{n'(n'-n)}{2nRL^2}h'^2r^2\cos^2\theta - \dfrac{n'(n'-n)}{nRL}\left(\dfrac{n'-n}{2nR}+\dfrac{1}{L}\right)h'^3r\cos\theta & \text{for } S' = R \ , \\[2em] & \hspace{4em}\text{(5-49b)} \\[1em] -\dfrac{n'(n'-n)}{4nRL^2}h'^2r^2 - \dfrac{n'(n'-n)}{2nRL^2}\left(\dfrac{n'}{n}\dfrac{L}{R}-1\right)h'^3r\cos\theta & \text{for } S' = \dfrac{n+n'}{n'}R \ . \\[2em] & \hspace{4em}\text{(5-49c)} \end{cases}$$

It is evident that the image of a point object placed at the center of curvature C is perfect in that all rays from the point object that are incident on the surface appear to diverge from the center of curvature after refraction; since the rays are incident normally on the surface, their angles of incidence and refraction are zero. Thus, the aplanatic points (R, R) are a Cartesian pair. It can be shown (see Problem 5.2) that for the conjugate pair $[(n+n')R/n, (n+n')R/n']$, not only the primary spherical aberration, but all orders of spherical aberration are zero, i.e., it is also a Cartesian pair. These pairs of aplanatic points are utilized for aplanatic imaging by contact magnifiers (see Problem 1.12). They are also used in the design of an aplanatic lens (see Section 5.10.5) and, in turn, a microscope objective for aplanatic imaging (see Problem 5.11). Owing to the symmetry of the refracting surface about its center of curvature C, the second Cartesian pair implies that the spherical surfaces of radii $(n'/n)R$ and $(n/n')R$ centered at C are perfect conjugates of each other. Moreover, the conjugates $(0, 0)$ imply that the refracting surface is perfectly imaged upon itself. Since spherical aberration, coma, and astigmatism are all zero in Eqs. (5-49a) and (5-49c), the corresponding conjugate planes are anastigmatic. A spherical refracting surface is Cartesian for the three conjugate pairs. The anastigmatic nature of a spherical surface for zero object distance yields an anastigmatic thin lens, called a field flattening lens, when placed at the image of a certain object. Its Petzval field curvature, which is independent of the object position, is used to cancel the Petzval curvature of the imaging system, as discussed in Section 5.11.

The aberration function for aplanatic points when the aperture stop is located at the surface can be obtained from Eqs. (5-49) by letting $L = S'$. The aplanatic points $S = 0 = S'$ must be excluded from such consideration since the aperture stop cannot lie in the image plane; since it cannot block the rays, it cannot be an aperture stop. The other two pairs of aplanatic points and the aberration coefficients associated with them can, of course, also be obtained from Eqs. (5-7b), (5-20b), (5-21b), (5-22c) and (5-23b). The aberration function for these pairs is given by

$$W(r,\theta;h') = \begin{cases} \dfrac{n'(n'-n)}{2nR^3}h'^2r^2\cos^2\theta + \dfrac{n'^3}{2R^3}\left(\dfrac{1}{n'^2}-\dfrac{1}{n^2}\right)h'^3r\cos\theta, \text{ for } S' = R \ , \\ \\ \hspace{6em} (5\text{-}50a) \\ \\ -\dfrac{n'^3(n'-n)}{4n(n'+n)^2R^3}h'^2r^2 + \dfrac{n'^6}{2(n'+n)^3R^3}\left(\dfrac{1}{n'^2}-\dfrac{1}{n^2}\right)h'^3r\cos\theta \ , \\ \\ \text{for } S' = \dfrac{n+n'}{n'}R \ . \hspace{6em} (5\text{-}50b) \end{cases}$$

Comparing Eqs. (5-49b) and (5-50a), and Eqs. (5-49c) and (5-50b), we note that spherical aberration and coma are, of course, zero for the object positions under consideration. However, this illustrates the fact that if a system is aplanatic for a certain position of the aperture stop, it is aplanatic for any position. Moreover, for an aplanatic system, the peak values of its astigmatism and field curvature do not change with a change in the position of its aperture stop. For example, if we let a_1 and a_2 be the radii of the exit pupils in Figures 5-2 and 5-6, respectively, we note that

$$\frac{a_1}{a_2} = \frac{S'}{L} \ . \hspace{6em} (5\text{-}51)$$

The peak value of astigmatism in Eq. (5-49b) is equal to $n'(n'-n)h'^2a_2^2/2nRL^2$ or $n'(n'-n)h'^2a_1^2/2nR^3$, which in turn is equal to the peak value of astigmatism in Eq. (5-50a). Similarly, the peak value of the field curvature in Eq. (5-49c) is given by $-n'(n'-n)h'^2a_2^2/4nRL^2$, or $-n'^3(n'-n)h'^2a_1^2/4n(n'+n)^2R^3$, which in turn is equal to the peak value of the field curvature in Eq. (5-50b). The peak value of distortion of the aplanatic system changes with a change in the position of the aperture stop, as may be seen by comparing the values obtained from Eqs. (5-49b) and (5-50a), and Eqs. (5-49c) and (5-50b). The manner in which the primary aberrations of an optical system change with a change in the position of its aperture stop is discussed in detail in Section 5.9.

5.5 CONIC REFRACTING SURFACE

So far we have determined the primary aberrations of a spherical refracting surface with an arbitrary location of its aperture stop. Now we consider the corresponding aberrations of a (spheric) *conic refracting surface*. We start with a description of the sag of a conic surface and determine the difference in its sag up to the fourth order from that of a spherical surface whose radius of curvature is the same as the vertex radius of curvature of the conic surface. For an on-axis point object, the path length of a chief ray from the object point to its image point does not change. However, the sag difference between the two surfaces contributes to a change in the optical path length of other rays, thereby contributing additional wave aberrations to them. For an off-axis point object, the optical path length of a chief ray also changes. The difference in the change in the optical path length of a ray with respect to that of the chief ray determines its additional aberration.

5.5.1 Sag of a Conic Surface

A *conic* is the locus of a point P which moves so that its distance from a fixed point F, called focus, bears a constant ratio e, called *eccentricity*, to its distance from a fixed straight line DE called the directrix. Consider a conic in the zx plane with an origin at its vertex V_0, as indicated in Figure 5-9. The z axis lies along its axis of symmetry. Let $(S, 0)$ be the coordinates of its *geometrical focus* F [which is not to be confused with its Gaussian focus, which depends on the refractive indices n and n' and its vertex radius of curvature according to Eq. (1-30) and which is the same for all conics shown in the figure]. Then, by the definition of a conic,

$$PF/PB = e \quad , \tag{5-52}$$

where $P(z, x)$ is a point on it. Substituting for PF and PB, we obtain

$$(z - S)^2 + x^2 = e^2 \left[z + (S/e) \right]^2$$
$$= (ze + S)^2 \quad , \tag{5-53}$$

where we have used the fact that $V_0 A = S/e$ by virtue of the limiting case of Eq. (5-52) in which P is replaced by V_0. Simplifying Eq. (5-53), we obtain

$$z^2 \left(1 - e^2 \right) - 2zS(1 + e) + x^2 = 0 \quad . \tag{5-54}$$

The vertex radius of curvature is given by

$$R = \left\{\left[1+(dz/dx)^2\right]^{3/2} \Big/ \left(d^2z/dx^2\right)\right\}_{z=0,x=0} \tag{5-55}$$

$$= S(1+e) \ . \tag{5-56}$$

Substituting Eq. (5-56) into Eq. (5-54), we obtain the equation of a conic of eccentricity e and vertex radius of curvature of R:

$$z^2\left(1-e^2\right)-2Rz+x^2 = 0 \ . \tag{5-57}$$

When $0 < e < 1$, Eq. (5-57) represents an *ellipse*,

$$\frac{(z-a)^2}{a^2}+\frac{x^2}{b^2} = 1 \ , \tag{5-58}$$

with its center at $(a, 0)$ and semimajor and semiminor axes given by

$$a = R\Big/\left(1-e^2\right) \tag{5-59}$$

and

$$b = R\left(1-e^2\right)^{1/2} \ , \tag{5-60}$$

respectively, as may be seen by comparing Eqs. (5-57) and (5-58). An ellipse with $a > b$ is referred to as a *prolate ellipse*. Its two foci, F_1 and F_2, lie symmetrically along the z axis on the same side of the vertex V_0. An ellipse with $a < b$ is referred to as an *oblate ellipse* and corresponds to an imaginary value of eccentricity so that e^2 is negative. The two foci in this case lie along an axis parallel to the x axis.

When $e > 1$, Eq. (5-57) represents a *hyperbola*,

$$\frac{(z-a)^2}{a^2}+\frac{x^2}{b'^2} = 1 \ , \tag{5-61}$$

where a is given by Eq. (5-59) and b' is given by

$$b' = R\left(e^2-1\right)^{1/2} \ . \tag{5-62}$$

The quantities a and b' are called the semilengths of the transverse and conjugate axes, respectively. The quantity $2a$ is equal to the distance between the vertices of the two branches of a hyperbola. Its two foci, F_1 and F_2, lie along the z axis on opposite sides of the vertex of a branch. The line joining F_1 and F_2 is called the transverse axis and the perpendicular bisector of the segment F_1F_2 is called the conjugate axis. When $e = 1$, Eq. (5-57) reduces to

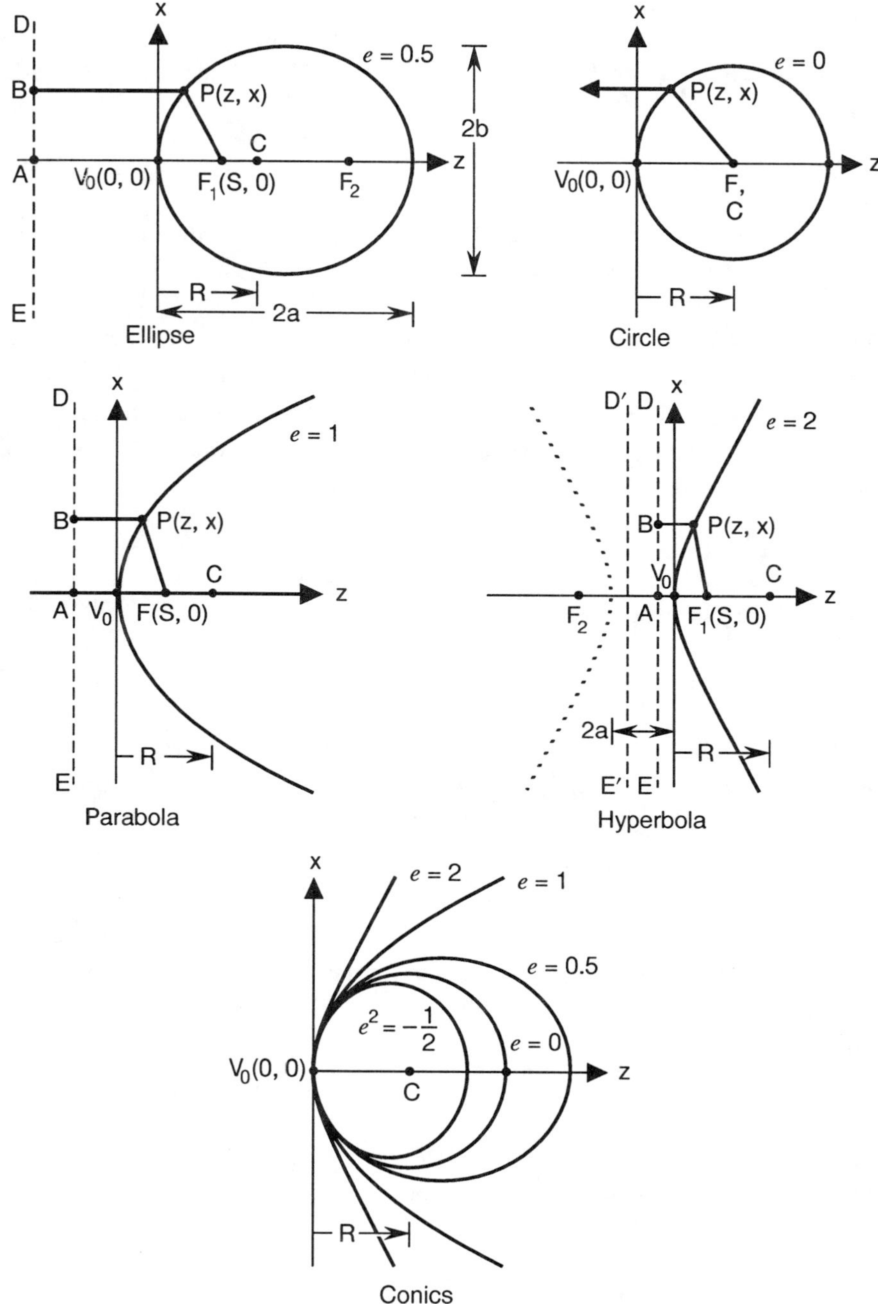

Figure 5-9. Conic surface of eccentricity e showing its vertex V_0, directrix DE, and symmetry axis z. The origin of the coordinate system lies at V_0. C and R are its vertex center and radius of curvature, respectively. An ellipse and a hyperbola have two geometrical foci each. In the case of a parabola, one focus lies at infinity, while in the case of a circle, the two foci coincide at its center and the directrix lies at infinity as indicated by the arrow.

$$x^2 = 2Rz \quad , \tag{5-63}$$

which is the equation of a *parabola* with its focus at $S = R/2$. When $e = 0$, Eq. (5-57) reduces to

$$(z - R)^2 + x^2 = R^2 \quad , \tag{5-64}$$

which is the equation of a *circle* of radius R centered at $(R, 0)$. The directrix in this case lies at infinity, as indicated by an arrow in the figure.

Both the ellipse and the hyperbola have two foci and two directrices. The parabola is a special case of an ellipse whose one focus lies at infinity. A circle is its special case where the two foci coincide at its center. A *conic of revolution* (referred to here simply as a *conic*) about the z axis may be obtained by replacing x^2 by r^2, where r is the distance of a point on it with coordinates (x, y, z) from its axis given by

$$r^2 = x^2 + y^2 \quad . \tag{5-65}$$

Thus, Eq. (5-57) may be generalized to

$$z^2\left(1 - e^2\right) - 2Rz + r^2 = 0 \quad . \tag{5-66}$$

The 3D surface thus obtained is called an *ellipsoid* for $0 < e < 1$, *hyperboloid* for $e > 1$, *paraboloid* for $e = 1$, and a *sphere* for $e = 0$. The *sag* of the conic is described by its z coordinate, which may be written from Eq. (5-66) as

$$z = \left\{ R \pm \left[R^2 - \left(1 - e^2\right)r^2 \right]^{1/2} \right\} \Big/ \left(1 - e^2\right) \quad . \tag{5-67}$$

We choose the negative sign in Eq. (5-67) since, for the conic, $z \to 0$ as $r \to 0$. Thus, we may write

$$z = \frac{R}{1 - e^2}\left[1 - \left(1 - \frac{1 - e^2}{R^2}r^2 \right)^{1/2} \right] \quad . \tag{5-68}$$

Multiplying the numerator and the denominator on the right-hand side of Eq. (5-68) by

$$\left[1 + \left(1 - \frac{1 - e^2}{R^2}r^2 \right)^{1/2} \right] \quad ,$$

we obtain

$$z = \frac{r^2/R}{1 + \left[1 - \left(1 - e^2\right)r^2/R^2 \right]^{1/2}} \quad . \tag{5-69}$$

Equation (5-69) describes the sag of a conic surface of eccentricity e and a vertex radius of curvature R. Compared to Eq. (5-68), it has the advantage that it does not give indeterminate sag for $e = 1$. The sag is often written in terms of a Schwarzschild's *conic constant* κ where $\kappa = -e^2$.

We will use a subscript c to denote the coordinates of a point on a conic surface. Thus, we describe the *sag of a conic surface* of eccentricity e and vertex radius of curvature R according to

$$z_c = \frac{r_c^2 / R}{1 + \left[1 - \left(1 - e^2 \right) r_c^2 / R^2 \right]^{1/2}} \ , \tag{5-70}$$

where, as illustrated in Figure 5-10, $\left(x_c, y_c, z_c \right)$ are the coordinates of a point $\overline{A}$ on it and

$$r_c = \left(x_c^2 + y_c^2 \right)^{1/2} \tag{5-71}$$

is the distance of the point from the z axis. The various conic surfaces are described by their values of e according to

$$e \ = \ 1 \qquad \text{Paraboloid} \tag{5-72a}$$

$$< \ 1 \qquad \text{Ellipsoid} \tag{5-72b}$$

$$> \ 1 \qquad \text{Hyperboloid} \tag{5-72c}$$

$$= \ 0 \qquad \text{Sphere} \quad . \tag{5-72d}$$

If we neglect the terms in r_c of an order higher than four, Eq. (5-70) becomes

$$z_c \ = \ \frac{r_c^2}{2R} + \left(1 - e^2 \right) \frac{r_c^4}{8R^3} \quad . \tag{5-73}$$

Thus, up to the fourth order in r_c, the sag of a spherical $\left(e = 0 \right)$ surface is larger than that of a conic surface by $e^2 r_c^4 / 8R^3$. (An exception is the oblate ellipsoid for which the reverse is true.)

5.5.2 On-Axis Point Object

Now, we consider the aberrations produced by a conic surface by comparing them with those produced by a spherical surface whose radius of curvature is equal to the vertex radius of curvature of the conic surface. The position of the Gaussian image of an object formed by a refracting surface depends on its vertex radius of curvature. Hence, the Gaussian image of an axial point object P_0 in Figure 5-11a or an off-axis point object

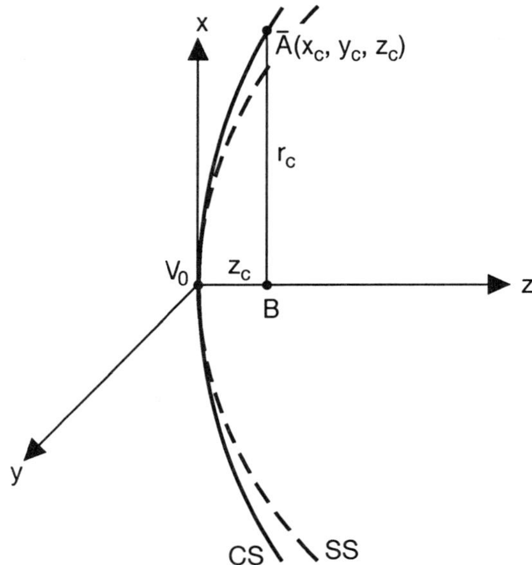

Figure 5-10. Sag of a conic surface _CS_. The origin of the coordinate system lies at the vertex V_0 of the conic. The axis about which the conic is rotationally symmetric is the z axis of the coordinate system. $z_c = V_0 B$ is the sag of a point _A_ on the conic. _SS_ is a spherical surface passing through V_0 and has the same radius of curvature as the vertex radius of curvature of the conic.

_P_ in Figure 5-11b lies at the same point P_0' or P', respectively, for both the conic and spherical surfaces.

Compared to a spherical surface, a conic surface introduces an additional aberration which for a ray from an axial point object P_0 passing through a point _A_ on the spherical surface in Figure 5-11a is given by

$$\Delta W_c\left(\overline{A}_0\right) \simeq \left(n' - n\right)\overline{A}_0 A \quad , \tag{5-74}$$

where

$$\begin{aligned}
\overline{A}_0 A &\simeq e^2 V_0 \overline{A}_0{}^4 \\
&\simeq e^2 r_c^4 / 8R^3
\end{aligned} \tag{5-75}$$

is approximately equal to the sag difference between a sphere and a conic of the same vertex radius of curvature at a height r_c from the optical axis. It represents the fact that the ray segment $\overline{A}_0 A$ lies in a medium of refractive index n' in the case of the conic surface and n in the case of the spherical surface. The refracted rays $\overline{A}_0 P_0'$ and $A P_0'$ for the two surfaces intersect the plane of the exit pupil _ExP_ at approximately the same point $Q(r, \theta)$, keeping in mind that in practice, $\overline{A}_0 A$ may only be on the order of a few

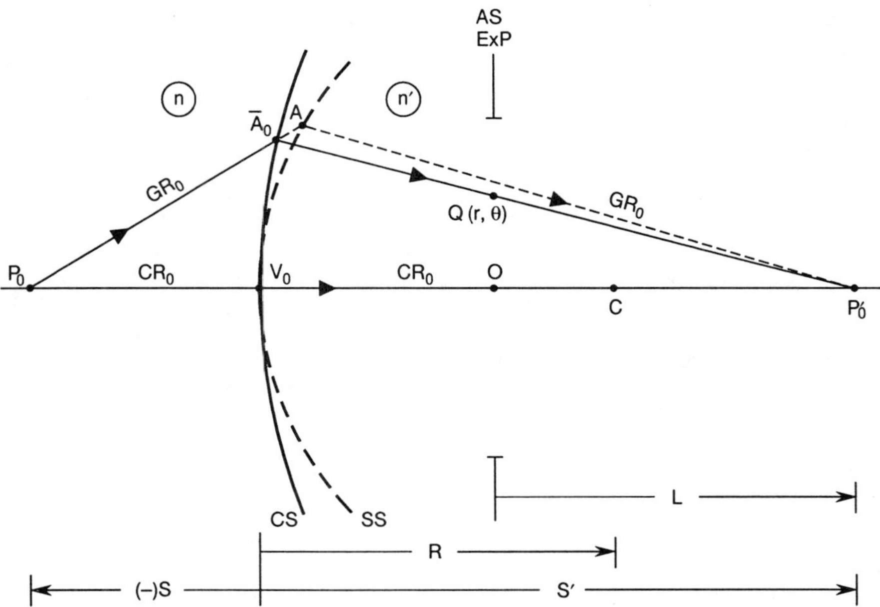

Figure 5-11a. Imaging of an on-axis point object P_0 by a conic refracting surface CS of a vertex radius of curvature R and center of curvature C. SS is a spherical surface of a radius of curvature R with its center of curvature at C. The Gaussian image, which is determined by the vertex radius of curvature, is located at P_0' for both surfaces.

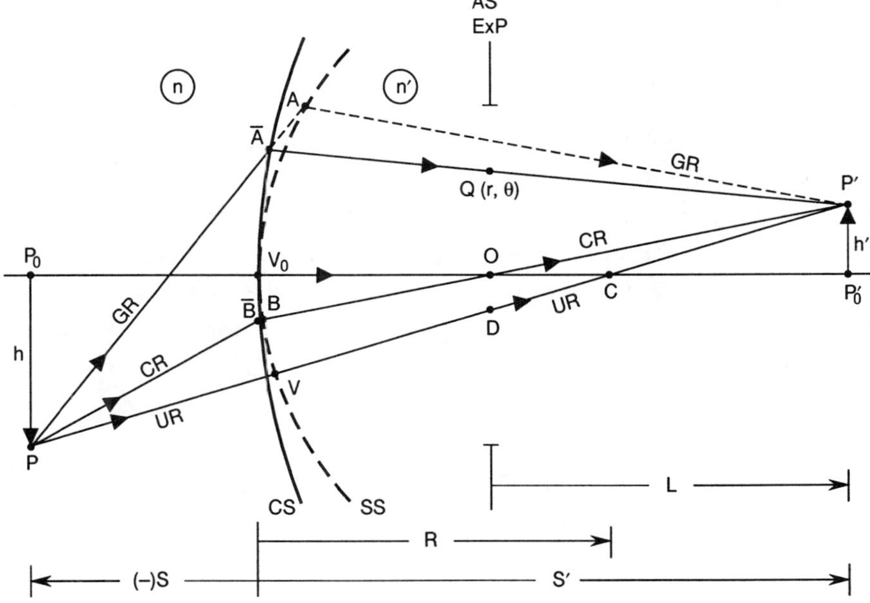

Figure 5-11b. Imaging of an off-axis point object P by a conic refracting surface of a vertex radius of curvature R and center of curvature C. The Gaussian image is located at P' for both the conic surface CS and spherical surface SS.

micrometers. According to Fermat's principle, any difference in the geometrical path lengths $\overline{A_0 A P_0'}$ and $\overline{A_0 P_0'}$ is of second order in $\overline{A_0 A}$ and, therefore, it is negligible. Replacing r_c by $V_0 A$ and substituting Eq. (5-75) into Eq. (5-74), we may write

$$\Delta W_c\left(\overline{A_0}\right) = \sigma V_0 \overline{A_0}^4 \ , \tag{5-76}$$

where

$$\sigma = \left(n' - n\right)e^2 / 8R^3 \ . \tag{5-77}$$

The aberration at a point Q in the plane of the exit pupil at a distance r from the optical axis is obtained by replacing r_c by $\left(S'/L\right)r$ i.e.,

$$\Delta W_c(Q) = \sigma\left(S'/L\right)^4 OQ^4 \ ,$$

or

$$\Delta W_c(r) = \sigma\left(S'/L\right)^4 r^4 \ . \tag{5-78}$$

The total aberration of a conic surface is obtained by adding the *conic contribution* to that of a spherical surface which is given by Eq. (5-27), i.e.,

$$\boxed{W_0(r) = \left(a_s + \sigma\right)\left(S'/L\right)^4 r^4} \ . \tag{5-79}$$

The spherical aberration is, of course, independent of the polar coordinate θ of the point Q.

5.5.3   Off-Axis Point Object

For an off-axis point object such as P in Figure 5-11b, the optical path length of the chief ray for a conic surface is also different from that for a spherical surface. The difference in the optical path lengths of the corresponding rays for the two surfaces lies in the fact that $\overline{A}A$ and $\overline{B}B$ lie in a medium of refractive index n' in the case of a conic surface and n in the case of a spherical surface. As in the case of an axial ray, the difference in the geometrical path lengths of the refracted rays for the two surfaces are of second order in $\overline{A}A$. Accordingly, the conic contribution to the aberration of a ray from the point object P and passing through point $\overline{A}$ on the conic surface is given by

$$\Delta W_c\left(\overline{A}\right) \simeq \left(n' - n\right)\left(\overline{A}A - \overline{B}B\right)$$

$$= \sigma\left(V_0 \overline{A}^4 - V_0 \overline{B}^4\right) \ . \tag{5-80}$$

Let (r, θ) be the polar coordinates of a point Q where the ray under consideration intersects the plane of the exit pupil. This point is approximately the same for the conic and spherical surfaces; the difference is greatly exaggerated in Figure 5-11b. Figure 5-12 shows a projection of the exit pupil on the refracting surface with P' as the center of projection. We note from the figure that

$$V_0\overline{A}^2 \;=\; \overline{AB}^2 \;+\; V_0\overline{B}^2 \;-\; 2\overline{AB}\,V_0\overline{B}\cos\theta \;. \tag{5-81}$$

Also, from Figure 5-11b

$$\overline{AB} \;\simeq\; (S'/L)r \tag{5-82a}$$

and

$$V_0\overline{B} \;\simeq\; gh' \;, \tag{5-82b}$$

where

$$g \;=\; \frac{S'-L}{L} \;. \tag{5-83}$$

Substituting Eqs. (5-82) and (5-83) into Eq. (5-81), squaring the result, and then substituting into Eq. (5-80), we obtain

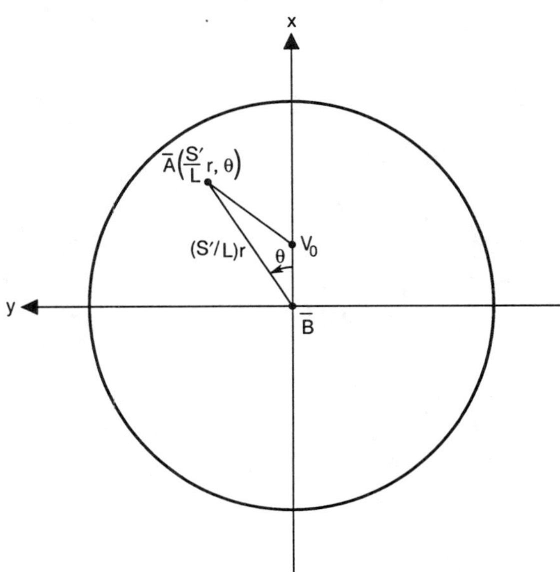

Figure 5-12. Projection of the exit pupil on the refracting surface as viewed from P' in Figure 5-11b. Point $\overline{B}$ which lies on the chief ray, forms the center of the projected pupil. In practice, the points A and $\overline{A}$ in Figure 5-11b are very close to each other. Hence, they are practically indistinguishable from each other in the above figure.

$$\Delta W_c(Q) = \sigma\left[\left(S'/L\right)^4 r^4 - 4\left(S'/L\right)^3 gh'r^3\cos\theta + 4\left(S'/L\right)^2 g^2 h'^2 r^2\cos^2\theta \right.$$

$$\left. + 2\left(S'/L\right)^2 g^2 h'^2 r^2 - 4\left(S'/L\right)g^3 h'^3 r\cos\theta\right] \quad . \tag{5-84}$$

Adding the conic contribution given by Eq. (5-84) to the aberration of a spherical surface given by Eq. (5-32), we obtain the total primary aberration function for a conic surface in the plane of the exit pupil. Thus,

$$W_c(Q) = W_s(Q) + \Delta W_c(Q) \quad ,$$

or

$$\boxed{\begin{aligned} W_c(r,\theta;h') &= a_{sc}r^4 + a_{cc}h'r^3\cos\theta + a_{ac}h'^2 r^2\cos^2\theta \\ &\quad + a_{dc}h'^2 r^2 + a_{tc}h'^3\cos\theta \quad , \end{aligned}} \tag{5-85}$$

$$a_{sc} = \left(S'/L\right)^4\left(a_s + \sigma\right) \tag{5-86a}$$

$$= a_{ss} + \sigma\left(S'/L\right)^4 \quad , \tag{5-86b}$$

where

$$a_{cc} = a_{cs} - 4\sigma g\left(S'/L\right)^3 \tag{5-87a}$$

$$= 4\left[da_{ss} - \sigma g\left(S'/L\right)^3\right] \quad , \tag{5-87b}$$

$$a_{ac} = a_{as} + 4\sigma g^2\left(S'/L\right)^2 \tag{5-88a}$$

$$= 4\left[d^2 a_{ss} + \sigma g^2\left(S'/L\right)^2\right] \quad , \tag{5-88b}$$

$$a_{dc} = a_{ds} + 2\sigma g^2\left(S'/L\right)^2 \tag{5-89a}$$

$$= 2\left[d^2 a_{ss} - \frac{n'(n'-n)}{8nRL^2} + \sigma g^2\left(S'/L\right)^2\right] \tag{5-89b}$$

$$= \frac{1}{2}\left[a_{ac} - \frac{n'(n'-n)}{2nRL^2}\right] \quad , \tag{5-89c}$$

and

$$a_{tc} = a_{ts} - 4\sigma g^3\left(S'/L\right) \tag{5-90a}$$

$$= 4\left[d^3 a_{ss} - \frac{n'(n'-n)d}{8nRL^2} - \sigma g^3\left(S'/L\right)\right] \quad . \tag{5-90b}$$

We note that the coefficient of Petzval curvature, represented by the second term on the right-hand side of Eq. (5-89c), does not change as we go from a spherical to a conic surface.

We also note that if the aperture stop is located at the surface so that $L = S'$, then $g = 0$. Hence, in that case, the aberrations of a conic surface differ from those of a spherical surface only in spherical aberration by σr^4, i.e.,

$$\boxed{W_c(r,\theta;h') \;=\; W_s(r,\theta;h') + \sigma r^4} \;. \tag{5-91}$$

When $L = S'$, the chief ray passes through the vertex V_0 and its optical path length is the same for both surfaces, i.e., the points B and $\overline{B}$ coincide with V_0 and, therefore, $\overline{B}B = 0$.

5.6 GENERAL ASPHERIC REFRACTING SURFACE

In Section 5.5 we showed how the difference in sag of a conic surface from that of a corresponding spherical surface contributes to the wave aberration of a ray. Following the same procedure, we can write the wave aberration of a ray for an arbitrary rotationally symmetric surface. A general rotationally symmetric aspheric surface with a vertex radius of curvature R is often described by its conic component and a series of higher-order terms (up to r^{12}) in the form

$$z_g \;=\; \frac{r^2/R}{1+\left[1-\left(1-e^2\right)r^2/R\right]^{1/2}} + a_4 r^4 + a_6 r^6 + a_8 r^8 + a_{10} r^{10} + a_{12} r^{12} \;, \tag{5-92}$$

where a_4, etc., are the *surface coefficients*. The sag of the surface up to the fourth order is given by

$$z_g \;=\; \frac{r^2}{2R} + \left(1-e'^2\right)\frac{r^4}{8R^3} \;, \tag{5-93a}$$

where

$$e'^2 \;=\; e^2 - 8a_4 R^3 \;. \tag{5-93b}$$

Thus, up to the fourth order, the general aspheric surface and a conic of eccentricity e' are equivalent. Hence, the primary aberrations of the general aspherical surface may be obtained from those of a conic surface, given above, by simply replacing e by e'.

5.7 SERIES OF COAXIAL REFRACTING (AND REFLECTING) SURFACES

Given the expressions for the wave aberrations of an arbitrary rotationally symmetric refracting surface with its aperture stop located at some arbitrary position, we now proceed to determine the aberrations of an imaging system consisting of a series of coaxial refracting surfaces. As an example, the procedure outlined in Section 5.7.1 below is applied in Section 5.7.2 to determine the field curvature wave aberration due to the Petzval curvature of a multisurface system. This procedure is also used in Sections 5.10, 5.11, and 5.12 to determine the aberrations of a thin lens, field flattener, and a plane-parallel plate, respectively. A relationship among the coefficients of Petzval curvature, field curvature, and astigmatism is obtained in Section 5.7.3.

5.7.1 General Imaging System

Consider an optical system consisting of a series of coaxial refracting and/or reflecting surfaces. Each surface produces primary aberrations with its own value of h' and L. The Gaussian image of a point object formed by the first surface acts as an object for the second surface, and so on. Similarly, the exit pupil ExP_1 for the first surface is the image of the system entrance pupil EnP formed by the surface, which in turn acts as the entrance pupil EnP_2 for the second surface, and so on. The aberration is calculated surface by surface, and the aberration of the system is obtained by adding the aberration contributions of all the surfaces. Since the aberration of a surface is calculated at a point on its exit pupil, the coordinates of a pupil point must be transformed using pupil magnification of a surface to obtain the aberration contribution of a surface at a point on the exit pupil of the system. Similarly, the image magnification of a surface can be used to obtain the system aberration in terms of the height of the image formed by the system. For example, if $W_1(x_1, y_1; h_1')$ represents the aberration at a point (x_1, y_1) in the plane of the exit pupil ExP_1 of the first surface for an image of height h_1', it can be converted to an aberration contribution at a conjugate point (x_2, y_2) in the plane of the exit pupil ExP_2 of the second surface and image height h_2' by letting

$$\left(x_1, y_1; h_1'\right) = \left(x_2/m_2, y_2/m_2; h_2'/M_2\right) \quad , \tag{5-94}$$

where m_2 and M_2 represent the transverse pupil and image magnifications, respectively, for the second surface. Thus, if $W_2(x_2, y_2; h_2')$ represents the aberration contribution of the second surface at the point (x_2, y_2) corresponding to an image height of h_2', the total aberration for a two-surface system will be given by

$$\boxed{W_s\left(x_2, y_2; h_2'\right) = W_1\left(\frac{x_2}{m_2}, \frac{y_2}{m_2}; \frac{h_2'}{M_2}\right) + W_2(x_2, y_2; h_2')} \quad . \tag{5-95}$$

An alternate approach for obtaining the system aberration is to calculate the peak values of the primary aberrations contributed by each surface and sum them term by term to obtain their peak values for the entire system. Thus, for example if A_{ci} is the peak value of coma contributed by the ith surface, then

$$A_c = \sum_{i=1}^{k} A_{ci} \tag{5-96}$$

is the peak value of coma for a system consisting of k surfaces. The coma aberration at a point (r, θ) in the plane of the exit pupil of the system is given by $A_c \rho^3 \cos\theta$, where $\rho = r/a$ is the normalized radial coordinate of the point.

5.7.2 Petzval Curvature and Corresponding Field Curvature Wave Aberration

We demonstrate the use of Eq. (5-95) by calculating the Petzval curvature of a multielement system in two different ways. The field curvature wave aberration contribution due to Petzval curvature can be calculated surface by surface or by first calculating the Petzval curvature produced by the whole system. If n_0, n_1, ..., n_k represent the refractive indices of the media separating a series of k refracting surfaces of vertex radii of curvature R_1, R_2, ..., R_k, then following Eq. (5-15), the radii of curvature R_{i1}, R_{i2}, ..., R_{ik}, of the Petzval image surfaces formed by them are given by

$$\frac{1}{n_1 R_{i1}} - \frac{1}{n_0 R_0} = \frac{1}{R_1}\left(\frac{1}{n_1} - \frac{1}{n_0}\right) , \qquad (5\text{-}96)$$

$$\frac{1}{n_2 R_{i2}} - \frac{1}{n_1 R_{i1}} = \frac{1}{R_2}\left(\frac{1}{n_2} - \frac{1}{n_1}\right) ,$$

$$\cdot$$
$$\cdot$$

and

$$\frac{1}{n_k R_{ik}} - \frac{1}{n_{k-1} R_{ik-1}} = \frac{1}{R_k}\left(\frac{1}{n_k} - \frac{1}{n_{k-1}}\right) , \qquad (5\text{-}97)$$

respectively. Adding these equations and letting the radius of curvature of the object surface $R_0 \to \infty$, we obtain the radius of curvature of the *Petzval image surface* produced by a system of k refracting surfaces according to

$$\boxed{\frac{1}{R_{ik}} = n_k \sum_{j=1}^{k} \frac{1}{R_j}\left(\frac{1}{n_j} - \frac{1}{n_{j-1}}\right) .} \qquad (5\text{-}98)$$

We note that it is independent of the object and image distances. The field curvature aberration of the system due to its Petzval curvature can be written immediately once R_{ik} is known. For a point object whose Gaussian image point is at a height h', the sag of the Petzval surface representing longitudinal defocus is given by $h'^2/2R_{ik}$. Hence, the field curvature aberration due to Petzval curvature is given by Eq. (5-17b) with S' replaced by L, i.e.,

$$\Delta W_d(r; h') = \frac{n_k}{4R_{ik}}\left(\frac{h'}{L}\right)^2 r^2$$

$$= a_{ps} h'^2 r^2 , \qquad (5\text{-}99\text{a})$$

where

$$a_{ps} = \frac{n_k}{4R_{ik} L^2} \qquad (5\text{-}99\text{b})$$

is the coefficient of Petzval curvature of the system, L is the distance of the image from the exit pupil of the system, and r is the radial distance of a point in the plane of the exit pupil from its center.

Example: Wave Aberration due to Petzval Curvature

We now derive Eq. (5-99) by application of Eq. (5-95), i.e., by adding the field curvature aberrations due to Petzval curvature produced by each surface of the imaging system. As indicated in Figure 5-13, let $h_1 \equiv h$ be the height of a point object $P_1 \equiv P$ placed at a distance $S_1 \equiv S$ from the first surface of the system. Let h_1' and S_1' be the height and the distance of the Gaussian image P_1', respectively, formed by the first surface. The entrance pupil EnP_1 of the system is also the entrance pupil EnP_1 for the first surface. Let ExP_1 be the image of EnP_1 formed by the first surface. Thus, ExP_1 is the exit pupil for the first surface. Let L_1 be the (axial) distance of the image P_1' from the exit pupil ExP_1. The field curvature aberration due to the Petzval curvature produced by the first surface is given by Eq. (5-17c) with S' replaced by L_1, i.e.,

$$\Delta W_{d1}(r_1;h_1') = -\frac{n_1(n_1-n_0)}{4n_0 R L_1^2} h_1'^2 r_1^2$$

$$= \frac{n_1^2}{4R_1}\left(\frac{1}{n_1}-\frac{1}{n_0}\right)\left(\frac{h_1'}{L_1}r_1\right)^2 \quad , \tag{5-100}$$

where r_1 is the radial distance of a point on ExP_1 from its center. Following Eq. (5-89c), it may also be written

$$\Delta W_{d1}(r_1;h_1') = \left(a_{d1}-\frac{1}{2}a_{a1}\right)h_1'^2 r_1^2 \quad , \tag{5-101}$$

where a_{d1} and a_{a1} are the defocus and astigmatism aberration coefficients for the first surface.

The image P_1' of an object formed by the first surface acts as an object for the second surface. Let h_2' and S_2' be the height and the image distance, respectively, of the image

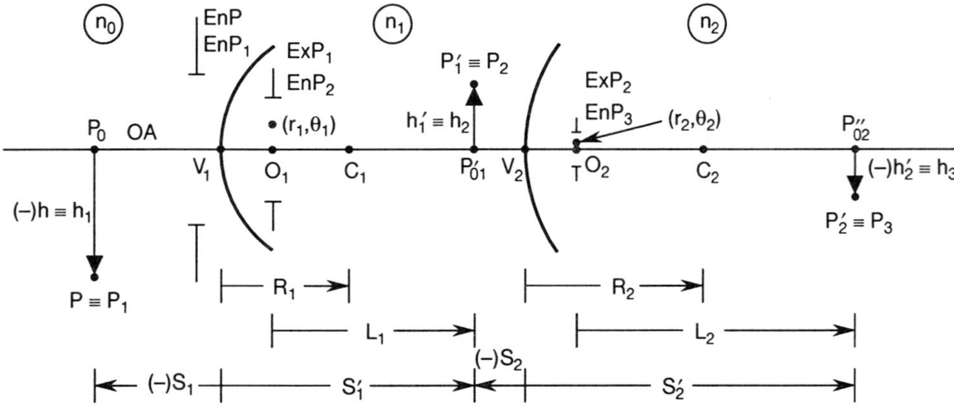

Figure 5-13. Imaging by a multisurface optical system.

P_2' formed by the second surface corresponding to an object $P_2 \equiv P_1'$ at a height $h_2 \equiv h_1'$ at a (numerically negative) distance S_2. Similarly, let ExP_2 be the image of $EnP_2 \equiv ExP_1$ formed by the second surface and L_2 be the distance of the image P_2' from it. Let r_2 be the radial distance of a point on ExP_2 from its center corresponding to the image of a point on EnP_2 at a distance r_1 from its center. Following Eq. (5-94), the aberration ΔW_{d1} can be written in terms of r_2 and h_2' according to

$$\Delta W_{d1}\left(r_1; h_1'\right) = \Delta W_{d1}\left(\frac{r_2}{m_2}; \frac{h_2'}{M_2}\right) \quad , \tag{5-102}$$

where

$$\left|m_2\right| = \frac{r_2}{r_1} = \frac{n_1\left(S_2' - L_2\right)}{n_2\left|\left(S_2 - L_1\right)\right|} \tag{5-103}$$

and

$$M_2 = \frac{h_2'}{h_1'} = \frac{n_1 S_2'}{n_2 S_2} \tag{5-104}$$

are the pupil and image magnifications for the second surface, respectively.

Equating the left-hand sides of the imaging relations

$$\frac{n_2}{S_2'} - \frac{n_1}{S_2} = \frac{n_2 - n_1}{R_2} \tag{5-105}$$

and

$$\frac{n_2}{S_2' - L_2} - \frac{n_1}{S_2 - L_1} = \frac{n_2 - n_1}{R_2} \tag{5-106}$$

for object and pupil imaging, respectively, we find that

$$m_2 M_2 L_1 = \left(n_1/n_2\right) L_2 \quad , \tag{5-107}$$

or

$$n_1 h_1'\left(r_1/L_1\right) = n_2 h_2'\left(r_2/L_2\right) \quad , \tag{5-108}$$

representing *Lagrange invariance* between object and image planes. Note that the angle $\left(r_1/L_1\right)$ is to be treated as a numerically negative quantity. Hence, substituting Eq. (5-108) into Eq. (5-110), the field curvature aberration due to Petzval curvature produced by the first surface may be written

$$W_{p1}(r_2;h_2') \equiv \Delta W_{d1}\left(\frac{r_2}{m_2};\frac{h_2'}{M_2}\right)$$

$$= \frac{n_2^2}{4R_1}\left(\frac{1}{n_1}-\frac{1}{n_0}\right)\left(\frac{h_2'r_2}{L_2}\right)^2 . \tag{5-109}$$

Continuing this procedure, we can write this aberration in terms of the radial coordinate r_k of a corresponding point on the exit pupil of the kth surface and the image height h_k' of the image formed by this surface. Thus

$$W_{p1}(r_k;h_k') = \frac{n_k^2}{4R_1}\left(\frac{1}{n_1}-\frac{1}{n_0}\right)\left(\frac{h_k'r_k}{L_k}\right)^2 , \tag{5-110}$$

where L_k is the distance of the image from the exit pupil for this surface. Similarly, the field curvature aberration due to Petzval curvature produced by the jth surface, where $j \le k$, can be written

$$W_{pj}(r_k;h_k') = \frac{n_k^2}{4R_j}\left(\frac{1}{n_j}-\frac{1}{n_{j-1}}\right)\left(\frac{h_k'r_k}{L_k}\right)^2 . \tag{5-111}$$

Adding the contributions of all the surfaces, the total field curvature aberration due to Petzval curvature produced by the system consisting of k surfaces is given by

$$W_{ps}(r_k;h_k') = \frac{n_k^2}{4}\left(\frac{h_k'r_k}{L_k}\right)^2\sum_{j=1}^{k}\frac{1}{R_j}\left(\frac{1}{n_j}-\frac{1}{n_{j-1}}\right)$$

$$= \frac{n_k}{4R_{ik}}\left(\frac{h_k'r_k}{L_k}\right)^2$$

$$= a_{ps}h_k'^2 r_k^2 , \tag{5-112a}$$

where

$$a_{ps} = \frac{n_k}{4R_{ik}L_k^2} \tag{5-112b}$$

is the coefficient of Petzval curvature of the system and R_{ik} is the radius of curvature of the Petzval image surface produced by the system and given by Eq. (5-98). Note that the subscript k in Eq. (5-112) refers to system parameters. For example, n_k is the refractive index of the image space, r_k is a radial coordinate in the plane of the exit pupil of the system, and L_k is the distance of the final image from the exit pupil. With appropriate change in notation, Eqs. (5-112) are identical to the corresponding Eqs. (5-99).

5.7.3 Relationship among Petzval Curvature, Field Curvature, and Astigmatism Wave Aberration Coefficients

Substituting Eq. (5-108) recursively into Eq. (5-101), the field curvature aberration due to Petzval curvature produced by the first surface in the plane of the exit pupil of the system can be written

$$\Delta W_{d1}\left(r_k; h_k'\right) = \left(a_{d1} - \frac{1}{2}a_{a1}\right)\left(\frac{n_k}{L_k}\right)^2\left(\frac{L_1}{n_1}\right)^2\left(h_k' r_k\right)^2 \quad . \tag{5-113}$$

Each surface contributes field curvature aberration due to Petzval curvature similar to that given by Eq. (5-113). Adding the contributions of all the surfaces, the field curvature aberration due to Petzval curvature for the whole system may be written

$$\Delta W_{ds}\left(r_k; h_k'\right) = \left(a_{ds} - \frac{1}{2}a_{as}\right)h_k'^2 r_k^2 \quad , \tag{5-114}$$

where a_{ds} and a_{as} represent the *field curvature* and *astigmatism coefficients* of the system. These coefficients are given by

$$a_{ds} = \left(\frac{n_k}{L_k}\right)^2 \sum_{j=1}^{k} a_{dj}\left(\frac{L_j}{n_j}\right)^2 \tag{5-115}$$

and

$$a_{as} = \left(\frac{n_k}{L_k}\right)^2 \sum_{j=1}^{k} a_{aj}\left(\frac{L_j}{n_j}\right)^2 \quad . \tag{5-116}$$

Since the left-hand sides of Eqs. (5-112a) and (5-114) represent the same quantity, comparing their right-hand sides, we obtain

$$a_{ds} - \frac{1}{2}a_{as} = \frac{n_k}{4R_{ik}L_k^2} = a_{ps} \quad . \tag{5-117}$$

Here, as in Eqs. (5-112), the subscript s represents the system, so that a_{ds} and a_{as} represent the coefficients of field curvature and astigmatism, respectively, for the system. Equation (5-117) was used in obtaining Eq. (4-49) from Eq. (4-48) when we discussed the sagittal and tangential image surfaces in Section 4.3.3.

5.8 ABERRATION FUNCTION IN TERMS OF SEIDEL SUMS OR SEIDEL COEFFICIENTS

Seidel sums and *Seidel coefficients* have been used in the literature to describe the primary aberration function of an optical system. (See Welford[2], Mouroulis and Macdonald,[3] and Born and Wolf.[4] Note, however that in References 2 and 3, the sign convention for wave aberration is opposite to ours.) In this section, we introduce these sums and coefficients and relate them to the aberration coefficients a_i and the peak aberration coefficients A_i.

Let the radius of the exit pupil of the system be a. Consider an object such that its Gaussian image has a maximum height of h'_{max}. The aberration at point (r, θ) in the plane of the exit pupil for a point object with a Gaussian image height h' is written in the form

$$W(\rho, \theta; h') = \frac{1}{8} S_I \rho^4 + \frac{1}{2} S_{II} \frac{h'}{h'_{max}} \rho^3 \cos\theta + \frac{1}{2} S_{III} \left(\frac{h'}{h'_{max}}\right)^2 \rho^2 \cos^2\theta$$

$$+ \frac{1}{4}\left(S_{III} + S_{IV}\right)\left(\frac{h'}{h'_{max}}\right)^2 \rho^2 + \frac{1}{2} S_V \left(\frac{h'}{h'_{max}}\right)^3 \rho\cos\theta \quad, \tag{5-118}$$

where

$$\rho = r/a \tag{5-119}$$

is the normalized radial coordinate of a pupil point and S_I, ..., and S_V are called the *Seidel sums* of the optical system representing contributions of its surfaces to spherical aberration, coma, astigmatism, field curvature, and distortion, respectively. The quantity S_{IV} is also called the *Petzval sum*. Within simple numerical factors, the Seidel sums represent the peak values of the primary aberrations. As pointed out in Section 5.7.1, the peak value of a primary aberration is obtained by summing the peak values contributed by the surfaces of the system. The Seidel sums may also be obtained by tracing two paraxial rays through the system, one from the axial object point through the edge of the entrance and, therefore, edge of the exit pupil and the other from a point on the edge of the object through the centers of the pupils.[2] When a Seidel sum is multiplied by the focal ratio of the image-forming light cone, it gives the spot radius in the Gaussian image plane in the case of S_I, sagittal coma in the case of S_{II}, radius of the circle of least confusion in the case of S_{III}, spot radius in the case of $S_{III} + S_{IV}$, and the image displacement in the case of S_V, as discussed in Chapter 4.

If the chief ray from a point object passing through the center of the entrance pupil makes an angle β with the optical axis, as indicated in Figure 5-14, then the aberration at a point $\left(r_{en}, \theta_{en}\right)$ in the plane of this pupil is written in the form

$$W\left(r_{en}, \theta_{en}; \beta\right) = -\frac{1}{4} B r_{en}^4 + F\beta r_{en}^3 \cos\theta_{en} - C\beta^2 r_{en}^2 \cos^2\theta_{en}$$

$$- \frac{1}{2} D\beta^2 r_{en}^2 + E\beta^3 r_{en}\cos\theta_{en} \quad, \tag{5-120}$$

where B, F, C, D, and E are called the *Seidel coefficients* and correspond to spherical aberration, coma, astigmatism, field curvature, and distortion, respectively.

The coordinates $\left(r_{en}, \theta_{en}\right)$ and (r, θ) of a ray in the planes of the entrance and exit pupils, respectively, are related to each other according to

$$r(\cos\theta, \sin\theta) = m r_{en}(\cos\theta_{en}, \sin\theta_{en}) \quad, \tag{5-121}$$

where m is the pupil magnification. If a_{en} represents the radius of the entrance pupil, then

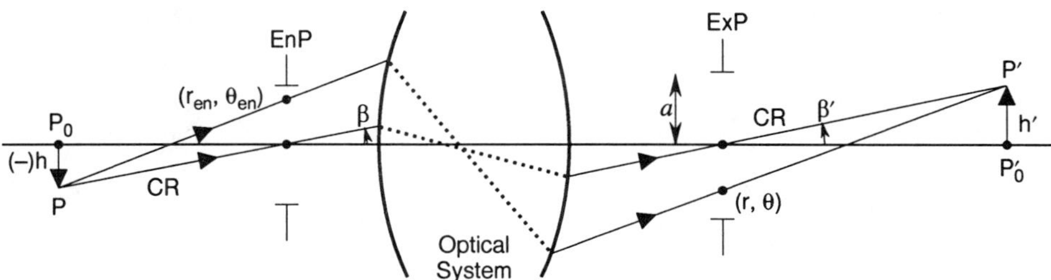

Figure 5-14. Description of the aberration of a system at its entrance pupil. As stated in Figure 1-26, a dotted line in the figure does not represent a ray, but merely a line joining its point of incidence on and its point of emergence from the system.

$$\rho = r/a = r_{en}/a_{en} \tag{5-122}$$

is the radial coordinate of a point in either pupil plane normalized by its radius. Accordingly, the aberration function in terms of the coordinates of the ray in the plane of the exit pupil may be written

$$W(\rho,\theta;\beta) = -\frac{1}{4}B\left(\frac{a}{m}\right)^4\rho^4 + F\left(\frac{a}{m}\right)^3\beta\rho^3\cos\theta - C\left(\frac{a}{m}\right)^2\beta^2\rho^2\cos^2\theta$$

$$-\frac{1}{2}D\left(\frac{a}{m}\right)^2\beta^2\rho^2 + E\left(\frac{a}{m}\right)\beta^3\rho\cos\theta \quad . \tag{5-123}$$

Comparing Eqs. (5-118), (5-123), and (5-32) or (5-85) with the aberration function in terms of the *peak aberration coefficients A_i* for a point object with a Gaussian image height of h',

$$W(\rho,\theta) = A_s\rho^4 + A_c\rho^3\cos\theta + A_a\rho^2\cos^2\theta + A_d\rho^2 + A_t\rho\cos\theta \quad , \tag{5-124}$$

we obtain

$$A_s = \frac{1}{8}S_I \tag{5-125a}$$

$$= -\frac{1}{4}\left(\frac{a}{m}\right)^4 B \tag{5-125b}$$

$$= a^4 a_s \quad , \tag{5-125c}$$

$$A_c = \frac{1}{2}\frac{h'}{h'_{max}}S_{II} \tag{5-126a}$$

$$= \beta\left(\frac{a}{m}\right)^3 F \tag{5-126b}$$

$$= h'a^3 a_c \quad , \tag{5-126c}$$

$$A_a = \frac{1}{2}\left(\frac{h'}{h'_{\text{max}}}\right)^2 S_{III} \tag{5-127a}$$

$$= -\beta^2\left(\frac{a}{m}\right)^2 C \tag{5-127b}$$

$$= h'^2 a^2 a_a \ , \tag{5-127c}$$

$$A_d = \frac{1}{4}\left(\frac{h'}{h'_{max}}\right)^2 \left(S_{III} + S_{IV}\right) \tag{5-128a}$$

$$= -\frac{1}{2}\beta^2\left(\frac{a}{m}\right)^2 D \tag{5-128b}$$

$$= h'^2 a^2 a_d \ , \tag{5-128c}$$

and

$$A_t = \frac{1}{2}\left(\frac{h'}{h'_{max}}\right)^3 S_V \tag{5-129a}$$

$$= \beta^3\left(\frac{a}{m}\right)E \tag{5-129b}$$

$$= h'^3 a a_t \ . \tag{5-129c}$$

It should be evident that we have dropped the second subscript on the aberration coefficients a_{ij} and that a_s in Eq. (5-125b) is what a_{ss} is in Eq. (5-32) or a_{sc} is in Eq. (5-85).

5.9 EFFECT OF CHANGE IN APERTURE STOP POSITION ON THE ABERRATION FUNCTION

By now we have discussed the aberrations of a single refracting surface and how to obtain the aberrations of a multisurface imaging system by adding the aberrations of its surfaces. Suppose we know the aberrations of a system for an object for a certain position of its aperture stop. In this section, we discuss how these aberrations change with a change in the position of its aperture stop, while adjusting its diameter so that the bundle of rays forming the image of the axial point object remains unchanged. Thus, the f-number of the image-forming light cone and the amount of light in the image do not change. In particular, we show, as expected, that the peak value of the spherical aberration of a system is independent of the position of its aperture stop. Moreover, if a system is free of spherical aberration, then its coma is also independent of the position of its aperture stop. However, if its spherical aberration is not zero, then its coma can be made zero by choosing an appropriate position of its aperture stop. It should be evident that the optical path of a ray does not change with a change in the position of the aperture stop. However, the chief ray which passes through its center does change and, therefore, the wave aberration of a ray with respect to it changes. The net effect is that the relative proportion of the various aberrations changes. We illustrate schematically how the coma or distortion of a system changes with a change in the position of its aperture stop. As an illustrative example, aberrations of a spherical refracting surface with its aperture stop not at the surface are obtained from its aberrations when its stop lies at the surface.

5.9.1 Change of Peak Aberration Coefficients

Consider, as indicated in Figure 5-15, an optical imaging system forming the image of an off-axis point object P (not shown in the figure). The image P' lies at a height h' from the optical axis of the system. Let the aperture stop of the system be located at a position so that the exit pupil of the system is located at ExP_1. Let the primary aberration function of the system be given by $W_{Q1}(x_1, y_1; h')$ representing the aberration of an image-forming ray PQ_1P' passing through a point $Q_1(x_1, y_1)$ in the plane of the exit pupil with respect to the chief ray PO_1P' passing through the center O_1 of the exit pupil.

Now, suppose we move the aperture stop to a new position along the optical axis so that the corresponding new exit pupil is located at ExP_2 with its center at O_2. A change in the stop position does not change the position of the image P'. The diameter of the aperture stop is adjusted so that the axial marginal ray does not change, with the consequence that the f-number of the axial image-forming light cone does not change. Thus, the exit pupils ExP_1 and ExP_2 subtend equal angles at the axial image point P_0' and

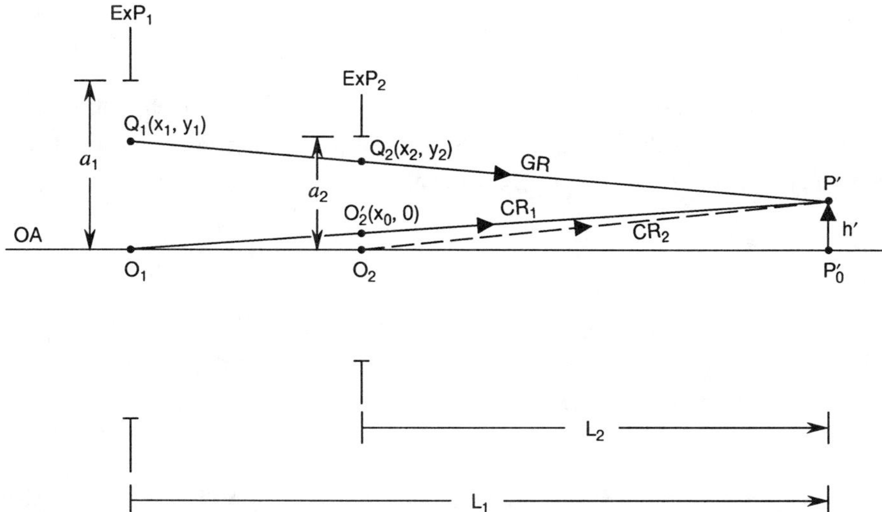

Figure 5-15. Exit pupils ExP_1 and ExP_2 corresponding to two positions of the aperture stop of an optical system forming a Gaussian image P' of an off-axis point object P (not shown). The chief rays CR_1 and CR_2 are for the pupils ExP_1 and ExP_2, respectively.

if a_1 and a_2 are their radii, and L_1 and L_2 are the axial distances of the Gaussian image plane from them, respectively, then

$$a_2/a_1 = L_2/L_1 \quad . \tag{5-130}$$

The chief ray PO_1P' intersects the plane of exit pupil ExP_2 at O_2' with rectangular coordinates $(x_0, 0)$ where, from triangles $O_1O_2O_2'$ and $O_1P_0'P'$

$$x_0 = \frac{h'}{L_1}(L_1 - L_2) \quad . \tag{5-131}$$

Note that O_1, P', and O_2' all lie in the tangential plane. From approximately similar triangles O_1Q_1P' and $O_2'Q_2P'$ in Figure 5-15 we note that

$$(x_1, y_1) \simeq \frac{L_1}{L_2}(x_2 - x_0, y_2) \quad , \tag{5-132}$$

where (x_2, y_2) are the coordinates of Q_2 with respect to O_2 as the origin.

The aberration of a ray PQ_1P' with respect to the chief ray PO_1P' represents the aberration at a point Q_1 with respect to the aberration at O_1 (which is zero by definition). It is also equal to the aberration of the ray PQ_1P' at Q_2 with respect to the aberration at O_2', where Q_2 represents the point of intersection of the ray with the plane of the exit pupil ExP_2. Thus, the aberration at Q_2 with respect to its value at O_2' may be obtained by substituting Eq. (5-132) into the expresson for $W_{Q_1}(x_1, y_1)$, i.e.,

$$W_{Q_2 \leftrightarrow O_2'}(x_2, y_2) \simeq W_{Q_1}\left[\frac{L_1}{L_2}(x_2 - x_0, y_2)\right] \quad . \tag{5-133}$$

However, we are interested in the aberration at Q_2 with respect to the aberration at O_2. It may be obtained by subtracting the aberration at O_2 with respect to O_2' from the aberration at Q_2 with respect to O_2'. Symbolically, we may write it in the form

$$W_{Q_2 \leftrightarrow O_2} = W_{Q_2 \leftrightarrow O_2'} - W_{O_2 \leftrightarrow O_2'} \quad , \tag{5-134}$$

where the second term on the right-hand side may be obtained from Eq. (5-133) by letting $(x_2, y_2) = (0, 0)$. Hence, the aberration function in the plane of the exit pupil ExP_2 with respect to the new chief ray PO_2P' is given by

$$W(x_2, y_2; h') = W_{Q_1}\left[\frac{L_1}{L_2}(x_2 - x_0, y_2)\right] - W_{Q_1}(-x_0 L_1/L_2, 0) \quad . \tag{5-135}$$

It is evident that the aberration function thus obtained has a value of zero associated with the new chief ray.

Let the primary aberration function at ExP_1 be given by

$$W_{Q_1}\left(x_1, y_1; h'\right) = a_{s1}\left(x_1^2 + y_1^2\right)^2 + a_{c1}h'x_1\left(x_1^2 + y_1^2\right)$$

$$+ a_{a1}h'^2 x_1^2 + a_{d1}h'^2\left(x_1^2 + y_1^2\right) + a_{t1}h'^3 x_1 \quad . \tag{5-136}$$

Substituting Eq. (5-136) into Eq. (5-135), we find that each aberration term of Eq. (5-136) of a certain order in pupil coordinates generates aberration terms of lower orders as well. Consider, for example, the spherical aberration term:

$$a_{s1}\left(x_1^2 + y_1^2\right)^2 \rightarrow a_{s1}\left(\frac{L_1}{L_2}\right)^4 \left\{\left[\left(x_2 - x_0\right)^2 + y_2^2\right] - x_0^4\right\}$$

$$= a_{s1}\left(\frac{L_1}{L_2}\right)^4 \left(r_2^4 - 4x_0 x_2 r_2^2 + 4x_0^2 x_2^2 + 2x_0^2 r_2^2 - 4x_0^3 x_2\right) \quad , \tag{5-137}$$

where

$$r_2^2 = x_2^2 + y_2^2 \quad . \tag{5-138}$$

The terms on the right-hand side of Eq. (5-137) represent, sequentially, spherical aberration, coma, astigmatism, field curvature, and distortion. The piston terms, which are proportional to x_0^4, cancel out, as expected. Substituting Eq. (5-136) into Eq. (5-135), we may write

$$W\left(x_2, y_2; h'\right) = a_{s2}\left(x_2^2 + y_2^2\right)^2 + a_{c2}h'x_2\left(x_2^2 + y_2^2\right) + a_{a2}h'^2 x_2^2$$

$$+ a_{d2}h'^2\left(x_2^2 + y_2^2\right) + a_{t2}h'^3 x_2 \quad , \tag{5-139}$$

where

$$a_{s2} = \left(L_1/L_2\right)^4 a_{s1} \quad , \tag{5-140}$$

$$a_{c2} = \left(\frac{L_1}{L_2}\right)^3 \left(a_{c1} - 4\gamma a_{s1}\right) \quad , \tag{5-141}$$

$$a_{a2} = \left(\frac{L_1}{L_2}\right)^2 \left(a_{a1} - 2\gamma a_{c1} + 4\gamma^2 a_{s1}\right) \quad , \tag{5-142}$$

$$a_{d2} = \left(\frac{L_1}{L_2}\right)^2 \left(a_{d1} - \gamma a_{c1} + 2\gamma^2 a_{s1}\right) \quad , \tag{5-143}$$

$$a_{t2} = \frac{L_1}{L_2}\left[a_{t1} - 2\gamma\left(a_{a1} + a_{d1}\right) + 3\gamma^2 a_{c1} - 4\gamma^3 a_{s1}\right] \quad , \tag{5-144}$$

and

$$\gamma = \left(L_1 - L_2\right)/L_2 \ . \tag{5-145}$$

Comparing Eqs. (5-136) and (5-139) and noting Eq. (5-130), we find that the old and the new peak aberration coefficients are related to each other according to

$$A_{s2} = A_{s1} \ , \tag{5-146}$$

$$A_{c2} = A_{c1} - 4\zeta A_{s1} \ , \tag{5-147}$$

$$A_{a2} = A_{a1} - 2\zeta A_{c1} + 4\zeta^2 A_{s1} \ , \tag{5-148}$$

$$A_{d2} = A_{d1} - \zeta A_{c1} + 2\zeta^2 A_{s1} \ , \tag{5-149}$$

and

$$A_{t2} = A_{t1} - 2\zeta\left(A_{a1} + A_{d1}\right) + 3\zeta^2 A_{c1} - 4\zeta^3 A_{s1} \ , \tag{5-150}$$

where

$$\zeta = \gamma h'/a_1 = \left(L_1 - L_2\right)h'/a_1 L_2 \ . \tag{5-151}$$

Equations (5-146) through (5-150) are called the *stop-shift equations*. They describe the change in the value of the peak aberrations due to a change in the position of the aperture stop. It is evident from them that an aberration of a certain order in pupil coordinates introduces aberrations of all lower orders as well. For example, a term in spherical aberration not only gives spherical aberration, but introduces coma, astigmatism, field curvature, and distortion as well.

From Eq. (5-146), we note that the *peak value of spherical aberration of a system is independent of the position of its aperture stop*. For a single refracting surface, such a result is evident from the discussion in Section 5.3.1. Indeed, such a result is to be expected for any system and for any order of spherical aberration since the axial bundle of rays was kept unchanged by adjusting the diameter of the aperture stop in its new position.

Equation (5-147) shows that if a system is free of spherical aberration, then the peak value of its coma is independent of the position of its aperture stop. It also shows that *if spherical aberration is not zero, its coma can be made zero by choosing an aperture stop position* corresponding to

$$\zeta = \frac{A_{c1}}{4A_{s1}} \qquad \text{or} \qquad \frac{L_1}{L_2} = 1 + \frac{a_{c1}}{4a_{s1}} \ . \tag{5-152}$$

The corresponding value of the astigmatism coefficient given by

$$A_{a2} = A_{a1} - A_{c1}^2/4A_{s1} \quad , \tag{5-153}$$

is minimum or maximum depending on whether A_{s1} is positive or negative, as may be seen from Eq. (5-148).

We also note that *if the system is aplanatic* (i.e., free of spherical aberration and coma) *for a certain position of the aperture stop, it is aplanatic for any position of the aperture stop.* Moreover, for such a system, Eqs. (5-148) and (5-149) show that *the peak values of its astigmatism and field curvature are also independent of the position of its aperture stop.* This was demonstrated for a single refracting surface in Section 5.4. However, the peak value of its distortion changes with a change in the position of its aperture stop unless the sum of the peak values of its astigmatism and field curvature is zero, as may be seen from Eq. (5-150). (Aplanatic systems are discussed in Sections 5.4, 5.10.5 and 6.4.1.) This equation shows that the *peak value of distortion depends on the position of the aperture stop unless spherical aberration, coma, and the sum of the peak values of astigmatism and field curvature are each zero, or the sum of the three terms depending on b on its right-hand side is zero.* An example of the latter is a spherical mirror with a collocated aperture stop compared with a stop located at its center of curvature, which is discussed in Sections 6.4.3 and 6.6.2, and Problem 6.2. As expected, *the coefficient of Petzval field curvature,* which depends on $2A_d - A_a$, *does not change when the position of the aperture stop is changed.*

For an anastigmatic system, i.e., one that is free of spherical aberration, coma, and astigmatism, the field curvature aberration is due only to the Petzval curvature, which, in turn, is independent of the position of its aperture stop. Its distortion depends on the position of its aperture stop unless its Petzval curvature is also zero. Of course, if all of the primary aberrations of a system are zero for a certain position of its aperture stop, then they are zero for any other position as well.

It should be noted that the optical path length of a ray, or its optical path difference with respect to another, does not change with a change in the position of the aperture stop. However, since the chief ray does change, the new aberration function merely describes the aberrations of the rays with respect to the new chief ray. Accordingly, the relative proportion of the various aberrations changes with a change in the position of the aperture stop. The position of the aperture stop also affects which and how many rays are transmitted by the system. Indeed, for high-quality imaging systems, a lens designer chooses the position of the aperture stop judiciously so that the rays with large aberrations are blocked by it without substantial loss in the amount of transmitted light. Those rays that are transmitted by the system intersect the image plane at certain points independent of the position of the aperture stop.

5.9.2 Illustration of the Effect of Aperture-Stop Shift on Coma and Distortion

We now illustrate the effect of a shift in the position of the aperture stop on coma and distortion of a system. Figure 5-16 illustrates how coma of a system is affected by a shift

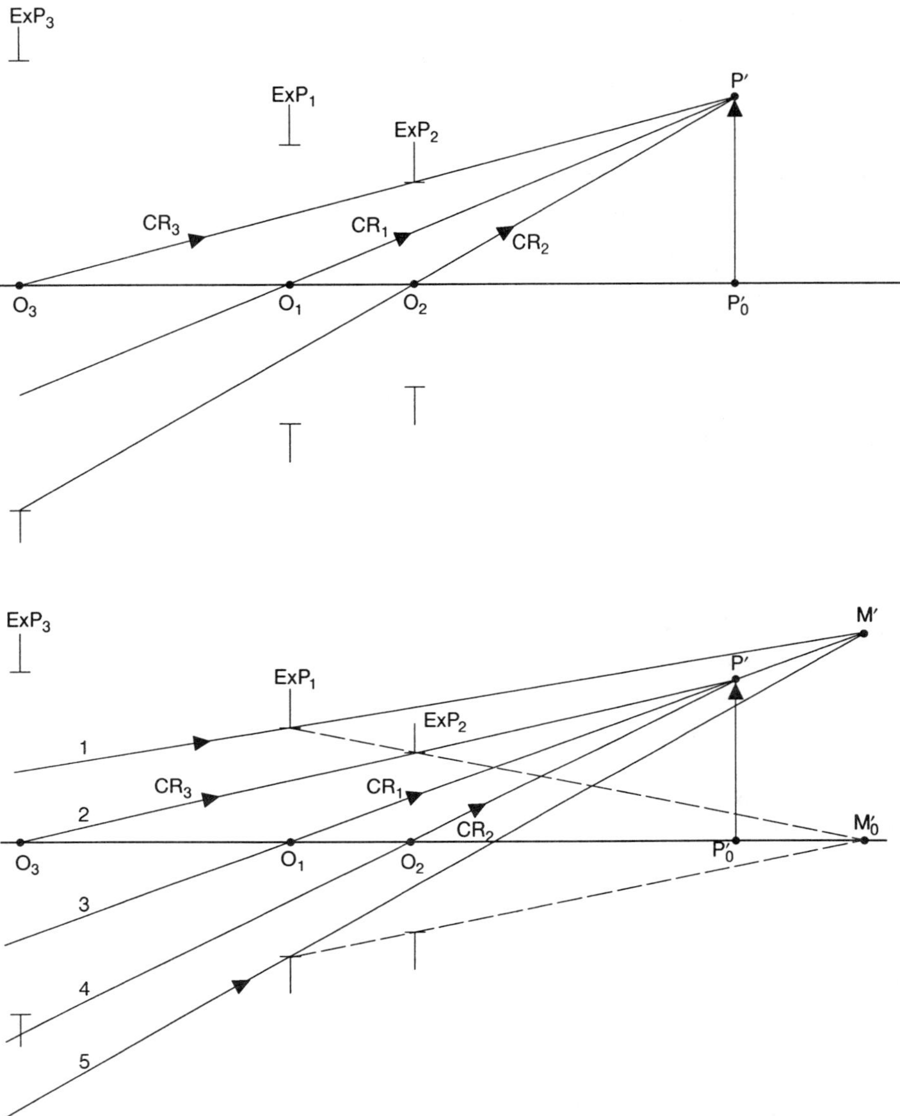

Figure 5-16. Effect of a shift in the position of the aperture stop of a system on its coma. (a) System with no spherical aberration or coma. (b) System with spherical aberration. Its coma depends on the position of the aperture stop, e.g., there is no coma for exit pupil ExP_1, there is negative coma for exit pupil ExP_2, and positive coma for ExP_3.

in its aperture stop. ExP_1, ExP_2, and ExP_3 represent three positions of its exit pupil corresponding to three positions of its aperture stop. Figure 5-16a shows all of the image rays meeting at the off-axis image point P' and there is neither spherical aberration nor coma regardless of the position of the exit pupil. For exit pupil ExP_1, Figure 5-16b shows that CR_1 is the chief ray and there is spherical aberration but no coma. Whereas the paraxial rays come to focus at P', the marginal rays meet at M'. For exit pupil ExP_2, ray 4 is the chief ray CR_2 and rays 3 and 5 that are symmetric about it do not meet it at the same point. Hence, there is coma that is negative $\left(a_c < 0\right)$ in the figure, since the marginal image formed by these rays lies below the point of intersection of the chief ray CR_2 with the marginal image plane. Similarly, ray 2 is the chief ray CR_3 for exit pupil ExP_3 and rays 1 and 3 that are symmetric about it do not meet it at the same point. There is positive coma in this case, since the marginal image formed by these rays lies above the point of intersection of the chief ray CR_3 with the marginal image plane.

Figure 5-17 shows how the distortion of a system is affected by the position of its aperture stop. The system is assumed to be free of spherical aberration and coma. It is free of distortion when ExP_1 is its exit pupil, since the chief ray CR_1 intersects the Gaussian image plane at the Gaussian image point P'. When ExP_2 is the exit pupil, the chief ray CR_2 intersects the Gaussian image plane at a point P'_p, which is below P' and there is a positive $\left(a_t > 0\right)$ or *pincushion distortion*. According to Eq. (4-63b) a negative ray aberration is obtianed for negative values of h' when the distortion wave aberration coefficient a_t is positive. Similarly, if the aperture stop and the entrance pupil lie at EnP_3, then the chief ray CR_3 intersects the Gaussian image plane at P'_b and there is a negative or *barrel distortion*. The manner in which distortion is affected by the position of the aperture stop depends on the location of the *tangential image P'_t* where the tangential rays (e.g., rays 2 and 3) intersect each other. P'_t lies to the right or left of P' depending on whether the sum of the astigmatism and field curvature coefficients is positive or negative (see Figure 4-12). If P'_t lies to the right of the Gaussian image plane, then ExP_2 corresponds to barrel distortion and EnP_3 corresponds to pincushion distortion. If P'_t lies in the Gaussian image plane, then distortion is not affected by the position of the aperture stop, i.e., distortion of an aplanatic system is independent of the position of its aperture stop if its tangential surface coincides with the Gaussian image plane. This happens when the coefficients of astigmatism and field curvature aberrations have equal magnitudes but opposite signs, as may be seen from Eq. (4-46).

5.9.3 Aberrations of a Spherical Refracting Surface with Aperture Stop Not at the Surface Obtained from Those with Stop at the Surface

As an application of the procedure outlined in Section 5.9.1, we determine the aberrations of a spherical refracting surface with an aperture stop not at the surface (see Section 5.3) from its aberrations when the stop is at the surface (see Section 5.2). The refracting surface has a radius of curvature R separating media of refractive indices n and n'. It forms the Gaussian image of an off-axis point object at P' at a height h' from the optical axis at an axial distance S' from the surface. When the stop is located at the surface, the primary aberration function is given by Eq. (5-19) where, according to Eqs. (5-20) through (5-23),

$$a_c = 4ba_s \quad , \tag{5-154}$$

$$a_a = 4b^2 a_s \quad , \tag{5-155}$$

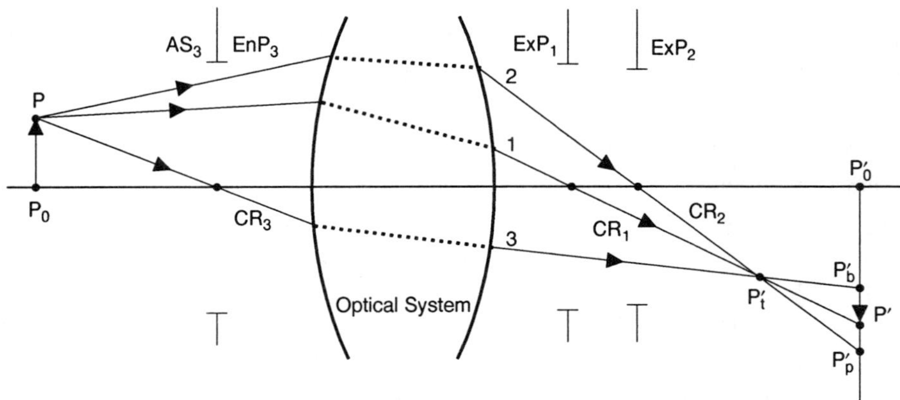

Figure 5-17. Effect of a shift in the position of the aperture stop of a system on its distortion. As stated in Figure 1-26, the system is assumed to be a multisurface one. Hence, a dotted line in the figure does not represent a ray but merely a line joining its point of incidence on and its point of emergence from the system.

$$a_d = 2b^2 a_s - \frac{n'(n'-n)}{4nRS'^2} \quad , \tag{5-156}$$

$$a_t = 4b^3 a_s - \frac{n'(n'-n)b}{2nRS'^2} \quad , \tag{5-157}$$

a_s is given by Eq. (5-7) or Eq. (5-8), and

$$b = \frac{R}{S'-R} \quad . \tag{5-158}$$

When the aperture stop is moved so that the Gaussian image plane lies at a distance L from the plane of the exit pupil, we let

$$L_1 = S' \tag{5-159}$$

$$L_2 = L \tag{5-160}$$

and

$$x_0 = \frac{n'}{S'}(S'-L) \tag{5-161}$$

in Eq. (5-139), and find that

$$W(x_2, y_2; h') = a_{ss}\left(x_2^2 + y_2^2\right)^2 + a_{cs}h'x_2\left(x_2^2 + y_2^2\right) + a_{as}h'^2 x_2^2$$
$$+ a_{ds}h'^2\left(x_2^2 + y_2^2\right) + a_{ts}h'^3 x_2 \quad , \tag{5-162}$$

where a_{ss}, a_{cs}, a_{as}, a_{ds}, and a_{ts} are given by Eqs. (5-33) through (5-37), respectively. Except for notation, Eq. (5-162) is the same as Eq. (5-32).

5.10   THIN LENS

In this section we consider a *thin lens* of refractive index n consisting of two spherical surfaces of radii of curvature R_1 and R_2, as illustrated in Figure 5-18, and determine its primary aberrations. Its optical axis is the line joining the centers of curvature C_1 and C_2 of its surfaces. Since the lens is thin, we neglect the spacing between its surfaces. We assume that the aperture stop is located at the lens, so that the entrance and exit pupils are also located there. The lens is located in air; therefore, the refractive index of the surrounding medium is 1. We apply the results of Sections 5.2 and 5.7 to obtain the imaging properties and aberrations of a thin lens. The *position* and *shape factors* of a lens are introduced and its Petzval curvature is determined. We show that the spherical aberration of a thin lens cannot be zero when both the object and its image are real. However, the coma of a thin lens can be zero for a certain position factor if its shape factor is appropriately chosen. The aplanatic points of the lens, i.e., conjugate points for which both spherical aberration and coma are zero, are discussed. The aberrations of a thin lens with conic surfaces are also considered. The spherical aberration of such a lens can be made zero by an appropriate choice of the eccentricities of its surfaces. A brief discussion of how the aberrations change as the aperture stop is moved away from the lens is also given.

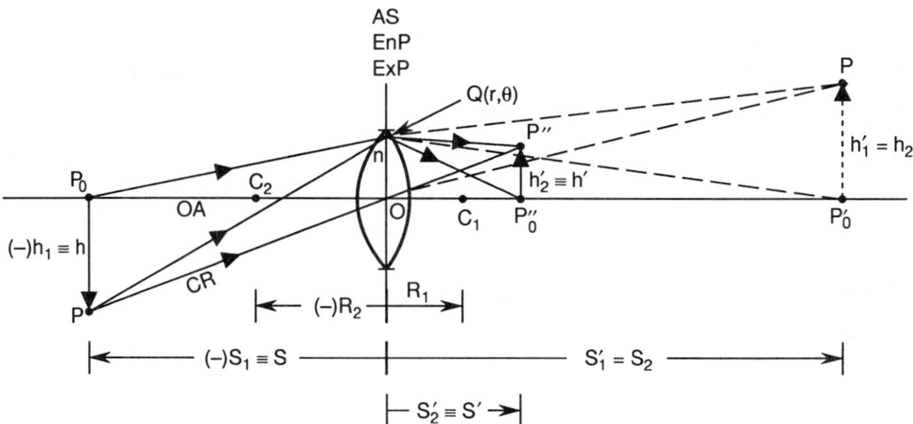

Figure 5-18. Imaging by a thin lens of refractive index n formed by two spherical surfaces of radii of curvature R_1 and R_2 with their centers of curvature at C_1 and C_2. Whereas R_1 is numerically positive, R_2 is negative. $P_0'P'$ is the Gaussian image of the object P_0P formed by the first surface. $P_0''P''$ is the image of the virtual object $P_0'P'$ formed by the second surface. The height h of the point object P is numerically negative, since it lies below the optical axis OA. The height h' of the final image P'' is numerically positive, since it lies above the optical axis. For a thin lens, its thickness is neglected. The aperture stop AS, entrance pupil EnP, and the exit pupil ExP are all located at the lens.

5.10.1 Imaging Relations

Consider a point object P located at a distance S_1 from the lens and at a height h_1 from its axis. The first surface forms the image of P at P' at a distance S_1' and at a height h_1'. Letting $n_1 = 1$ and $n_1' = n$ in Eqs. (5-5) and (5-9), we obtain

$$\frac{n}{S_1'} - \frac{1}{S_1} = \frac{n-1}{R_1} \tag{5-163}$$

and

$$M_1 = h_1'/h_1 = S_1'/nS_1 \quad , \tag{5-164}$$

respectively, where M_1 is the magnification of the image formed by the first surface. Since the image at P' is a *virtual object* for the second surface, letting $n_2 = n$, $n_2' = 1$, $S_2 = S_1'$, and $h_2 = h_1'$ in Eqs. (5-5) and (5-9), we obtain the distance S_2' and height h_2' of the final image P'' according to

$$\frac{1}{S_2'} - \frac{n}{S_1'} = \frac{1-n}{R_2} \tag{5-165}$$

and

$$\begin{aligned} M_2 &= h_2'/h_1' \\ &= n\,S_2'/S_1' \quad , \end{aligned} \tag{5-166}$$

respectively, where M_2 is the magnification of the image formed by the second surface. Adding Eqs. (5-163) and (5-165), we obtain

$$\frac{1}{S_2'} - \frac{1}{S_1} = (n-1)\left(\frac{1}{R_1} - \frac{1}{R_2} \right) \quad . \tag{5-167}$$

Letting the object distance $S_1 = S$ and the final image distance $S_2' = S'$, Eq. (5-167) may be written

$$\frac{1}{S'} - \frac{1}{S} = \frac{1}{f'} \quad , \tag{5-168}$$

where f' is the image-space focal length of the lens given by

$$\frac{1}{f'} = (n-1)\left(\frac{1}{R_1} - \frac{1}{R_2} \right) \tag{5-169}$$

as may be seen by letting $S \to -\infty$. Equation (5-168) is the Gaussian imaging equation and Eq. (5-169) is called the *lens maker's formula* for a thin lens. Similarly, letting the object height $h_1 = h$, the final image height $h_2' = h'$, and combining Eqs. (5-164) and (5-166) we obtain the magnification M of the image formed by the lens:

$$M = M_1 M_2 \tag{5-170}$$

$$= h'/h \tag{5-171a}$$

$$= S'/S \quad . \tag{5-171b}$$

5.10.2 Thin Lens with Spherical Surfaces and Aperture Stop at the Lens

The primary aberrations of a thin lens with an aperture stop located at the lens can be determined by applying Eq. (5-19) to its two refracting surfaces. It is not difficult to see that the aberration of a ray produced by the lens is equal to the sum of its aberrations produced by its two surfaces. For example, the aberration of a ray from the point object P incident on the lens at a point Q is given by (see Figure 5-18)

$$W(Q) = [PQP''] - [POP''] \quad . \tag{5-172}$$

The aberrations produced by the first and the second surfaces are given by

$$W_1(Q) = [PQP'] - [POP'] \tag{5-173a}$$

and

$$W_2(Q) = [P'QP''] - [P'OP''] \quad , \tag{5-173b}$$

respectively. Their sum may be written

$$W_1(Q) + W_2(Q) = [PQ + QP'] - [PO + OP'] + [P'Q + QP''] - [P'O + OP''] \quad . \tag{5-174}$$

Since P' is a virtual object for the second surface; i.e., since the rays $P'Q$ and $P'O$ are virtual, the optical path lengths $[P'Q]$ and $[P'O]$ are numerically negative. For example, $[P'Q] = -nP'Q$. Thus, the optical path length $[P'Q]$ for the second surface is equal to the optical path length $[QP']$ for the first surface in magnitude but opposite in sign, i.e., $[P'Q] = -[QP']$. Similarly, $[P'O] = -[OP']$. Hence, Eq. (5-174) reduces to

$$W_1(Q) + W_2(Q) = [PQ + QP''] - [PO + OP'']$$

$$= [PQP''] - [POP''] \quad . \tag{5-175}$$

Comparing Eqs. (5-172) and (5-175), we obtain the desired result

$$\boxed{W(Q) = W_1(Q) + W_2(Q)} \tag{5-176}$$

Since the aperture stop lies at the thin lens, it is also the exit pupil for the two surfaces. We determine the aberration contributed by each surface at a point (r, θ) in the plane of this pupil and add their contributions according to Eq. (5-95). The aberration contributed by the first surface may be written

$$W_1(r, \theta; h_1') = a_{s1} r^4 + a_{c1} h_1' r^3 \cos\theta + a_{a1} h_1'^2 r^2 \cos^2\theta + a_{d1} h_1'^2 r^2 + a_{t1} h_1'^3 r \cos\theta \quad . \tag{5-177}$$

Or, writing h_1' in terms of h_2' according to Eq. (5-166)

$$W_1\left(r,\theta;h_2'/M_2\right) = a_{s1}r^4 + \left(a_{c1}/M_2\right)h_2'r^3\cos\theta + \left(a_{a1}/M_2^2\right)h_2'^2r^2\cos^2\theta$$

$$+ \left(a_{d1}/M_2^2\right)h_2'^2r^2 + \left(a_{t1}/M_2^3\right)h_2'^3r\cos\theta \quad . \tag{5-178}$$

Similarly, the aberration contributed by the second surface may be written

$$W_2\left(r,\theta;h_2'\right) = a_{s2}r^4 + a_{c2}h_2'r^3\cos\theta + a_{a2}h_2'^2r^2\cos\theta + a_{d2}h_2'^2r^2 + a_{t2}h_2'^3r\cos\theta \ . \tag{5-179}$$

Adding Eqs. (5-178) and (5-179), we obtain the aberrations of a thin lens

$$W\left(r,\theta;h_2'\right) = W_1\left(r,\theta;h_2'/M_2\right) + W_2\left(r,\theta;h_2'\right) \quad , \tag{5-180}$$

or

$$\boxed{W\left(r,\theta;h'\right) = a_s r^4 + a_c h'r^3\cos\theta + a_a h'^2r^2\cos^2\theta + a_d h'^2r^2 + a_t h'^3 r\cos\theta \quad ,} \tag{5-181}$$

where

$$a_s = a_{s1} + a_{s2} \quad , \tag{5-182a}$$

$$a_c = \left(a_{c1}/M_2\right) + a_{c2} \quad , \tag{5-182b}$$

$$a_a = \left(a_{a1}/M_2^2\right) + a_{a2} \quad , \tag{5-182c}$$

$$a_d = \left(a_{d1}/M_2^2\right) + a_{d2} \quad , \tag{5-182d}$$

$$a_t = \left(a_{t1}/M_2^3\right) + a_{t2} \quad , \tag{5-182e}$$

and we have replaced h_2' by h'. Now we determine the various aberration coefficients.

Letting $n_1 = 1$ and $n_1' = n$ in Eq. (5-7b), we obtain the contribution of the first surface to spherical aberration

$$a_{s1} = -\frac{n(n-1)}{8}\left(\frac{1}{R_1} - \frac{1}{S_1'}\right)^2\left(\frac{n}{R_1} - \frac{1+n}{S_1'}\right) \quad . \tag{5-183}$$

Letting

$$\boxed{\begin{aligned} p &= -\frac{2f'}{S_1} - 1 \\[2mm] &= 1 - \frac{2f'}{S_2'} \end{aligned}}$$

$$\tag{5-184a}$$

$$\tag{5-184b}$$

and

$$q = \frac{R_2 + R_1}{R_2 - R_1} \; , \tag{5-185}$$

Eq. (5-183) becomes

$$a_{s1} = -\frac{1}{64n^2 f'^3}\left(p + \frac{q+n}{n-1}\right)^2\left[p\left(n^2 - 1\right) + q + n^2\right] \; . \tag{5-186}$$

The quantities p and q are called the *position* and *shape factors* of a thin lens, respectively. Several examples of these factors are illustrated in Figures 5-19 and 5-20. Both *positive* and *negative lenses* (in the sense of the sign of their image-space focal length f') are considered in these figures. The names associated with the different lens shapes are also indicated in Figure 5-20. We note from Eq. (5-169) that the focal length of a thin lens does not change if the radii of curvature of its two surfaces are changed so that different shape factors. Similarly, the lenses in Figure 5-20b have the same focal length but different shape factors. The process of changing the shape factor of a lens while keeping its focal length fixed is called *lens bending*. It is used in controlling aberrations, as discussed in Section 5.10.4.

Similarly, letting $n_2 = n$ and $n_2' = 1$ in Eq. (5-7b), we obtain the contribution of the second surface to spherical aberration

$$a_{s2} = \frac{n-1}{8n^2}\left(\frac{1}{R_2} - \frac{1}{S_2'}\right)^2\left(\frac{1}{R_2} - \frac{n+1}{S_2'}\right) \tag{5-187a}$$

$$= \frac{1}{64n^2 f'^3}\left(p + \frac{q-n}{n-1}\right)^2\left[p\left(n^2 - 1\right) + q - n^2\right] \; . \tag{5-187b}$$

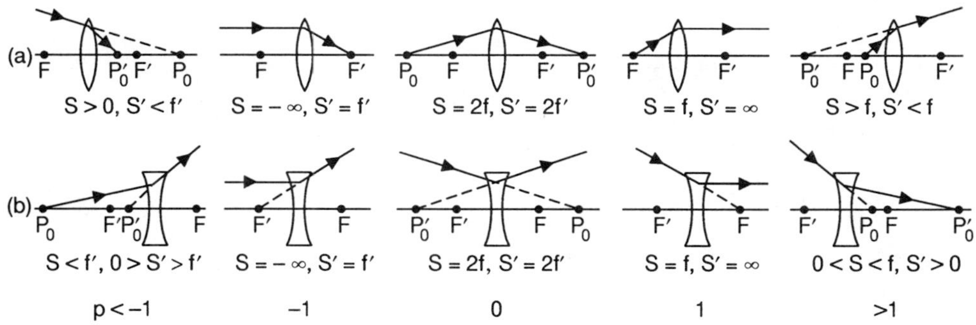

Figure 5-19. *Position factor* $1 < p < -1$ of a thin lens. (a) Positive lens, i.e., $f' > 0$. (b) Negative lens, i.e., $f' < 0$. F and F' are the object- and image-space focal points of a lens of image-space focal length f'. P_0 and P_0' represent an axial point object and its point image, respectively. S and S' are the object and image distances from the center of the lens. Note that $f = -f'$.

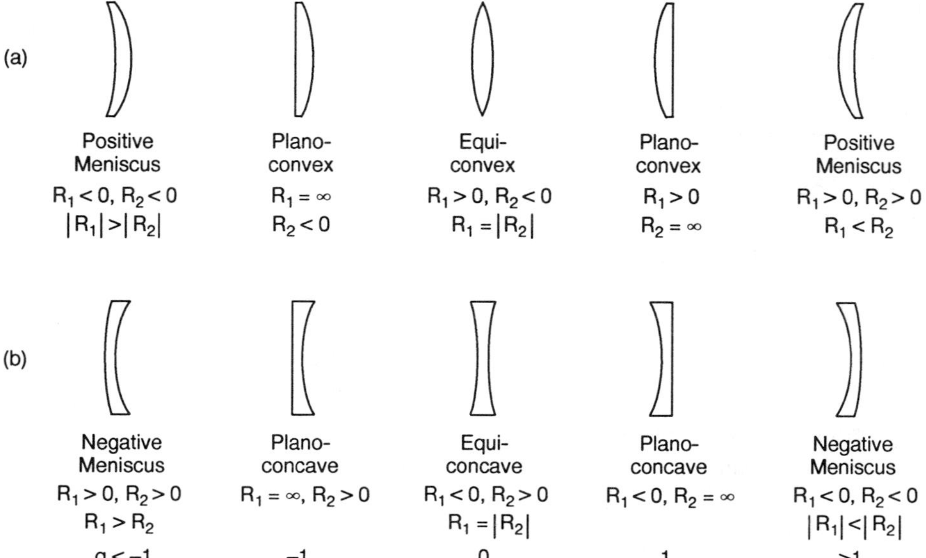

Figure 5-20. *Shape factor* $1 < q < -1$ *of a thin lens with spherical surfaces of radii of* curvature R_1 and R_2. **(a) Positive lens. (b) Negative lens.**

The *spherical aberration* of the lens is obtained by adding the contributions of its two surfaces according to Eq. (5-182a), i.e.,

$$a_s = a_{s1} + a_{s2}$$

$$= -\frac{1}{32n(n-1)f'^3}\left[\frac{n^3}{n-1} + (3n+2)(n-1)p^2 + \frac{n+2}{n-1}q^2 + 4(n+1)pq\right] \ . \tag{5-188}$$

Similarly, following Eq. (5-20a), the *coma* of a thin lens given by Eq. (5-182b) may be written

$$a_c = \left(a_{c1}/M_2\right) + a_{c2} \tag{5-189a}$$

$$= \left(4b_1 a_{s1}/M_2\right) + 4b_2 a_{s2} \ , \tag{5-189b}$$

where [following Eq. (5-11b)]

$$b_1 = \frac{R_1}{S_1' - R_1} \tag{5-190a}$$

and

$$b_2 = \frac{R_2}{S_2' - R_2} \ . \tag{5-190b}$$

Substituting for a_{s1} and a_{s2}, we obtain

$$\frac{a_{c1}}{M_2} = \frac{n-1}{2S_2'}\left(\frac{1}{R_1} - \frac{1}{S_1'}\right)\left(\frac{n}{R_1} - \frac{1+n}{S_1'}\right)$$

$$= -\frac{1}{8n^2 f'^2 S_2'}\left(p+\frac{q+n}{n-1}\right)\left[p(n^2-1)+q+n^2\right] \tag{5-191}$$

and

$$a_{c2} = -\frac{1}{8n^2 f'^2 S_2'}\left(p+\frac{q-n}{n-1}\right)\left[p(n^2-1)+q-n^2\right] \ . \tag{5-192}$$

Adding Eqs. (5-191) and (5-192), we obtain

$$\boxed{a_c = -\frac{1}{4nf'^2 S'}\left[(2n+1)p+\frac{n+1}{n-1}q\right]} \ , \tag{5-193}$$

where we have replaced S_2' by S'. We note that like spherical aberration, coma also depends on both the position and shape factors of the lens.

Following Eq. (5-21), the *astigmatism* of a thin lens given by Eq. (5-182c) may be written

$$\begin{aligned}
a_a &= \left(a_{a1}/M_2^2\right)+a_{a2}\\
&= \left(4b_1^2 a_{s1}/M_2^2\right)+4b_2^2 a_{s2}\\
&= -\frac{n-1}{2nS_2'^2}\left(\frac{n}{R_1}-\frac{1+n}{S_1'}\right)-\frac{1-n}{2n^2 S_2'^2}\left(\frac{1}{R_2}-\frac{n+1}{S_2'}\right)
\end{aligned}$$

or

$$\boxed{a_a = -\frac{1}{2f'S'^2}} \ . \tag{5-194}$$

We note that it does not depend on the refractive index or the shape factor of the lens.

Following Eq. (5-22c), the *field curvature* of a thin lens given by Eq. (5-182d) may be written

$$\begin{aligned}
a_d &= \left(a_{d1}/M_2^2\right)+a_{d2}\\
&= \frac{n}{4S_2'^2}\left(\frac{1}{R_1}-\frac{1}{S_1'}\right)\left(\frac{1}{n^2}-1\right)+\frac{1}{4S_2'^2}\left(\frac{1}{R_2}-\frac{1}{S_2'}\right)\left(1-\frac{1}{n^2}\right) \ ,
\end{aligned}$$

or

$$\boxed{a_d = -\frac{n+1}{4nf'S'^2}} \tag{5-195a}$$

or

$$\boxed{a_d = \frac{1}{2}a_a - \frac{1}{4nf'S'^2}} \ . \tag{5-195b}$$

Like astigmatism, field curvature of the lens also does not depend on its shape factor.

Following Eqs. (5-23b) and (5-182e), we find that a thin lens does not produce any *distortion*, since

$$a_t = \left(a_{t1}/M_2^3\right) + a_{t2}$$

$$= \frac{1}{2S_2'^3}\left(\frac{1}{n^2} - 1\right) + \frac{1}{2S_2'^3}\left(1 - \frac{1}{n^2}\right)$$

$$= 0 \ . \tag{5-196}$$

Although the lens surfaces produce distortion, the lens as a whole does not, i.e., the distortion contributions of the two surfaces cancel each other. This is understandable for a thin lens with a collocated aperture stop, since the chief ray CR in that case passes through its center undeviated, intersecting the Gaussian image plane at the image point P'', as shown in Figure 5-18. Distortion is generally nonzero when the aperture stop does not lie at the lens, as pointed out in Section 5.10.7.

Substituting Eqs. (5-188) and (5-193)–(5-196) into Eq. (5-181), we obtain the aberration function for a thin lens of focal length f', forming a distortion-free image of a point object at a distance S' from it at a height h' from its optical axis. Note that its spherical aberration, coma, and astigmatism and field curvature vary with its focal length as f'^{-3}, f'^{-2}, and f'^{-1}, respectively.

5.10.3 Petzval Surface

The Petzval radius of curvature of a thin lens can be obtained by applying Eq. (5-15) to refraction by its two surfaces, or equivalently, by using Eq. (5-98) and letting $n_0 = 1$, $n_1 = n$, and $n_2 = 1$. The radius of curvature of the Petzval image surface produced by the first refracting surface (for a planar object) is given by

$$\frac{1}{R_{i1}} = \frac{1-n}{R_1} \ . \tag{5-197}$$

The second refracting surface images the first Petzval surface into a second surface, with a radius of curvature R_{i2} given by

$$\frac{1}{R_{i2}} - \frac{1}{n R_{i1}} = \frac{1}{R_2}\left(1 - \frac{1}{n}\right) \ ,$$

or

$$\frac{1}{R_{i2}} = \frac{1-n}{n}\left(\frac{1}{R_1} - \frac{1}{R_2}\right) \tag{5-198a}$$

$$= -\frac{1}{nf'} \ . \tag{5-198b}$$

Thus, the radius of curvature R_p of the Petzval image surface is given by

$$R_p \equiv R_{i2}$$

$$= -nf' \quad .$$

$$(5\text{-}199)$$

Equation (5-199) may also be obtained directly by substituting Eq. (5-195b) into Eq. (5-117), where $n_k = 1$ and $L_k = S'$.

Note that the Petzval image surface for the lens corresponds to contributions of field curvature aberrations of the type given by Eq. (5-17) for the two refracting surfaces. Also, the radius of curvature of the Petzval surface does not depend on the object or the image distance; it depends only on the refractive index and the focal length of the lens. Its value is numerically negative for a positive lens; i.e., the Petzval surface is curved toward the lens with its center of curvature lying to its left, as illustrated in Figure 5-21a. The radius of curvature of the virtual Petzval surface for a negative lens is numerically positive as, illustrated in Figure 5-21b; it lies to the left of the lens and is curved toward it.

5.10.4 Spherical Aberration and Coma

Now we consider *lens bending,* i.e., how to determine the value of its shape factor q, to control its aberrations. From Eqs. (5-188) and (5-193) we note that the spherical aberration and coma of a thin lens depend on its position and shape factors. For a given position factor p, the value of the shape factor which minimizes the spherical aberration is given by the condition

Substituting Eq. (5-201) into Eq. (5-188), we obtain the corresponding minimum spherical aberration

$$a_{smin} = -\frac{1}{32 f'^3}\left[\left(\frac{n}{n-1}\right)^2 - \frac{n}{n+2} p^2\right] \quad . \tag{5-202}$$

We note from Eq. (5-188) that for a given value of p, a_s as a function of q follows a parabola with its vertex at (q_{min}, a_{smin}). For different values of p, the parabolas have the same shape but different vertices.

It is evident from Eqs. (5-184) that when both an object and its image are real,

$$-1 \le p \le 1 \quad , \quad \text{or} \quad p^2 \le 1 \quad . \tag{5-203}$$

As indicated in Figure 5-19, the case $p = -1$ corresponds to an object at infinity and the image at the image-space focal plane of the lens. Similarly, $p = 1$ corresponds to an

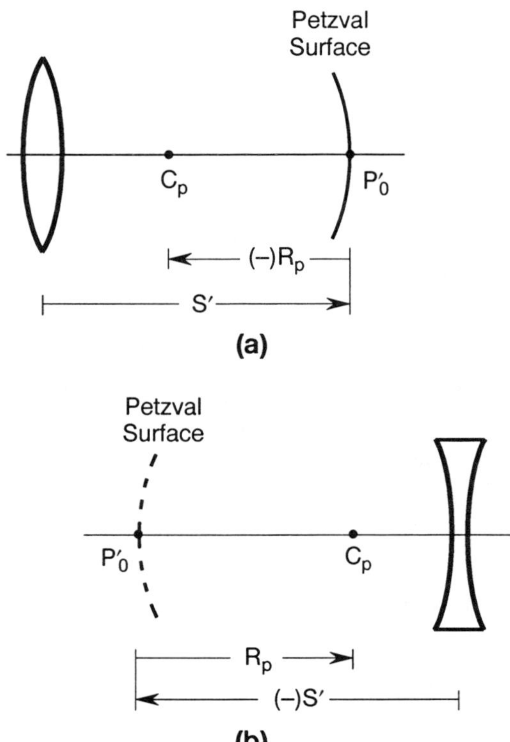

Figure 5-21. Petzval surface of a thin lens. (a) Real for a positive lens. (b) Virtual for a negative lens. C_p is the center of curvature of the Petzval surface.

object at the object-space focal plane and the image at infinity. The case $p = 0$ corresponds to an object and its image lying at distances $2f$ and $2f'$, respectively. For spherical aberration to be zero, Eq. (5-202) yields

$$p^2 = \frac{n(n+2)}{(n-1)^2} \tag{5-204a}$$

$$> 1 \; . \tag{5-204b}$$

Hence, *spherical aberration of a thin lens (with spherical surfaces) cannot be zero when both the object and its image are real.*

For a thin lens with a refractive index $n = 1.5$, Eqs. (5-188), (5-201), and (5-202) reduce to

$$a_s = -\frac{1}{24 f'^3}\left(6.75 + 3.25 p^2 + 7q^2 + 10 pq\right) \; , \tag{5-205}$$

$$q_{min} = -(5/7)p \; , \tag{5-206}$$

and

$$a_{smin} \;=\; -\,\frac{1}{32 f'^3}\left(9 - \frac{3}{7}p^2\right) \;, \tag{5-207}$$

respectively. Figure 5-22 shows how spherical aberration varies with q for $p = 0$. The minimum value of spherical aberration corresponds to $q_{min} = 0$, i.e., an *equiconvex lens.* As pointed out earlier, the variation of spherical aberration with q for other values of p follows the same parabola except that the location of its vertex (q_{min}, a_{smin}) depends on p. The vertices of the parabolas follow the lower parabolic curve in Figure 5-22, which represents a_{smin} as a function of q obtained by substituing Eq. (5-206) into Eq. (5-207). The solid dots on this curve indicate various values of p. The minimum value of spherical aberration approaches zero for $|p| = \sqrt{21}$. It changes its sign for larger values of $|p|$.

It follows from Eq. (5-193) that the coma of a thin lens is zero if its position and shape factors are related to each other according to

$$q \;=\; -\,\frac{(2n+1)(n-1)p}{n+1} \;. \tag{5-208}$$

For $n = 1.5$, Eqs. (5-193) and (5-208) reduce to

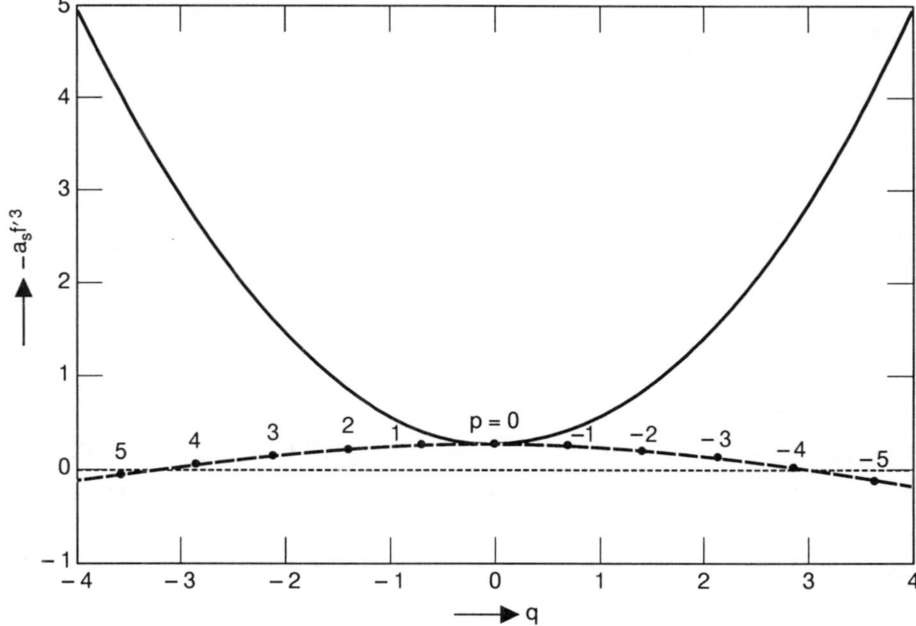

Figure 5-22. Spherical aberration of a thin lens as a function of its shape factor q. The variation of its minimum value with q is indicated by the lower parabolic curve. Several values of p are indicated on this curve.

$$a_c = -\frac{1}{6f'^2 S'}(4p+5q) \tag{5-209}$$

and

$$q = -0.8p \quad, \tag{5-210}$$

respectively. For $p=-1$, the values of q giving minimum spherical aberration $\left(q_{min} = 0.71\right)$ and zero coma ($q = 0.8$) are approximately the same. Thus, a lens designed for zero coma for parallel incident light will have practically the minimum amount of spherical aberration. *Since spherical aberration of a thin lens varies as* f'^{-3}, *it is possible to make it zero for a combination of two or more lenses having focal lengths of different signs.* A doublet designed to correct for spherical aberration can at the same time be corrected for coma, without correcting its astigmatism (see Problems 5.7 and 5.12).

5.10.5 Aplanatic Lens

We now design a lens to be aplanatic and determine the conjugate pairs, called *aplanatic points*, for which spherical aberration and coma are zero. We have already seen that the coma of a thin lens is zero if its position and shape factors are related to each other according to Eq. (5-208). Substituting this value of q into Eq. (5-188) for the coefficient of spherical aberration, we obtain

$$a_s = -\frac{n^3}{32n(n-1)f'^3}\left[(n+1)^2 - (n-1)^2 p^2\right] \tag{5-211}$$

$$= 0 \; if \; p = \pm\frac{n+1}{n-1} \quad. \tag{5-212}$$

Substituting the value of p for $a_s = 0$ into Eq. (5-208) for zero coma, we obtain

$$q = \mp(2n+1) \quad. \tag{5-213}$$

The object and image distances that satisfy Eq. (5-212) are the aplanatic points of a thin lens whose shape factor is given by Eq. (5-213). Or, a thin lens whose shape factor is given by Eq. (5-213) is aplanatic for conjugate points given by Eq. (5-212). Substituting the positive value of p from Eq. (5-212) into Eqs. (5-184), we obtain the conjugate points

$$(S, S') = \left[-\frac{n-1}{n}f', -(n-1)f'\right] \quad. \tag{5-214}$$

The magnification of the image is n. Choosing the negative value of p simply exchanges the object and image points, i.e., it yields

$$(S, S') = \left[(n-1)f', \frac{n-1}{n}f'\right] \quad. \tag{5-215}$$

The magnification of the image in this case is $1/n$. We note that if the object is real, the image is virtual. Similarly, if the object is virtual, the image is real.

Substituting Eq. (5-213) into Eq. (5-185), we find for the negative value of q that

$$\frac{R_2}{R_1} = \frac{n}{n+1} \quad . \tag{5-216}$$

Combining Eq. (5-216) with Eq. (5-169), we may write the radii of curvature in terms of the focal length:

$$R_1 = -\frac{n-1}{n} f' \tag{5-217a}$$

and

$$R_2 = -\frac{n-1}{n+1} f' \quad . \tag{5-217b}$$

Similarly, for the positive value of q,

$$\frac{R_2}{R_1} = \frac{n+1}{n} \quad , \tag{5-218}$$

$$R_1 = \frac{n-1}{n+1} f' \quad , \tag{5-219a}$$

and

$$R_2 = \frac{n-1}{n} f' \quad . \tag{5-219b}$$

The lens represented by Eqs. (5-219) is simply a lens represented by Eqs. (5-217) but turned around so that its front surface becomes the rear surface and vice versa. It should be noted that according to Eqs. (5-212) and (5-213), a positive value of p goes with a negative value of q. Similarly, a negative value of p goes with a positive value of q. Accordingly, Eqs. (5-214) and (5-217) go together, and Eqs. (5-215) and (5-219) go together.

From Eqs. (5-214) and (5-217), we note that

$$S = S_1 = R_1 \quad , \tag{5-220}$$

i.e., the axial point object lies at the center of curvature of the first surface of the lens. Thus, as discussed in Section 5.4, the object lies at an aplanatic point of the first surface. According to Eq. (5-50a), this surface contributes only astigmatism and distortion. The image by the first surface also lies at its center of curvature, which is at a distance

$$S_2 = R_1 = \frac{n+1}{n} R_2 \tag{5-221}$$

from the second surface, where we have used Eq. (5-216) in the last step. Thus, the image formed by the first surface lies at an aplanatic point for the second surface, as may be seen from the discussion in Section 5.4. Hence, imaging by the lens is aplanatic, and the final image lies at

$$
\begin{aligned}
S' &\equiv S_2' \\
&= -(n-1)f' \\
&= (n+1)R_2 \quad .
\end{aligned}
\tag{5-222}
$$

According to Eq. (5-50b), this surface contributes only field curvature and distortion. The astigmatism of the lens given by Eq. (5-194) is contributed by its first surface only. Similarly, its field curvature given by Eq. (5-195a) is contributed by its second surface only. The distortion contributions of its two surfaces cancel each other so that the lens is distortion free, as expected for a thin lens with an aperture stop at the lens (see Problem 5.10).

Equations (5-215) and (5-219) can be discussed in a similar manner. The aplanatic pointsof a thin aplanatic lens are illustrated in Figure 5-23 for a real and a virtual object. It is evident that an aplanatic lens can only make a diverging beam less divergent or a converging beam more convergent. It cannot, for example, convert a diverging beam into a converging beam.

5.10.6 Thin Lens with Conic Surfaces

When an aperture stop is located at a refracting surface, its primary aberration function for a conic surface is different from that for a corresponding spherical surface only in spherical aberration according to Eqs. (5-86) through (5-90); the other aberrations are the same for the two surfaces. Of course, the Gaussian imaging equations are identical for the two surfaces. The difference in the coefficients of spherical aberration is given by

$$
a_{sc} - a_{ss} = \sigma \quad ,
\tag{5-223}
$$

where σ is given by Eq. (5-77). Letting e_1 and e_2 be the eccentricities of the two conic surfaces of a thin lens with vertex radii of curvature R_1 and R_2, and noting that $n_1 = 1$, $n_1' = n$, $n_2 = n$, and $n_2' = 1$, the coefficient of its spherical aberration may be written

$$
a_{sc} = a_s + \sigma_1 + \sigma_2 \quad ,
\tag{5-224}
$$

where

$$
\sigma_1 = (n-1)e_1^2 / 8R_1^3
\tag{5-225}
$$

and

$$
\sigma_2 = -(n-1)e_2^2 / 8R_2^3 \quad .
\tag{5-226}
$$

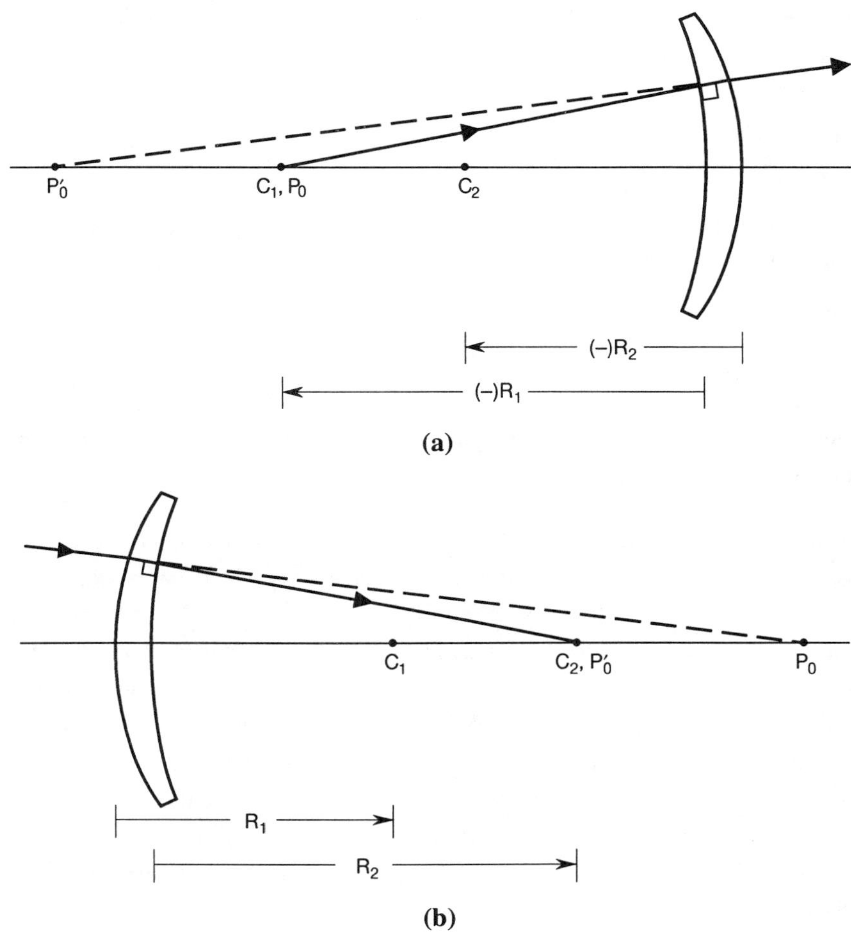

Figure 5-23. Aplanatic points P_0 and P_0' of a thin aplanatic lens with its centers of curvature at C_1 and C_2. (a) Real object point P_0 at C_1 and virtual image point P_0'. The object ray is incident normally to the first surface. (b) Virtual object point P_0 and real image P_0' at C_2. The image ray is refracted normally to the second surface.

or

$$\boxed{a_{sc} = a_s + \frac{n-1}{8}\left(\frac{e_1^2}{R_1^3} - \frac{e_2^2}{R_2^3}\right)}\ .$$

$(5\text{-}227)$

It is evident that spherical aberration of the lens can be made zero by an appropriate choice of the eccentricities of its two surfaces (see Problem 5.9).

5.10.7 Thin Lens with Aperture Stop Not at the Lens

So far we have discussed aberrations of a thin lens with a collocated aperture stop. Now we consider its aberrations when its aperture stop is located at a position such that the Gaussian image $P_0'P'$ lies at a distance L from its exit pupil, as illustrated in Figure 5-24. The peak values of its primary aberrations change according to Eqs. (5-146) through (5-150) with

$$\zeta = \frac{S' - L}{a_1 L} h' \quad , \tag{5-228}$$

where a_1 is the radius of the lens (i.e., the radius of the exit pupil when the aperture stop is located at the lens). The radius a_2 of the exit pupil located at distance L from the Gaussian image is given by

$$a_2 = (L/S')a_1 \quad . \tag{5-229}$$

This value keeps the f-number of the imaging-forming light cone as well as the amount of light in the image unchanged.

Given the peak aberration values for a lens with an aperture stop collocated with it, the peak aberration values for an aperture stop that is not collocated can be obtained from Eqs. (5-146) through (5-150) by substituting Eq. (5-228) into them. We noted earlier that when the aperture stop is collocated with the lens, its spherical aberration and coma depend on its shape factor, but its astigmatism and field curvature do not. Moreover, its distortion is zero. However, with a noncollocated stop, its spherical aberration does not change, and its astigmatism and field curvature depend on its shape factor due to its spherical aberration (which is nonzero for a real object and a real image) and coma. Its coma can be made zero by an appropriate choice of the position of its aperture stop. Its distortion is generally nonzero since the chief ray is no longer the undeviated ray it was in Figure 5-18.

5.11   FIELD FLATTENER

A *field flattener* is a thin lens, typically planoconvex or planoconcave, placed in the image plane of an optical system to flatten the curvature of its image surface. An example of such a lens is a planoconvex lens placed at the image plane of a Schmidt camera to flatten the Petzval curvature of its spherical mirror (see Section 6.3.1). Since the *field-flattening lens* is placed at the image plane of an optical system, the distances of the object and image for it are (practically) zero. Consequently, the expressions obtained in Section 5.10 for the aberrations of a thin lens cannot be used. A stop placed in the image plane cannot act as an aperture stop since it cannot control the focused imaging beams. To obtain the aberrations of a field flattener, we start with Eq. (5-49a) for the aberrations of a single refracting surface for zero object distance and add them for its two surfaces. Since the object distance for the two lens surfaces is zero, each surface is anastigmatic, i.e., neither surface introduces any spherical aberration, coma, or astigmatism. Hence, the field-flattening lens is also anastigmatic. The Petzval curvature, which is independent of the object location, is the same as that determined in Section 5.10.3. We now show that a field flattener introduces not only Petzval curvature but distortion as well.

5.11.1 Imaging Relations

Consider a field-flattening lens placed at the image plane of an optical system as illustrated in Figure 5-25. Let its image-space focal length be f' given by

$$\frac{1}{f'} = (n-1)\left(\frac{1}{R_1} - \frac{1}{R_2}\right) \ ,$$

(5-230)

where n is its refractive index and R_1 and R_2 are the radii of curvature of its two surfaces. Let the exit pupil of the imaging system under consideration be at a (numerically negative) distance s_1 from the lens; the exit pupil ExP of the system is the entrance pupil EnP_1 for the lens. Its image ExP_1 by the first surface of the lens is the exit pupil for the surface and lies at a distance s_1' given by [see Eq. (5-163)]

$$\frac{n}{s_1'} = \frac{n-1}{R_1} + \frac{1}{s_1} \ .$$

(5-231)

It is also the entrance pupil EnP_2 for the second surface. Its image ExP_2 by the second surface is the exit pupil for that surface as well as the lens and lies at a distance s_2' given by [see Eq. (5-165)]

$$\begin{aligned}
\frac{1}{s_2'} &= \frac{1-n}{R_2} + \frac{n}{s_1'} \\
&= \frac{1-n}{R_2} + \frac{n-1}{R_1} + \frac{1}{s_1} \ ,
\end{aligned}$$

(5-232)

where we have substituted for s_1' from Eq. (5-231).

The image $P_0'P'$ formed by the system is the object for the lens and, in particular, for its first surface. Since the object lies at the surface, its image formed by the surface also

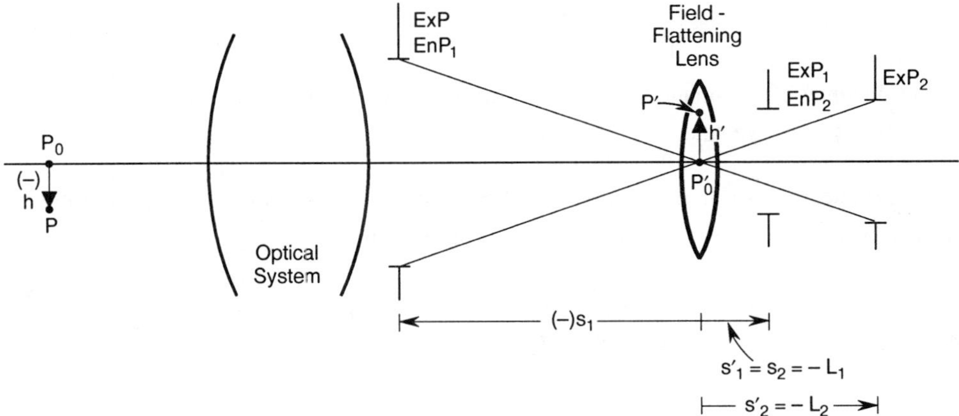

Figure 5-25. Field-flattening lens placed at the image formed by a certain optical system for flattening its curvature.

lies at it with a unity magnification. This image being the object for the second surface, its image by it also lies at it with unity magnification. Thus, if h' is the height of the image formed by the system, the heights h_1' and h_2' of the images formed by the two lens surfaces, respectively, are also equal to h', i.e.,

$$h_1' = h_2' = h' \ . \tag{5-233}$$

5.11.2 Aberration Function

Letting $n_1 = 1$ and $n_1' = n$ in Eq. (5-49a), the aberration contributed by the first surface at a point (x_1, y_1) on its exit pupil ExP_1 may be written

$$
\begin{aligned}
W_1(x_1, y_1; h_1') &= a_{d1}h_1'^2 \left(x_1^2 + y_1^2\right) + a_{t1}h_1'^3 x_1 \\[2mm]
&= \frac{n(n-1)}{2R_1 L_1^2}\left[-\frac{h'^2}{2}\left(x_1^2 + y_1^2\right) + \frac{R_1 + L_1}{R_1}h'^3 x_1\right] \ ,
\end{aligned}
\tag{5-234}
$$

where

$$L_1 = -s_1' \tag{5-235}$$

is the distance of the image of height h_1' from the exit pupil ExP_1. This distance is numerically negative since the image lies (at the lens) to the left of the exit pupil.

Similarly, letting $n_2 = n$ and $n_2' = 1$ in Eq. (5-49a), the aberration contributed by the second surface at a point (x_2, y_2) on its exit pupil ExP_2 may be written

$$
\begin{aligned}
W_2(x_2, y_2; h_2') &= a_{d2}h_2'^2 \left(x_2^2 + y_2^2\right) + a_{t2}h_2'^3 x_2 \\[2mm]
&= \frac{(n-1)}{2nR_2 L_2^2}\left[\frac{h'^2}{2}\left(x_2^2 + y_2^2\right) - \frac{R_2 + L_2}{R_2}h'^3 x_2\right] \ ,
\end{aligned}
\tag{5-236}
$$

where

$$L_2 = -s_2' \tag{5-237}$$

is the distance of the image of height h_2' from the exit pupil ExP_2. This distance is also numerically negative since the image lies (at the lens) to the left of the exit pupil.

The aberration function for the field-flattening lens is obtained by combining the aberration contributions of its two surfaces (see Section 5.7.1):

$$W_s(x_2, y_2; h_2') = W_1\left(\frac{x_2}{m_2}, \frac{y_2}{m_2}; \frac{h_2'}{M_2}\right) + W_2(x_2, y_2; h_2') \ , \tag{5-238}$$

where

$$
\begin{aligned}
m_2 &= ns_2'/s_1' \\
&= n L_2/L_1
\end{aligned}
\tag{5-239}
$$

and

$$M_2 = h_2'/h_1'$$

$$= 1 \tag{5-240}$$

are the pupil and image magnifications, respectively, for the second surface. Substituting Eqs. (5-234) and (5-236) along with Eqs. (5-239) and (5-240) into Eq. (5-238), we obtain

$$W_s\left(x_2, y_2; h_2'\right) = a_{ds}h'^2\left(x_2^2 + y_2^2\right) + a_{ts}h'^3 x_2 \quad , \tag{5-241}$$

where

$$a_{ds} = -1/4nf'L_2^2 \tag{5-242}$$

and

$$a_{ts} = \frac{n-1}{2L_2}\left(\frac{R_1 + L_1}{R_1^2 L_1} - \frac{R_2 + L_2}{nR_2^2 L_2}\right) \quad . \tag{5-243}$$

Substituting for L_1 and L_2 from Eqs. (5-235) and (5-237) in terms of Eqs. (5-231) and (5-232) into the quantity in parenthesis in Eq. (5-243), we may write

$$a_{ts} = \frac{1}{2f'L_2}\left(\frac{1}{nR_1} + \frac{1}{R_2} - \frac{1}{ns_1}\right) \quad . \tag{5-244}$$

It is evident from Eq. (5-241) that the field-flattening lens introduces both field curvature and distortion. Noting that the sag of an image point at a height h' on a spherical image surface of radius of curvature R is equal to $h'^2/2R$ and that it represents the longitudinal defocus with respect to the Gaussian image point, comparing the field curvature term of Eq. (5-241) with Eq. (5-99b), we find that the image observed on a spherical surface of radius of curvature $R_p = -nf'$ is defocus free. This image surface is, of course, the Petzval surface discussed in Section 5.10.3. Moreover, comparing the distortion term of Eq. (5-241) with Eq. (3-21), we note that the image point is displaced by a height

$$\Delta h' = L_2 a_{ts} h'^3$$

$$= \frac{1}{2f'}\left(\frac{1}{nR_1} + \frac{1}{R_2} - \frac{1}{ns_1}\right)h'^3 \quad . \tag{5-245}$$

Since the field-flattening lens does not introduce any spherical aberration, coma, or astigmatism, it is anastigmatic.

The Petzval field curvature produced by the lens is used to cancel or flatten the curvature of an image surface produced by a certain system. Thus, an image surface of radius of curvature R_i is flattened if a lens of focal length $f' = R_i / n$ is placed at the image plane. The lens does introduce some distortion.

5.12   PLANE-PARALLEL PLATE

5.12.1 Introduction

A *plane-parallel plate*, as its name implies, is a plate with two surfaces that are parallel to each other. It is a thick lens whose two surfaces have infinite radii of curvature. Unlike a lens, a plane parallel plate is not used for imaging per se, but is often used in imaging systems as a beam splitter or a window. The imaging and aberration equations for such a plate cannot be obtained from those for a thin lens by letting the radii of curvature of its two surfaces approach infinity, since its thickness is neglected by its definition. However, as discussed below, they can be obtained by applying the imaging equations (5-5) and (5-9), and aberration equation (5-32) for a spherical surface to its two surfaces and combining the results thus obtained. We show that the distance between an object and its image formed by the plate is independent of the object position, and the aberration produced by it approaches zero as the object distance approaches infinity. Thus, as illustrated in Figure 5-26a, a plane-parallel plate placed in the path of a converging beam not only displaces its focus from P_1 by a certain amount (which depends only on the thickness and the refractive index of the plate) to P_2, but also introduces aberrations into it. In the case of a collimated beam, it only shifts the beam without introducing any aberration.

Figure 5-26b shows a *right-angle reflecting prism* as an example of a plane-parallel plate. It is used in optical systems to deviate the path of a beam by 90°. Its diagonal face acts like a mirror because the rays incident on it undergo a total internal reflection. The "unfolded" path of the rays, called a *tunnel diagram*, illustrates that the prism ABC is equivalent to a plane-parallel plate ABCD in terms of their optical path lengths.

5.12.2 Imaging Relations

Consider, as indicated in Figure 5-27, a circular plane-parallel plate of radius a, thickness t, and refractive index n forming the image of a point object P lying at a distance S from its front surface and at a height h from its axis. Let the aperture stop of the plate be located at its front surface.

Using Eqs. (5-5) and (5-9), we determine the location of the image of the point object P. For the first surface $n_1 = 1$, $n_1' = n$ and $R_1 = \infty$. Accordingly, it forms the image of P at P' such that

$$S_1' = nS_1$$

$$\equiv nS \tag{5-246}$$

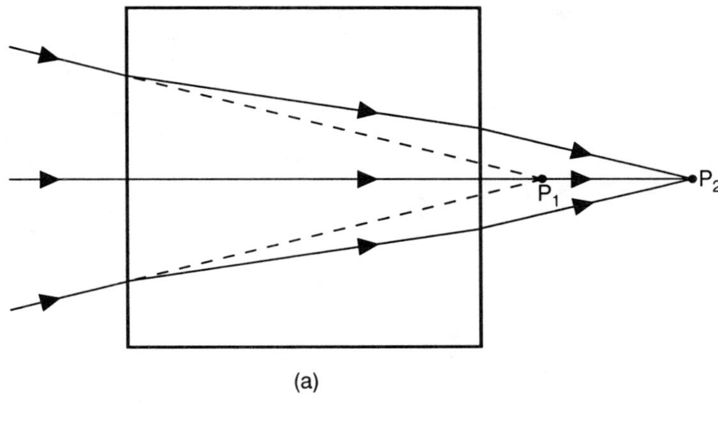

(a)

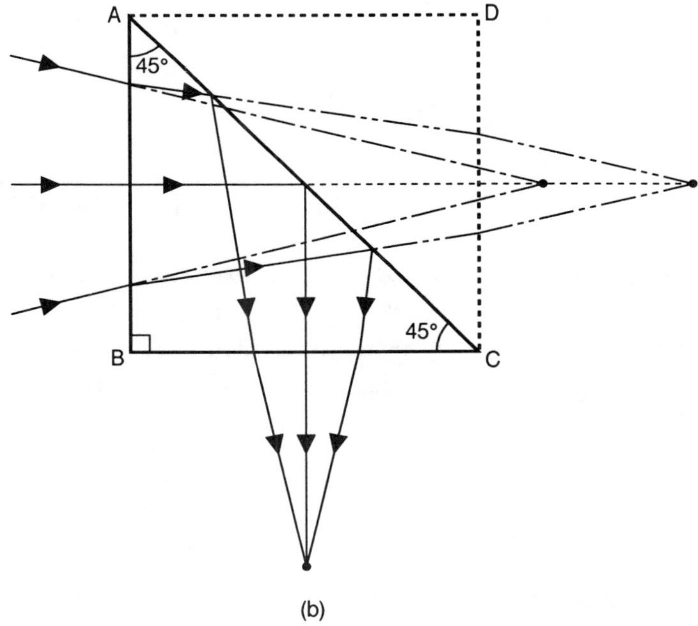

(b)

Figure 5-26. (a) Plane-parallel plate placed in the path of a converging beam of light. Rays incident on the plate converging toward P_1 converge toward P_2 after refraction by it. (b) A right-angle reflecting prism placed in the path of a converging beam. The optical path lengths of the rays for the prism are equivalent to those for a plane-parallel plate, where the virtual portion ADC of the equivalent plate is obtained by a reflection of its real portion ABC by the reflecting surface AC.

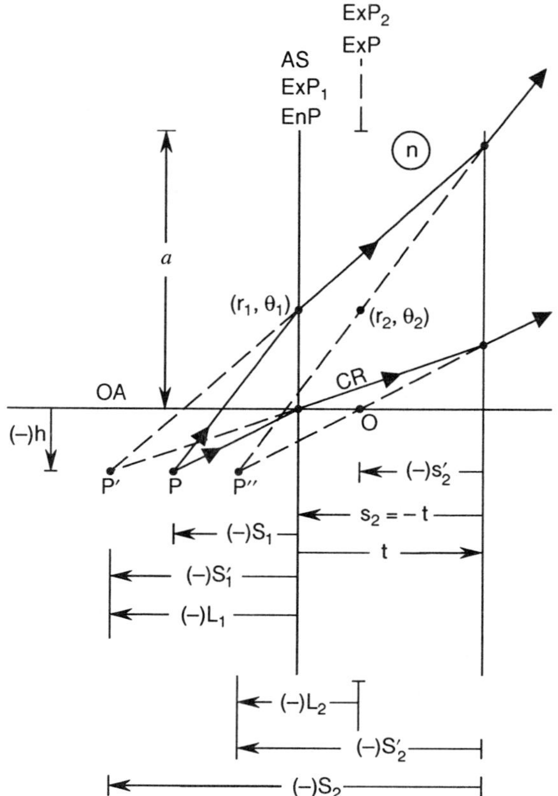

Figure 5-27. Imaging of a point object P by a plane-parallel plate of refractive index n and thickness t. P' is the image of P formed by the first surface, and P'' is the image of P' formed by the second surface of the plate. The aperture stop AS and, therefore, the entrance pupil EnP of the plate are located at the first surface. A negative sign in parentheses indicates a numerically negative quantity.

and

$$M_1 \;=\; h'_1 / h_1 \;=\; n_1 S'_1 / n'_1 S_1 \;=\; 1 \;\;, \tag{5-247}$$

where $h_1 \equiv h$. For the second surface, $n_2 = n$, $n'_2 = 1$, $R_2 = \infty$, and $S_2 = S'_1 - t$. Hence, it forms the image of P' at P'' such that

$$
\begin{aligned}
S'_2 &= S_2 / n \\
     &= \left(S'_1 - t \right) / n \\
     &= S - \frac{t}{n}
\end{aligned}
\tag{5-248}
$$

and

$$M_2 \;=\; h'_2 / h'_1 \;=\; n_2 S'_2 / n'_2 S_2 \;=\; 1 \;\;. \tag{5-249}$$

Noting that S'_2 is numerically negative, the displacement PP'' of the final image from the object may be written

$$PP'' = -S_1 - \left(-S_2' - t\right)$$

or

$$\boxed{PP'' = t\left(1 - 1/n\right)} \quad . \tag{5-250}$$

Thus, the *image displacement PP" is independent of the object distance S*; it depends only on the thickness t and refractive index n of the plate.

Since the aperture stop is located at the first surface, the entrance pupil EnP of the system is also located there. Moreover, the entrance and exit pupils EnP_1 and ExP_1 for this surface are also located at the surface. The entrance pupil EnP_2 for the second surface is ExP_1. The exit pupil ExP_2 for this surface is the image of EnP_2 formed by it. Thus, letting $n_2 = n$, $n_2' = 1$, $s_2 = -t$, and $R_2 = \infty$, we find from Eqs. (5-5) and (5-9) that ExP_2 is located at a distance $s_2' = -t/n$ from the second surface and its magnification $m_2 = 1$. As expected from Eq. (5-250), ExP_2 lies at a distance $t(1 - 1/n)$ from the first surface. Of course, ExP_2 is also the exit pupil ExP of the system. It is evident that for the first surface, the distance L_1 of the image P' from ExP_1 is equal to its distance S_1' from the surface. For the second surface, distance L_2 of the image P'' from ExP_2 is given by

$$L_2 = S_2' - s_2' \quad , \tag{5-251}$$

since L_2, S_2', and s_2' are all numerically negative. Substituting for S_2' and s_2', we find that

$$L_2 = S \quad . \tag{5-252}$$

5.12.3 Aberration Function

Now we determine the primary aberration function of the plate. We start with the aberration $W_1\left(r_1, \theta_1; h_1'\right)$ contributed by the first surface at a point $\left(r_1, \theta_1\right)$ in the plane of ExP_1. Letting $n_1 = 1$, $n_1' = n$, and $R_1 = \infty$, Eq. (5-7b) yields

$$a_{s1} = \frac{n\left(n^2 - 1\right)}{8S_1'^3} \quad . \tag{5-253}$$

Moreover, Eq. (5-29b) reduces to $d_1 = -1$, and since $S_1' = L_1$, Eq. (5-33) reduces to $a_{ss} = a_{s1}$. Since $R_1 = \infty$, implying that the Petzval surface coincides with the Gaussian image plane, the Petzval contributions to field curvature and distortion represented by the second terms on the right-hand side of Eqs. (5-36) and (5-37), respectively, are zero. Hence, for the first surface, Eq. (5-32) may be written

$$W_1\left(r_1, \theta_1; h_1'\right) = a_{s1}\left(r_1^4 - 4h_1'r_1^3\cos\theta_1 + 4h_1'^2r_1^2\cos^2\theta_1 + 2h_1'^2r_1^2 - 4h_1'^3r_1\cos\theta_1\right) \quad . \tag{5-254}$$

Next, we determine the aberration $W_2(r_2,\theta_2;h_2')$ contributed by the second surface at a point (r_2,θ_2) in the plane of ExP_2. Letting $n_2 = n$, $n_2' = 1$, and $R_2 = \infty$, Eq. (5-7b) yields for this surface

$$a_{s2} = -\frac{n^2-1}{8n^2 S_2'^3} \quad . \tag{5-255}$$

Once again, Eq. (5-29b) reduces to $d_2 = -1$ and the Petzval contributions to field curvature and distortion are zero. Hence, for the second surface, Eq. (5-32) may be written

$$W_2(r_2,\theta_2;h_2') = a_{ss2}\left(r_2^4 - 4h_2'r_2^3\cos\theta_2 + 4h_2'^2 r_2^2\cos^2\theta_2 + 2h_2'^2 r_2^2 - 4h_2'^3 r_2\cos\theta_2\right) , \tag{5-256}$$

where

$$a_{ss2} = \left(S_2'/L_2\right)^4 a_{s2} \quad . \tag{5-257}$$

Finally, since m_2 and M_2 are both unity, following Eq. (5-95), the aberration of the plane-parallel plate at a point (r,θ) in the plane of its exit pupil can be written

$$W(r,\theta;h) = W_1(r,\theta;h) + W_2(r,\theta;h) \quad , \tag{5-258}$$

where we have written h in place of h_2', since they are equal to each other. Substituting Eqs. (5-254) and (5-256) into (5-258), we may write

$$\boxed{W(r,\theta;h) = a_s\left(r^4 - 4hr^3\cos\theta + 4h^2 r^2\cos^2\theta + 2h^2 r^2 - 4h^3 r\cos\theta\right)} \quad , \tag{5-259}$$

where

$$a_s = a_{s1} + \left(S_2'/L_2\right)^4 a_{s2} \quad . \tag{5-260}$$

Substituting Eqs. (5-246), (5-248), (5-252), (5-253), and (5-255) into Eq. (5-260), we obtain

$$a_s = \frac{n(n^2-1)}{8S_1'^3}\left(1 - \frac{nS_2'}{S_1'}\right) , \tag{5-261}$$

or

$$a_s = \frac{(n^2-1)t}{8n^3 S^4} \quad . \tag{5-262}$$

Note that the *aberration increases linearly with the plate thickness t*. Moreover, as expected, the aberration reduces to zero for a collimated incident beam $(S \to -\infty)$.

5.13 CHROMATIC ABERRATIONS

5.13.1 Introduction

So far in this chapter, we have discussed the imaging relations and the monochromatic aberrations of an imaging system. Although the wavelength of the object radiation was not explicitly stated, the refractive indices used in the expressions for imaging and aberrations were for a certain wavelength. Now, the refractive index of a transparent substance decreases with increasing wavelength. Accordingly, a thin lens, for example, made of such a substance will have a shorter focal length for a shorter wavelength. Consequently, an axial point object emanating white light will be imaged at different distances along the axis depending on the wavelength; i.e., the image will not be a "white" point. Similarly, light of each wavelength will form an image of a finite object and each image will have a different size. The axial and transverse extents of the image of a multiwavelength point object are called *longitudinal* and *transverse chromatic aberrations*, respectively. They describe a chromatic change in the position and magnification of the image, and that is the subject of this section.

We start this section with a discussion of the chromatic aberrations of a single refracting surface and apply the results to obtain the chromatic aberrations of a thin lens, a doublet, and finally, a general system consisting of a series of refracting surfaces. The chromatic aberrations of a plane-parallel plate are considered as an example of the general theory. An *achromatic doublet* consisting of two thin lenses that are separated or in contact so that its focal length is independent of wavelength is discussed. Numerical examples are given to illustrate the concepts. A brief discussion of how an achromatic doublet can be designed so that it is aplanatic is also given.

Since the refractive index of a transparent substance depends on the wavelength, the optical path length of a ray passing through it also depends on the wavelength. Accordingly, the aberrations of a refracting system also vary with the wavelength. For example, the variation of spherical aberration with wavelength, called *spherochromatism*, can be calculated in the case of a thin lens by substituting the appropriate value of the refractive index in Eq. (5-188). An example of spherochromatism is considered in Section 6.6 where the variation of spherical aberration of a Schmidt plate with wavelength or refractive index is discussed. However, this variation is generally small, especially for a narrow spectral bandwidth.

5.13.2 Single Refracting Surface

First we consider, as indicated in Figure 5-28, the chromatic aberrations of a single refracting surface of a vertex radius of curvature R separating media of refractive indices n and n'. The distance S' and height h' of the image P' of a point object P lying at a distance S and height h are given by the relations (see Section 5.2)

$$\frac{n'}{S'} - \frac{n}{S} = \frac{n'-n}{R} \tag{5-263}$$

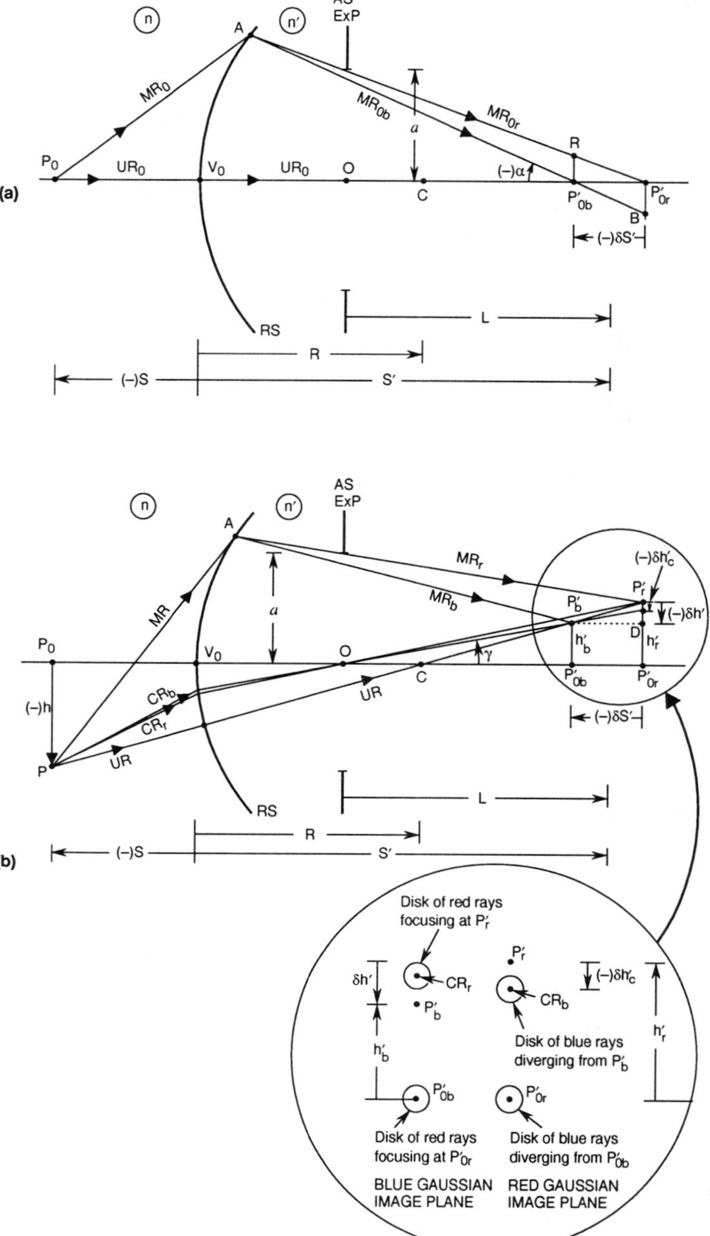

Figure 5-28. Longitudinal and transverse chromatic aberrations $\delta S'$ and $\delta h'$ of a refracting surface RS. The distance S' and height h' of the image P' of an off-axis point object P depend on the wavelength of object radiation. The subscripts b and r denote blue and red light. UR and MR are the undeviated and marginal rays, respectively. (a) On-axis imaging. (b) Off-axis imaging. The change in a certain quantity represents the difference in its value for the blue and red rays, e.g., the change in the image distance is $\delta S' = S'_b - S'_r$, where S'_b and S'_r are the image distances for the blue and red rays.

and

$$M = h'/h$$

$$= nS'/n'S \quad ,\tag{5-264}$$

where M is the transverse magnification of the image. Let δ represent a small change in a certain quantity corresponding to a small change in the wavelength. Since the object distance S is independent of the wavelength, differentiating both sides of Eq. (5-263), we obtain

$$\frac{\delta n'}{S'} - \frac{n'}{S'^2}\,\delta S' - \frac{\delta n}{S} = \frac{\delta n' - \delta n}{R} \quad .\tag{5-265}$$

Substituting for S from Eq. (5-263), we find that

$$\frac{\delta S'}{S'} = \left(\frac{\delta n}{n} - \frac{\delta n'}{n'}\right)\left(\frac{S'}{R} - 1\right) \quad .\tag{5-266}$$

Similarly, since the object height h is independent of wavelength, differentiating both sides of Eq. (5-264), we obtain

$$\frac{\delta M}{M} = \delta h'/h'$$

$$= \frac{\delta n}{n} - \frac{\delta n'}{n'} + \frac{\delta S'}{S'}$$

$$= \left(\frac{\delta n}{n} - \frac{\delta n'}{n'}\right)\frac{S'}{R} \quad ,\tag{5-267}$$

where in the last step we have used Eq. (5-266). Note that the fractional chromatic variation of magnification is independent of the object (or image) height. The quantities δn and $\delta n'$ represent the difference in the refractive index of the object and image spaces, respectively, for the blue and red light. The blue and red light represent, in general, the shortest and the longest wavelengths of the object radiation spectrum. The chromatic change $\delta S' = S'_b - S'_r$ in the position of the axial image represents the distance between the axial Gaussian images for the blue and red light. It is called the *longitudinal chromatic aberration*, or simply, the *axial color*. The chromatic change $\delta h' = h'\delta M / M$ in the image height represents the difference $h'_b - h'_r$ in the heights of the blue and red chief rays in the blue and red Gaussian image planes, respectively. It is called the *transverse chromatic aberration of the Gaussian images.*

From a practical standpoint, the quantity of interest is the size of the image of a point object in a given Gaussian image plane. For example, the image of an on-axis point object in the red Gaussian image plane consists of a bright red Gaussian image point P'_{0r} at the center surrounded by blue rays. The radius $P'_{0r}B$ of the blue disk of rays is given by (see Figure 5-28a)

$$\boxed{\begin{aligned} r_i &= \alpha\,\delta S' \\ &= (a/L)\,|\delta S'| \end{aligned}} \quad ,$$

(5-268)

where a is the radius of the exit pupil and L is the distance of the image from it. Similarly, the image in the blue Gaussian image plane consists of a bright blue Gaussian image point P'_{0b} at the center surrounded by red rays. The radius $P'_{0b}R$ of the red disk is approximately the same as that of the blue disk. For a given angular size of the light cone forming a Gaussian image point, the ratio a/L is fixed, i.e., if the position of the exit pupil is changed so that L changes, its diameter (in practice, the diameter of the aperture stop) is also changed so that a/L does not change. Hence, the size of the blue or red image disk does not change as the position of the exit pupil is changed. The radius r_i of the image disk is called the *transverse chromatic aberration of an image point in a given image plane* and its value is independent of a stop shift.

In the case of an off-axis object point P, its image in the red Gaussian image plane consists of a red Gaussian image point and a displaced disk of blue rays. The radius of the blue disk is approximately the same as that for the on-axis image. The displacement of the blue disk represents the difference in the heights of the blue and red chief rays in this image plane. We note from Figure 5-28b that the displacement, called the *lateral color* representing chromatic aberration of the chief ray in a given image plane, is given by

$$\boxed{\begin{aligned} \delta h'_c &= \delta h' - \gamma\,\delta S' \\ &= \delta h' - (h'/L)\,\delta S' \end{aligned}} \quad ,$$

(5-269)

where γ is the angle the blue chief ray CR_b makes with the optical axis in image space. It differs from $\delta h'$, which is the difference in the heights of the blue and red chief rays in the blue and red Gaussian image planes, repectively. Like $\delta S'$ and $\delta h'$, $\delta h'_c$ is also numerically negative in Figure 5-28b.

We note that the value of $\delta h'_c$ changes as the value of L changes. This is to be expected since the chief ray changes as the position of the exit pupil is changed. As an example, when the exit pupil lies at the center of curvature of the surface, $\delta h'_c$ must approach zero since the undeviated ray UR becomes the chief ray for both blue and red light. From similar triangles $CP'_{0b}P'_b$ and $P'_bDP'_r$ in Figure 5-28b, we find that

$$\delta h' = \frac{h'}{S' - R}\,\delta S' \quad .$$

(5-270)

Substituting Eq. (5-270) into Eq. (5-269), we obtain

$$\delta h'_c = h'\left(\frac{1}{S' - R} - \frac{1}{L}\right)\delta S' \quad .$$

(5-271)

Hence, $\delta h'_c = 0$ as $L \to S' - R$, i.e., when the exit pupil lies at the center of curvature. The values of the lateral colors $\delta h'_{c1}$ and $\delta h'_{c2}$ corresponding to two exit pupil locations so that the image lies at disatnces L_1 and L_2 from them are related to each other according to

$$\delta h'_{c2} = \delta h'_{c1} - \zeta r_i \quad , \tag{5-272}$$

where ζ is given by Eq. (5-151). Equation (5-272) represents the *stop-shift equation* for the lateral color.

It is evident from Eq. (5-269) that if the longitudinal aberration $\delta S'$ is zero [it can not happen for a single surface (unless $S' = R$) or even a thin lens (unless $S' = 0$)], then $\delta h'_c$ is equal to $\delta h'$ independent of the position of the exit pupil. In this respect, it is similar to coma, which, as discussed in Section 5.9.1 [see Eq. (5-147)], is independent of the position of the exit pupil when spherical aberration is zero.

5.13.3 Thin Lens

The chromatic aberrations of an image formed by a thin lens of focal length f' and refractive index n can be obtained by applying the results for a single refracting surface successively to its two surfaces. Or, we can obtain them from the imaging and magnification equations of a thin lens, namely, Eqs. (5-168) through (5-171). Since the image-space focal length f' of the lens depends on its refractive index n, the image distance S' and height h' also depend on it, i.e., the image is accompanied by both the longitudinal and transverse chromatic aberrations.

Differentiating Eqs. (5-168) and (5-171) with respect to the refractive index, we obtain

$$\frac{\delta S'}{S'^2} = \frac{\delta f'}{f'^2}$$

$$= -\frac{1}{f'V} \tag{5-273}$$

and

$$\frac{\delta M}{M} = \frac{\delta h'}{h'} \tag{5-274a}$$

$$= \frac{\delta S'}{S'} \tag{5-274b}$$

$$= -\frac{S'}{f'V} \quad , \tag{5-274c}$$

respectively, where

$$V = \frac{(n-1)}{\delta n} \qquad (5\text{-}275\text{a})$$

is called the *dispersive constant* of the lens material. Thus, for a change δn in the refractive index, there is a corresponding change $\delta f'$ in the focal length, $\delta S'$ in the image distance, and $\delta h'$ in the image height. It is evident from Eq. (5-273) that the smaller the value of δn, larger the value of V, smaller the change in focal length, and the smaller the values of chromatic aberrations.

It is common practice to consider n as the refractive index for the yellow line of helium $(\lambda = 0.5876\,\mu m)$ called the d line, and δn as the difference $n_F - n_C$ between the refractive indices for the Fraunhofer lines F and C, i.e., for the blue $(\lambda = 0.4861\,\mu m)$ and red $(\lambda = 0.6563\,\mu m)$ lines of hydrogen. Glass manufacturers often give the refractive index data as a six-digit number. For example, BK7 glass is specified as #517642. The first three digits define its refractive index according to $n_d - 1 = 0.517$ and the remaining three digits define its dispersive constant according to

$$V = \frac{n_d - 1}{n_F - n_C} \qquad (5\text{-}275\text{b})$$

$$= 64.2 \quad . \qquad (5\text{-}275\text{c})$$

The dispersive constant of a glass defined according to Eq. (5-275b) is called its *Abbe number*.

The refractive indices of the available lens materials and their Abbe numbers from Schott Optical Glass are given in Figure 5-30, called an n_d/V_d diagram. Each glass in this diagram is identified by a point whose position is called its *optical position*. The Abbe numbers of glasses vary from about 20 to 90. The glasses with $n_d > 1.60$, $V_d > 50$ or $n_d < 1.60$, $V_d > 55$ are called *crowns* and are indicated by the letter K; others are called *flints* and are indicated by the letter F. The simple crown (*kron* in German) glasses (soda-lime-silicate glasses) have low dispersion, and simple flint glasses (lead-alkali-silicate glasses) have high dispersion. The addition of barium oxide (BaO) yields a low dispersion with a relatively high refractive index. The borosilicate crown glasses contain boron oxide (B_2O_3) instead of calcium oxide used in the normal soda-lime-silicate glass. The addition of boron oxide yields a low refractive index and low dispersion. The light and heavy flint glasses contain low and high lead and barium amounts, respectively. Use of fluorine instead of oxygen also lowers the refractive index and dispersion. The barium flint glasses contain both barium oxide and lead oxide; he crown flint glasses contain calcium oxide and lead oxide, resulting in average dispersions. Use of rare earths such as lanthanum (La) yields glasses of high refractive index and high Abbe numbers. The terms heavy and light crowns or flints are also used, e.g., barium heavy flint (BaSF) or phosphorus heavy crown (PSK) (The letter S is for *schwer* in German, meaning "heavy" or "dense"). The barium crown glasses contain a large proportion of boron oxide and

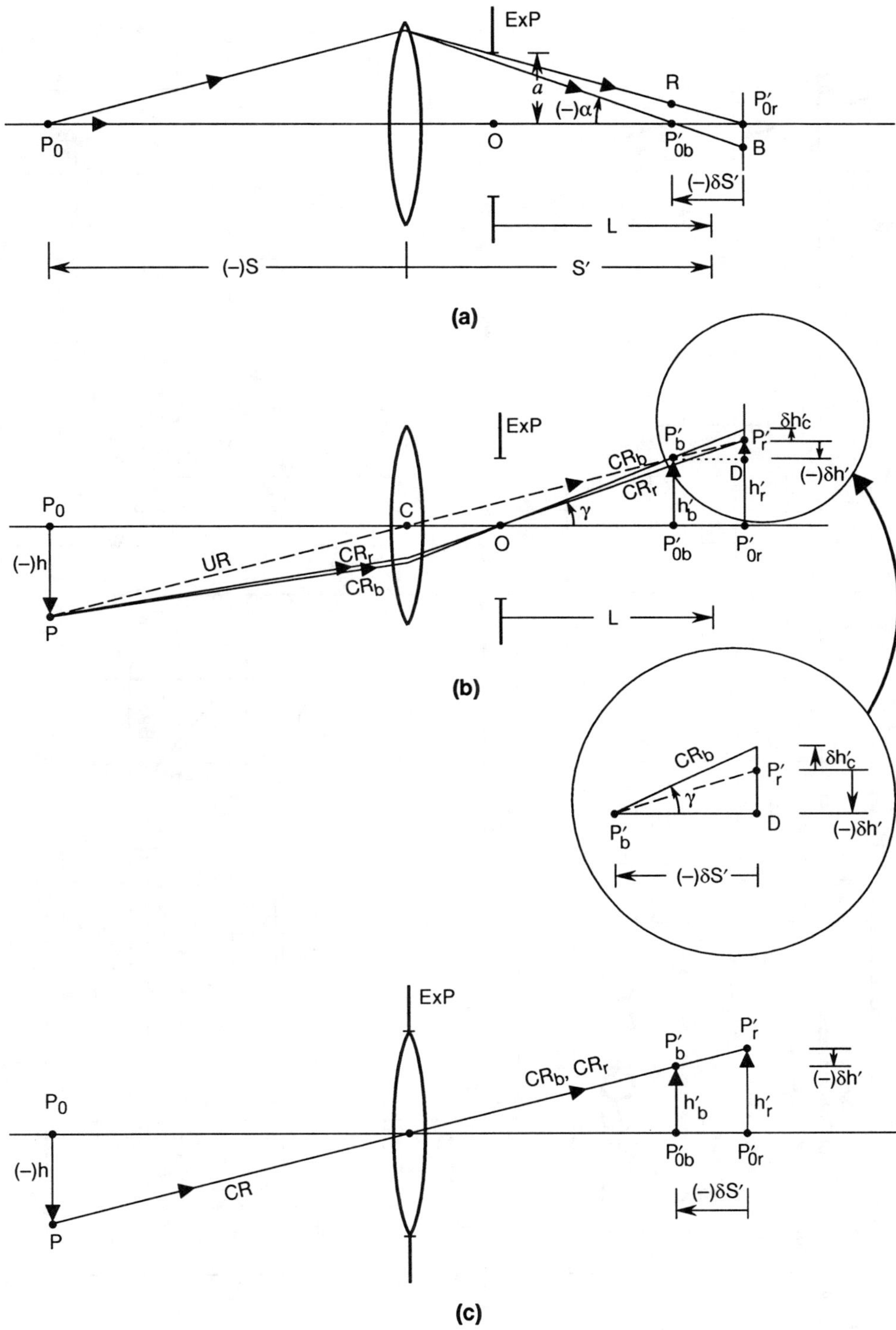

Figure 5-29. Chromatic aberrations of a thin lens. (a) On-axis imaging. (b) Off-axis imaging. (c) Off-axis imaging with exit pupil at the lens.

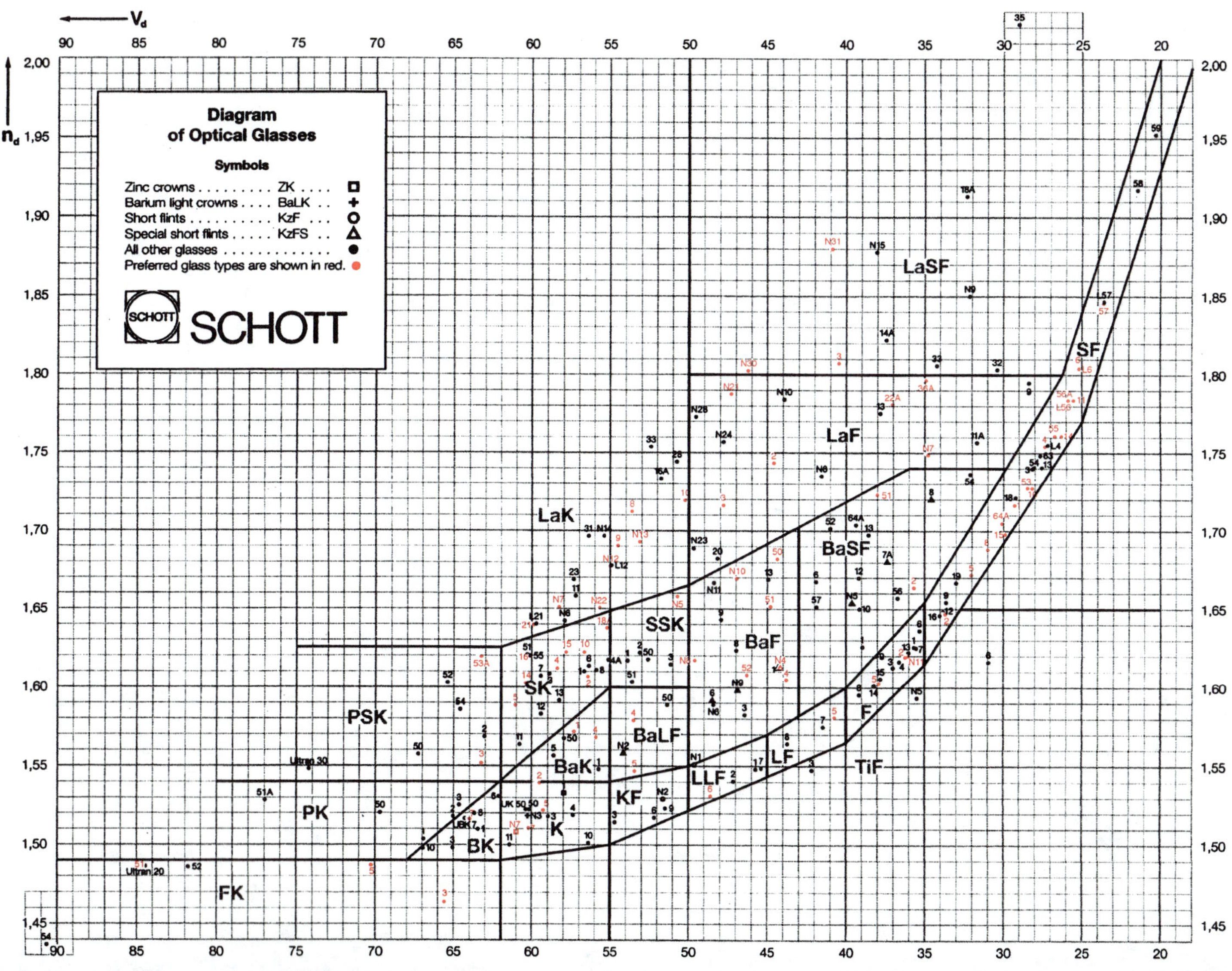

Figure 5-30. Refractive indices and Abbe numbers of various glass materials available from Schott Optical Glass, Inc.

barium oxide, while their silicon dioxide (SiO_2) content is low. The K group in the diagram includes the barium light crowns (BaLK) and the zinc crown (ZK). The glasses given in the diagram are for use in the visible light. The materials for use with infrared radiation have been discussed by McCarthy,[5] whose publications are listed in the references.

The radius of the blue or the red disk of rays in the red or the blue Gaussian image plane, respectively, is again given by Eq. (5-268), as may be seen from Figure 5-29a. Moreover, from Figure 5-29b, we can show that the (numerically positive) displacement $\delta h'_c$ of the blue disk from the red Gaussian image point P'_r of an off-axis point object P is given by Eq. (5-269). Substituting

$$\delta h' = \left(h' / S'\right)\delta S' \qquad\qquad (5\text{-}276a)$$

from Eqs. (5-274a) and (5-274b) (or from similar triangles $CP'_{0b}P'_b$ and $P'_b DP'_r$ in Figure 5-29b), Eq. (5-269) becomes

$$\delta h'_c = h'\left(\frac{1}{S'} - \frac{1}{L'}\right)\delta S' \quad . \qquad\qquad (5\text{-}276b)$$

The lateral color $\delta h'_c$ approaches zero when the exit pupil lies at the lens as in Figure 5-29c, i.e., as $L \to S'$. The chief ray in this case passes through the center of the lens undeviated regardless of its wavelength. Since the chief rays of different colors are coincident, they intersect an image plane at the same point. In a given image plane, rays (other than the chief ray) of different colors are not in sharp focus due to longitudinal chromatic aberration.

As a numerical example, Figure 5-31 shows how the focal length of a thin lens made of BK7 glass varies with wavelength. The variation of its refractive index is also shown in the figure. We note that the refractive index decreases as the wavelength increases. Hence, according to Eq. (5-272b), the focal length increases as the wavelength increases.

5.13.4 General System: Surface-by-Surface Approach

Now we consider an imaging system consisting of k refracting surfaces as illustrated in Figure 5-32. Let L be the distance of the image from its exit pupil ExP. If S_i and S'_i represent the object and image distances corresponding to the ith surface of a vertex radius of curvature R_i separating media of refractive indices n_{i-1} and n_i, the imaging Eq. (5-262) may be written

$$\frac{n_i}{S'_i} - \frac{n_{i-1}}{S_i} = \frac{n_i - n_{i-1}}{R_i} \quad . \qquad\qquad (5\text{-}277)$$

The image formed by the ith surface is the object for the $(i+1)$th surface. If t_i represents the axial spacing between two adjacent surfaces i and $i + 1$, the object distance for the $(i+1)$th surface is given by

$$S_{i+1} = -t_i + S_i' \quad . \tag{5-278}$$

Differentiating Eq. (5-277) with respect to wavelength, we obtain

$$n_i \frac{\delta S_i'}{S_i'^2} = \left(\frac{\delta n_{i-1}}{n_{i-1}} - \frac{\delta n_i}{n_i} \right) K_i + n_{i-1} \frac{\delta S_{i-1}'}{S_i^2} \quad , \tag{5-279}$$

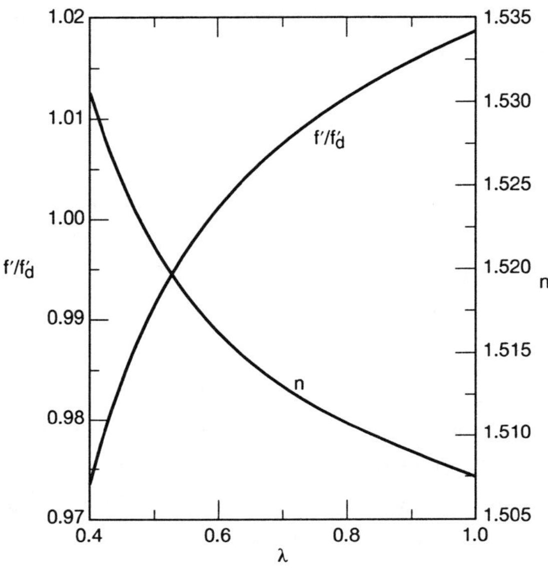

Figure 5-31. Variation of refractive index and focal length of a thin lens made of BK7 glass #517642 with wavelength. The focal length is normalized by its value for the *d* line. The wavelength is in micrometers.

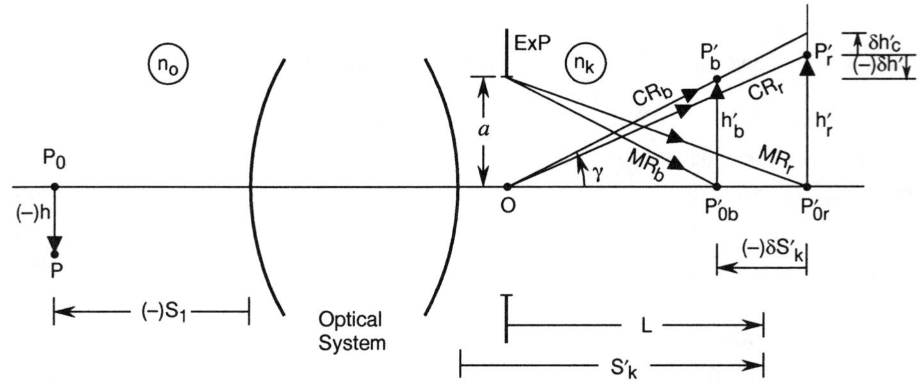

Figure 5-32. Chromatic aberrations of a general imaging system.

where

$$K_i = n_i\left(\frac{1}{R_i} - \frac{1}{S_i'}\right) \tag{5-280a}$$

$$= n_{i-1}\left(\frac{1}{R_i} - \frac{1}{S_i}\right) \ , \tag{5-280b}$$

and, following Eq. (5-278), we have let

$$\delta S_i = \delta S_{i-1}' \ . \tag{5-281}$$

Equation (5-279) gives a recursive relation for determination of the longitudinal chromatic aberration, or the chromatic change $\delta S_k'$ in the position of the image.

The transverse magnification of the image formed by the ith surface is given by

$$M_i = h_i'/h_i$$

$$= h_i'/h_{i-1}'$$

$$= \frac{n_{i-1}S_i'}{n_i S_i} \ , \tag{5-282}$$

where h_i' is the height of the image formed by the ith surface. Of course, the height h_{i-1}' of the image formed by the $(i-1)$th surface is the height of the object for the ith surface. The magnification of the overall system is equal to the product of the magnifications produced by each surface. It may be written

$$M = h_k'/h_o$$

$$= \frac{h_1'}{h_o}\frac{h_2'}{h_1'}\cdots\frac{h_k'}{h_{k-1}'}$$

$$= \prod_{i=1}^{k} M_i$$

$$= \frac{n_o}{n_k}\frac{S_1'S_2'\cdots S_k'}{S_1 S_2 \cdots S_k} \ , \tag{5-283}$$

where $h_o = h$ is the height of the object point and n_o is the refractive index of the object space. Differentiating both sides of Eq. (5-283) with respect to wavelength, we obtain

$$\frac{\delta M}{M} = \delta h'_k / h'_k$$

$$= \frac{\delta n_o}{n_o} - \frac{\delta n_k}{n_k} + \sum_{i=1}^{k} \left(\frac{\delta S'_i}{S'_i} - \frac{\delta S_i}{S_i} \right) \quad , \tag{5-284}$$

where $\delta S_1 = 0$. Substituting Eq. (5-281) into Eq. (5-284), we may write

$$\frac{\delta M}{M} = \frac{\delta n_o}{n_o} - \frac{\delta n_k}{n_k} + \frac{\delta S'_k}{S'_k} + \sum_{i=1}^{k-1} \delta S'_i \left(\frac{1}{S'_i} - \frac{1}{S_{i+1}} \right) \quad . \tag{5-285}$$

If the refractive indices n_o and n_k of the object and image spaces, respectively, are equal, then the first two terms on the right-hand side of Eq. (5-285) cancel each other. In practice, the imaging system lies typically in air, in which case the two terms are individually equal to zero. We note, in particular, that if the system is designed so that its longitudinal chromatic aberration $\delta S'_k$ is zero, its transverse chromatic aberration $\delta h'_k$ is generally not equal to zero.

The radius of the blue or red disk of rays in the red or the blue Gaussian image plane, respectively, is given by Eq. (5-268), where $\delta S'$ is equal to $\delta S'_k$, as may be seen from Figure 5-32. Similarly, we can show that the displacement $\delta h'_c$ of the blue disk from the red Gaussian image point, i.e., the lateral color representing the difference in the heights between the blue and red chief rays in an image plane is given by Eq. (5-269). The stop-shift equation for the lateral color is also given by Eq. (5-272).

Next, we apply the equations derived in this section to obtain the chromatic aberrations of a plane-parallel plate.

Example: Chromatic Aberrations of a Plane-Parallel Plate

As in Section 5.12, we consider a plane-parallel plate of thickness t and refractive index n forming the image of a point object P lying at a distance S from its front surface and at a height h from its axis (see Figure 5-33). To determine the chromatic aberrations of its image P'', we note that, for imaging by the first surface, $n_o = 1$, $n_1 = n$, and $R_1 = \infty$. Substituting in Eqs. (5-279) and (5-280), we obtain

$$K_1 = -1/S \tag{5-286a}$$

$$= -n/S'_1 \tag{5-286b}$$

and

$$n \frac{\delta S'_1}{S'^2_1} = \frac{\delta n}{nS} \quad ,$$

or

$$\delta S'_1 = S \, \delta n \quad . \tag{5-287}$$

For the second surface, $n_1 = n$, $n_2 = 1$, $R_2 = \infty$, $S_2 = S'_1 - t = nS - t$,

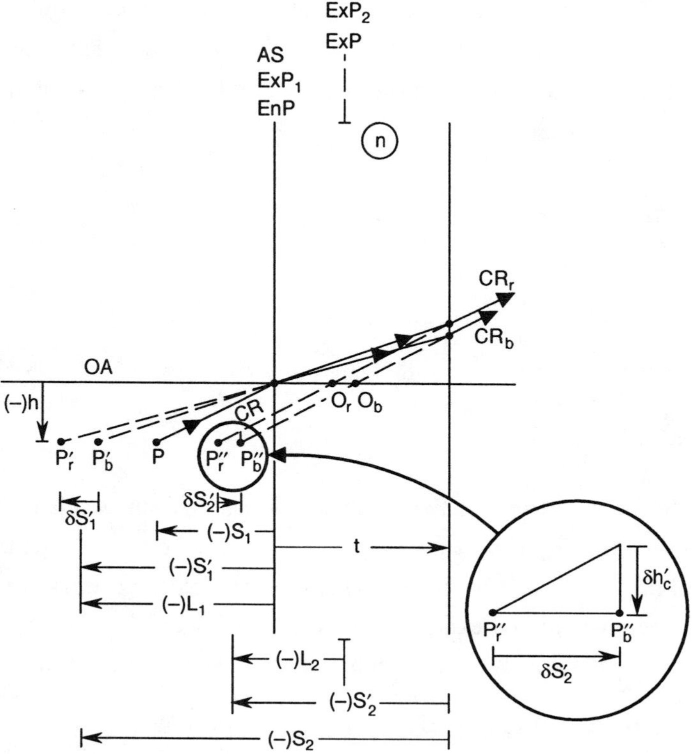

Figure 5-33. Chromatic aberrations of a plane-parallel plate.

$$K_2 = -n/S_2 \tag{5-288a}$$

$$= -1/S'_2 \quad , \tag{5-288b}$$

and

$$\frac{\delta S'_2}{S'^2_2} = -\frac{\delta n}{S_2} + n\frac{\delta S'_1}{S^2_2} \quad . \tag{5-289}$$

Substituting for the various quantities, we find that the longitudinal chromatic aberration is given by

$$\delta S'_2 = \frac{t}{n^2}\delta n \quad . \tag{5-290}$$

This result can, of course, be obtained very simply from Eq. (5-248).

According to Eq. (5-284) with $k = 2$, the transverse chromatic aberration is given by

$$\frac{\delta h'}{h'} = \frac{\delta S'_2}{S'_2} + \delta S'_1\left(\frac{1}{S'_1} - \frac{1}{S_2}\right)$$

$$= 0 \quad . \tag{5-291}$$

It is not surprising that $\delta h'$ is zero, since the image magnification is unity regardless of the refractive index of a ray due to zero refracting power of the plate. The lateral color representing the difference in the heights of the blue and red chief rays in the final image plane is given by Eq. (5-269):

$$\delta h'_c = -\frac{h'}{L_2}\, \delta S'_2 \quad,$$

or

$$\delta h'_c = -\frac{h'}{L_2}\frac{t}{n^2}\, \delta n \quad, \tag{5-292}$$

where L_2 is the (numerically negative) distance of the final image plane from the exit pupil of the plate. From Eq. (5-252), we note that $L_2 = S$. Of course, the exit pupil, which is the image of the first surface by the second, also has chromatic aberrations. That is why the centers of the blue and red exit pupil are shown in Figure 5-33 to lie on the optical axis at O_b and O_r, respectively. Its impact on Eq. (5-292) is a second-order effect.

5.13.5 General System: Use of Principal and Focal Points

Just as we obtained in Section 1.3.5 an imaging equation in terms of the positions of the focal points and principal points of a multielement imaging system, similarly, we can obtain a relationship between the chromatic aberrations and the chromatic displacements of these points. To obtain a relationship between the longitudinal chromatic aberration and the displacements of the focal points and the principal points with a change in wavelength, it is convenient to use the Newtonian imaging equation (1-78)

$$zz' = ff' \quad, \tag{5-293}$$

where z is the object distance from the object-space focal point F, z' is the image distance from the image-space focal point F', and f and f' are the object-space and image-space focal lengths of the imaging system, respectively, as illustrated in Figure 5-34. The two focal lengths are related to each other according to

$$\frac{n'}{f'} = -\frac{n}{f} \quad, \tag{5-294}$$

where n and n' are the refractive indices of the object and image spaces, respectively. Substituting for f from Eq. (5-294) into Eq. (5-293), we may write

$$zz' = -\frac{n}{n'}\, f'^2 \quad. \tag{5-295}$$

Taking a logarithmic differentiation of Eq. (5-295) with respect to wavelength, we obtain

$$\frac{\delta z}{z} + \frac{\delta z'}{z'} = \frac{\delta n}{n} - \frac{\delta n'}{n'} + \frac{2}{f'}\delta f' \quad . \tag{5-296}$$

Let u be the distance of the object from the vertex V of the first surface of the system. Similarly, let u' be the distance of the image from the vertex V' of its last surface. Also, let v and v' be the distances of the principal points H and H' from the vertices V and V' of the first and the last surfaces of the system, respectively. Then

$$z = u - f - v \tag{5-297}$$

and

$$z' = u' - f' - v' \quad . \tag{5-298}$$

Differentiating Eqs. (5-297) and (5-298) with respect to wavelength, we obtain

$$\delta z = \delta u - \delta f - \delta v \tag{5-299}$$

and

$$\delta z' = \delta u' - \delta f' - \delta v' \quad . \tag{5-300}$$

The transverse and longitudinal magnifications M_t and M_l of the image are given by [see Eqs. (1-77) and (1-71)]

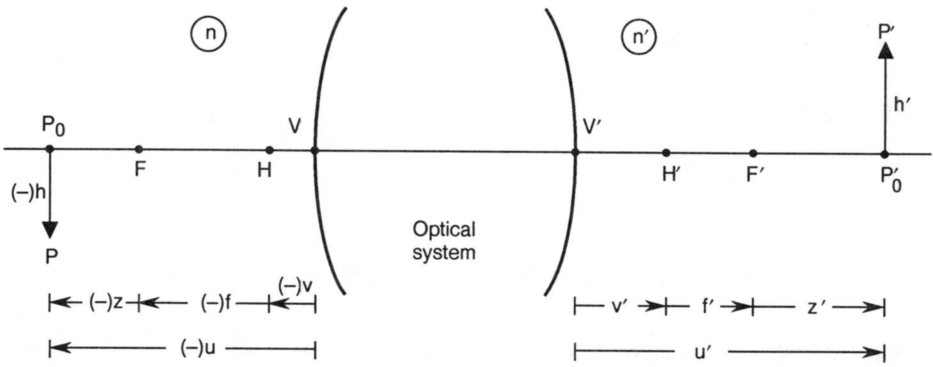

Figure 5-34. General imaging system showing the location of its principal and focal points H, H', and F and F', respectively. Also shown are the object and image locations.

$$M_t = h'/h \tag{5-301a}$$

$$= -f/z \tag{5-301b}$$

$$= -z'/f' \tag{5-301c}$$

and

$$M_l = (n'/n) M_t^2 \quad , \tag{5-302}$$

respectively. Thus,

$$z = -f/M_t$$

$$= \frac{nf'}{n'M_t} \tag{5-303}$$

and

$$z' = -f'M_t \quad . \tag{5-304}$$

Substituting Eqs. (5-299), (5-300), (5-303), and (5-304) into Eq. (5-296) we obtain

$$\frac{\delta u - \delta f - \delta v}{nf'/n'M_t} + \frac{\delta u' - \delta f' - \delta v'}{-f'M_t} = \frac{\delta n}{n} - \frac{\delta n'}{n'} + \frac{2}{f'} \delta f' \quad ,$$

or

$$M_l \left(\delta u - \delta f - \delta v \right) - \left(\delta u' - \delta f' - \delta v' \right) = f'M_t \left(\frac{\delta n}{n} - \frac{\delta n'}{n'} \right) + 2M_t \, \delta f' \quad . \tag{5-305}$$

Now, we write δf in terms of f' and $\delta f'$. From Eq. (5-294)

$$f = -\frac{n}{n'} f' \tag{5-306}$$

and by differentiation

$$\delta f = -\frac{\delta n}{n'} f' + \frac{nf'}{n'^2} \delta n' - \frac{n}{n'} \delta f' \quad . \tag{5-307}$$

Substituting Eq. (5-307) into Eq. (5-305) and rearranging the terms, we obtain

$$\delta u' - M_l \delta u = \delta v' - M_l \delta v + \left(1 - M_t\right)^2 \delta f' - f'M_t \left(1 - M_t\right)\left(\frac{\delta n}{n} - \frac{\delta n'}{n'} \right) \quad . \tag{5-308}$$

In practice, the object position is fixed (unless it is an image formed by a nonachromatic preceding system). Hence, $\delta u = 0$. Moreover, if the refractive indices of the object and image spaces are the same, then the last term on the right-hand side of Eq. (5-308) is also zero. In practice, it is zero for a system in air since $n = n' = 1$ and, therefore, $\delta n = \delta n' = 0$. Hence, under such conditions, Eq. (5-308) reduces to

$$\delta u' = \delta v' - M_t^2 \delta v + (1 - M_t)^2 \delta f' \quad . \tag{5-309}$$

Thus, the longitudinal chromatic aberration $\delta u'$ can be determined for any value of the image magnification M_t from the change $\delta f'$ of the image-space focal length f' and the displacements δv and $\delta v'$ of the principal points H and H', respectively. The displacements of the principal and focal points are determined in the usual manner by tracing blue and red rays incident on the system parallel to its optical axis.

To determine the transverse chromatic aberration, we consider Eqs. (5-301) in the form

$$h'/h = -z'/f' \tag{5-310}$$

and take its logarithmic differentiation. Thus,

$$\frac{\delta h'}{h'} - \frac{\delta h}{h} = \frac{\delta z'}{z'} - \frac{\delta f'}{f'}$$

$$= -\frac{\delta u' - \delta f' - \delta v'}{f' M_t} - \frac{\delta f'}{f'}$$

$$= -\frac{1}{f'} \left[\frac{\delta u' - \delta v'}{M_t} + \left(1 - \frac{1}{M_t}\right) \delta f' \right]$$

$$= M_t \frac{\delta v}{f'} - (M_t - 1)\frac{\delta f'}{f'} \quad , \tag{5-311}$$

where we have used Eqs. (5-300), (5-304), and (5-309). Generally, the object height h will be fixed (unless it is the image formed by a nonachromatic preceding system) and, therefore, $\delta h = 0$. Hence, Eq. (5-311) reduces to

$$\boxed{\delta h' = \frac{h'}{f'}\left[M_t \delta v - (M_t - 1)\delta f'\right] \quad .} \tag{5-312}$$

The lateral color $\delta h_c'$ representing the transverse chromatic aberration of the chief rays in a given image plane can be obtained from Eq. (5-269) by substituting Eq. (5-312) into it, and replacing $\delta S'$ by $\delta u'$. As a simple example of a general system, the chromatic aberrations of a thick lens are considered in Problem 5.15, where the conditions for an *achromatic singlet* are derived.

Example: Achromatic Doublet

Consider two thin lenses of image-space focal lengths f_1' and f_2' separated by a distance t as in Figure 1-37. The focal length f of the combination is given by Eq. (1-123), i.e.,

$$\frac{1}{f'} = \frac{1}{f_1'} + \frac{1}{f_2'} - \frac{t}{f_1'f_2'} \quad . \tag{5-313}$$

Differentiating Eq. (5-313) with respect to wavelength, we find that a *doublet* consisting of two separated thin lenses is *achromatic* with respect to its focal length $\left(\delta f' = 0\right)$ if

$$t = \frac{f_1'V_1 + f_2'V_2}{V_1 + V_2} \quad , \tag{5-314}$$

where V_1 and V_2 are the dispersive constants of the lenses. Thus, the focal length f' of the doublet is the same for the two wavelengths used to define V for which the difference in the refractive index is δn.

If the two lenses are made of the same material $\left(V_1 = V_2\right)$, then Eq. (5-314) reduces to

$$t = \frac{1}{2}\left(f_1' + f_2'\right) \quad . \tag{5-315}$$

Thus, a doublet made of lenses of the same material is achromatic if their spacing is equal to half the sum of their image-space focal lengths. Substituting Eq. (5-315) into Eq. (5-313), we obtain the focal length of the achromatic doublet,

$$\frac{1}{f'} = \frac{1}{2}\left(\frac{1}{f_1'} + \frac{1}{f_2'}\right) \quad . \tag{5-316}$$

Since the lenses are made of the same material, both f_1' and f_2' vary with the wavelength in the same manner. Hence, Eq. (5-315) can be satisfied at one wavelength only. Accordingly, the focal length of the doublet given by Eq. (5-316) is independent of the wavelength to first order in δn.

The longitudinal chromatic aberration can be determined from Eq. (5-309) with $\delta f' = 0$, i.e., from

$$\delta u' = \delta v' - M_t^2 \, \delta v \quad . \tag{5-317}$$

Since f' is fixed, the image-space focal point F' and the principal point H' are displaced by the same amount $\delta v'$. Now F' lies at a distance

$$t_2 = f'\left(1 - \frac{t}{f_1'}\right) \tag{5-318}$$

from the center of the second lens [see Eq. (1-124)]. Differentiating with respect to wavelength, we find that the image-space principle point H' and the focal point F' are displaced by

$$\delta v' \equiv \delta t_2$$

$$= f't\,\frac{\delta f_1'}{f_1'^2}$$

$$= -\frac{f't}{f_1'V} \quad , \tag{5-319}$$

where V is the dispersive constant of the lenses. Similarly, considering the distance $f\left(1 - t/f_2'\right)$ of the object-space focal point F from the center of the first lens and noting that the object-space focal length f and the image-space focal length f' are related to each other according to $f = -f'$, we find that the object-space principal point H and focal point F are displaced by an amount

$$\delta v = -f't\,\frac{\delta f_2'}{f_2'^2}$$

$$= \frac{f't}{f_2'V} \quad . \tag{5-320}$$

Substituting Eqs. (5-319) and (5-320) into Eq. (5-317) and using Eq. (5-273), we obtain the longitudinal chromatic aberration

$$\delta u' = f't\left(\frac{\delta f_1'}{f_1'^2} + M_t^2\,\frac{\delta f_2'}{f_2'^2}\right)$$

$$= -\frac{f't}{V}\left(\frac{1}{f_1'} + \frac{M_t^2}{f_2'}\right) \quad . \tag{5-321}$$

We note that it is not zero unless $M_t^2 = -f_2'/f_1'$.

The transverse chromatic aberration can be obtained from Eq. (5-312) with $\delta f' = 0$, i.e., from

$$\delta h' = \frac{h'}{f'}\,M_t\,\delta v$$

$$= \frac{h'tM_t}{Vf_2'} \quad , \tag{5-322}$$

where in the last step we have made use of Eqs. (5-273) and (5-320). The transverse magnification M_t of an object lying at infinity is zero. Hence, its transverse chromatic aberration is also zero. Its longitudinal chromatic aberration is given by $-f't/Vf_1'$ according to Eq. (5-321).

A numerical example of an *achromatic doublet* consisting of two separated thin lenses using BK7 glass is shown in Figure 5-35a. It is a *Huygens eyepiece* consisting of two planoconvex thin lenses of focal lengths 7.5 and 15 cm, respectively, with a separation of 11.25 cm. The object-space focal point F_2 of the second lens coincides with the image-space principal point H' of the eyepiece. Similarly, the image-space focal point F_1' of the first lens coincides with the object-space principal point H of the eyepiece. An eyepiece is used with a telescope or a microscope objective. The objective forms the image of an object in the object-space focal plane (passing through F) of the eyepiece which, in turn, forms the image at infinity for comfortable viewing by a human eye, as illustrated in Figure 5-35b. The focal length of the doublet is 10 cm and its variation with wavelength is shown Figure 5-35c. Its minimum value is 10 cm, corresponding to the d line. Its value increases as the wavelength deviates from that of the d line. It is evident from the paraboliclike variation that the focal length has the same value at two different wavelengths. However, it does not have the same value for the wavelengths used in defining the dispersive constant V. Figure 5-35d shows how the back focal distance t_2, i.e., the distance of the focal point F' from the center of the second lens, varies with wavelength. We note that t_2 is numerically negative, indicating that F' lies to the left of the second lens. The magnitude of the distance is 5 cm for the d line and it decreases as the wavelength increases.

In order that the longitudinal chromatic aberration of the doublet for an object at infinity be zero, the position of F' must be independent of the wavelength, i.e., δt_2 obtained from Eq. (5-318) must be zero. Substituting for f' from Eq. (5-313), Eq. (5-318) may be written

$$\frac{1}{t_2} = \frac{1}{f_1' - t} + \frac{1}{f_2'} \qquad\qquad (5\text{-}323)$$

$$= \frac{1}{\dfrac{1}{(n-1)\kappa_1} - t} + (n-1)\kappa_2 \quad , \qquad\qquad (5\text{-}324)$$

where κ for a lens in terms of the radii of curvature R_1 and R_2 of its two surfaces is given by

$$\kappa_i = \left(\frac{1}{R_1} - \frac{1}{R_2}\right)_i \quad , \quad i = 1, 2 \quad . \qquad\qquad (5\text{-}325)$$

Differentiating Eq. (5-324), we find that the variation of t_2 with respect to n is equal to zero if the value of t is given by

$$f_2' = -f_1'\left(1 - t/f_1'\right)^2 \ .$$ (5-326)

It is clear from Eq. (5-326) that the image-space focal lengths f_1' and f_2' of the two lenses must be of opposite signs. Moreover, this equation requires a spacing between the lenses that is different from that given by Eq. (5-315). Thus, *for an object lying at infinity, the transverse and longitudinal chromatic aberrations of an air-spaced doublet cannot be simultaneously equal to zero.*

If the two thin lenses are in contact $(t = 0)$, then the doublet, called a *thin-lens doublet*, is achromatic with respect to its focal length, according to Eq. (5-314), if the ratio of their focal lengths is given by

$$\frac{f_1'}{f_2'} = -\frac{V_2}{V_1} \ .$$ (5-327)

Since, for zero spacing, Eq. (5-313) reduces to

$$\frac{1}{f'} = \frac{1}{f_1'} + \frac{1}{f_2'} \ ,$$ (5-328)

the two focal lengths are given by

$$f_1' = \frac{f'(V_1 - V_2)}{V_1}$$ (5-329)

and

$$f_2' = \frac{f'(V_2 - V_1)}{V_2} \ .$$ (5-330)

Thus, an achromatic thin-lens doublet is obtained by combining a positive lens of low dispersion (small δn or large V) and a negative lens of high dispersion.

By the definition of a thin lens, the principal points of a thin-lens doublet coincide at its center. Hence, the focal points of an achromatic thin-lens doublet for different colors also coincide. Accordingly, the longitudinal and transverse chromatic aberrations are zero, regardless of the value of the object distance. It should be noted, however, that the focal length of a thin-lens doublet can be made the same at only two selected wavelengths for which the difference δn in the refractive indices is used in defining V. This may be seen as follows. Substituting the expressions for the focal length and the Abbe number from Eqs. (5-272b) and (5-275b), respectively, into Eq. (5-327), we obtain

$$\frac{(n_{d2} - 1)\kappa_2}{(n_{d1} - 1)\kappa_1} = -\frac{n_{d2} - 1}{n_{F2} - n_{C2}} \frac{n_{F1} - n_{C1}}{n_{d1} - 1}$$

or

$$\frac{\kappa_2}{\kappa_1} = -\frac{n_{F1} - n_{C1}}{n_{F2} - n_{C2}} \ .$$ (5-331)

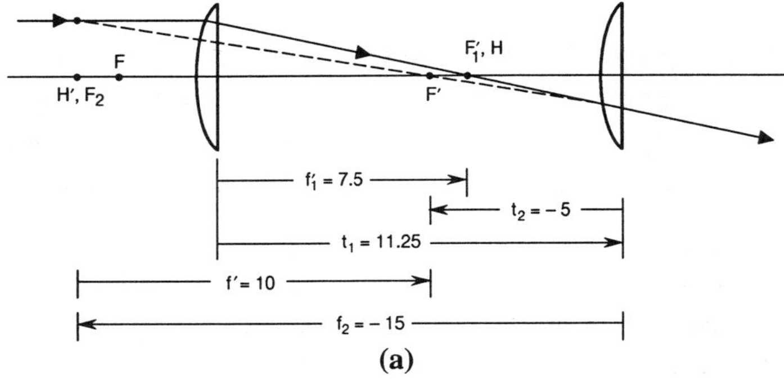

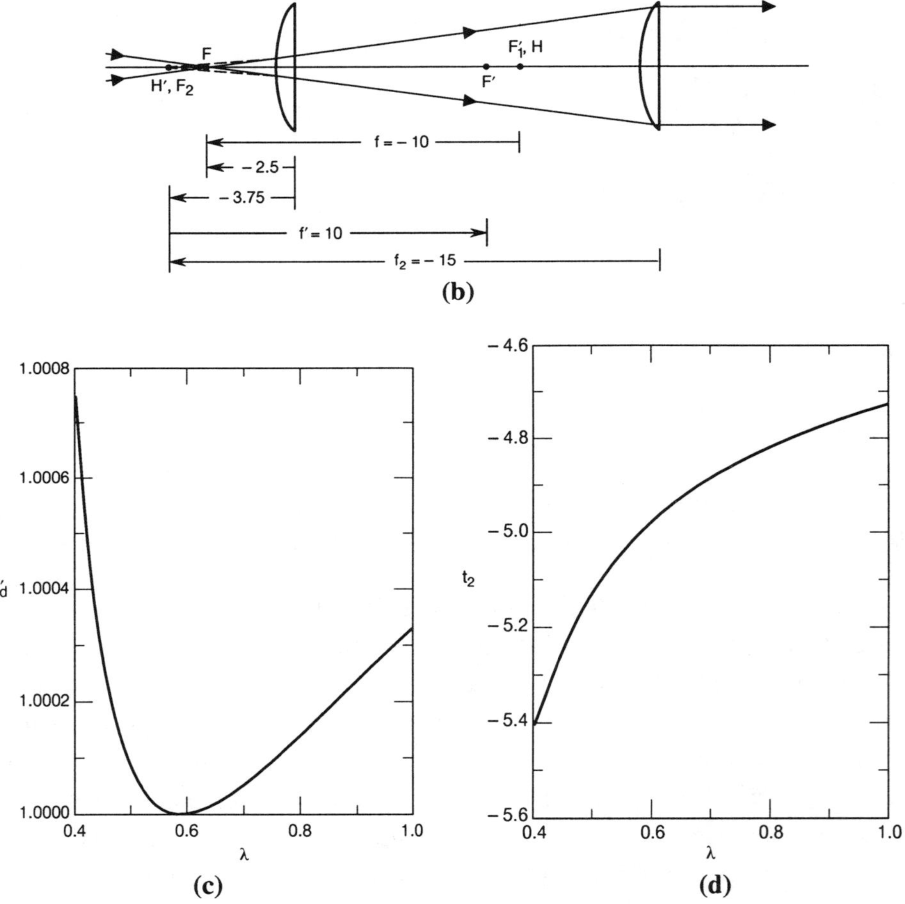

Figure 5-35. Achromatic air-spaced doublet. (a) Schematic of a Huygens eyepiece of focal length 10 cm. The two thin lenses are made of BK7 glass. (b) The eyepiece forms an image at infinity of the image at F formed by the objective (not shown). (c) Variation of focal length of the doublet with wavelength. (d) Variation of back focal distance t_2 with wavelength. The wavelength is in mircometers and t_2 is in centimeters.

The focal lengths f_F' and f_C' of the doublet for the F and C lines are equal to each other according to Eq. (5-328) if

$$\frac{1}{f_{F1}'} + \frac{1}{f_{F2}'} = \frac{1}{f_{C1}'} + \frac{1}{f_{C2}'} \quad , \tag{5-332}$$

or

$$\left(n_{F1} - 1\right)\kappa_1 + \left(n_{F2} - 1\right)\kappa_2 = \left(n_{C1} - 1\right)\kappa_1 + \left(n_{C2} - 1\right)\kappa_2 \quad , \tag{5-333}$$

which reduces to Eq. (5-331). It should be evident from the foregoing that f_F' and f_C', which are equal to each other, cannot be equal to the focal length of the doublet for a third wavelength. The focal lengths of the doublet for another pair of wavelengths will be equal to each other provided the ratio of the differences in the refractive indices for them is equal to that given by Eq. (5-331). The residual chromatic aberration at wavelengths other than λ_F and λ_C is called the *secondary spectrum*.

A numerical example of an achromatic thin-lens doublet made of BK7 and SF2 glasses is shown in Figure 5-36a. It is a *cemented doublet* in that the contact surface between the two thin lenses is common. Thus, the radius of curvature of the second surface of the first lens is the same as that of the first surface of the second lens. The focal length of the doublet is 10 cm for the d line. How it varies with focal length is shown in Figure 5-36b. We note again from the parabolic-like variation that the focal length has the same value for two different wavelengths. However, compared to the thick-lens doublet shown in Figure 5-34, there is a built-in design feature of equal focal lengths for the F and C lines.

We note from Eqs. (5-329) and (5-330) that since V_1 and V_2 are positive, f_1' and f_2' have opposite signs. Moreover, the specification of f' and the dispersive constants of the lens materials specifies their focal lengths f_1' and f_2'. However, the focal length of a thin lens depends on the difference in the curvatures of its surfaces, while its spherical aberration and coma depend on the curvatures through its shape factor (see Section 5.10). This degree of freedom can be utilized to make the achromatic thin-lens doublet aplanatic as well, i.e., the radii of curvature of its four surfaces can be chosen so that its spherical aberration and coma are both equal to zero.

We start by assuming that the aperture stop lies at the doublet. Once an aplanatic doublet is designed, it will remain aplanatic regardless of the position of its aperture stop, as discussed in Section 5.9. The aberrations of its two lenses combine to give its aberrations in the same manner as the aberrations of the two surfaces of a thin lens discussed in Section 5.10.2. Thus, the spherical aberration coefficient a_s of the doublet is given by Eq. (5-182a), where a_{s1} and a_{s2} are the corresponding coefficients of its two lenses obtained according to Eq. (5-188). Given the object distance, we determine the position factors p_1 and p_2 for the two lenses. Substituting for the various quantities, we obtain a quadratic equation for a_s in terms of the shape factors q_1 and q_2 of the two lenses. Similarly, the coma coefficient a_c of the doublet is given by (5-182b), where a_{c1} and a_{c2} are the corresponding coefficients of the two lenses according to Eq. (5-193)

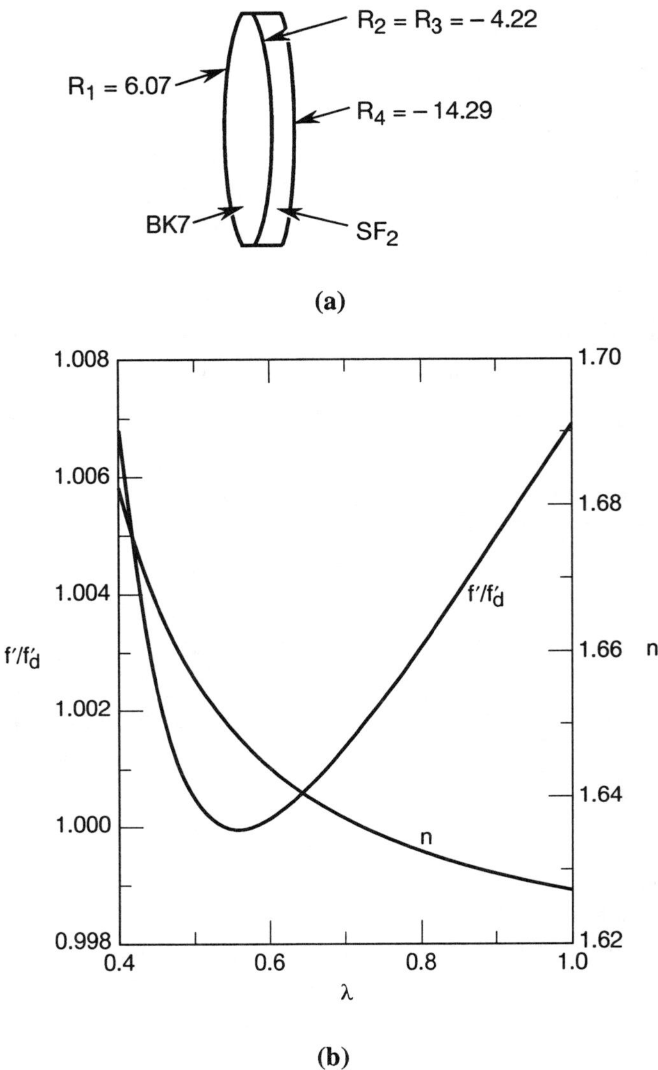

Figure 5-36. Achromatic thin-lens doublet. (a) Cemented doublet of 10-cm focal length made of BK7 and SF2 glass lenses. The focal lengths of the two lenses are 4.82 cm and – 9.29 cm, respectively. (b) Variation of focal length with wavelength. The variation of the refractive index n of SF_2 glass is also shown in the figure.

and M_2 is the magnification of the image formed by the second lens. Thus, we obtain a linear equation for a_c in terms of q_1 and q_2. Letting a_s and a_c be equal to zero and solving the two equations simultaneously, we obtain a pair of solutions for q_1 and q_2. The equations for the focal length and shape factor of a lens can be solved simultaneously to obtain its radii of curvature. It is possible that of the two solutions thus obtained, one is more practical than the other from a fabrication standpoint.

The astigmatism and field curvature coefficients of the achromatic aplanatic thin-lens doublet can be obtained from Eqs. (5-182c) and (5-182d), respectively, where the corresponding coefficients for its lenses are given by Eqs. (5-194) and (5-195). These coefficients do not depend on the shape factors of the lenses. The distortion coefficient of the doublet given by Eq. (5-182e) is zero since the corresponding coefficients for the lenses are zero according to Eq. (5-196). As discussed in Section 5.10.2, the peak values of astigmatism and the field curvature aberrations of an aplanatic system do not change with a change in the position of its aperture stop. However, the distortion of the system does change according to Eq. (5-150). Thus, an aplanatic doublet whose aperture stop is not collocated with it will have distortion unless the sum of its astigmatism and field curvature coefficients is zero.

In general, the radii of curvature of the four surfaces of an *achromatic aplanatic doublet* will be different from each other. However, if one chooses the contact surface of the lenses to be common, then the design is overconstrained and the achromatic thin-lens doublet, in general, will not be aplanatic. An *aplanatic cemented doublet* can be designed provided one is free to choose the material for one of the lenses. With an appropriate refractive index and Abbe number, the required focal length and cancellation of the spherical aberration and coma of the other lens can be achieved (see Problem 5.12).

Although spherical aberration and coma of a system of thin lenses in contact can be made zero, its Petzval curvature and astigmatism are generally nonzero. Whereas the Petzval curvature is independent of the position of the aperture stop, its astigmatism is not (unless its spherical aberration and coma are each equal to zero). It can be shown that astigmatism cannot be zero when the aperture stop is located at the lens system unless its focal length is infinity (see Problem 5.7).

5.13.6 Chromatic Aberrations as Wave Aberrations

The chromatic aberrations of a refracting surface or a system represent the variation of image distance and height with the wavelength of the object radiation, and can be written as wave aberrations. The longitudinal chromatic aberration or axial color represents chromatic longitudinal defocus; hence it can be written as a defocus wave aberration. The wavefronts for different wavelengths are spherical but their radii of curvature are longer for the longer wavelengths. If the red wavefront is chosen as the reference sphere, then the defocus wave aberration corresponding to an axial color of dS' is given by [see Eq. (3-15a)]

$$W_d(r) = -\frac{n_i}{2}\frac{\delta S'}{R^2}r^2 \quad , \tag{5-334}$$

where n_i is the refractive of the image space.

Similarly, in the case of lateral color, the wavefronts are spherical but their centers of curvature lie at a higher height from the optical axis for the longer wavelength. Again choosing the red wavefront as the reference sphere, the wavefront tilt aberration due to a lateral color of $\delta h'$ is given by [see Eq. (3-21)]

$$W_t(r, \theta) = n_i \frac{\delta h'}{R} r \cos \theta \quad . \tag{5-335}$$

The chromatic defocus or tilt aberration of a system can be calculated from its axial or lateral color, or by adding the contributions of its elements.

5.14   SYMMETRICAL PRINCIPLE

The symmetrical principle states that the aberrations that are odd in field angle or image height, e.g., coma, distortion, and lateral color, are zero for a system that is symmetric about its aperture stop when it images an object with a magnification of -1, as illustrated in Figure 5-37. Because of the symmetry of the system, the refractive indices of the object and image spaces must be equal. Hence, according to Eq. (1-64), an image of magnification -1 is formed when the object and image distances are equal in magnitude but opposite in sign. From Eq. (1-66), the object and image distances from the respective principal points are $2f$ and $2f'$, as illustrated in Figure 5-37. Again, because of the symmetry, the left half of the system must form the image of the object either in the plane of the aperture stop or at infinity. The intermediate image in the plane of the aperture stop is not a practical case because the two can not coexist; otherwise the aperture stop can not function as the aperture stop. Therefore, the intermediate image must be formed at infinity, which, in turn, implies that the object must be in the object-space focal plane of the left half of the system. The intermediate image at infinity is the object for the right half of the system. Thus, the final image must be formed in the image-space focal plane of the right half. Because of the symmetry, the entrance and exit pupils have the same size, i.e., the pupil magnification is unity. Hence, the entrance pupil lies in the object-space principal plane and the exit pupil lies in the image-space principal plane.

Let h be the height of a point object P in the object-space focal plane of the left half of the system from the optical axis. Let $W_1(x, y, h)$ be the contribution of the left half of

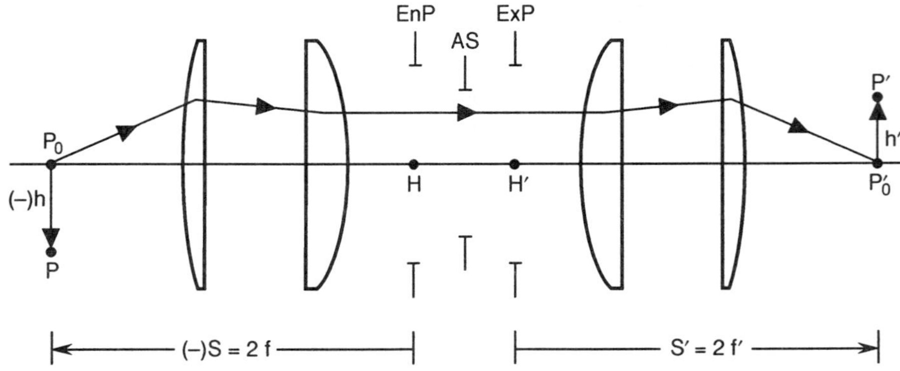

Figure 5-37. Symmetrical optical system imaging an object with a magnification of -1. The object lies in the object-space focal plane of the left half of the system and the image lies in the image-space focal plane of the right half.

the system to the aberration at a point (x, y) of the aperture stop in forming the image of P at infinity. The aberration contributed by the right half of the system in imaging an object lying at infinity is equal to its contribution when imaging an object lying at P' in its image-space focal plane, since the two objects are conjugates of each other. This contribution is given by $W_2(x, y, -h)$ as may be seen by folding the system about its aperture stop. The right half of the system is identical to the left half when folded, but $f' = -f$ and P' lies at a height $h' = -h$ because of the -1 magnification between it and P. Hence the aberration of the system forming the image P' of P with a magnification of -1 is given by

$$W(x, y, h) = W_1(x, y, h) + W_2(x, y, -h) \ . \tag{5-336}$$

Since the functional dependence of W_1 and W_2 on x, y, and h is the same, it is evident that those aberrations that depend on the odd powers of h (e.g., coma, distortion, and lateral color) cancel each other. Accordingly, the aberration function of the system depends on h through its even powers only. Since the two halves of the system contribute equally to an even aberration term, its value is equal to twice the value contributed by either half of the system.

A simple example of a symmetrical system is that of two identical thin lenses with an aperture stop placed halfway between them. When an object is placed in the front focal plane of the first lens so that it forms the image at infinity, it lies at a distance that is twice the object-space focal length of the system regardless of the spacing between the lenses. The final image with a magnification of -1 lies in the back focal plane of the second lens which is at a distance that is twice the image-space focal length of the system. Of course, the object and image distances for the system are measured from its object- and image-space principal points, respectively.

5.15 PUPIL ABERRATIONS AND CONJUGATE-SHIFT EQUATIONS

5.15.1 Introduction

If the expressions for the primary aberrations of a system are known, their values for any position of the object can be calculated by substituting the value of the object distance. However, in Section 5.9 we saw how the image aberrations for a certain position of the aperture stop can be obtained from those for another by using the stop-shift equations. Similarly, it is possible to obtain the aberrations for one position of the object from those for another.[2,6] While the stop-shift equations help the lens designer to place the stop at an appropriate position to minimize the most detrimental aberrations or vignette the rays with such aberrations, the conjugate-shift equations are not that useful since the optical systems are typically used for a narrow range of object distances. For example, the astronomical telescopes are used for imaging objects lying at infinity; cameras are used for objects at relatively long or short distances using a lens appropriate for the distance; and microscopes are used for objects at very small distances. Any variation in the object distance, i.e., the depth of field, is accommodated by the depth of focus of the system. Nevertheless, the conjugate-shift equations are useful, for example, in knowing that a system cannot simultaneously image perfectly two objects lying at different distances, but certain aberrations can be made zero simultaneously for more than one object position.

Just as in the case of a stop shift we adjusted the stop size so that the axial marginal ray did not change, similarly we adjust the size of the object as its position is changed so that the chief ray from its edge point does not change. The conjugate-shift equations are obtained as follows. By interchanging the roles of the object and the entrance pupil, we obtain the aberrations of *pupil imagery* in the same manner as those for the *object imagery*. A conjugate shift is applied to these aberrations just as a stop shift was applied to the image aberrations, to obtain the pupil aberrations for a different object position. By comparing the aberrations thus obtained with their nominal form for the new object position, we obtain the effects of conjugate shifts on the aberrations of object imagery. It is found that although all primary aberrations can not be made zero and invariant with object position, conditions can be obtained under which certain aberrations can be corrected and made invariant, or corrected for more than one object position.

5.15.2 Pupil Aberrations

From Eqs. (5-85) through (5-90), the peak values of the primary aberrations for object imagery by a conic refracting surface of eccentricity e (see Figure 5-38) are given by

$$A_s = \left[a_{ss} + \sigma \left(S'/L \right)^4 \right] a^4 \quad , \tag{5-337a}$$

$$A_c = 4 \left[d a_{ss} - \sigma g \left(S'/L \right)^3 \right] h' a^3 \quad , \tag{5-337b}$$

$$A_a = 4 \left[d^2 a_{ss} + \sigma g^2 \left(S'/L \right)^2 \right] h'^2 a^2 \quad , \tag{5-337c}$$

$$A_d = \frac{1}{2} A_a - \frac{n'(n'-n)}{4nR} \frac{h'^2 a^2}{L^2} \quad , \tag{5-337d}$$

and

$$A_t = 4 \left[d^3 a_{ss} - \frac{n'(n'-n)d}{8nRL^2} - \sigma g^3 \left(S'/L \right) \right] h'^3 a \quad , \tag{5-337e}$$

where

$$a_{ss} = -\left(\frac{S'}{L} \right)^4 \frac{n'(n'-n)}{8n^2} \left(\frac{1}{R} - \frac{1}{S'} \right)^2 \left(\frac{n'}{R} - \frac{n+n'}{S'} \right) \quad , \tag{5-338}$$

$$d = \frac{R - S' + L}{S' - R} \quad , \tag{5-29b}$$

$$g = \frac{S' - L}{L} \quad , \tag{5-83}$$

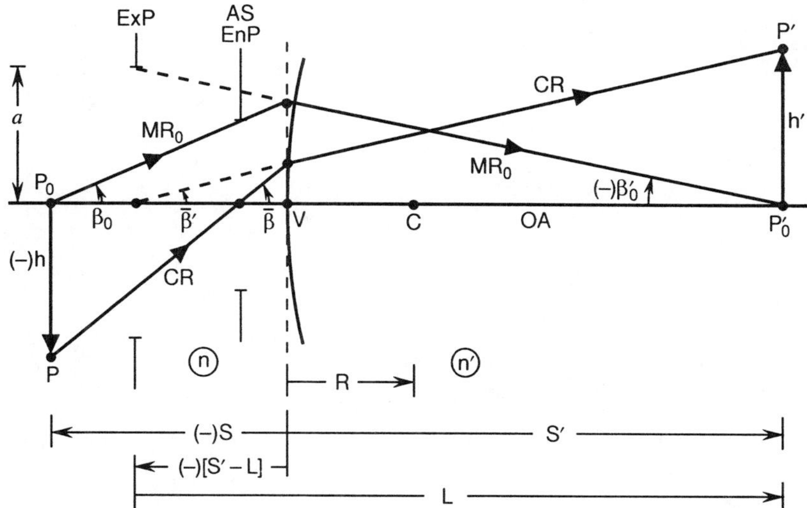

Figure 5-38. Imaging by a refracting surface of eccentricity e and vertex radius of curvature R and vertex center of curvature C separating media of refractive indices n and n'. An object of height h from the optical axis VC lies at a distance S from the surface and its image lies at a height h' at a distance S'. The aperture stop AS is also the entrance pupil EnP and its image by the refracting surface is the exit pupil ExP. The distance of the image from the exit pupil is L. The slope angles of the axial marginal ray MR_0 are β_0 and β'_0 in the object and image spaces, respectively. Similarly, the slope angles of the edge chief ray CR in these spaces are $\overline{\beta}$ and $\overline{\beta}'$.

and

$$\sigma = \frac{(n'-n)e^2}{8R^3} \ .$$
(5-77)

For simplicity of notation, we have omitted from the aberration coefficients the subscript c representing the conic.

When the roles of the entrance pupil and the object are interchanged, the entrance and exit pupils become the object and image, and the object and image become the entrance and exit pupils, respectively. Moreover, the axial marginal ray becomes the edge chief ray and viceversa. The aberrations for the pupil imagery can be obtained from those for the image by changing $L \to -L$, $S' \to S'-L$, and interchanging a and h'. Thus, the peak values of the primary aberrations for pupil imagery can be written

$$\overline{A}_s = \left(\overline{a}_{ss} + \sigma g^4\right)h'^4 \ ,$$
(5-339a)

$$\overline{A}_c = 4\left[\overline{d}\,\overline{a}_{ss} - \sigma g^3(S'/L)\right]ah'^3 \ ,$$
(5-339b)

$$\overline{A}_a = 4\left[\overline{d}^2\overline{a}_{ss} + \sigma g^2(S'/L)^2\right]a^2h'^2 \ ,$$
(5-339c)

$$\overline{A}_d = \frac{1}{2}\overline{A}_a - \frac{n'(n'-n)}{4nR}\frac{a^2h'^2}{L^2} \quad , \tag{5-339d}$$

and

$$\overline{A}_t = 4\left[\overline{d}^3\overline{a}_{ss} - \frac{n'(n'-n)\overline{d}}{8nRL^2} - \sigma g(S'/L)^3\right]ah'^3 \quad , \tag{5-339e}$$

where

$$\overline{a}_{ss} = -\frac{n'(n'-n)}{8n^2}g^4\left(\frac{1}{R} - \frac{1}{S'-L}\right)^2\left(\frac{n'}{R} - \frac{n+n'}{S'-L}\right) \quad , \tag{5-340a}$$

or

$$\overline{a}_{ss} + \sigma g^4 = -\frac{S'-L}{4S'}\frac{A_t}{ah'^3} - \frac{n'(n'-n)g^2(R-S'+L)}{8nRS'L} \quad , \tag{5-340b}$$

and

$$\overline{d} = \frac{S'-R}{R-S'+L} = \frac{1}{d} \quad . \tag{5-341}$$

Note that the quantity g transforms into $\overline{g} = -S'/L$ and the ratio S'/L transforms into $-g$ when the roles of the object and the entrance pupil are interchanged. We also note that the aspheric contribution to astigmatism is the same for the exit pupil and the image. The second term in the field curvature coefficient representing the aberration due to Petzval curvature is also the same, as expected. It should be evident that the expressions for the aberrations of the exit pupil given above hold only if the aperture stop lies in the object space so that it is also the entrance pupil of the system, because only then can the exit pupil be its image as formed by the whole system.

As an example, if we let $L \to S'$ in Eqs. (5-339), we obtain the aberrations of the exit pupil when it lies at the surface, in agreement with the aberrations given by Eq. (5-49a) for an aplanatic point lying on a spherical surface. Similarly, if we let $S'-L \to (n+n')/n'R$, we obtain the aberrations given by Eq. (5-49c) for the other aplanatic point of the spherical surface. In both cases, the sign of L must be changed because of the interchange of the roles of the object and the entrance pupil. The aberrations given by Eq. (5-49b) for the third aplanatic point of the surfcae, namely, an object at its center of curvature, cannot be obtained from Eqs. (5-339) for the pupil aberrations since the exit pupil cannot lie in the plane of the image.

Using the slope angles of the axial marginal ray and the edge chief ray, we can write the aberrations of the exit pupil in terms of the aberrations of the image in a convenient form:

$$\overline{A}_s = -\frac{S'-L}{4aS'}h'A_t - \frac{n'(n'-n)g^2(R-S'+L)}{8nR^2S'L}h'^4 \quad , \tag{5-342a}$$

$$\overline{A}_c = A_t + \frac{1}{2}H\Delta\!\left(\overline{\beta}^2\right) \quad , \tag{5-342b}$$

$$\overline{A}_a = A_a + \frac{1}{2}H\Delta\!\left(\beta_0\overline{\beta}\right) \quad , \tag{5-342c}$$

$$\overline{A}_d = A_d + \frac{1}{4}H\Delta\!\left(\beta_0\overline{\beta}\right) \quad , \tag{5-342d}$$

and

$$\overline{A}_t = A_c + \frac{1}{2}H\Delta\!\left(\beta_0^2\right) \quad , \tag{5-342e}$$

where

$$H = nh\beta_0 = n'h'\beta_0' \tag{5-343}$$

is the Lagrange invariant in terms of the image parameters, β_0 and β_0' are the slope angles of the incident and refracted axial marginal rays, $\overline{\beta}$ and $\overline{\beta}'$ are the slope angles of the incident and refracted edge chief rays, and $\Delta(x)$ represents the difference between the quantity x after and before refraction, e.g., $\Delta\!\left(\beta_0^2\right) = \beta_0'^2 - \beta_0^2$. In terms of the image distance S' and distane L of the image from the exit pupil, the slope angles are given by

$$\beta_0 = -a_{en}/L_o \tag{5-344a}$$

$$= -aS'/SL$$

$$= -\frac{aS'}{nL}\left(\frac{n'}{S'}-\frac{n'-n}{R}\right) \quad , \tag{5-344b}$$

$$\beta_0' = -a/L \quad , \tag{5-345}$$

$$\overline{\beta} = h/L_o \tag{5-346}$$

$$= \frac{h'(S'-L)}{nL}\left[\frac{n'}{S'-L}-\frac{n'-n}{R}\right] \quad ,$$

and

$$\overline{\beta}' = h'/L_o \tag{5-347}$$

where a_{en} is the radius of the entrance pupil and L_o is the distance of the object P_0P from it.

When the aperture stop is shifted, the peak aberration coefficients change according to the stop-shift equations, namely, Eqs. (5-146) through (5-150). For pupil imagery, the object is the effective stop. Hence, when the object is shifted so that its image lies at a distance L^* from the exit pupil (compared to a distance L of the image of the object in its original position), its size is changed so that the edge chief ray does not change, as illustrated in Figure 5-39. The new peak aberration coefficients (indicated by an *) are given by

$$\overline{A}_s^* = \overline{A}_s \quad , \tag{5-348a}$$

$$\overline{A}_c^* = \overline{A}_c - 4\overline{\zeta}\,\overline{A}_s \quad , \tag{5-348b}$$

$$\overline{A}_a^* = \overline{A}_a - 2\overline{\zeta}\,\overline{A}_c + 4\overline{\zeta}^2\overline{A}_s \quad , \tag{5-348c}$$

$$\overline{A}_d^* = \overline{A}_d - \overline{\zeta}\,\overline{A}_c + 2\overline{\zeta}^2\overline{A}_s \quad , \tag{5-348d}$$

and

$$\overline{A}_t^* = \overline{A}_t - 2\overline{\zeta}\left(\overline{A}_a + \overline{A}_d\right) + 3\overline{\zeta}^2 A_c - 4\overline{\zeta}^3\overline{A}_s \quad , \tag{5-348e}$$

where

$$\overline{\zeta} = \frac{L - L^*}{L^*}\,\frac{a}{h'} \quad . \tag{5-349}$$

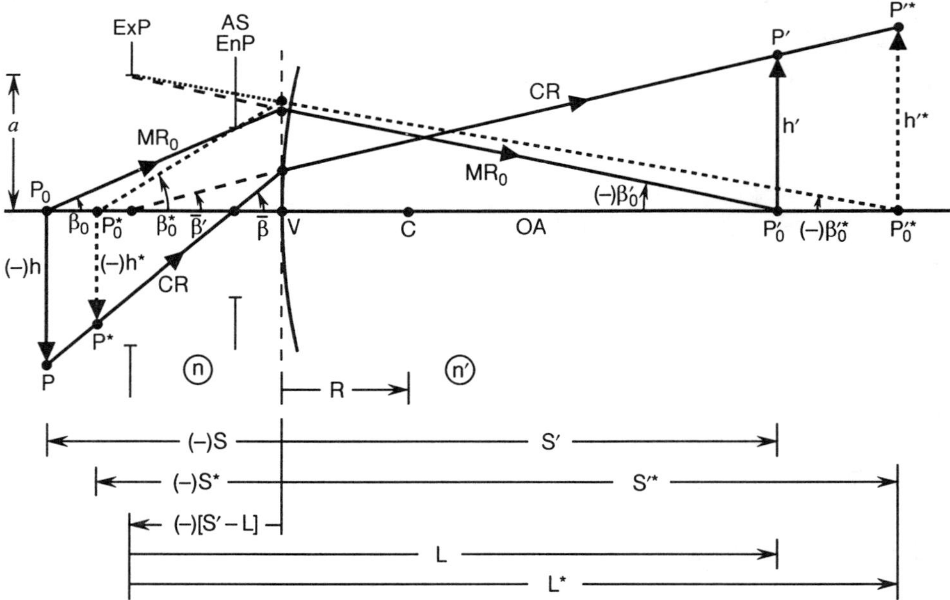

Figure 5-39. When the object in Figure 5-38 is shifted from P_0P to $P_0^*P^*$, its size is changed as illustrated so that the edge chief ray CR does not change. The new axial marginal ray is shown with dashed lines. The parameters corresponding to the new object are shown with an *.

5.15.3 Conjugate-Shift Equations

To obtain the image aberration coefficients for the new image position (also indicated by an *), we note that the object size is changed so that the edge chief ray does not change, as illustrated in Figure 5-39. Accordingly, the slope angles $\overline{\beta}$ and $\overline{\beta}'$ of the chief ray in the object and image spaces do not change. Moreover, the Lagrange invariant H also does not change, i.e., $nh\beta_0 = nh^*\beta_0^*$ since $h/h^* = L_o/L_o^* = \beta_0^*/\beta_0$ from Figure 5-39, where L_o^* is the distances of the object $P_0^*P^*$ from the entrance pupil. The Lagrange invariant associated with the chief ray is given by

$$\overline{H} = na_{en}\overline{\beta} = na_{en}\,h/L_o = -nh\beta_0 = -H \quad , \tag{5-350}$$

that is, it is equal in magnitude but opposite in sign to the Lagrange invariant associated with the marginal ray. The slope angles β_0^* and $\beta_0'^*$ of the incident and refracted axial marginal rays for the new object position are given by

$$\beta_0^* = \beta_0 - \overline{\zeta}\,\overline{\beta} \tag{5-351a}$$

and

$$\beta_0'^* = \beta_0' - \overline{\zeta}\,\overline{\beta}' \quad . \tag{5-351b}$$

Equation (5-351b) is easy to obtain from Figure 5-39. Equation (5-351a) is obtained by noting that

$$\begin{aligned}
\beta_0^* - \beta_0 &= -a_{en}\left(\frac{1}{L_o^*} - \frac{1}{L}\right) \\
&= \frac{a_{en}}{h}\frac{L_o - L_o^*}{L_o^*}\overline{\beta}
\end{aligned} \tag{5-352}$$

and

$$\begin{aligned}
\frac{L_o - L_o^*}{L_o^*} &= \frac{S - S^*}{S^* - t} \\
&= \frac{S' - S'^*}{S'^* - t'}\frac{S}{S'}\frac{t'}{t} \\
&= \frac{L - L^*}{L^*}\frac{h}{h'}\frac{a}{a_{en}} \quad ,
\end{aligned} \tag{5-353}$$

where t and t' are the distances of the entrance and exit pupils from the refracting surface and we have made use of the equations for imaging and magnification, e.g.,

$$\frac{n'}{S'} - \frac{n}{S} = \frac{n'}{S'^*} - \frac{n}{S^*} = \frac{n'}{t'} - \frac{n}{t} \quad . \tag{5-354}$$

It should be noted that $\bar{\zeta}$ given by Eq. (5-349) is also given by

$$\bar{\zeta} = \frac{L_o - L_o^*}{L_o^*}\frac{a_{en}}{h} \quad ,$$
(5-355)

that is $\bar{\zeta}$ is invariant under refraction. Hence it has the same value for each surface in a multisurface system.

By replacing S' by S'^* and L by L^* in Eq. (5-337a), the spherical aberration for the new conjugates can be written

$$A_s^* = \left(\frac{S'^*}{L^*}\right)^4\left[-\frac{n'(n'-n)}{8n^2}\left(\frac{1}{R}-\frac{1}{S'^*}\right)^2\left(\frac{n'}{R}-\frac{n+n'}{S'^*}\right)+\sigma\right]a^4 \quad .$$
(5-356a)

Following Eqs. (5-342b) through (5-342e), the other aberrations for the new image position can also be obtained from:

$$\bar{A}_c^* = A_t^* +\frac{1}{2}H\Delta(\bar{\beta}^2) \quad ,$$
(5-356b)

$$\bar{A}_a^* = A_a^* +\frac{1}{2}H(\beta_0^*\bar{\beta}) \quad ,$$
(5-356c)

$$\bar{A}_d^* = A_d^* +\frac{1}{4}H\Delta(\beta_0^*\bar{\beta}) \quad ,$$
(5-356d)

and

$$\bar{A}_t^* = A_c^* +\frac{1}{2}H\Delta(\beta_0^{*2}) \quad .$$
(5-356e)

The spherical aberration given by Eq. (356a) can also be written in terms of the aberrations for the old conjugates in the form

$$A_s^* = A_s -\bar{\zeta}\left[A_c + \frac{1}{8}H\Delta(\beta_0^2)\right] + \bar{\zeta}^2\left[A_a + A_d +\frac{3}{8}H\Delta(\beta_0\bar{\beta})\right]$$

$$- \bar{\zeta}^3\left[A_t + \frac{3}{8}H\Delta(\bar{\beta}^2)\right] + \bar{\zeta}^4\bar{A}_s \quad .$$
(5-357a)

Comparing Eqs. (5-356b) through (5-356e) with the corresponding Eqs. (5-348b) through (5-348e), we obtain coma, astigmatism, field curvature, and distortion for the new conjugates:

$$A_c^* = A_c -2\bar{\zeta}\left[A_a + A_d + \frac{1}{4}H\Delta(\beta_0\bar{\beta})\right] + \bar{\zeta}^2\left[3A_t + H\Delta(\bar{\beta}^2)\right]-4\bar{\zeta}^3\bar{A}_s \quad ,$$
(5-357b)

$$A_a^* = A_a -2\bar{\zeta}\left[A_t +\frac{1}{4}H\Delta(\bar{\beta}^2)\right] + 4\bar{\zeta}^2\bar{A}_s \quad ,$$
(5-357c)

$$A_d^* = A_d - \overline{\zeta}\left[A_t + \frac{1}{4}H\Delta\left(\overline{\beta}^2\right)\right] + 2\overline{\zeta}^2\overline{A}_s \quad , \tag{5-357d}$$

and

$$A_t^* = A_t - 4\overline{\zeta}\,\overline{A}_s \quad . \tag{5-357e}$$

Equations (5-357) are the *conjugate-shift equations*, describing the peak aberration coefficients for one object position in terms of those for another. Note, however, they also involve the spherical aberration $\overline{A}_s$ of the exit pupil.

The peak aberration coefficients obtained above for a single surface can be generalized to a system consisting of more than one surface by summing them over all the surfaces. The $\Delta(x)$ terms now represent the difference between the quantity x in the image and object spaces of the system since the intermediate terms cancel each other.

5.15.4 Invariance of Image Aberrations

The conditions under which a certain primary aberration can be made zero and invariant with object position can be obtained from Eqs. (5-357). However, all primary aberrations cannot be made zero and invariant simultaneously for a system of finite power. Certain combinations of aberrations can, however, be made invariant. From Eq. (5-357e), we note that distortion is invariant with a conjugate shift if the spherical aberration $\overline{A}_s$ of the pupil is zero. If, in addition, distortion is given by $A_t = -H\Delta\left(\overline{\beta}^2\right)\big/4$, then Eq. (5-357c) shows that its astigmatism is also invariant. This distortion is zero if $\Delta\left(\overline{\beta}^2\right) = 0$, i.e., if the entrance and exit pupils are located at the nodal points of the system so that $\overline{\beta}' = \overline{\beta}$.

Now consider a system for which the pupils lie at its nodal points and the refractive indices of the object and image spaces are equal to unity so that the nodal points coincide with the principal points of the system. In this case, the chief ray in the image space is parallel to the chief ray in the object space (see Figure 5-40), $L = S'$, and

$$\Delta\left(\beta_0\overline{\beta}\right) = \overline{\beta}'\left(\beta_0' - \beta_0\right) = h'\beta_0'\left(\frac{1}{S'} - \frac{1}{S}\right) = h'\beta'\frac{1}{f'} = HK \quad . \tag{5-358}$$

where $K = 1/f'$ is the power of the system. We note from Eq. (5-353b) that coma is invariant if the system satisfies three conditions:

$$\overline{A}_s = 0\,, A_t = 0\,, \text{and } A_a + A_d = -H^2K/4 \quad . \tag{5-359}$$

Thus, a system with zero and invariant distortion, and, therefore, invariant astigmatis (and field curvature), cannot have invariant coma if $A_a + A_d = 0$ (giving a flat tangential field) and the power is non zero. Similarly, a system with zero and invariant distortion and zero and invariant astigmatism, cannot have invariant coma if it has a flat Petzval surface and non zero power.

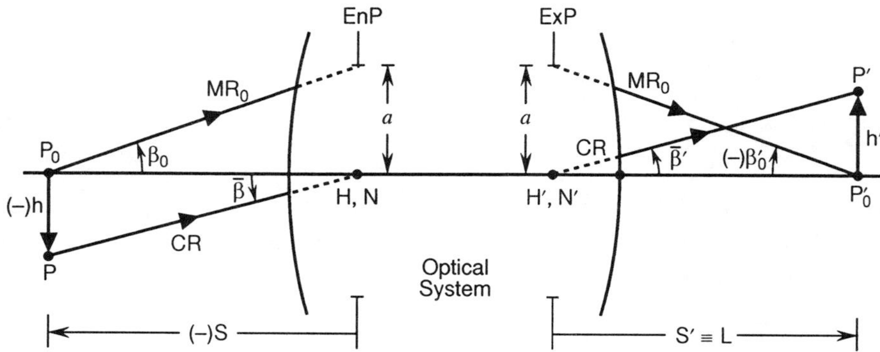

Figure 5-40. Imaging by a system for which the entrance pupil lies in the object-space principal plane and the exit pupil lies in the image-space principal plane. The refractive indices of the object and image spaces are equal and, therefore, the nodal points coincide with the corresponding principal points. Accordingly, the chief ray in the image space is parallel to the chief ray in the object space, i.e., $\overline{\beta}' = \overline{\beta}$.

In order that spherical aberration be invariant, Eq. (5-357a) shows that the following four conditions must be satisfied:

$$\overline{A}_s = 0 \quad , \tag{5-360a}$$

$$8A_t + 3H\Delta\left(\overline{\beta}^2\right) = 0 \quad , \tag{5-360b}$$

$$8\left(A_a + A_d\right) + 3H\Delta\left(\beta_0\overline{\beta}\right) = 0 \quad , \tag{5-360c}$$

and

$$8A_c + H\Delta\left(\beta_0^2\right) = 0 \quad . \tag{5-360d}$$

As discussed in Section 5.9.1, the peak value of spherical aberration is independent of the position of the aperture stop. For pupil imagery, the object is the effective aperture stop. As the object is moved in the plane of the entrance pupil, its size reduces to a point. $\overline{A}_s = 0$ implies that this point object is imaged with zero spherical aberration. Hence, A_s can be invariant only if it is zero. Moreover, since $\Delta\left(\beta_0^2\right)$ varies with conjugates, being zero only for an angular magnification of ± 1, $A_s = 0$ implies that coma can be zero only at this magnification, corresponding to the incompatibility of the sine and Herschel conditions at any other magnification, as discussed in Section 3.7.3.

5.15.5 Simultaneous Correction of Aberrations for Two or More Object Positions

Equation (5-357a) shows that, in principle, since the spherical aberration of a system depends on $\overline{\zeta}$ as $\overline{\zeta}^4$, it can be zero simultaneously for four object positions. Similarly, Eqs. (5-357b), (5-357c), and (5-357e) show that coma can be zero for three object positions, astigmatism for two, and distortion for one, respectively. Moreover, if distortion is zero and invariant, then the number of object positions for which the other aberrations are zero is reduced by one. For example, as discussed in Section 5.4, a spherical surface has three aplanatic points, i.e., spherical aberration and coma are zero for three object positions. For two of them, astigmatism is also zero.

Consider a system for which spherical aberration, coma, and astigmatism are zero for two object positions. In that case, the stop-shift equations show that they are zero for any position of the aperture stop. Hence, the aperture stop may be chosen at a nodal point of the system without any loss of generality, resulting in $\Delta(\overline{\beta}^2) = 0$ and $\Delta(\beta_0\overline{\beta}) = HK$. Letting $A_a = 0 = A_a^*$ in Eq. (5-357c), we find that

$$\overline{\zeta} = A_t / 2\overline{A}_s \quad , \tag{5-361}$$

where $\overline{A}_s$ is the spherical aberration of the pupil when it is located at the nodal point. Letting $A_c = 0 = A_c^*$ in Eq. (5-357b) and substituting Eq. (5-361), we obtain

$$A_t^2 = \left(4A_d + H^2K\right)\overline{A}_s \quad . \tag{5-362}$$

Finally, letting $A_s = 0 = A_s^*$ in Eq. (5-357a) and substituting Eqs. (5-361) and (5-362), we obtain

$$A_t HK = 2\overline{A}_s \Delta(\beta_0^2) \quad . \tag{5-363}$$

Substituting Eq. (5-361) into Eq. (5-357e), we find that $A_d^* = A_d$, as expected for zero astigmatism since A_d in that case reduces to the aberration coefficient due to Petzval curvature which is independent of the object position. Equations (5-362) and (5-363) are the conditions under which spherical aberration, coma, and astigmatism of a system with air for object and image spaces are corrected simultaneously for two object positions. Equation (5-362) shows that if the Petzval curvature is zero, then K and $\overline{A}_s$ must have opposite signs. Moreover, substituting Eq. (5-361) into Eq. (5-357e), we find that $A_t^* = -A_t$, i.e., distortions for the two object positions are equal in magnitude but opposite in sign. Also the angular magnifications for the two conjugates are reciprocal of each other, as may be seen by considering the example of aplanatic points of a spherical surface.

Wynne[6] also discusses the aplanatic conditions for three object positions and shows that a system cannot have flat field anastigmatism correction for two object positions and be aplanatic for a third. He also obtains the conjugate-shift equations for chromatic aberrations.

REFERENCES

1.   H. A. Buchdahl, *Optical Aberration Coefficients*, Oxford, London (1954); reprinted by Dover, New York (1968).

2.   W. T. Welford, *Aberrations of the Symmetrical Optical System*, Academic Press, New York (1974).

3.   P. Mouroulis and J. Macdonald, *Geometrical Optics and Optical Design*, Oxford, New York (1997).

4.   M. Born and E. Wolf, *Principles of Optics*, Pergamon, New York (1985).

5   D. E. McCarthy, "The reflection and transmission of infrared materials, Part 1, Spectra from 2 μm to 50 μm," *Appl. Opt.* **2**, 591-595 (1963); "Part 2, Bibliography," *Appl. Opt.* **2**, 596–603 (1963); "Part 3, Spectra from 2 μm to 50 μm," *Appl. Opt.* **4**, 317-320 (1965); "Part 4, Bibliography," *Appl. Opt.* **4**. 507-511 (1965); "Part 5, Spectra from 2 μm to 50 μm," *Appl. Opt.* **7**, 1997–2000 (1965); "Part 6, Bibliography," *Appl. Opt.* **7**, 2221–2225 (1965).

6.   C. G. Wynne, "Primary aberrations and conjugate change," *Proc. Phys. Soc.* **65B**, 429-437 (1952).

PROBLEMS

5.1   As an example of a *Cartesian refracting surface,* (a) show that a collimated beam incident on an ellipsoidal surface separating media of refractive indices n and n' is focused perfectly at its right-hand side geometrical focus provided its eccentricity is given by $e = n/n'$. (b) Also show that the primary spherical aberration given by Eq. (5-79) reduces to zero. Of course, for the ellipsoidal surface obtained in part (a), higher orders of spherical aberration are also zero.

5.2   Consider a *glass sphere* of radius R and refractive index n imaging a point object lying at a distance R/n from its center. Show that its image observed from the other side of the center is perfect by showing that (a) all object rays incident on the lens surface intersect at its Gaussian image point after being refracted by it, (b) the optical path lengths of the rays from the object point to its Gaussian image point are equal to each other. The two points form a *Cartesian pair* in that one is a perfect image of the other. (c) It is shown in Section 5.4 that the conjugate pair is aplanatic. Show explicitly that the pair satisfies the *sine condition* of Eq. (3-97).

5.3   Show that the spherical aberration of a refracting surface is given by Eq. (5-6) even if r represents the chord $V_0 Q$ of a surface point Q instead of its distance from the axis. (Hint: In Figure 5-1, show that the cosine of the angle QV_0C is equal to $V_0 Q/2R$ and determine $P_0 Q$ and QP_0' from the triangles $P_0 Q V_0$ and $V_0 Q P_0'$, respectively.)

5.4   In Gaussian optics, the angle that a ray makes with the optical axis, or with a surface normal at its point of incidence, is assumed to be small so that its sine is approximately equal to itself. The primary aberrations of a system represent the next level of approximation in which $\sin\theta = \theta - \theta^3/3!$. Show that a ray refracted at a point Q in Figure 5-1, according to Snell's law in this approximation, intersects the Gaussian image plane at a distance $r_i = 4S'a_s r^3/n'$ below the optical axis, a result that is obtained by the substitution of Eq. (5-6) into Eq. (3-13).

5.5   By considering P' as the object point and P as its Gaussian image point in Figure 5-2, show from Eq. (5-19) that the primary aberrations of the ray PQP' do not change when the roles of P and P' are interchanged.

5.6   Show that for an object at infinity, (a) the spherical aberration of a *glass sphere* of radius of curvature R and refractive index n is given by $a_s = (n-1)\left(n^2 - 3n + 1\right)/4n^3 R^3$. Assuming that the aperture stop is located at the lens center, (b) determine the aberration coefficients of other primary aberrations also, and (c) calculate the peak values of the aberrations for $R = 3$ cm and $n = 1.5$.

5.7   Show that the radius of curvature of the *Petzval surface* of a system consisting of a series of m thin lenses of refracting indices n_j, and focal lengths f_j', where $j = 1, 2,$ ..., m, is given by $R_p^{-1} = \sum_{j=1}^{m} \left(n_j f_j'\right)^{-1}$. Also show that the peak value of its astigmatism for an object at a distance S and a height h is given by $A_a = -(ah/S)^2/2f'$, where a is the radius of the aperture stop located at the lens system and f' is the focal length of the system given by $f'^{-1} = \sum\limits_{j=1}^{m} f_j'^{-1}$.

5.8 Design a thin lens for focusing a *parallel beam of light* with minimum spherical aberration at a distance of 10 cm from it. (a) Determine the radii of curvature of its surfaces if its refractive index is 1.5. (b) Determine the peak value of spherical aberration for an He-Ne beam 4 mm in diameter. (c) Repeat problem (b) if the lens is turned around. (d) How does the aberration change if the diameter of the beam is increased to 1 cm?

5.9 The shape factor of a *thin lens* for zero coma is given by Eq. (5-208). (a) For an object at infinity, determine the radii of curvature of its two spherical surfaces in terms of its focal length f' and refractive index n. (b) Show that its corresponding spherical aberration is given by $a_s = -n^3/8(n^2-1)^2 f'^3$. (c) Its spherical aberration can be made zero by making its surfaces nonspherical. Determine the eccentricities of its conic surfaces so that the spherical aberration is zero. Assume that the surfaces have equal eccentricities. (d) Calculate its primary aberration coefficients for $f' = 10$ cm and $n = 1.5$.

5.10 (a) Show that astigmatism contributed by the first surface of an *aplanatic lens* given by Eq. (5-50a) represents the astigmatism of the lens given by Eq. (5-194). Similarly, show that field curvature contribution by the second surface given by Eq. (5-50b) represents the field curvature of the lens given by Eq. (5-195a) Also show that the distortion contributions of its two surfaces cancel each other so that the lens is distortion free. (b) Design an aplanatic thin lens of focal length 15 cm and refractive index 1.5, i.e., calculate the radii of curvature of its two surfaces. (c) Determine its aplanatic points and sketch them on a diagram showing the centers of curvature of its surfaces. (d) Calculate its astigmatism and field curvature coefficients.

5.11 In an *oil immersion microscope*, an object placed on a slide is surrounded by oil, which in turn is in contact with a hyperhemispherical lens (called an *Amici lens*) so that the object lies at an aplanatic point of the spherical surface. The oil is chosen so that its refractive index is as close as possible to that of the lens. The magnified image is further magnified by an *aplanatic lens* whose first surface has its center of curvature at the image formed by the first lens. If the radius of curvature of the spherical surface of the first lens is 1.2 cm, determine (a) the location of the object with respect to its vertex, (b) the radii of curvature of the surfaces of the second lens assuming it to be placed at a distance of 5 mm from the vertex of the first, and (c) the magnification of the final image. Assume that the lenses and the oil have a refractive index of 1.5. The oil is used to increase the angle of the light cone from the object point transmitted by the system as well as to increase its resolution by a factor equal to the refractive index of the oil. [See J. R. Benford, "Microscope objectives," in *Applied Optics and Optical Engineering*, ed. R. Kingslake, Vol. III, pp. 145–182, Academic Press, New York (1965).]

5.12 Consider a *thin-lens doublet* (two thin lenses in contact) focusing a parallel beam of light incident at an angle of 5° from its axis. The refractive index of each lens is 1.5. The radii of curvature of the first lens are 9.2444 cm and − 15.5197 cm. For the second lens, they are − 9.5618 cm and − 15.3120 cm. (a) Calculate the peak values of the primary aberrations of the system assuming a beam of 2 cm diameter. (Spherical aberration and coma should come out to be practically zero, i.e., the doublet is *aplanatic.*) (b) Calculate the radii curvature of the sagittal, tangential, best, and Petzval image surfaces.

5.13 Design a *thin-lens doublet* of focal length 15 cm that is both *achromatic* and *aplanatic* for a parallel beam of light incident on it using borosilicate crown glass #517645 and dense flint glass #617366. Determine the peak values of its astigmatism, field curvature, and distortion wave aberrations for light incident at an angle of 5° from its axis if its aperture stop is 2 cm in diameter placed at a distance of 1.5 cm from it.

5.14 Consider a *plane-parallel plate* placed in the path of a converging beam. The plate has a refractive index of 1.5, a thickness of 1 cm, and a diameter of 4 cm. In the absence of the plate, the beam comes to a focus at a distance of 8 cm from its front surface at a height of 0.5 cm from its axis. (a) Calculate the position of the focus in the presence of the plate. (b) Also calculate the peak values of the primary aberrations of the system. (c) Determine its chromatic aberrations for $\delta n = 0.008$ and illustrate by a diagram.

5.15 Consider the *thick lens* of refractive index n, thickness t, and surfaces of radii of curvature R_1 and R_2 discussed in Section 1.4.4. (a) Show that its back focal distance t_2 can be written

$$\frac{1}{t_2} = (n-1)\left[\frac{1}{R_1 - bt} - \frac{1}{R_2}\right] ,$$

where $b = (n-1)/n$. (b) By letting $\partial t_2/\partial n = 0$, show that the position of its focal point is achromatic if its thickness is given by

$$R_2 = \frac{(R_1 - bt)^2}{R_1 - b^2 t} .$$

Show that the corresponding focal length may be written

$$f' = \frac{b(t/R_1) - 1}{b^2 t} R_1^2 .$$

(c) Show that it is achromatic with respect to its focal length if its thickness is given by

$$t = \frac{n^2 (R_1 - R_2)}{n^2 - 1} \quad ,$$

or that the distance between the centers of curvature of its two surfaces is given by t/n^2. Show that the corresponding focal length in this case is given by

$$\frac{1}{f'} = \frac{n-1}{n+1} \left(\frac{1}{R_1} - \frac{1}{R_2} \right) \quad ,$$

i.e., it is longer by a factor of $n + 1$ compared with that of a corresponding thin lens.

5.16 Consider a *concentric lens* (see Problem 1.14) made of BK7 glass, with radii of curvature 5 cm and 4 cm, placed in a converging beam of image-forming light of a certain system so that the axial image is concentric with the lens. (a) Show that the lens introduces only astigmatism and distortion. Neither surface of the lens introduces any spherical aberration, coma, or axial color. (b) Determine the peak values of the wave and ray aberrations introduced by each surface in the final image plane for a 1-cm aperture stop placed at the first surface of the lens for an image of height 0.5 cm. (c) Calculate the radius of curvature of the tangential image surface. (d) Calculate the lateral color introduced by each surface and show that their contributions cancel each other. Rosin suggested the use of a concentric lens for controlling the tangential field curvature of an imaging system. However, he incorrectly stated that the lens did not introduce any distortion [see S. Rosin, "Concentric lens," *J. Opt. Soc. Am.* **49**, 862–864 (1959)].

5.17 Consider the *Mangin mirror* of Problem 1.5 imaging an object so that the image distance is S'. Show that its longitudinal chromatic aberration (or axial color) is given by

$$\delta S' = \left[S'^2 \left(2f_s' - R_1 \right) / n\, R_1 f_s' \right] \delta n \quad .$$

For an aperture stop located at the mirror, its lateral color is zero.

CHAPTER 6

CALCULATION OF PRIMARY ABERRATIONS: REFLECTING AND CATADIOPTRIC SYSTEMS

6.1   **Introduction** **367**

6.2   **Conic Reflecting Surface** **367**

    6.2.1   Conic Surface 367

    6.2.2   Imaging Relations 370

    6.2.3   Aberration Function 370

6.3   **Petzval Surface** **375**

6.4   **Spherical Mirror** **377**

    6.4.1   Aberration Function and Aplanatic Points for Arbitrary Location of Aperture Stop 377

    6.4.2   Aperture Stop at the Mirror Surface 379

    6.4.3   Aperture Stop at the Center of Curvature of Mirror 381

6.5   **Paraboloidal Mirror** **384**

6.6   **Catadioptric Systems** **385**

    6.6.1   Introduction...385

    6.6.2   Schmidt Camera 385

    6.6.3   Bouwers-Maksutov Camera 394

6.7   **Beam Expander** **398**

    6.7.1   Introduction...398

    6.7.2   Gaussian Parameters 398

    6.7.3   Aberration Contributed by Primary Mirror 400

    6.7.4   Aberration Contributed by Secondary Mirror 401

    6.7.5   System Aberration 402

6.8   **Two-Mirror Astronomical Telescopes** **402**

    6.8.1   Introduction...402

    6.8.2   Gaussian Parameters 403

    6.8.3   Petzval Surface 408

    6.8.4   Aberration Contributed by Primary Mirror 408

    6.8.5   Aberration Contributed by Secondary Mirror 410

    6.8.6   System Aberration 412

    6.8.7   Classical Cassegrain and Gregorian Telescopes........413

    6.8.8   Aplanatic Cassegrain and Gregorian Telescopes 416

    6.8.9   Afocal Telescope 416

    6.8.10   Couder Anastigmatic Telescopes 417

    6.8.11   Schwarzschild Telescope...418

    6.8.12   Dall-Kirkham Telescope..420

6.9   **Astronomical Telescopes Using Aspheric Plates** **422**

    6.9.1   Introduction..422

    6.9.2   Aspheric Plate in a Diverging Object Beam 422

    6.9.3   Aspheric Plate in a Converging Image Beam 425

    6.9.4   Aspheric Plate and a Conic Mirror 426

    6.9.5   Aspheric Plate and a Two-Mirror Telescope.............................428

References **431**

Problems **432**

Chapter 6

Calculation of Primary Aberrations: Reflecting and Catadioptric Systems

6.1 INTRODUCTION

In Chapter 5, we discussed with examples how to determine the aberrations of an imaging system consisting of refracting surfaces imaging a point object. In this chapter, we consider imaging systems with reflecting surfaces, i.e., catoptric or mirror system*s*. Catadioptric systems, i.e., those consisting of reflecting and refracting elements are also discussed. We start with a system consisting of a single reflecting surface. Although its aberrations may be derived by using the technique used in Sections 5.2 through 5.5 for a refracting surface (and it is quite instructive to do so), detailed derivations are not given here. Instead, we obtain the results for a conic reflecting surface from those for a corresponding refracting surface by substituting the refractive index associated with the reflected rays equal to the negative of the refractive index associated with the incident rays. The aberrations of a spherical mirror with the aperture stop located at its center of curvature are discussed and the results are utilized to describe catadioptric systems such as the Schmidt and Bouwers-Maksutov cameras. Next a beam expander with two confocal paraboloidal mirrors is discussed. It is shown that such a system is anastigmatic, i.e., it is free of spherical aberration, coma, and astigmatism. Finally, the aberrations of a two-mirror system imaging an astronomical object are discussed and the aberration properties of telescopes such as classical and aplanatic Cassegrain and Gregorian, Couder, and Schwarzschild are described. Finally, the aberrations of aspheric plates used in astronomical telescopes are discussed.

6.2 CONIC REFLECTING SURFACE

In this section, we discuss conic reflecting surfaces briefly and give expressions for their primary aberrations as obtained from the corresponding expressions for a refracting surface by substituting for the refractive index associated with the reflected rays equal to the negative of the refractive index associated with the incident rays. Of course, these expressions can also be derived in the same manner as we did for a refracting surface in Sections 5.2 through 5.5.

6.2.1 Conic Surface

A *Cartesian reflecting surface* is one for which light rays from a given point object pass through the same image point after reflection from it. The image is thus aberration free. The Cartesian surfaces for reflection are the conics of revolution. As discussed in Section 5.4.1, a conic represents the locus of a point P such that its distance from a fixed point F, called *geometrical focus*, bears a constant ratio e, called *eccentricity*, to its distance from a fixed straight line called the directrix.

As discussed in Section 5.5.1, a conic of eccentricity e with its origin at its vertex, a vertex radius of curvature R, and its axis of symmetry along the z axis is described by

$$z^2\left(1 - e^2\right) - 2Rz + r^2 = 0 \quad , \tag{6-1}$$

where (x, y, z) are the coordinates of a point on it whose radial distance from the optical axis is given by

$$r^2 = x^2 + y^2 \quad . \tag{6-2}$$

For $0 < e < 1$, Eq. (6-1) represents an *ellipsoid* as illustrated in Figure 6-1a centered at (0, 0, a) with semimajor and semiminor axes given by

$$a = R\big/\left(1 - e^2\right) \tag{6-3a}$$

and

$$b = R\big/\left(1 - e^2\right)^{1/2} \quad , \tag{6-3b}$$

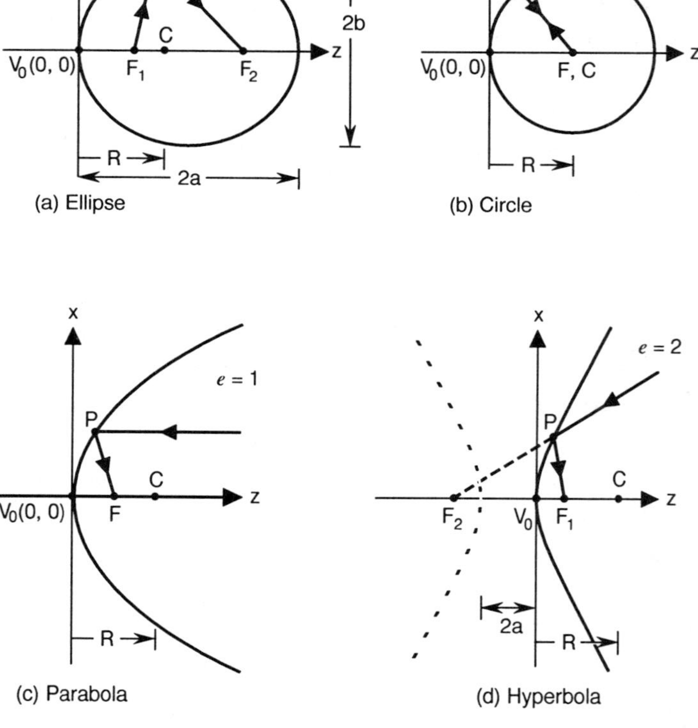

Figure 6-1. Conic reflecting surfaces with a vertex center of curvature C. (a) Ellipse. (b) Circle. (c) Parabola. (d) Hyperbola. The 3-D surfaces are obtained by rotating the figures shown here about the z axis.

respectively. The ellipsoid has two geometrical foci, F_1 and F_2, located at $[0, 0, R/(1 \pm e)]$, and two directrices. The two foci are perfect conjugate points in that all the rays from one of them pass through the other after reflection by the surface. This may be seen by considering a point $P(x, y, z)$ on the surface and showing that

$$F_1 P + PF_2 \;=\; 2a \;\; , \tag{6-4}$$

i.e., by showing that the optical path length $[F_1 PF_2]$ is independent of the location of the point P on the surface. Indeed, this is how an ellipse is often defined. It can also be shown that the incident ray $F_1 P$ and the reflected ray PF_2 make equal angles with the surface normal at the point P; i.e., the ray PF_2 is indeed the reflected ray. Note that the conjugate points F_1 and F_2 lie on the same side of the vertex of an ellipsoidal mirror. Such a mirror is an example of a Cartesian reflecting surface for the conjugate foci.

For $e = 0$, Eq. (6-1) represents a *sphere* of radius of curvature R centered at $(0, 0, R)$, as shown in Figure 6-1b. The two foci are now collocated at the center. Thus, a spherical mirror forms the image of a point object located at its center of curvature without any aberrations, i.e., it is a Cartesian reflecting surface for the conjugates lying at its center of curvature.

For $e = 1$, Eq. (6-1) represents a *paraboloid* with one focus at $(0, 0, R/2)$ and the other at $(0, 0, \infty)$, as illustrated in Figure 6-1c. Thus, a paraboloidal mirror focuses an on-axis collimated beam perfectly at a distance $R/2$ from the vertex along its axis. It is an example of a Cartesian reflecting surface for which one of the conjugates lies at infinity and the other at its geometrical focus, both conjugates lying on its optical axis.

For $e > 1$, Eq. (6-1) represents a *hyperboloid* with semilengths of the transverse and conjugate axes given by

$$a \;=\; R/\left(1 - e^2\right) \tag{6-5a}$$

and

$$b' \;=\; R/\left(e^2 - 1\right)^{1/2} \;\; , \tag{6-5b}$$

respectively. The hyperboloid has two foci, F_1 and F_2, and two directrices. It consists of two branches as shown in Figure 6-1d. Whether the mirror is concave or convex to the incident light, the optical path length $[F_1 PF_2]$ is constant given by

$$F_1 P - PF_2 \;=\; -2a \;\; , \tag{6-6}$$

where the optical path length PF_2 is negative since it is virtual. Indeed, this is how a hyperboloid is defined; i.e., it is the locus of a point whose distances from two fixed points (called foci) differ by a constant. Note that the foci of a hyperboloidal mirror lie on opposite sides of its vertex.

It should be noted that the Gaussian focus of any conic surface with a vertex radius of curvature R lies at a distance $R/2$ from its vertex. It coincides with a geometrical focus only in the case of a paraboloidal mirror.

6.2.2 Imaging Relations

As indicated in Figure 6-2, consider a spherical reflecting surface (mirror) of radius of curvature R imaging a point object. Let the aperture stop of the mirror be located at the surface so that the entrance and exit pupils EnP and ExP, respectively, are also located there. The line joining the vertex V_0 and the center of curvature C of the surface defines its optical axis. Consider an axial point object P_0 at a distance S from the vertex as in Figure 6-2a. Let P_0' be its Gaussian image at a distance S' from the vertex. The object and image distances are related to each other according to (see Section 1.3.7)

$$\frac{1}{S'} + \frac{1}{S} = \frac{2}{R} = \frac{1}{f'} \ ,$$

$$(6\text{-}7)$$

where

$$f' = R/2 \qquad\qquad (6\text{-}8)$$

is the focal length of the mirror. All of the quantities, R, S, S' and f' are numerically negative in Figure 6-2 (For the sign convention, see Section 1.3.2.). If an off-axis point object P is located at a height h, as in Figure 6-2b, its Gaussian image P' lies at a height h' given by

$$M = h'/h = -S'/S \ , \qquad\qquad (6\text{-}9)$$

where M is the magnification of the image. The Gaussian imaging properties of a conic reflecting surface with a vertex radius of curvature R are identical to those of a spherical mirror with the same radius of curvature.

6.2.3 Aberration Function

The wave aberration $W_0(r)$ of an axial object ray incident at a point Q on the mirror at a zonal height r with respect to the chief ray CR_0 incident at the center V_0 of the exit pupil is given by the difference in the optical path lengths of the two rays traveling from P_0 to P_0', i.e.,

$$\begin{aligned}
W_0(r) &= \left[P_0 Q P_0' \right] - \left[P_0 V_0 P_0' \right] \\
&= \left(n P_0 Q - n' \left| Q P_0' \right| \right) - \left(n P_0 V_0 - n' \left| V_0 P_0' \right| \right) \\
&= n \left[\left(P_0 Q + \left| Q P_0' \right| \right) + \left(S + S' \right) \right] \ ,
\end{aligned}$$

$$(6\text{-}10)$$

where n is the refractive index of the medium associated with the incident rays. The path length segments $Q P_0'$ and $V_0 P_0'$ are treated as numerically negative quantities since they correspond to rays traveling backward. The refractive index n' associated with the reflected rays is equal to $-n$ for the same reason.

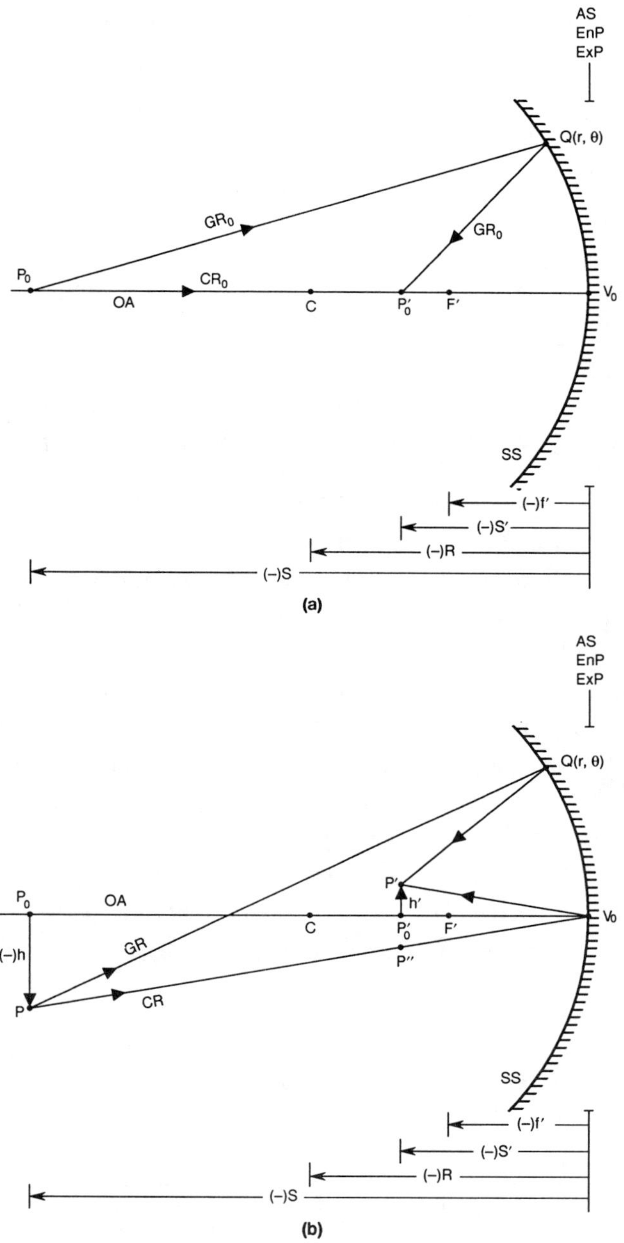

Figure 6-2. (a) Axial imaging by a spherical reflecting surface SS of radius of curvature R , center of curvature C, vertex V_0, and focal point F'. The aperture stop is located at the surface. Accordingly, the entrance and exit pupils EnP and ExP, respectively, are also located there. The axial point object P_0 and its Gaussian image P'_0 lie at at distances S and S' from the vertex, respectively. (b) Off-axis imaging. The point object P and its Gaussian image P' lie at heights h and h', respectively, from the optical axis OA. The quantities f', R, S, S' and h are all numerically negative. GR is a general ray and CR is the chief ray.

Following the same procedure as in Section 5.2.1 for a spherical refracting surface, and noting that R is numerically negative in Figure 6-2, we find that up to the fourth order in r,

$$\boxed{W_0(r) = a_s r^4 \quad,}$$
(6-11)

where

$$\boxed{a_s = \frac{n}{4R}\left(\frac{1}{R} - \frac{1}{S'}\right)^2 .}$$
(6-12)

Comparing Eqs. (5-5) and (6-7), and Eqs. (5-7b) and (6-12), we note that the results for a reflecting surface can be obtained from those for a refracting surface if we let $n' = -n$.

Similarly, following the treatments of Sections 5.2.2, 5.3, and 5.5, we can obtain the aberrations of a spherical mirror for an off-axis point object with the aperture stop at the surface (Figure 6-2), the aperture stop not at the surface (Figure 6-3), and the aberrations of a conic mirror (Figure 6-4), respectively. Thus, if we let $n' = -n$ in Eqs. (5-86) through (5-90), we obtain the primary aberration function of a conic mirror for a point object P at a height h from the optical axis with a Gaussian image P' at a height h', and at a distance L from the exit pupil ExP, representing the optical path difference $\left[P\overline{A}P'\right] - \left[P\overline{B}P'\right]$ between a general ray $P\overline{A}P'$ and the chief ray $P\overline{B}P'$, up to the fourth order in object (or image) and pupil coordinates. It also represents the difference in the optical path lengths of a ray and a chief ray from the object point P up to the reference sphere. The reference sphere passes through the center of the exit pupil with its center of curvature at the Gaussian image point P'. Thus, the wave aberration of the ray represents its optical path length from the wavefront passing through O to the refrence sphere.

The wave aberration at a point $Q(r, \theta)$ where the ray intersects the reference sphere, which is approximately the point of its projection in the plane of the exit pupil, is given by

$$\boxed{W_c(r,\theta;h') = a_{sc} r^4 + a_{cc} h' r^3 \cos\theta + a_{ac} h'^2 r^2 \cos^2\theta + a_{dc} h'^2 r^2 + a_{tc} h'^3 r \cos\theta \quad,}$$
(6-13)

$$a_{sc} = (S'/L)^4 (a_s + \sigma)$$
(6-14a)

$$= a_{ss} + \sigma(S'/L)^4 \quad,$$
(6-14b)

$$a_{cc} = a_{cs} - 4\sigma g(S'/L)^3$$
(6-15a)

$$= 4\left[da_{ss} - \sigma g(S'/L)^3\right] \quad,$$
(6-15b)

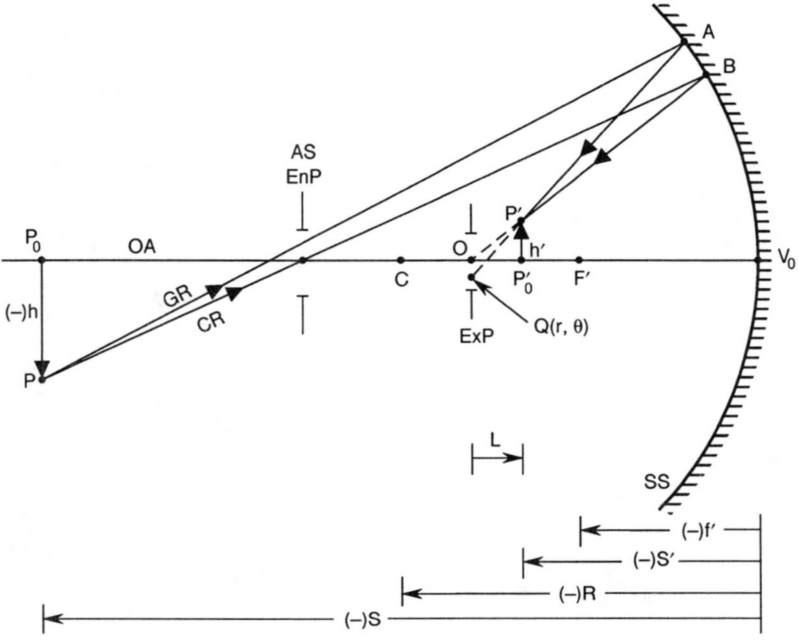

Figure 6-3. Off-axis imaging by a spherical reflecting surface SS. The aperture stop AS is not located at the surface. The image $P_0'P'$ of an object P_0P lies to the right of the exit pupil ExP and, therefore, its distance L from the pupil is numerically positive in the figure.

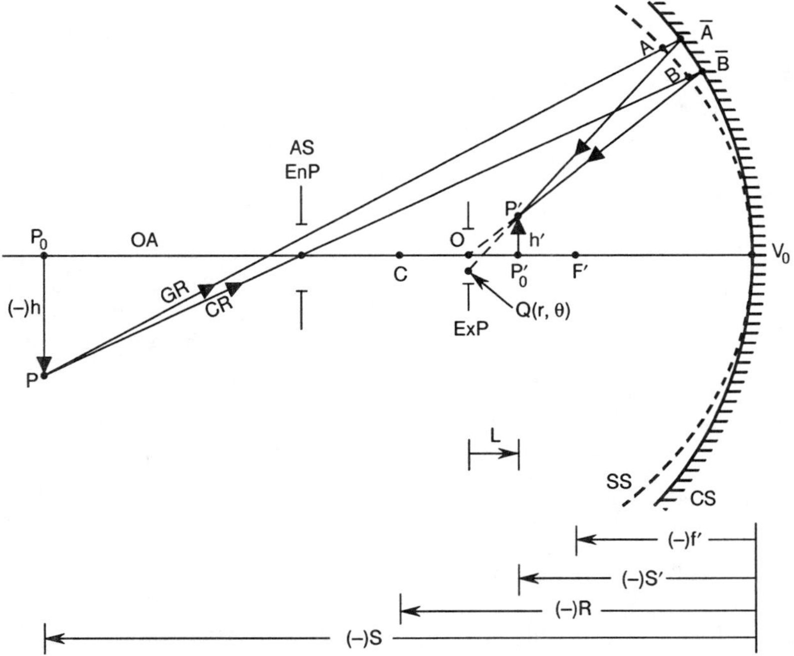

Figure 6-4. Off-axis imaging by a conic reflecting surface CS. The aperture stop is not located at the surface.

$$a_{ac} = a_{as} + 4\sigma g^2 (S'/L)^2 \tag{6-16a}$$

$$= 4\left[d^2 a_{ss} + \sigma g^2 (S'/L)^2 \right] \;, \tag{6-16b}$$

$$a_{dc} = a_{ds} + 2\sigma g^2 (S'/L)^2 \tag{6-17a}$$

$$= 2\left[d^2 a_{ss} - \frac{n}{4RL^2} + \sigma g^2 (S'/L)^2 \right] \tag{6-17b}$$

$$= \frac{1}{2}\left(a_{ac} - \frac{n}{RL^2} \right) \;, \tag{6-17c}$$

and

$$a_{tc} = a_{ts} - 4\sigma g^3 (S'/L) \tag{6-18a}$$

$$= 4\left[d^3 a_{ss} - \frac{nd}{4RL^2} - \sigma g^3 (S'/L) \right] \;. \tag{6-18b}$$

From Eqs. (5-29b), (5-33), (5-77), and (5-83), the quantities $d, a_{ss}, \sigma,$ and g are given by

$$d = \frac{R - S' + L}{S' - R} \;, \tag{6-19}$$

$$a_{ss} = (S'/L)^4 a_s \;, \tag{6-20}$$

$$\sigma = -ne^2 / 4R^3 \;, \tag{6-21}$$

and

$$g = \frac{S' - L}{L} \;, \tag{6-22}$$

respectively. The second term on the right-hand side of Eq. (6-17c) represents the field curvature aberration coefficient due to Petzval curvature of the reflecting surface. We note that when the aperture stop is located at the conic surface so that $L = S'$, then $g = 0$. Hence, in that case the aberrations of a conic surface differ from those of a corresponding spherical surface only in spherical aberration by σr^4. The other primary aberrations of the two surfaces are identical with each other. Generally, a mirror will be used in air. Hence, the refractive index n will be 1 when the rays are incident on the mirror from left to right and -1 when they are incident from right to left.

As examples of systems with reflecting surfaces, we consider the primary aberrations of a spherical mirror, a paraboloidal mirror, Schmidt and Bouwers-Maksutov cameras, a beam expander consisting of two confocal paraboloidal mirrors, and two-mirror astronomical telescopes.

6.3 PETZVAL SURFACE

Given the aberrations of a reflecting surface (and those of a refracting surface) the aberrations of a multisurface reflecting or a catadioptric system can be calculated by following the procedure outlined in Section 5.7.1. In this section, we discuss the Petzval curvature of systems that are considered in later sections.

The radius of curvature R_{ik}, or R_p, of the Petzval surface of a system consisting of k refracting surfaces of radii of curvature R_j, $j = 1, 2, ..., k$, separating media of refractive indices $n_0, n_1, ..., n_k$ respectively, is given by Eq. (5–98):

$$\frac{1}{R_{ik}} = n_k \sum_{j=1}^{k} \frac{1}{R_j} \left(\frac{1}{n_j} - \frac{1}{n_{j-1}} \right) \ . \tag{6-23}$$

The rays on each surface are incident from left to right and the radius of curvature R_j of a surface, including the Petzval surface, is numerically positive or negative, depending on whether its center of curvature lies to the right or the left of its vertex, i.e., depending on whether it is convex or concave to the light incident on it. If the jth surface of a system is a reflecting one, then $n_{j-1} = 1$ and $n_j = -1$ when rays are incident on it from left to right. However, if they are incident from right to left as, for example, on the secondary mirror of a two-mirror system, then $n_{j-1} = -1$ and $n_j = 1$.

For a system consisting of a single mirror of radius of curvature R, the refractive indices are $n_0 = 1$ and $n_1 = -1$. Thus, Eq. (6-23) reduces to

$$\frac{1}{R_{i1}} = -\frac{1}{R}(-1-1) \ ,$$

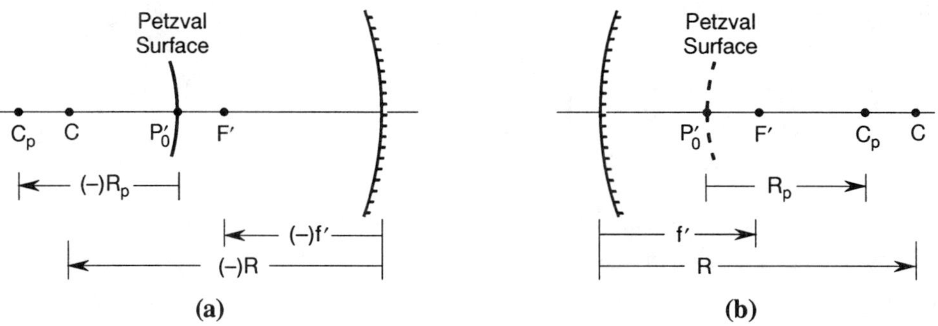

Figure 6-5. Petzval surface of a mirror. (a) Concave mirror with a real Petzval image surface. (b) Convex mirror with a virtual Petzval image surface. C and F' are the center of curvature and the focal point of the mirror. P'_0 is the axial image point and C_p is the center of curvature of the Petzval surface. $P'_0 C_p$ is the radius of curvature of the Petzval surface.

or

$$R_p \;=\; R/2 \;=\; f' \;\;. \tag{6-24}$$

For a concave (converging or a positive) mirror with its center of curvature to the left of its vertex, R is numerically negative (see Figure 6-5a). Therefore, R_p is also numerically negative, or the Petzval surface is convex to the light rays incident on it with a radius of curvature equal to the focal length of the mirror. For a convex (diverging or a negative) mirror with its center of curvature to the right of its vertex, R is numerically positive (see Figure 6-5b), and f' and R_p are also numerically positive. The Petzval surface is virtual and concave to the light incident on it with its center of curvature to the right of its vertex. (Actually, the light rays are diverging from the image surface.) When the object is at infinity so that the image lies in the focal plane of the mirror, the Petzval surface is concentric with the mirror, regardless of whether the mirror is concave or convex.

For a system consisting of two mirrors with radii of curvature R_1 and R_2, the refractive indices have the values $n_0 = 1$, $n_1 = -1$, and $n_2 = 1$ (a second reflection makes n_2 positive). Thus, Eq. (6-23) reduces to

$$\frac{1}{R_{i2}} \;=\; \frac{1}{R_1}(-1-1) + \frac{1}{R_2}(1+1) \;\;,$$

or

$$\frac{1}{R_p} \;=\; 2\left(-\frac{1}{R_1} + \frac{1}{R_2}\right)$$

$$=\; -\frac{1}{f_1'} + \frac{1}{f_2'} \;\;, \tag{6-25}$$

where f_1' and f_2' are the focal lengths of the mirrors. Similarly, we find that R_p for a system consisting of k mirrors with radii of curvature R_j, $j = 1, 2, ..., k$, is given by

$$\boxed{\;\frac{1}{R_p} \;=\; 2(-1)^k \sum_{j=1}^{k} (-1)^j \frac{1}{R_j} \;\;.\;} \tag{6-26}$$

Now we consider a catadioptric system consisting of a concave mirror and a thin lens of refractive index n, as in a Schmidt camera (discussed in Section 6.6.2) where the lens is used as a field flattener. Let the radius of curvature of the mirror be R_1 and those of the lens surfaces be R_2 and R_3. The refractive indices are given by $n_0 = 1, n_1 = -1$, $n_2 = -n$, and $n_3 = -1$. Note that n_2 and n_3 are negative because the rays are incident on the lens from right to left. Thus, Eq. (6-23) reduces to

$$\frac{1}{R_{i3}} \;=\; -\left[\frac{1}{R_1}(-1-1) + \frac{1}{R_2}\left(\frac{1}{-n}+1\right) + \frac{1}{R_3}\left(-1-\frac{1}{-n}\right)\right] \;\;,$$

or

$$\frac{1}{R_p} = \frac{2}{R_1} - \frac{n-1}{n}\left(\frac{1}{R_2} - \frac{1}{R_3}\right)$$

$$= \frac{1}{f_m'} - \frac{1}{nf_l'} \quad , \tag{6-27}$$

where f_m' and f_l' are the focal lengths of the mirror and the lens, respectively. A planar (flat) Petzval surface is obtained as $R_p \to \infty$, or when $f_l' = f_m'/n$. The focal length of the concave mirror in a Schmidt camera is negative. Hence, the focal length of the field-flattening lens is also negative, implying that the focus of the lens lies to its left. Thus, it is a positive or a converging lens since light is being incident on it from right to left. In practice, one chooses a planoconvex lens so that its planar surface $(R_3 = \infty)$ lies against the image surface (see Figure 6-14). In that case, the radius of curvature of its curved surface is given by

$$R_2 = (n-1)R_1/2n \quad , \tag{6-28}$$

which is numerically negative, or the surface is convex to the light rays incident on it.

The radius of curvature R_{ik} of the Petzval surface of a system can also be determined from its astigmatism and defocus aberration coefficients according to Eq. (5–117), i.e.,

$$\boxed{\frac{1}{R_{ik}} = \left(2L_k^2/n_k\right)\left(2a_d - a_a\right)} \quad , \tag{6-29}$$

where L_k is the radius of curvature of the reference sphere with respect to which the aberration of the system is defined; it is equal to the distance of the Gaussian image plane from the plane of the exit pupil of the system. The quantitiy n_k is the refractive index of the image space of the system. In the case of a single mirror in air, substituting Eq. (6-17c) into Eq. (6-29) and letting $n_k = -1$, we obtain Eq. (6-24).

6.4 SPHERICAL MIRROR

In this section, we start our discussion with the aberrations and aplanatic points of a spherical mirror when its aperture stop is located at some arbitrary position. We then consider its aberrations when its aperture stop is located at the mirror surface or at its center of curvature. It is shown that field curvature and distortion are zero with the former location, while coma, astigmatism, and distortion are zero for the latter location. The latter location is utilized in Schmidt and Bouwers-Maksutov cameras which are discussed in Section 6.6.

6.4.1 Aberration Function and Aplanatic Points for Arbitrary Location of Aperture Stop

Consider, as indicated in Figure 6-3, a spherical mirror of radius of curvature R and focal length f' imaging an object P_0P lying at a distance S from its vertex V_0. The aperture stop of the mirror is located at a position so that the Gaussian image $P_0'P'$ lies at a distance L from its exit pupil ExP. Let the height of a point object P from the optical axis of the mirror be h. Its Gaussian image P' lies at a distance S' and a height h', given by Eqs. (6-7) and (6-9), respectively. For a spherical mirror, the eccentricity $e = 0$, and, therefore, according to Eq. (6-21), $\sigma = 0$. Hence, letting $n = 1$ for a mirror in air with light incident from left to right, its primary aberration function obtained from Eq. (6-13) may be written

$$\boxed{W_s\left(r,\theta;h'\right) = a_{ss}\,r^4 + a_{cs}\,h'r^3\cos\theta + a_{as}\,h'^2 r^2\cos^2\theta + a_{ds}\,h'^2 r^2 + a_{ts}\,h'^3 r\cos\theta\ ,} \tag{6-30}$$

where

$$a_{ss} = \left(S'/L\right)^4 a_s \tag{6-31a}$$

$$= \frac{S'^2\left(S' - R\right)^2}{4R^3 L^4}\ , \tag{6-31b}$$

$$a_{cs} = 4d a_{ss} \tag{6-32a}$$

$$= -\frac{S'^2\left(S' - R\right)\left(S' - R - L\right)}{R^3 L^4}\ , \tag{6-32b}$$

$$a_{as} = 4d^2 a_{ss} \tag{6-33a}$$

$$= \frac{S'^2\left(S' - R - L\right)^2}{R^3 L^4}\ , \tag{6-33b}$$

$$a_{ds} = 2d^2 a_{ss} - \frac{1}{2RL^2} \tag{6-34a}$$

$$= \frac{1}{2RL^2}\left[\frac{S'^2\left(S' - R - L\right)^2}{R^2 L^2} - 1\right]\ , \tag{6-34b}$$

and

$$a_{ts} = 4\left(d^3 a_{ss} - \frac{d}{4RL^2}\right) \tag{6-35a}$$

$$= -\frac{S'^2\left(S' - R - 2L\right) + L^2\left(S' + R\right)}{R^3 L^4}\left(S' - R + L\right)\ . \tag{6-35b}$$

As defined in Section 5.4, an optical imaging system that is free of spherical aberration and coma is called an *aplanatic system*. Conjugate points that are free of these aberrations are called *aplanatic points*. From Eqs. (6-31b) and (6-32b), it is easy to see that both spherical aberration and coma coefficients are zero when either $S' = 0$ or $S' = R$. The corresponding values of the object distance S are 0 and R, respectively.

Thus, $(0,0)$ and (R,R) are the aplanatic pairs of points for a spherical mirror. The image of a point object located at the surface or at its center of curvature is perfect and collocated with it, i.e., the axial conjugate pair is Cartesian. The other aberration coefficients corresponding to these points are given by

$$a_{as} = \begin{cases} 0 \text{ for } S' = 0 \ , & (6\text{-}36a) \\[2ex] 1/RL^2 \text{ for } S' = R \ , & (6\text{-}36b) \end{cases}$$

$$a_{ds} = \begin{cases} -1/2RL^2 \text{ for } S' = 0 \ , & (6\text{-}37a) \\[2ex] 0 \text{ for } S' = R \ , & (6\text{-}37b) \end{cases}$$

and

$$a_{ts} = \begin{cases} \dfrac{R+L}{R^2 L^2} \text{ for } S' = 0 \ , & (6\text{-}38a) \\[3ex] -\dfrac{2(R-L)}{R^2 L^2} \text{ for } S' = R \ . & (6\text{-}38b) \end{cases}$$

Hence, the aberration function for the aplanatic points may be written

$$W_s\left(r,\theta;h'\right) = \begin{cases} -\dfrac{1}{2RL^2}h'^2 r^2 + \dfrac{R+L}{R^2 L^2}h'^3 r\cos\theta \text{ for } S' = 0 \ , & (6\text{-}39a) \\[3ex] \dfrac{1}{RL^2}h'^2 r^2 \cos^2\theta - \dfrac{2(R-L)}{R^2 L^2}h'^3 r\cos\theta \text{ for } S' = R \ . & (6\text{-}39b) \end{cases}$$

We note that the three aplanatic points of a spherical refracting surface discussed in Section 5.4 reduce to only two for a corresponding reflecting surface. Moreover, the spherical surfaces that were perfect conjugates of each other for a refracting surface reduce to the reflecting surface itself in the case of a mirror. Of course, the results derived above can be obtained from those of Section 5.4 by letting $n = 1$ and $n' = -1$.

6.4.2 Aperture Stop at the Mirror Surface

We now assume that the aperture stop is located at the mirror so that the entrance and exit pupils EnP and ExP, respectively, are also located there, as illustrated in Figure 6-2. Therefore, the distance L of the image from the exit pupil is equal to S'. Following Eqs. (6-12) and (6-20), the spherical aberration of the mirror is given by

$$\begin{aligned} a_{ss} &= \frac{1}{4R}\left(\frac{1}{R}-\frac{1}{S}\right)^2 \\[2ex] &= \frac{1}{4R}\left(\frac{1}{R}-\frac{1}{S'}\right)^2 \ . \end{aligned} \qquad (6\text{-}40)$$

According to Eq. (6-19), the quantity d for the present case is given by

$$d = R/(S' - R) \ . \tag{6-41}$$

Following Eqs. (6-15) to (6-18), the other primary aberrations are given by

$$a_{cs} = 4d\,a_{ss}$$

$$= \frac{S' - R}{R^2 S'^2} \tag{6-42}$$

$$a_{as} = 4d^2 a_{ss}$$

$$= \frac{1}{R S'^2} \ , \tag{6-43}$$

$$a_{ds} = \frac{1}{2}\left(a_{as} - \frac{1}{R S'^2}\right)$$

$$= 0 \ , \tag{6-44}$$

and

$$a_{ts} = 4d^3 a_{ss} - \frac{d}{R S'^2}$$

$$= 0 \ . \tag{6-45}$$

Hence, the primary aberration function for a spherical mirror with a collocated aperture stop may be written

$$\boxed{W_s(r,\theta;h') = \frac{1}{4R}\left(\frac{1}{R} - \frac{1}{S'}\right)^2 r^4 + \frac{S' - R}{R^2 S'^2} h' r^3 \cos\theta + \frac{1}{R S'^2} h'^2 r^2 \cos^2\theta \ .} \tag{6-46}$$

We note that both the field curvature and distortion aberration coefficients are zero when the aperture stop is located at the mirror. A zero distortion implies that the chief ray CR in Figure 6-2b actually passes through the Gaussian image point P' (which lies along the undeviated ray PCP' in the Gaussian image plane at a height h' from the optical axis) after reflection by the mirror. This may also be seen directly by considering a point P'' on the incident chief ray lying below P'. From similar triangles $P_0 V_0 P$ and $P_0' V_0 P''$, $P_0' P'' / P_0 P$ is equal to S'/S, which is equal to $-h'/h$ from Eq. (6-9). Thus, $P_0' P'' = -h'$. The reflected chief ray will intersect the image plane at the same height above the axis as P'' is below, i.e., it passes through the image point P'.

If the object is located at infinity (e.g., a star), as in Figure 6-6, then

$$S' = R/2$$

$$= f' \tag{6-47}$$

and

$$d = -2 \ . \tag{6-48}$$

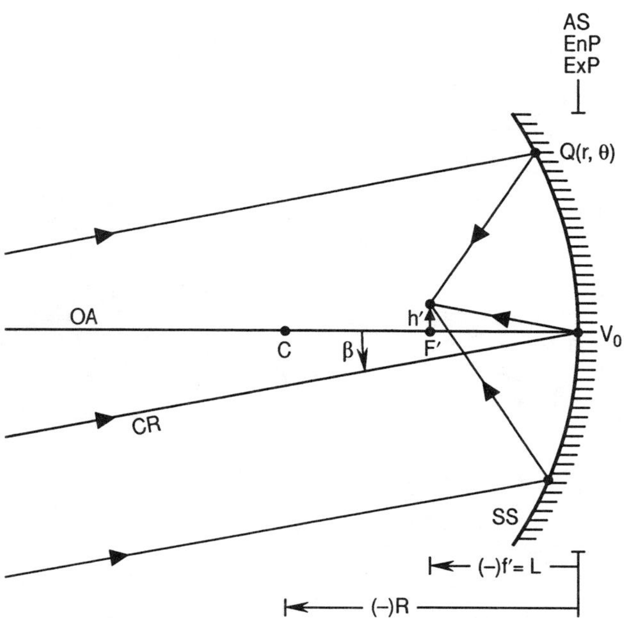

Figure 6-6. Spherical concave mirror imaging a point object lying at infinity at an angle β from its optical axis.

If it lies at an angle β from the optical axis, then

$$h' = -\beta f' \quad . \tag{6-49}$$

Substituting Eqs. (6-47) through (6-49) into Eq. (6-46), we obtain the primary aberration function for a spherical mirror for an object at infinity at an angle β from its optical axis:

$$\boxed{\begin{aligned} W_s(r,\theta;\beta) &= \frac{1}{4R^3} + \frac{1}{R^2}\beta r^3 \cos\theta + \frac{1}{R}\beta^2 r^2 \cos\theta \\ &= \frac{1}{32f'^3}r^4 + \frac{1}{4f'^2}\beta r^3 \cos\theta + \frac{1}{2f'}\beta^2 r^2 \cos^2\theta \quad . \end{aligned}} \tag{6-50}$$

6.4.3 Aperture Stop at the Center of Curvature of Mirror

If the aperture stop is located at the center of curvature C of the mirror, as indicated in Figure 6-7, then the entrance pupil EnP is also located there. The exit pupil ExP, which is the image of the aperture stop by the mirror, is also located there, as may be seen by letting $s = R$ in Eq. (6-7). Thus, $s' = R$, and the pupil magnification is given by $m = -s'/s = -1$. The chief ray CR passes through C and is, therefore, incident normally on the mirror. Accordingly, it is reflected by the mirror upon itself. The distance L of the image $P_0' P'$ of an object $P_0 P$ from the exit pupil is numerically positive since it lies to the right of the exit pupil. Accordingly, we may write

$$L = S' - R \quad . \tag{6-51}$$

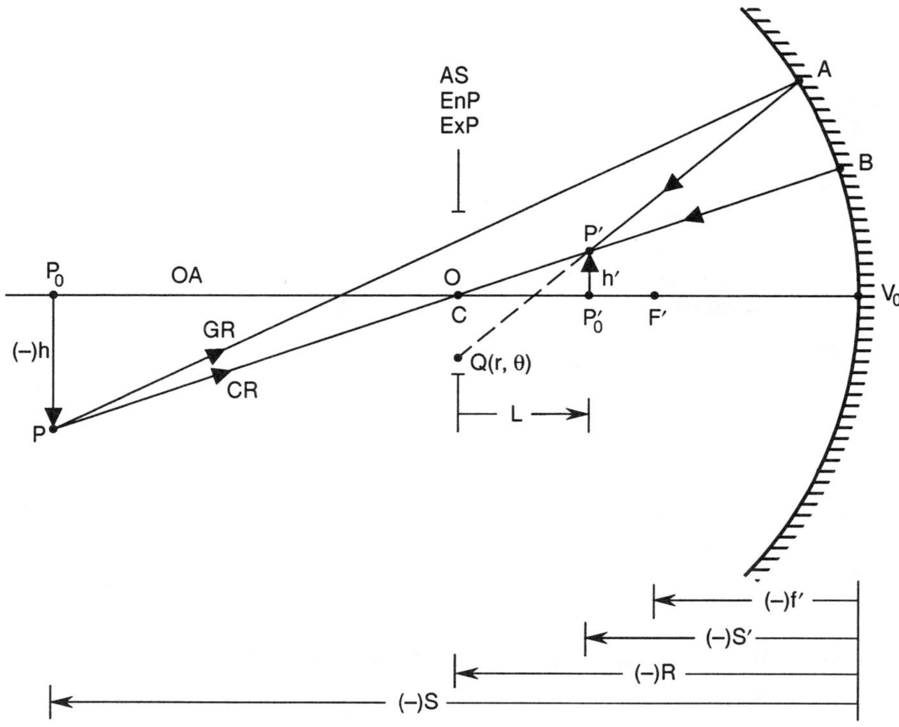

Figure 6-7. Imaging by a spherical concave mirror with the aperture stop located at its center of curvature.

Substituting Eqs. (6-12) and (6-51) into Eq. (6-20), we obtain

$$a_{ss} = \frac{S'^2}{4R^3(S'-R)^2} \quad . \tag{6-52}$$

Substituting Eq. (6-51) into Eq. (6-19), we find that

$$d = 0 \quad . \tag{6-53}$$

Letting $d = 0$ in Eqs. (6-15) through (6-18), we obtain

$$a_{cs} = 0 \quad , \tag{6-54}$$

$$a_{as} = 0 \quad , \tag{6-55}$$

$$a_{ds} = -\frac{1}{2R(S'-R)^2} \quad , \tag{6-56}$$

and

$$a_{ts} = 0 \quad . \tag{6-57}$$

Once again, zero distortion implies that the chief ray actually passes through the Gaussian image point P' after reflection by the mirror. This is indeed the case since the undeviated ray PC on which P' lies is also the chief ray.

Thus, coma, astigmatism, and distortion of a spherical mirror with the aperture stop located at its center of curvature are zero. A concave mirror has a negative spherical aberration but a positive field curvature aberration. If the image is observed on a spherical surface of radius of curvature $R/2$, a surface that is convex to the light incident on it, lying at a distance S' from the mirror, then the field curvature coefficient given by Eq. (6-56) also vanishes. The spherical image surface is, of course, the Petzval image surface discussed in Section 6.3. Equation (6-13) for the aberration function reduces to

$$\boxed{W_s(r;h') = \frac{S'^2 r^4}{4R^3(S'-R)^2} - \frac{h'^2 r^2}{2R(S'-R)^2}}$$

(6-58)

It may be noted with the aid of Figure 6-7 that in going from Eq. (6-46) to Eq. (6-58), the maximum value of r has been reduced by a factor of $S/(S-R)$ or $-S'/(S'-R)$. Hence, the peak value of spherical aberration has not changed due to a change in the position of the aperture stop, as expected from Section 5.9.1. We may add that the aberration function for a spherical mirror with its aperture stop located at its center of curvature given by Eq. (6-58) can be obtained from its aberration function given by Eq. (6-46) when the aperture stop lies at the mirror by using Eqs. (5-146) through (5-151) (see Problem 6.3).

For a point object lying at infinity (see Figure 6-6 except that now we are considering an aperture stop located at the center of curvature of the mirror), the image distance is given by Eq. (6-47), i.e., $S' = R/2 = f'$. Therefore, Eq. (6-51) yields

$$L = -R/2$$

(6-59)

and the spherical image surface of radius of curvature $R/2$ is concentric with the mirror. If the object lies at an angle β from the optical axis, its Gaussian image lies at a height h' given by

$$h' = -\beta R/2 = -\beta f' \ .$$

(6-60)

Hence, the primary aberration function of Eq. (6-58) reduces to

$$\boxed{\begin{aligned} W_s(r;\beta) &= \frac{r^4}{4R^3} - \frac{\beta^2 r^2}{2R} \\ &= \frac{r^4}{32 f'^3} - \frac{\beta^2 r^2}{4f'} \end{aligned}}$$

(6-61)

We note that the spherical aberration is the same as for a mirror with the aperture stop at its surface, as expected, since $|S'/L| = 1$. It can be eliminated by placing, at the center of curvature of the mirror, a glass plate whose thickness varies as r^4. This, of course, is the principle of the Schmidt camera, as discussed in Section 6.6.

It is not difficult to see why all aberrations, except spherical, vanish when the aperture stop is located at the center of curvature of a spherical mirror and the image is observed on the Petzval surface. Since the exit pupil is also located at the center of curvature, the chief ray corresponding to an off-axis point object passes through it. Moreover, since the mirror is spherical, any line passing through its center of curvature forms the optical axis. Hence, every point object is like an on-axis object; therefore, the only aberration that arises (with respect to its Petzval image) is spherical aberration. The Petzval curvature, corresponding to the second term on the right-hand side of Eq. (6-17b), is nonzero. It has the implication that an image aberrated by spherical aberration alone is formed on a spherical surface of radius of curvature f'. Since f' is numerically negative, the center of curvature of the surface lies to the left of its vertex. The surface, of course, is the Petzval image surface passing through the axial image point P_0'. It is concentric with the mirror when the object is at infinity. Its radius of curvature f' is independent of the object location.

6.5 PARABOLOIDAL MIRROR

For a paraboloidal $(e = 1)$ mirror imaging an object at infinity, we note from Eqs. (6-12) and (6-21) that

$$a_s = 1/4R^3 \tag{6-62}$$

and

$$\sigma = -1/4R^3 \ , \tag{6-63}$$

respectively. Hence, substituting these relations into Eq. (6-14a), we find that its spherical aberration

$$a_{sc} = 0 \ . \tag{6-64}$$

This is true regardless of the value of L, i.e., its spherical aberration is zero regardless of the location of its aperture stop. Since its spherical aberration is zero, it follows from Eq. (5-123) that the peak value of its coma is also independent of the position of its aperture stop. This fact can be shown explicitly from Eq. (6-15) by letting $S' = R/2$ and determining the peak aberration (see Problem 6.4).

When the aperture stop is located at the paraboloidal mirror so that $S' = L$, then Eqs. (6-20) and (6-22) yield $a_{ss} = a_s$ and $g = 0$, respectively. Hence, Eqs. (6-15) through (6-18) show that in this case its primary aberrations, other than spherical aberration, are identical with those for a spherical mirror. Thus, the image of an object lying at infinity at an angle β from the axis of the mirror is given by

$$\boxed{\begin{aligned} W_p(r,\theta;\beta) &= \frac{1}{R^2}\beta r^3 \cos\theta + \frac{1}{R}\beta^2 r^2 \cos^2\theta \\[6pt] &= \frac{1}{4f'^2}\beta r^3 \cos\theta + \frac{1}{2f'}\beta^2 r^2 \cos^2\theta \ . \end{aligned}}$$

$$(6-65a)$$

$$(6-65b)$$

i.e. it suffers only from coma and astigmatism. Of course, the image of an axial point object ($\beta = 0$) at infinity is aberration free. We note from Eq. (5-148) that since both spherical aberration and coma are not zero, it is possible to find a position of the aperture stop that yields zero astigmatism. It can be shown that when the aperture stop is located at the focal plane of the paraboloidal mirror, its astigmatism is zero (see Problem 6.4).

The primary aberration function of a conic mirror with an aperture stop located at its conic focus is considered in Problem 6.6. A spherical mirror with its aperture stop located at its center of curvature is a special case of this problem.

6.6 CATADIOPTRIC SYSTEMS

6.6.1 Introduction

We have seen that a paraboloidal mirror forms an aberration-free image of a point object only when it lies on its axis at an infinite distance from it. A spherical mirror gives spherical aberration even when the aperture stop is located at its center of curvature. Spherical aberration of such a mirror can be compensated by placing an aspheric plate at its center of curvature as in a Schmidt camera, or by using a meniscus lens concentric with the mirror as in a Bouwers-Maksutov camera. Such cameras are examples of catadioptric systems. The image of an extended object formed by these cameras is formed free of primary aberrations on a spherical surface that is concentric with the mirror. However, because of the variation of the refractive index of the plate and the lens with the wavelength of object radiation, complete correction of spherical aberration takes place only at one wavelength.

6.6.2 Schmidt Camera

An optical system consisting of a spherical mirror and a transparent plate of nonuniform thickness placed at its center of curvature to compensate for its spherical aberration is called a *Schmidt camera*. The plate is appropriately called the *Schmidt plate*. The aperture stop and, therefore, the entrance and exit pupils of the system, are located at the center of curvature of the mirror. Accordingly, as discussed in Section 6.3.3, the mirror introduces only spherical aberration and Petzval field curvature.

For an axial object at infinity, a nonaxial ray QA intersects the axis after reflection at an axial point F'' that is slightly closer to the mirror vertex than the paraxial focus F', as indicated in Figure 6-8. This may be seen from the isosceles triangle CAF'', where $CF'' = AF''$. Since,

$$CF'' + AF'' > CA = 2|f'| \ ,$$

therefore,

$$CF'' > |f'| = CF',$$

where f' is the focal length of the mirror. From Eq. (6-61) the optical path difference between a ray of zone r and an axial (chief) ray is given by

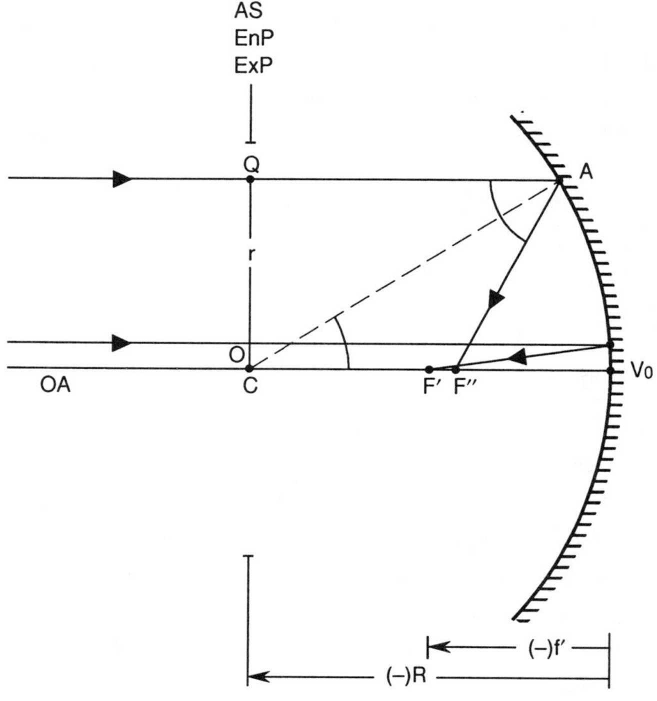

Figure 6-8. Imaging by a concave spherical mirror with the aperture stop located at its center of curvature C. Rays of different zones from an axial object at infinity intersect the axis of the mirror after reflection at different points, such as F' and F'' thus forming an image aberrated by spherical aberration. The ray intersecting the axis at F'' has a zone of $\sqrt{3}a/2$, where a is the radius of the aperture stop.

$$W(r) = \frac{r^4}{4R^3}$$

$$= \frac{r^4}{32 f'^3} \; . \tag{6-66}$$

Since R is numerically negative, so is $W(r)$, implying that the optical path length $[QAF']$ of the nonaxial ray to the focus F' is shorter than that of the axial ray $[CV_0F']$. The negative sign is consistent with the ray QA intersecting the optical axis at a point F'' to the right of F' after reflection by the mirror. This may be seen by comparing the wavefront and the Gaussian reference sphere both passing through C at the exit pupil and noting that the former is less curved than the latter by virtue of the negative spherical aberration. In order that the two optical path lengths be equal, the optical path length of the nonaxial ray must be increased. This is indeed what happens in the case of a paraboloidal mirror. Since its sag compared to that of a spherical mirror is less by $r^4/64|f'^3|$ according to Eq. (5-73), a ray of zone r travels an extra optical path length of $r^4/32|f'^3|$. However, as we have seen, although spherical aberration is zero for a paraboloidal mirror, its other aberrations are not.

If a plate of refractive index n and thickness $t(r)$ is placed at the center of curvature with a flat surface normal to the axis of the mirror, the additional optical path length introduced by the plate is given by $(n-1)t(r)$. All object rays transmitted by the system travel equal optical path lengths and converge to a common focus F' if $t(r)$ is given by

$$W(r) + (n-1)t(r) = 0 \quad ,$$

or

$$\boxed{t(r) = -\frac{r^4}{32(n-1)f'^3}} \quad \cdot \tag{6-67}$$

From a value of zero at its center, the thickness of the plate increases proportional to the fourth power of the zonal radius. In practice, a plane-parallel plate of small thickness t_0 would be added to it so that it can be fabricated. The shape of the plate is illustrated in Figure 6-9. It should be noted that in obtaining Eq. (6-67), we have neglected the fact that a ray incident parallel to the optical axis at a zone r on the plate is incident at a zone that is slightly larger than r; i.e., we have neglected the divergence of the incident beam produced by the plate. We also note that the Schmidt camera is an anastigmat since its spherical aberration, coma, and astigmatism are all zero. Its distortion is also zero, as discussed in Section 6.4.3. The only nonzero primary aberration is the Petzval curvature of the spherical mirror. Hence, the image observed on a spherical surface concentric with the mirror passing through F' is free of all primary aberrations.

Although spherical aberration of the mirror is corrected by the use of a Schmidt plate, complete correction can take place at only one wavelength. Since the refractive index of the plate varies with the wavelength of object radiation, the change in the optical path length, or the angular deviation of a ray produced by the plate, also varies with it. Hence, spherical aberration of the system varies with the wavelength. Such a variation is called *spherochromatism*. To reduce spherochromatism, we proceed as follows.

Consider a ray corresponding to a refractive index n and passing through the plate at a zone r. Its wave aberration produced by the plate is given by $(n-1)t(r)$. Following Eq. (3-11), its angular deviation produced by the plate is given by

$$\psi = (n-1)\frac{dt}{dr} \quad \cdot \tag{6-68}$$

Substituting Eq. (6-67) into Eq. (6-68) we obtain

$$\psi = -\frac{r^3}{8f'^3} \quad \cdot \tag{6-69}$$

From Eq. (6-68), the *angular dispersion* of the rays is given by

$$\Delta\psi = \Delta n\frac{dt}{dr} \quad , \tag{6-70}$$

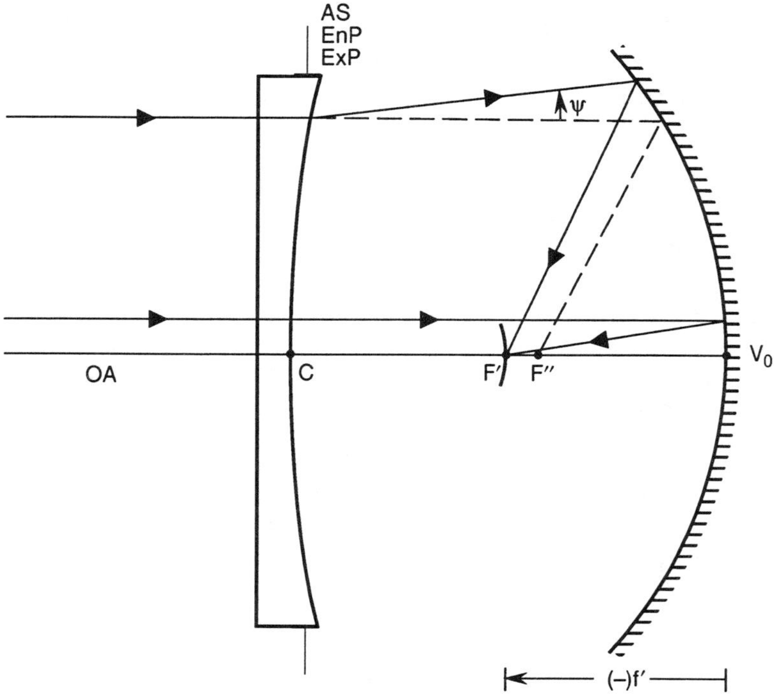

Figure 6-9. Schematic of a Schmidt camera consisting of a concave spherical mirror and a transparent plate placed at its center of curvature C. The spherical aberration of the mirror is precorrected by the plate so that the system forms an image free of this aberration. The dashed lines indicate the path of a ray in the absence of the plate. The thickness of the plate is minimum at its center. The Petzval image surface is illustrated by the spherical surface concentric with the mirror passing through F'.

where Δn is the variation of the refractive index of the plate across the spectral bandwidth of the object radiation. Substituting for dt/dr from Eq. (6-68) into Eq. (6-70), we obtain

$$\Delta\psi = \frac{\Delta n}{n-1}\psi \quad , \tag{6-71a}$$

showing that the angular dispersion $\Delta\psi$ of a ray produced by the plate is proportional to its angular deviation ψ. For a plate of radius a, the maximum value of ψ is $-a^3/8f'^3$, and occurs for the marginal rays. The corresponding maximum value of $\Delta\psi$ is given by

$$[\Delta\psi]_{max} = -\frac{\Delta n}{n-1}\frac{a^3}{8f'^3} \quad . \tag{6-71b}$$

To reduce spherochromatism, we must reduce the maximum value of ψ. To do so, we add to the plate a very thin planoconvex lens. Such a lens will reduce the focus distance so that the rays are now focused at a point F'' instead of F', as in Figure 6-10. A planoconvex lens introduces thickness to the plate varying as r^2. Thus, the plate thickness may be written

$$t(r) = t_0 - \frac{r^4}{32(n-1)f'^3} + \frac{br^2}{n-1} \quad , \tag{6-72}$$

where b is a constant chosen to minimize spherochromatism. Comparing the defocus aberration br^2 introduced by the plate with Eq. (3-15), we find that the distance between F' and F'' is given by $2bf'^2$. F'' lies on the right-hand side of F', as in Figure 6-10, if b is numerically negative.

For simplicity, we write Eq. (6-72) in the form

$$t(r) = t_0 - \frac{a^4}{32(n-1)f'^3}\left(\rho^4 - c\rho^2\right) \quad , \tag{6-73}$$

where

$$c = 32bf'^3/a^2 \tag{6-74}$$

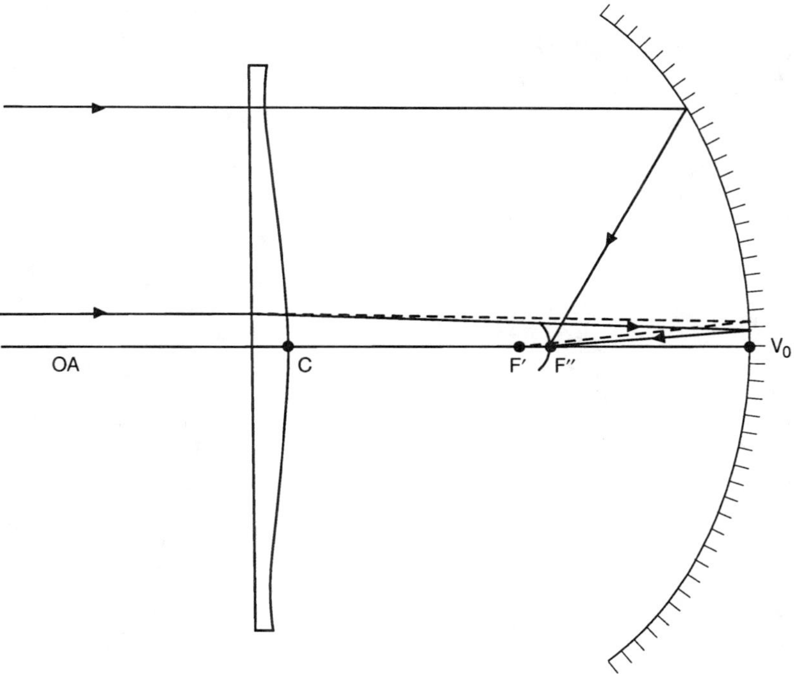

Figure 6-10. Schmidt camera with a plate introducing minimum chromatic aberration. The dashed lines indicate the path of a ray in the absence of the Schmidt plate. All rays passing through the plate and reflected by the concave spherical mirror are focused at F'', where the ray passing through the neutral zone of the plate is focused by the mirror. The thickness of the plate is maximum at its center.

and

$$\rho = r/a \ . \tag{6-75}$$

The thickness variations of plates with different values of c are illustrated in Figure 6-11. We note that the depth of material removal, starting with a plane-parallel plate, is minimum when $c = 1$. However, we are interested in minimizing the maximum absolute value of the angular deviation of rays in the range $0 \le \rho \le 1$. As shown next, this requires that $c = 1.5$.

Substituting Eq. (6-72) into Eq. (6-68), we find that the angular deviation of a ray is now given by

$$\psi = -\frac{a^3}{16 f'^3}\left(2\rho^3 - c\rho\right) \ . \tag{6-76}$$

We want to determine the value of c that minimizes the maximum absolute value of ψ as ρ varies from 0 to 1. This problem is similar to the balancing of spherical aberration with defocus to minimize the spot radius, as discussed in Section 4.3.1. It is clear that c must be positive since a negative value increases the absolute value of ψ for any ray. Its maximum absolute value in the range $0 \le \rho \le 1$ occurs either at its stationary point $\rho_1 = \sqrt{c/6}$ obtained by letting $\partial\psi/\partial\rho = 0$, or at $\rho_2 = 1$. (Zero is also a possible extremal value of ρ, but it corresponds to $\psi = 0$, which is a minimum absolute value of ψ.) The corresponding values of ψ are given by:

$$\psi_1 = \frac{a^3 c^{3/2}}{24\sqrt{6} f'^3} \tag{6-77a}$$

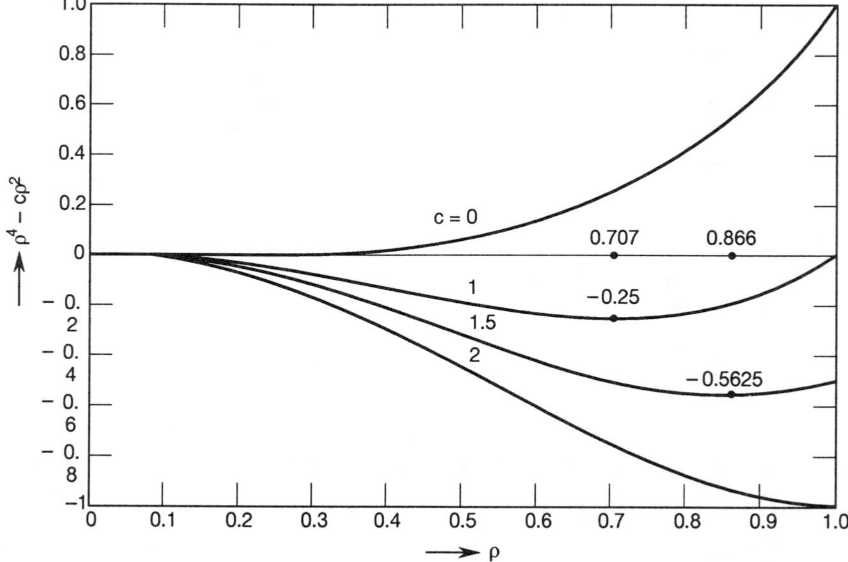

Figure 6-11. Thickness variation of a Schmidt plate for different values of c. A minimum thickness variation is obtained when $c = 1$.

and

$$\psi_2 = -\frac{a^3}{8f'^3}\left(1 - c/2\right) \ ,\qquad\qquad (6\text{-}77b)$$

respectively.

As shown in Figure 6-12, $|\psi_1|$ increases as c increases. However, $|\psi_2|$ first decreases, approaches zero when $c = 2$, and then increases as c increases. The value of c that minimizes the maximum value of $|\psi|$ is that for which $|\psi_1|$ and $|\psi_2|$ are equal, i.e., the point of intersection of the two curves in Figure 6-12. Letting $|\psi_1| = |\psi_2|$, we obtain from Eqs. (6-77)

$$\frac{c^{3/2}}{3\sqrt{6}} = \left|1 - c/2\right| \ .$$

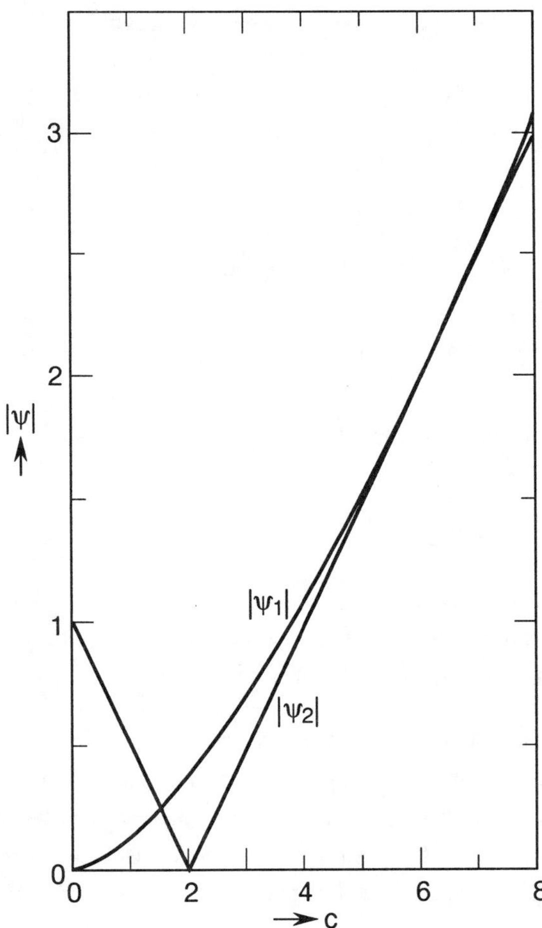

Figure 6-12. Optimization of plate thickness for minimum chromatic aberration. ψ **is in units of** $a^3/8\left|f'^3\right|$.

Or, squaring both sides, we may write

$$c^3 - (27/2)c^2 + 54c - 54 = 0 \quad .$$
(6-78)

Equation (6-78) has three solutions, namely $c = 6$, 6, and $3/2$. Substituting these values of c in Eq. (6-77a) or (6-77b), we find that the maximum absolute value of ψ is minimum and given by $-a^3/32f'^3$ when $c = 3/2$. This may be seen from Figure 6-12 also. We note that the maximum angular deviation and, therefore, spherochromatism, is reduced by a factor of 4 when $c = 1.5$ compared with its value when $c = 0$. Substituting $c = 1.5$ into Eq. (6-73), we find that the plate thickness required for eliminating spherical aberration introduced by the mirror and minimizing the spherochromatism introduced by the plate is given by

$$\boxed{t(r) = t_0 - \frac{a^4}{32(n-1)f'^3}\left(\rho^4 - \frac{3}{2}\rho^2\right) \quad .}$$
(6-79)

We note from Eqs. (6-68) and (6-79) that $\psi \sim \partial t/\partial r = 0$ for $\rho = \sqrt{3}/2$. Since a ray incident normal to the plate at this value of r passes through it undeviated, as indicated in Figure 6-10, it is called the *neutral zone* of the plate. As may be seen from Figure 6-11, the variation in thickness of the plate and the material removal are maximum at this zone. This variation is more than twice the variation for a minimum-thickness plate; compare the numbers -0.5625 and -0.25 in the figure which occur at zones of $\rho = 0.867$ and $\rho = 0.707$, respectively.

Substituting Eq. (6-79) into (6-70), we obtain the angular dispersion of the rays

$$\Delta\psi = -\frac{\Delta n a^3}{8(n-1)f'^3}\left(\rho^3 - \frac{3}{4}\rho\right) \quad .$$
(6-80a)

Its maximum value occurs for rays with $\rho = 1/2$ and 1 given by

$$[\Delta\psi]_{max} = -\frac{\Delta n}{32(n-1)}\frac{a^3}{f'^3} \quad .$$
(6-80b)

This is 1/4 of the corresponding value given by Eq. (6-71b) for $c = 0$. The dependence of angular dispersion on the value of c [obtained by substituting Eq. (6-73) into Eq. (6-70)] is illustrated in Figure 6-13. We note that it is minimum when $c = 1.5$.

The spherical Petzval image surface can be flattened by placing a thin lens of positive focal length near it, i.e., by canceling the Petzval curvature of the mirror with that of the lens, as discussed in Section 6.3. For a planoconvex field-flattening lens of refractive index n, the radius of curvature of its curved surface is given by $-(n-1)f'/n$. Its planar surface is placed against the image or film as illustrated in Figure 6-14.

The spherochromatism of a Schmidt camera can be avoided if a reflecting aspheric corrector[1-3] is used in place of the refracting Schmidt plate. The corrector is tilted with respect to the mirror axis so that the beam reflected by it may be incident on the mirror.

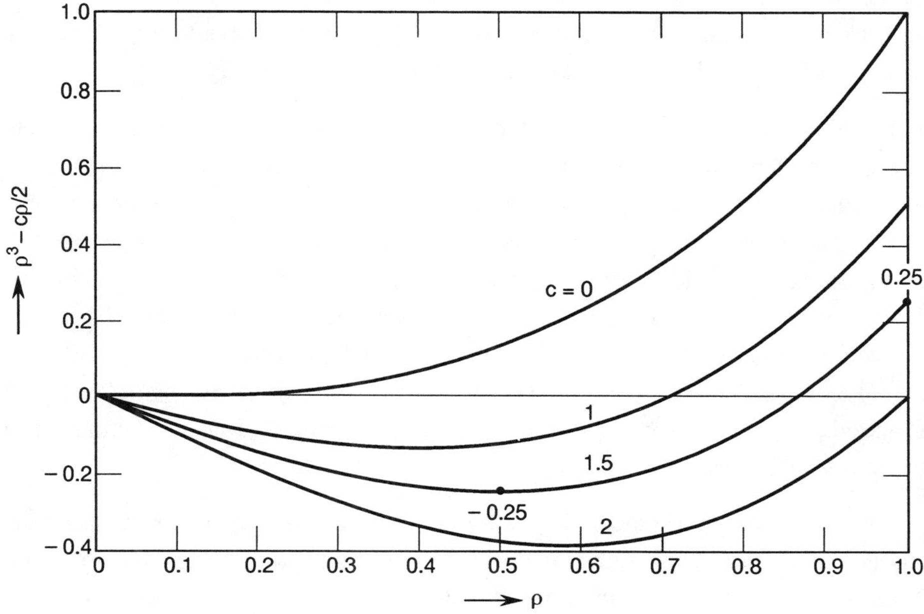

Figure 6-13. Dependence of angular dispersion on the value of *b*. It is minimum when $c = 1.5$.

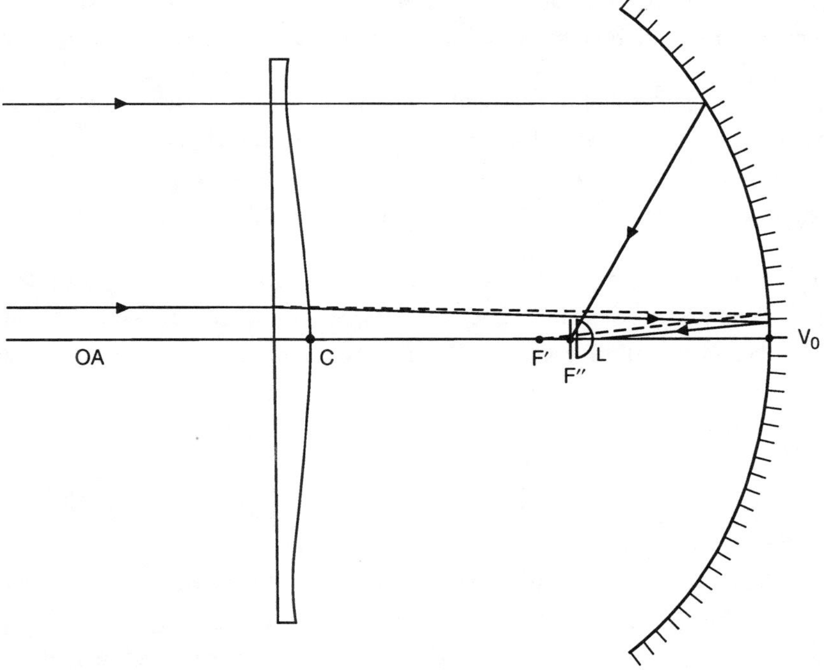

Figure 6-14. Schmidt camera with a planoconvex field-flattening lens *L*. The spherical Petzval image surface illustrated in Figures 6-9 and 6-10 has been flattened by the lens.

The surface figure of the corrector has elliptical symmetry (instead of the circular symmetry of the Schmidt plate) and is correspondingly more difficult to fabricate. As a result, the reflecting corrector has not found widespread application.

It was shown in Section 6.4 that a spherical mirror with an aperture stop located at its center of curvature gives only spherical aberration and field curvature. The Schmidt plate compensates for the spherical aberration and, therefore, the image of an extended object observed on a spherical surface concentric with the mirror is free of primary aberrations. Strictly speaking, the lens component of the plate also introduces small amounts of primary aberrations (see Problem 6.5). The spherical aberration contributed by it can be made zero by slightly adjusting the value of r^4 term in the plate thickness $t(r)$. Its Petzval field curvature can be corrected by the field-flattening lens. Similarly, the *field-flattening lens* introduces small amounts of distortion, higher-order monochromatic aberrations, and chromatic aberrations.

For an on-axis point object, the mirror contributes some *secondary* or *sixth-order spherical aberration*. This can be made zero by introducing an r^6 term in the plate. For off-axis point objects, the Schmidt plate given by Eq. (6-73) also introduces sixth-order aberrations. One of these is the *oblique or lateral spherical aberration* varying as $\beta^2 r^4$, where β is the field angle of a point object. The other is a *sixth-order* (or secondary) *astigmatism* (called *wings* by Schwarzschild) varying as $\beta^2 r^4 \cos^2 \theta$. The ratio of the peak values of these two aberrations is $1:4n$ (see Linfoot[4]). As a result, the geometrical spot diagrams (discussed for primary aberrations in Chapter 4) for off-axis point objects have a much larger width in the tangential direction than that in the sagittal direction.

It should be noted that as the field angle β increases, the size of the focal surface also increases, which, in turn, obscures the ray bundle incident on the mirror. For a field of view of radius β, the *linear obscuration* of the on-axis beam incident on the mirror is given by $\epsilon = 2\beta F$, where F is the focal ratio of the system.

The Schmidt camera can be generalized to using a conic mirror in place of the spherical. This is discussed in Section 6.9.2. The aberrations of a conic mirror with a corrector plate placed at its geometrical focus, where the aperture stop is also located,[5] are considered in Problem 6.6.

6.6.3 Bouwers-Maksutov Camera

In a Schmidt camera, we have seen how the spherical aberration of a spherical mirror is compensated by an aspheric Schmidt plate placed at its center of curvature. Now, we consider an alternative approach introduced independently by Bouwers[6] and Maksutov[7] in which a spherical meniscus corrector lens is used to compensate for the spherical aberration of the mirror. An advantage of the meniscus lens over the Schmidt plate is its simpler fabrication because of its spherical surfaces. We refer to an imaging system consisting of a spherical mirror and a meniscus corrector as the *Bouwers-Maksutov camera*.

We start by considering a thin meniscus lens whose two surfaces are concentric with the mirror as shown in Figure 6-15. If the aperture step is placed at their common center of curvature, then the lens, the mirror, and the system as a whole all have their exit pupils lying at the common center of curvature. As discussed in Section 6.3.3, the system has no unique axis and, therefore, there are no off-axis aberrations. There is no unique position for the placement of the meniscus lens as long as it is concentric with the mirror. Its thickness and refractive index are such that they compensate for the spherical aberration of the mirror. The image of an extended object at infinity formed on a surface concentric with the mirror is free of primary aberrations.

The spherical aberration of a thin lens of refractive index n imaging an object lying at infinity is given by Eq. (5-188) with the position factor $p = -1$. The shape factor q of the lens given by Eq. (5-185) in terms of its radii of curvature R_1 and R_2 can be written in terms of its image-space focal length f_i' and its radius of curvature R_1 by using Eq. (5-169) for the focal length. Thus, substituting

$$q = -1 + 2f_i'(n-1)/R_1 \tag{6-81}$$

and

$$p = -1 \tag{6-82}$$

into Eq. (5-188), we obtain

$$a_s = \frac{-1}{8(n-1)f_i'^3}\left[\frac{n^2}{n-1} - \frac{f_i'}{R_1}(2n+1) + \frac{f_i'^2}{R_1^2}\frac{(n-1)(n+2)}{n}\right] . \tag{6-83}$$

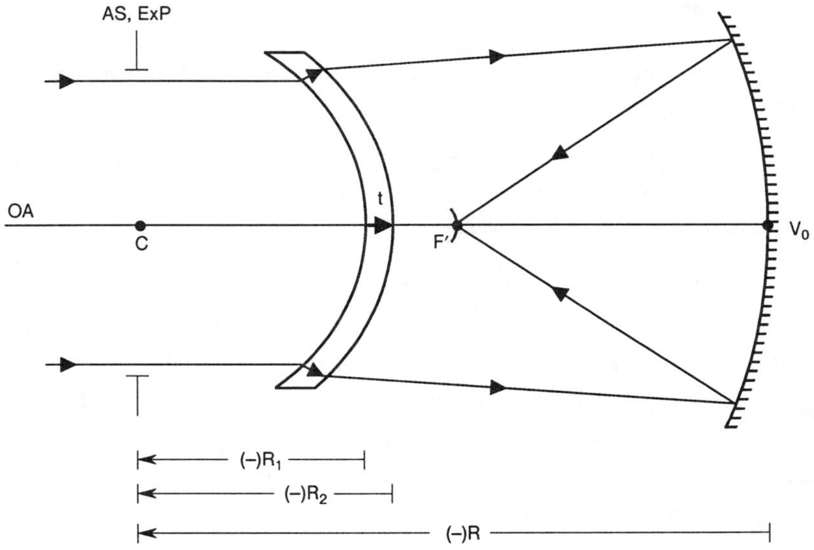

Figure 6-15. Bouwers-Maksutov camera showing a concave spherical mirror and a concentric meniscus corrector lens. F' **is the Gaussian focus of the mirror.**

From Eq. (1-123) for the focal length of a *thick lens* given later as Eq. (6-93), we can show that the focal length of a concentric lens is given by (see Problem 1.13)

$$f_l' = -\frac{nR_1R_2}{t(n-1)} \quad , \tag{6-84}$$

where its (numerically positive) thickness t is given by

$$t = R_1 - R_2 \quad . \tag{6-85}$$

Note that in Figure 6-15, R_1 and R_2 are numerically both negative. Substituting Eq. (6-84) into Eq. (1-126), we find that $t_2 = f_l' + R_2$, implying that the principal point H' in Figure 1-38 lies at the center of curvature of the second surface. Hence, the concentric lens behaves as a thin lens of long negative focal length f_l' placed at the common center of curvature of its two surfaces. The numerical value of its focal length is much larger than the values of its radii of curvature. Hence, the first two terms on the right-hand side of Eq. (6-83) may be neglected in comparison with the third, and we may write

$$a_s = -\frac{1}{8f_l'R_1^2}\frac{n+2}{n} \tag{6-86}$$

$$= \frac{(n-1)(n+2)}{8n^2}\frac{t}{R_1^3R_2} \quad , \tag{6-87}$$

where in the last step we have substituted for the focal length according to Eq. (6-84). We note that the spherical aberration introduced by the concentric lens depends on its thickness t explicitly. However, Eq. (6-87) gives an approximate value of its spherical aberration not only because we neglected two terms of Eq. (6-83), but also because that equation is for a thin lens.

Adding the spherical aberrations of the corrector lens and the mirror, we obtain the spherical aberration of the system

$$\boxed{a_{ss} = \frac{(n-1)(n+2)}{8n^2}\frac{t}{R_1^3R_2} + \frac{1}{4R^3} \quad .} \tag{6-88}$$

Note that in obtaining Eq. (6-88), we have neglected the fact that a ray incident parallel to the axis of the system at a zone r of the corrector is incident at a slightly larger value of r on the mirror (just as we did in the case of a Schmidt plate). This is equivalent to assuming that the image formed by the meniscus lens lies at a sufficiently large distance that it is practically infinity for the mirror. For a given value of the radius of curvature R of the mirror, the spherical aberration given by Eq. (6-88) can be made zero by a suitable choice of R_1, t, and n.

As in the case of a Schmidt camera, here too the spherical aberration is zero at a wavelength corresponding to a refractive index n. At other wavelengths, spherical aberration will be nonzero. To estimate the chromatic aberration introduced by the meniscus lens, we proceed as follows: parallel rays of a given wavelength corresponding to a refractive index n incident on the lens are brought to a virtual focus by it at a distance f_l' from it. Rays of another wavelength corresponding to a refractive index $n + \Delta n$ are brought to a virtual focus at a distance $f_l' + \Delta f_l'$. The relationship between $\Delta f_l'$ and Δn may be obtained by differentiating Eq. (6-84). Thus,

$$\Delta f_l' = -\frac{f_l'}{nV} \quad , \qquad (6\text{-}89a)$$

where

$$V = (n-1)/\Delta n \qquad (6\text{-}89b)$$

is the dispersive constant (Abbe number) of the lens material. Since the concentric lens behaves like a thin lens of focal length f_l' placed at the common center of curvature of its surfaces (see Problem 1.13), the focal length f_s' of the lens-mirror system is given by [see Eq. (1-147)]

$$\frac{1}{f_s'} = \frac{1}{f_m'} + \frac{1}{f_l'} \quad , \qquad (6\text{-}90)$$

where f_m' is the focal length of the mirror. We find from Eq. (6-90) that a change of $\Delta f_l'$ in f_l' produces a change of $\Delta f_s'$ in f_s' given by

$$\Delta f_s' = \left(f_s'/f_l' \right)^2 \Delta f_l' \quad . \qquad (6\text{-}91)$$

Substituting Eq. (6-89a) into Eq. (6-91), we obtain the longitudinal chromatic aberration of the image:

$$\Delta f_s' = -\frac{f_s'^2}{nVf_l'} \quad . \qquad (6\text{-}92)$$

The chromatic aberration can be reduced by using an achromatic meniscus lens made of two different materials that are cemented together. An alternative approach is to make the focal length of the meniscus lens invariant with respect to its refractive index. The focal length of a thick lens is given by Eq. (1-123):

$$\frac{1}{f_l'} = (n-1)\left(\frac{1}{R_1} - \frac{1}{R_2} \right) + \frac{t(n-1)^2}{nR_1R_2} \quad . \qquad (6\text{-}93)$$

Differentiating both sides with respect to n and equating the result to zero, we obtain (see Problem 5.15).

$$\boxed{\; t = \frac{n^2}{n^2-1}(R_1 - R_2) \;} \quad . \qquad (6\text{-}94)$$

When Eq. (6-94) is satisfied, the transverse chromatic aberration of the lens for an object at infinity is zero [see Eq. (5-312)]. Comparing Eqs. (6-85) and (6-94), we find that the achromatic lens for $n = 1.5$ is 1.8 times as thick as the one with chromatic aberration. The achromatic lens is no longer concentric; the spacing between the centers of curvature of its surfaces is given by

$$\Delta z = t - \left(R_1 - R_2 \right)$$

$$= t/n^2 \ , \tag{6-95}$$

so that the center of curvature of the second surface lies closer to the mirror. It is evident that the smaller the value of t, the more concentric the two surfaces are. Substituting for $R_1 - R_2$, in terms of t according to Eq. (6-94) into Eq. (6-93), we obtain the focal length of the *achromatic meniscus corrector*:

$$f_l' = \frac{n^2}{(n-1)^2} \frac{R_1 R_2}{t} \ . \tag{6-96a}$$

Or, substituting for t, from Eq. (6-94) we may write

$$\frac{1}{f_l'} = \frac{n-1}{n+1} \left(\frac{1}{R_1} - \frac{1}{R_2} \right) \ , \tag{6-96b}$$

showing that the focal length of the thick-lens achromatic corrector is longer by a factor of $n + 1$ compared with that of a corresponding thin lens.

6.7 BEAM EXPANDER

6.7.1   Introduction

As an example of a two-mirror system, we consider the aberrations of a beam expander consisting of two confocal paraboloidal mirrors, M_1 and M_2, with the aperture stop located at M_2 as illustrated in Figure 6-16. Since the two mirrors have a common focus, a collimated beam incident on the system is focused by the first mirror and recollimated by the second; hence the term "beam expander" (or "reducer" when used in reverse). We show that the expanded beam is free of spherical aberration, coma, and astigmatism, thus making the beam expander an anastigmat. It does have some field curvature and distortion. A beam expander such as that shown in Figure 6-16 is an afocal telescope called a *Mersenne telescope*.

6.7.2   Gaussian Parameters

Since the aperture stop is located at the concave secondary mirror M_2, its image by the convex primary mirror M_1 is the entrance pupil EnP of the system. Accordingly, the exit pupil ExP_1 for M_1 (which would be the image of $EnP \equiv EnP_1$ by M_1) is located at M_2. Moreover, the exit pupils ExP of the system and ExP_2 for M_2 are also located at M_2. Thus, the exit pupils for the two mirrors as well as for the system are located at M_2. A collimated beam incident at an angle β from the optical axis is made divergent

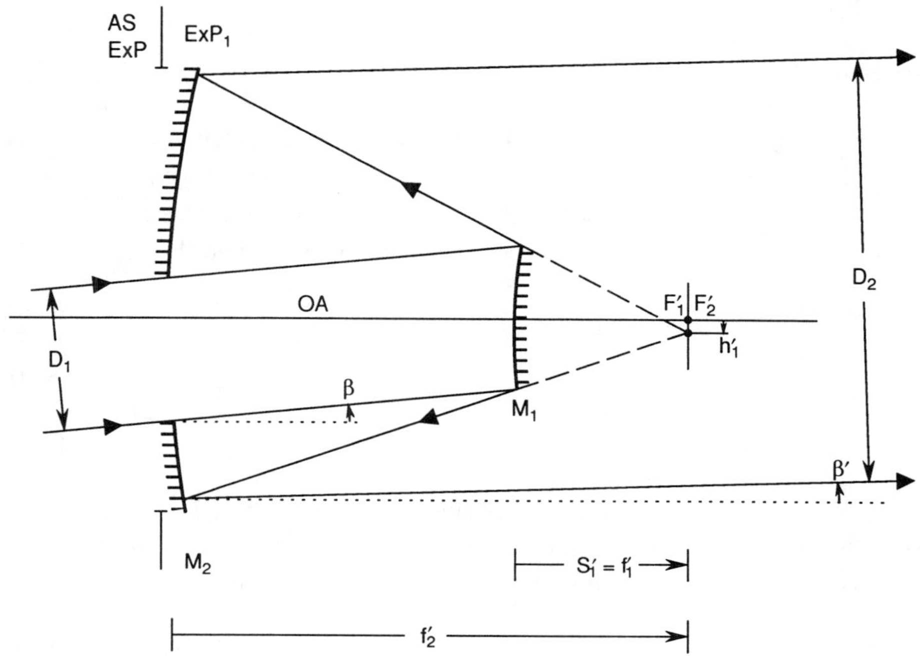

Figure 6-16. Schematic of a beam expander system consisting of two confocal para-boloidal mirrors M_1 and M_2 with their focal points F_1' and F_2'. The aperture stop is located at M_2. Its image by M_1 is the entrance pupil (not shown) of the system. The dotted lines shown are parallel to the optical axis OA.

by the convex mirror M_1 with a virtual focus on the common focal plane such that

$$S_1' = f_1' \tag{6-97}$$

and

$$h_1' = -\beta f_1' \ , \tag{6-98}$$

where $f_1' = R_1/2$ is its focal length. Since a collimated beam is equivalent to a point object at infinity, the paraboloidal mirror M_1 does not introduce any spherical aberration.

Since the exit pupil for M_1 lies at the concave mirror M_2, we note that the distance of the image formed by M_1 from its exit pupil is

$$L_1 = f_2' \ , \tag{6-99}$$

where f_2' is the focal length of mirror M_2. According to Eqs. (6-19) and (6-22), the parameters d_1 and g_1 are given by

$$d_1 = -\left(f_1' + f_2'\right)/f_1' \tag{6-100}$$

and

$$g_1 = \left(f_1' - f_2'\right)/f_2' \quad, \tag{6-101}$$

respectively.

The divergent beam (produced by M_1) incident on the secondary mirror is equivalent to a point object lying in its focal plane, i.e., $S_2 = f_2'$. Hence, the image formed by it lies at infinity $\left(S_2' = \infty\right)$; i.e., the beam exiting from the beam expander is also collimated. The direction of the exit beam with respect to the optical axis is given by

$$\beta' = -h_1'/f_2' \tag{6-102a}$$

$$= \beta\left(f_1'/f_2'\right) \tag{6-102b}$$

$$= \beta/M \quad, \tag{6-102c}$$

where $M = \beta'/\beta = f_2'/f_1' = D_2/D_1$ is the angular demagnification as well as the beam expansion ratio of the beam expander. Moreover, since the exit pupil for M_2 is coincident with it, $L_2 = S_2'$. Accordingly, Eqs. (6-9), (6-19), and (6-22) yield

$$h_2'/S_2' = -h_1'/f_2' \quad, \tag{6-103}$$

$$d_2 \to R_2/S_2' = 2f_2'/S_2' \quad,$$

or

$$d_2 h_2' = -2h_1' \quad, \tag{6-104}$$

and

$$g_2 = 0 \quad, \tag{6-105}$$

respectively.

6.7.3 Aberration Contributed by Primary Mirror

Since the rays from an object lying at infinity are incident on the paraboloidal mirror from left to right, $n_1 = 1$, and Eqs. (6-12) and (6-21) yield

$$a_{s1} = 1/32 f_1'^3 \tag{6-106a}$$

$$= -\sigma_1 \quad. \tag{6-106b}$$

Hence, we find from Eq. (6-20) that for the primary mirror M_1,

$$a_{ss1} = f_1'/32 f_2'^4 \quad. \tag{6-107}$$

Substituting for the various quantities in Eqs. (6-13) through (6-18), we obtain the primary aberration function introduced by M_1

$$W_1(r,\theta;h_1') = a_{cc1}h_1'r^3\cos\theta + a_{ac1}h_1'^2r^2\cos^2\theta + a_{dc1}h_1'^2r^2 + a_{tc1}h_1'^3r\cos\theta \quad , \quad (6\text{-}108)$$

where

$$a_{cc1} = -1/4f_2'^3 \quad , \tag{6-109a}$$

$$a_{ac1} = 1/2f_2'^3 \quad , \tag{6-109b}$$

$$a_{dc1} = \left(f_2'^{-1} - f_1'^{-1}\right)/4f_2'^2 \quad , \tag{6-109c}$$

$$a_{tc1} = \left(\frac{1}{4f_1'^2 f_2'} - \frac{3}{4f_2'^3} + \frac{1}{2f_1'f_2'^2}\right) \quad , \tag{6-109d}$$

and (r,θ) are the coordinates of a point in the plane of the exit pupil located at M_2. We note that the spherical aberration $a_{sc1} = 0$, as expected for a paraboloidal mirror imaging an object lying at infinity.

6.7.4   Aberration Contributed by Secondary Mirror

Since the rays are incident on the paraboloidal secondary mirror M_2 from right to left, therefore, $n_2 = -1$. Moreover, $L_2 = S_2'$. Hence, the quantity a_{ss} for an image formed at infinity is given by

$$a_{ss2} = a_{s2} \tag{6-110a}$$

$$= -1/32\,f_2'^3 \tag{6-110b}$$

$$= -\sigma_2 \quad . \tag{6-110c}$$

Substituting for the various quantities in Eqs. (6-13) through (6-18), we obtain the primary aberration function introduced by M_2:

$$W_2(r,\theta;h_2') = a_{cc2}h_2'r^3\cos\theta + a_{ac2}h_2'^2r^2\cos^2\theta$$

$$= 4d_2 a_{ss2}h_2'r^3\cos\theta + 4d_2^2 a_{ss2}h_2'^2r^2\cos^2\theta \quad .$$

Or, writing it as a function of h_1' by using Eq. (6-104),

$$W_2(r,\theta;h_1') = \frac{1}{4f_2'^3}\,h_1'r^3\cos\theta - \frac{1}{2f_2'^3}\,h_1'^2r^2\cos^2\theta \quad . \tag{6-111}$$

Thus, as pointed out in Section 6.5 [see Eq. (6-65)] for a paraboloidal mirror with a collocated aperture stop and image (instead of the object) lying at infinity, the only nonzero aberrations contributed by M_2 are coma and astigmatism. These two aberrations are equal in magnitude but opposite in sign to the corresponding aberrations contributed by the primary mirror.

6.7.5   System Aberration

Adding the aberration contributions of the two mirrors given by Eqs. (6-108) and (6-111) and substituting for h_1' from Eq. (6-102a), we obtain the primary aberration function of the beam expander system as a function of the exit angle β' of the beam:

$$W(r,\theta;\beta') = W_1(r,\theta;\beta') + W_2(r,\theta;\beta') \quad ,$$

or

$$W(r,\theta;\beta') = \frac{1}{4}\left[\beta'^2\left(\frac{1}{f_2'} - \frac{1}{f_1'}\right)r^2 - \beta'^3\left(M^2 + 2M - 3\right)r\cos\theta\right] \quad . \tag{6-112}$$

Thus, we find that the beam expander is an anastigmatic system; i.e., it does not introduce spherical aberration, coma, or astigmatism. The only aberrations introduced by it are the field curvature and distortion terms of mirror M_1. The field curvature term represents an aberration due to the Petzval curvature of the system, as may be seen from Eqs. (6-25), (6-112), and (5-99). The consequence of the nonzero field curvature is that the output beam is actually not quite collimated, but it is focused at a distance of $2/\beta'^2\left(f_2'^{-1} - f_1'^{-1}\right)$ from the exit pupil of the system, as may be seen by comparing the field curvature aberration with the sag of a spherical surface. Similarly, following Eq. (3-21), the consequence of the nonzero distortion is that the output beam makes an angle of $\beta' - \beta'^3\left(M^2 + 2M - 3\right)/4$ with the optical axis.

6.8 TWO-MIRROR ASTRONOMICAL TELESCOPES

6.8.1   Introduction

Although the Schmidt and Bouwers-Maksutov cameras provide good imagery over a wide field of view, they are not suitable when the specrtral band is wide due to the chromatic aberrations of the corrector plate. Moreover, for large telescopes, the fabrication of a correspondingly large corrector plate may become impractical. The next obvious step is to consider telescopes with two mirrors such that the aberrations of one mirror are canceled by those of the other. In this section, we consider astronomical imaging by a reflecting system consisting of two conic mirrors. We start with a discussion of the Gaussian imaging relations, determine the aberration contribution of each of the two mirrors, and appropriately add them to obtain expressions for the system aberrations. The general results thus obtained are applied to discuss the aberrations of classical telescopes such as the Cassegrain, Gregorian, Ritchey-Chrétien, and Schwarzschild. Reflecting telescopes have the advantage that the images formed by them do not suffer from chromatic aberrations. However, since one mirror obscures a portion of the other, the beam of light that forms the final image is *annular,* resulting in a decrease in the amount of light in the image.

The strict definition of a telescope is an optical system that is afocal, i.e., for an object lying at infinity on one side, it forms the image at infinity on the other side. This definition held as long as the image was observed by humans, since the eye is most relaxed when it sees an object lying at infinity. However, with the advent of photographic film and lately solid-state optical detectors, the definition of a telescope has evolved to a system that is focal so that the image is formed at a finite distance on the film or a detector array often called the focal plane array.

There are many papers written on the subject of two-mirror telescopes. Some of these are listed under the references.[8-12] Also listed is a paper on *unobscured two-mirror systems* in which the beam of light forming the final image is circular.[13] However, such systems do not have an axis of rotational symmetry and are more difficult to fabricate and assemble. Although our discussion is limited to two-mirror systems, it is easier to design systems with three mirrors that have zero primary aberrations.[14]

6.8.2   Gaussian Parameters

Consider, as indicated in Figure 6-17, a two-mirror astronomical telescope imaging an object lying at infinity. Tha axial image formation is illustrated by the solid-line rays and the off-axis by the dashed-line rays. We assume that the aperture stop of the system lies at its primary mirror M_1. Consequently, the entrance pupil *EnP* of the system and the entrance and exit pupils EnP_1 and ExP_1 for M_1 also lie at M_1. ExP_1 is also the entrance pupil EnP_2 for the secondary mirror M_2. Its image by the secondary mirror M_2 is the exit pupil ExP_2 for M_2 as well as the exit pupil *ExP* of the system. The reason for choosing the aperture stop location at the primary mirror is that generally this mirror is the larger element and hence is more difficult to fabricate. For a finite (or nonzero) field of view, this mirror would have to be oversized if the aperture stop were not placed at it. The Gaussian parameters of the system will be determined by using Eqs. (6-8) and (6-11) through (6-22). Its aberrations will be determined by calculating the aberrations of its two mirrors and adding them according to the procedure outlined in Section 5.7.1.

Let R_1 and R_2 be the vertex radii of curvature of the primary and secondary mirrors M_1 and M_2, respectively. Their corresponding focal lengths are given by $f_1' = R_1/2$ and $f_2' = R_2/2$. The mirrors are rotationally symmetric conics of eccentricity e_1 and e_2. They are coaxial so that the system is rotationally symmetric about the optical axis that passes through their vertices. Let the vertex-to-vertex spacing from mirror M_1 to mirror M_2 be (a numerically negative quantity) t.

Applying Eq. (6-8) to the primary mirror M_1, we note that for an object at infinity, $S_1 = -\infty$ and, therefore, the image is formed at its focus F_1', called the *prime focus*, at a (numerically negative) distance $S_1' = f_1'$ from M_1. This image is the object for the secondary mirror M_2 and lies at a distance

$$S_2 = f_1' - t \tag{6-113}$$

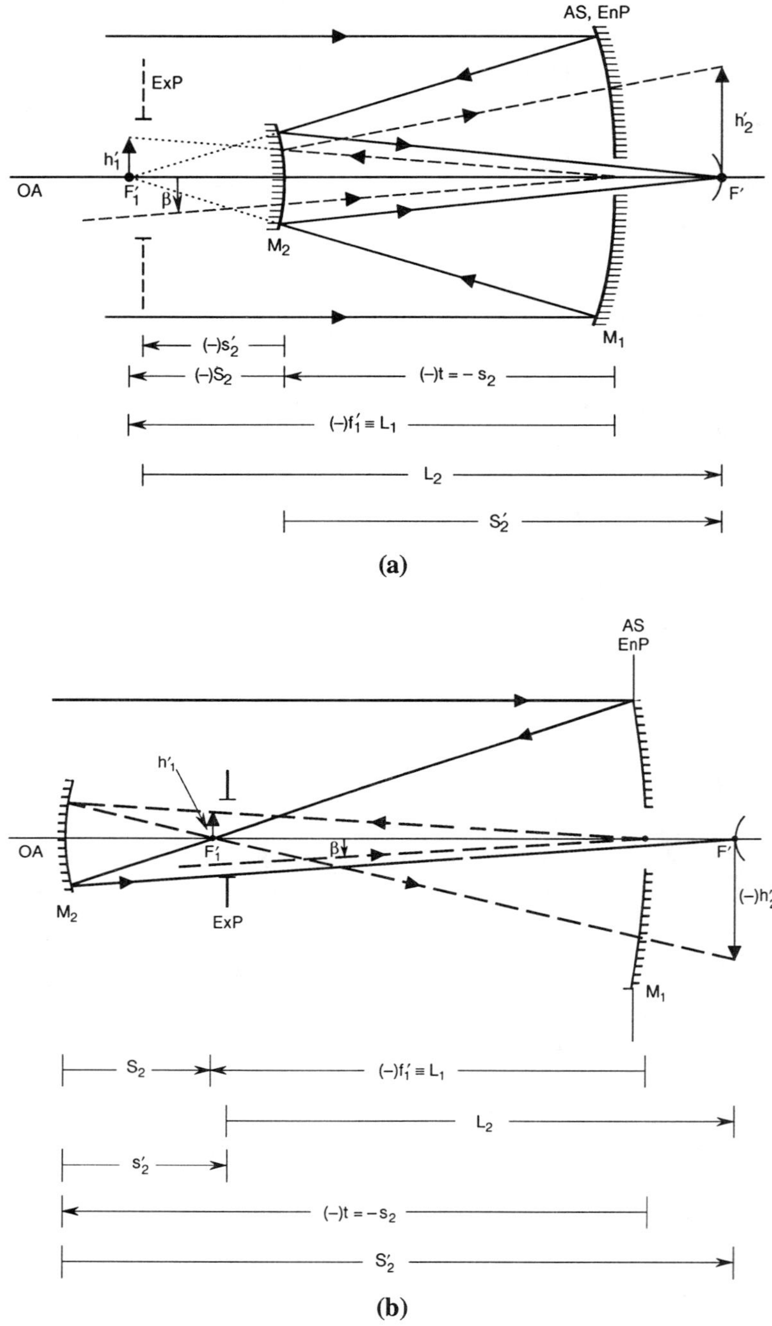

Figure 6-17. Astronomical telescope consisting of two conic mirrors M_1 and M_2. (a) Cassegrain and (b) Gregorian forms are shown in this figure. The aperture stop of the telescope lies at its primary mirror M_1. The spherical image surface passing through the focal point F' is an illustration of the Petzval image surface. The axial image formation is illustrated by the solid line rays and off-axis by the dashed line rays.

from it. In Figure 6-17a, F_1' lies inside the focus of the secondary mirror, i.e., $\left|S_2\right|<\left|f_2'\right|$, but in Figure 6-17b, it lies outside, i.e., $S_2 > f_2'$. In both cases, a real image is formed by M_2 that lies at the telescope (or Cassegrain) focus F' at a distance S_2' given by

$$\frac{1}{S_2'} = \frac{1}{f_2'} - \frac{1}{S_2} \tag{6-114a}$$

$$= \left(f_1'-t-f_2'\right)/f_2'\left(f_1'-t\right) \tag{6-114b}$$

$$= f_1'/f'\left(f_1'-t\right) \tag{6-114c}$$

$$= f_1'/f_2'\left(f'+f_1'\right) \ , \tag{6-114d}$$

where

$$\boxed{f' = \frac{f_1'f_2'}{f_1'-f_2'-t}} \tag{6-115}$$

is the focal length of the system [see Eq. (1-142)]. S_2' locates the image formed by the system and $S_2'+t$ gives the distance of the image from the primary mirror, called the *working distance*.

For an object lying at infinity at an angle β from the optical axis of the system, the height h_1' of its image formed by M_1 is given by

$$h_1' = -\beta f_1' \ . \tag{6-116}$$

The image formed by M_1 is the object for M_2. The height h_2' of the final image formed by M_2 (and, therefore, by the system) is given by

$$M_2 = h_2'/h_1' \tag{6-117a}$$

$$= -S_2'/S_2 \ . \tag{6-117b}$$

Substituting for S_2 and S_2' from from Eqs. (6-113) and (6-114c), respectively, we obtain

$$M_2 = -f'/f_1' \ . \tag{6-117c}$$

The magnification M_2 of the image formed by the secondary mirror is called the *secondary magnification*. From Eqs. (6-116) and (6-117), we obtain

$$\boxed{h_2' = \beta f'} \ . \tag{6-118}$$

The exit pupil *ExP* of the system is the image of the entrance pupil EnP_2 or *EnP* by M_2. Its location (using small letters) is given by

$$\frac{1}{s_2'}+\frac{1}{s_2} = \frac{1}{f_2'} \ , \tag{6-119}$$

where $s_2 = -t$. Thus, the exit pupil is located at a distance

$$s_2' = tf_2'/(t + f_2') \qquad (6\text{-}120)$$

from M_2. Since f_2' is numerically negative in Figure 6-17a, so is s_2' in this figure. The magnification of the exit pupil is given by

$$m_2 = -s_2'/s_2$$

$$= s_2'/t \qquad (6\text{-}121\text{a})$$

$$= f_2'/(t + f_2') \ . \qquad (6\text{-}121\text{b})$$

It is numerically positive in Figure 6-17a, but negative in Figure 6-17b. Substituting

$$f_2' = f'(f_1' - t)/(f' + f_1') \qquad (6\text{-}122)$$

from Eqs. (4-114c) and (4-114d), we may write

$$\boxed{m_2 = f'(f_1' - t)/f_1'(f' + t) \ .} \qquad (6\text{-}123)$$

The diameter of the exit pupil is given by

$$D_{ex} = |m_2| D_1 \ , \qquad (6\text{-}124)$$

where D_1 is the diameter of the primary mirror, since it is also the entrance pupil EnP.

The focal ratio (or the f-number) of the image-forming light cone is equal to the radius of curvature of the reference sphere (with respect to which the aberration of the system is defined) divided by the diameter of its exit pupil. The reference sphere passes through the center of the exit pupil and it is centered at the Gaussian image of the point object for which the aberration is under consideration. Thus, the radius of curvature of the reference sphere is the distance of the final image from the exit pupil. In our notation, this distance is L_2 and is given by

$$L_2 = S_2' - s_2' \ . \qquad (6\text{-}125)$$

From Eqs. (6-114c) and (6-123), we obtain

$$S_2' = m_2(f' + t) \ . \qquad (6\text{-}126\text{a})$$

Also

$$s_2' = m_2 t \qquad (6\text{-}126\text{b})$$

by the definition of m_2. Hence, Eq. (6-125) may be written

$$L_2 = m_2 f' \ . \qquad (6\text{-}127)$$

L_2 is numerically positive in Figure 6-17 since the final image lies to the right of the exit pupil ExP. The focal ratio of the image-forming light cone is given by

$$\boxed{F \;=\; L_2/D_{ex}} \tag{6-128a}$$

or

$$\boxed{F \;=\; |f'|/D_1 \;\;.} \tag{6-128b}$$

Since the object is at infinity, we expect the focal ratio of the image-forming light cone to be equal to that of the system, as discussed in Section 2.6.3. This may also be seen by determining the principal point as in Figure 1-42.

To obtain the aberrations contributed by the two mirrors, we need their parameters d and g corresponding to Eqs. (6-19) and (6-22). For the primary mirror, since the aperture stop of the system is collocated with it, the exit pupil for it is also located there. Hence, the distance of the image formed by it from its exit pupil is given by

$$L_1 \;=\; S_1' \;=\; f_1' \;=\; R_1/2 \;\;, \tag{6-129}$$

so that

$$d_1 \;=\; \left(R_1 - S_1' + L_1\right)\big/\left(S_1' - R_1\right) \;=\; -2 \tag{6-130}$$

and

$$g_1 \;=\; \left(S_1' - L_1\right)\big/L_1 \;=\; 0 \;\;. \tag{6-131}$$

Note that L_1 in Figure 6-17 is numerically negative since the image formed by mirror M_1 lies to the left of its exit pupil (which is at the mirror). The exit pupil for the primary mirror is also the entrance pupil for the secondary mirror. The exit pupil for the secondary mirror is the image of its entrance pupil formed by it. This is also the exit pupil of the system and the final image lies at a distance L_2 from it. Hence, using Eqs. (6-121), (6-122), (6-125), and (6-126), we obtain

$$
\begin{aligned}
d_2 &= \left(R_2 - S_2' + L_2\right)\big/\left(S_2' - R_2\right) \\[2ex]
&= \frac{2f_2' - s_2'}{S_2' - 2f_2'} \\[2ex]
&= \frac{2 - t/\left(t + f_2'\right)}{\left[\left(f' + f_1'\right)/f_1'\right] - 2} \\[2ex]
&= \frac{\left[2f'f_1'/\left(f' - f_1'\right)\right] - t}{f' + t}
\end{aligned}
\tag{6-132}
$$

and

$$g_2 = \left(S_2' - L_2\right)/L_2$$
$$= s_2'/L_2$$
$$= t/f' \quad . \tag{6-133}$$

6.8.3   Petzval Surface

As discussed in Section 6.3, the radius of curvature of the *Petzval surface* of a two-mirror system is given by Eq. (6-25), i.e.,

$$\boxed{R_p = f_1'f_2'/\left(f_1' - f_2'\right)} \quad . \tag{6-134}$$

This surface is shown passing through F' in Figure 6-17. As pointed out later in Section 6.8.7, it is curved toward the primary mirror in the case of a Cassegrain telescope, and away from it in the case of a Gregorian telescope.

Now we are ready to determine the aberration function of the system. Using Eq. (6-13), we first determine the aberration contributed by the primary mirror, then by the secondary mirror, and finally combine the two contributions to obtain the system aberration. The aberrations are written in terms of the focal lengths f_1' of the primary mirror and f' of the system.

6.8.4   Aberration Contributed by Primary Mirror

The aberration contributed by the primary conic mirror at a point $\left(r_1, \theta_1\right)$ on its exit pupil ExP_1 (i.e., in its own plane) is given by

$$W_{c1}(r_1, \theta_1; h_1') = a_{sc1}r_1^4 + a_{cc1}h_1'r_1^3\cos\theta_1 + a_{ac1}h_1'^2r_1^2\cos^2\theta_1$$
$$+ a_{dc1}h_1'^2r_1^2 + a_{tc1}h_1'^3r_1\cos\theta_1 \quad . \tag{6-135}$$

The coefficient of spherical aberration is given by

$$a_{sc1} = \left(S_1'/L_1\right)^4\left(a_{s1} + \sigma_1\right) \quad , \tag{6-136}$$

where, letting $n_1 = 1$ in Eqs. (6-12) and (6-21),

$$a_{s1} = \frac{1}{4R_1}\left(\frac{1}{R_1} - \frac{1}{S_1'}\right)^2$$
$$= 1/32\,f_1'^3 \tag{6-137}$$

and

$$\sigma_1 = -e_1^2/32f_1'^3 \quad . \tag{6-138}$$

Since $S_1' = L_1 = f_1'$, we find that

$$a_{sc1} = \left(1 - e_1^2\right)/32f_1'^3 \tag{6-139}$$

and

$$a_{ss1} = \left(S_1'/L_1\right)^4 a_{s1}$$

$$= 1/32\, f_1'^3 \quad . \tag{6-140}$$

Since $g_1 = 0$, the aberration coefficients of the conic primary mirror are the same as those for a corresponding spherical mirror. Thus, the coma coefficient is given by

$$\begin{aligned} a_{cc1} &= a_{cs1} \\ &= 4d_1 a_{ss1} \\ &= -1/4\, f_1'^3 \quad . \end{aligned} \tag{6-141}$$

Similarly, the coefficient of astigmatism is given by

$$\begin{aligned} a_{ac1} &= a_{as1} \\ &= 4d_1^2 a_{ss1} \\ &= 1/2\, f_1'^3 \quad . \end{aligned} \tag{6-142}$$

The primary mirror does not contribute any field curvature or distortion since

$$\begin{aligned} a_{dc1} &= a_{ds1} \\ &= \frac{1}{2}\left(a_{ac1} - \frac{1}{R_1 L_1^2}\right) \\ &= 0 \end{aligned} \tag{6-143}$$

and

$$\begin{aligned} a_{tc1} &= a_{ts1} \\ &= 4\left(d_1^3 a_{ss1} - \frac{d}{4R_1 L_1^2}\right) \\ &= 0 \quad . \end{aligned} \tag{6-144}$$

Substituting Eqs. (6-139), and (6-141) through (6-144) into Eq. (6-135), the aberration contributed by the primary mirror at a point (r_1, θ_1) in its plane may be written

$$W_{c1}(r_1,\theta_1;h_1') = \frac{1-e_1^2}{32 f_1'^3} r_1^4 - \frac{h_1'}{4 f_1'^3} r_1^3 \cos\theta_1 + \frac{h_1'^2}{2 f_1'^3} r_1^2 \cos^2\theta_1 \quad . \tag{6-145}$$

This aberration is at a point on the entrance pupil EnP of the system. The aberration at a point (r_2, θ_2) on the exit pupil ExP of the system may be obtained by letting

$$(x_1, y_1) = \left(x_2/m_2,\, y_2/m_2\right) \tag{6-146}$$

or

$$r_1 = r_2 / |m_2| \tag{6-147}$$

and

$$r_1 \cos\theta_1 = r_2 \cos\theta_2 / m_2 \quad , \tag{6-148}$$

where, in general

$$(x, y) = r(\cos\theta, \sin\theta) \quad . \tag{6-149}$$

Also, from Eqs. (6-117a) and (6-117c), h_1' may be written in terms of h_2'. Thus, Eq. (6-145) for the aberration contributed by the primary mirror when referred to the exit pupil ExP and in terms of the image height h_2' may be written

$$W_{c1}(r_2,\theta_2;h_2') = \frac{1-e_1^2}{32 m_2^4 f_1'^3} r_2^4 + \frac{h_2'}{4 m_2^3 f' f_1'^2} r_2^3 \cos\theta_2 + \frac{h_2'^2}{2 m_2^2 f'^2 f_1'} r_2^2 \cos^2\theta_2 \quad . \tag{6-150}$$

6.8.5 Aberration Contributed by Secondary Mirror

The aberration contributed by the secondary conic mirror at a point (r_2,θ_2) in the plane of the exit pupil ExP of the system may be written

$$W_{c2}(r_2,\theta_2;h_2') = a_{sc2} r_2^4 + a_{cc2} h_2' r_2^3 \cos\theta_2 + a_{ac2} h_2'^2 r_2^2 \cos^2\theta_2$$

$$+ a_{dc2} h_2'^2 r_2^2 + a_{tc2} h_2'^3 r_2 \cos\theta_2 \quad . \tag{6-151}$$

The coefficient of spherical aberration is given by

$$a_{sc2} = \left(S_2'/L_2 \right)^4 (a_{s2} + \sigma_2) \quad , \tag{6-152}$$

where, keeping in mind that the refractive index associated with the rays incident on the secondary mirror is $n_2 = -1$,

$$\left(S_2'/L_2 \right) = (f'+t)/f' \quad , \tag{6-153}$$

$$a_{s2} = -\frac{1}{4R_2}\left(\frac{1}{R_2} - \frac{1}{S_2'} \right)^2$$

$$= -\frac{1}{32 f_2'^3}\left(\frac{f'-f_1'}{f'+f_1'} \right)^2 \tag{6-154}$$

and

$$\sigma_2 = e_2^2 / 32 f_2'^3 \quad . \tag{6-155}$$

Hence

$$a_{sc2} = \left(\frac{f'+t}{f'}\right)^4 \frac{1}{32 f_2'^3}\left[e_2^2 - \left(\frac{f'-f_1'}{f'+f_1'}\right)^2\right] \ . \tag{6-156}$$

Substituting for f_2' in terms of f' and f_1' according to Eq. (6-122), Eq. (6-156) can be written

$$a_{sc2} = \frac{\left(f'+f_1'\right)^3\left(f_1'-t\right)}{32 m_2^4 f'^3 f_1'^4}\left[e_2^2 - \left(\frac{f'-f_1'}{f'+f_1'}\right)^2\right] \ . \tag{6-157}$$

Now

$$\begin{aligned}
a_{ss2} &= \left(S_2'/L_2\right)^4 a_{s2} \\[2mm]
&= \left(\frac{f'+t}{f'}\right)^4 \frac{-1}{32 f_2'^3}\left(\frac{f'-f_1'}{f'+f_1'}\right)^2 \\[2mm]
&= -\frac{f'+t}{32 m_2^3 f'^4 f_1'^3}\left(f'+f_1'\right)\left(f'-f_1'\right)^2 \tag{6-158a} \\[2mm]
&= -\frac{\left(f'+t\right)^2}{32 m_2^2 f'^5 f_1'^2 \left(f_1'-t\right)}\left(f'+f_1'\right)\left(f'-f_1'\right)^2 \ . \tag{6-158b}
\end{aligned}$$

Therefore, the coma coefficient is given by

$$\begin{aligned}
a_{cc2} &= 4\left[d_2 a_{ss2} - \sigma_2 g_2\left(S_2'/L_2\right)^3\right] \\[2mm]
&= -\frac{1}{4 m_2^3 f'}\left\{\frac{1}{f_1'^2} - \frac{1}{f'^2} + \frac{t\left(f'+f_1'\right)^3}{2 f'^3 f_1'^3}\left[e_2^2 - \left(\frac{f'-f_1'}{f'+f_1'}\right)^2\right]\right\} \ . \tag{6-159}
\end{aligned}$$

The coefficient of astigmatism is given by

$$\begin{aligned}
a_{ac2} &= 4\left[d_2^2 a_{ss2} + \sigma_2 g_2^2\left(S_2'/L_2\right)^2\right] \\[2mm]
&= -\frac{f'+f_1'}{8 m_2^2 f'^5 f_1'^2 \left(f_1'-t\right)}\left\{\left[t\left(f'-f_1'\right)-2 f' f_1'\right]^2 - e_2^2 t^2\left(f'+f_1'\right)^2\right\} \ . \tag{6-160}
\end{aligned}$$

The field curvature coefficient is given by

$$\begin{aligned}
a_{dc2} &= \frac{1}{2}\left(a_{ac2} + \frac{1}{R_2 L_2^2}\right) \\[2mm]
&= \frac{1}{2}\left(a_{ac2} + \frac{1}{2 m_2^2 f'^2 f_2'}\right) \ . \tag{6-161}
\end{aligned}$$

The distortion coefficient is given by

$$a_{tc2} = 4\left(d_2^3 a_{ss2} + \frac{d_2}{4R_2 L_2^2} - \sigma_2 g_2^3 \frac{S_2'}{L_2} \right) \ .$$
(6-162)

6.8.6 System Aberration

Substituting Eqs. (6-157), (6-159), and (6-163) to (6-165) into Eq. (6-151) and combining the result obtained with Eq. (6-150), we obtain the *primary aberration function of the conic system*

$$W_{cs}(r_2,\theta_2;h_2') = W_{c1}(r_2,\theta_2;h_2') + W_{c2}(r_2,\theta_2;h_2') \ ,$$

or

$$\boxed{ \begin{aligned} W_{cs}(r_2,\theta_2;h_2') &= a_{scs}r_2^{\ 4} + a_{ccs}h_2'r_2^3 \cos\theta_2 + a_{acs}h_2'^2 r_2^2 \cos^2\theta_2 \\ &\quad + a_{dcs}h_2'^2 r_2^2 + a_{tcs}h_2'^3 r_2 \cos\theta_2 \ , \end{aligned} }$$
(6-163)

where

$$a_{scs} = \frac{1}{32m_2^4 f_1'^3}\left\{ 1 - e_1^2 + \frac{(f_1'-t)(f'+f_1')^3}{f'^3 f_1'}\left[e_2^2 - \left(\frac{f'-f_1'}{f'+f_1'}\right)^2 \right] \right\} \ ,$$
(6-164)

$$a_{ccs} = \frac{1}{4m_2^3 f'^3}\left\{ 1 - \frac{t(f'+f_1')^3}{2f'f_1'^3}\left[e_2^2 - \left(\frac{f'-f_1'}{f'+f_1'}\right)^2 \right] \right\} \ ,$$
(6-165)

$$a_{acs} = \frac{1}{2m_2^2 f'^4 (f_1'-t)}\left\{ -f_1'(f'+t) + \frac{t^2(f'+f_1')^3}{4f'f_1'^2}\left[e_2^2 - \left(\frac{f'-f_1'}{f'+f_1'}\right)^2 \right] \right\} \ ,$$
(6-166)

$$\begin{aligned} a_{dcs} &= a_{dc2} \\ &= \frac{1}{2}\left(a_{ac2} + \frac{1}{2m_2^2 f_2' f'^2} \right) \\ &= \frac{1}{2}a_{acs} + \frac{1}{4m_2^2 f'^2}\left(\frac{1}{f_2'} - \frac{1}{f_1'} \right) \ . \end{aligned}$$
(6-167)

and

$$a_{tcs} = a_{tc2} \ .$$
(6-168)

We note that only spherical aberration of the conic system depends on the eccentricities (e_1 and e_2) of both mirrors. Its coma and astigmatism depend on the eccentricity e_2 of the secondary mirror only. This is because the aperture stop of the system is at the primary mirror and, therefore, its eccentricity does not affect its coma and astigmatism.

Letting $h_2' = \beta f'$, we may also write the aberration function in terms of the field angle β of the point object under consideration. By using Eqs. (6-146) to (6-148), the aberration function in the plane of the entrance pupil (i.e., in the plane of the primary mirror) can also be obtained.

Now we consider some special cases of a two-mirror system.

6.8.7 Classical Cassegrain and Gregorian Telescopes

The classical Cassegrain and Gregorian telescopes consist of a concave paraboloidal primary mirror. Thus, the focal length of the mirror $f_1' < 0$ and its eccentricity is

$$e_1 = 1 \ . \tag{6-169}$$

The image of an axial point object at infinity formed by this mirror is aberration free and lies at a distance f_1' from it. This aberration-free image is the object for the secondary mirror. According to Eq. (6-157) or (6-164), the image formed by the secondary mirror is also aberration free, provided its eccentricity is given by

$$e_2 = \left(f' - f_1'\right)/\left(f' + f_1'\right) \tag{6-170a}$$

$$= \left(M_2 + 1\right)/\left(M_2 - 1\right) \ . \tag{6-170b}$$

Thus, the spherical aberration of the telescope is zero when the eccentricities of its mirrors are given by Eqs. (6-169) and (6-170). Now, $e_2 > 1$ corresponds to a hyperboloidal secondary mirror if the telescope focal length f' is positive, and $e_2 < 1$ corresponds to an ellipsoidal mirror if it is negative. A telescope consisting of a paraboloidal primary mirror and a hyperboloidal secondary mirror is called a *Cassegrain telescope*. Similarly, a telescope consisting of a paraboloidal primary mirror and an ellipsoidal secondary mirror is called a *Gregorian telescope*. The point at which the primary mirror forms the image and the point at which the secondary mirror forms the final image are the two geometrical foci of the hyperboloidal surface in the case of a Cassegrain telescope and of the ellipsoidal surface in the case of a Gregorian. These two points lie on opposite sides of the secondary mirror in the case of a Cassegrain telescope, but on the same side in the case of a Gregorian. In the Cassegrain telescope, the secondary mirror is convex $\left(f_2' < 0\right)$ to the light incident on it while it is concave $\left(f_2' > 0\right)$ in the Gregorian. These facts may be seen from the expressions for the object and image distances S_2 and S_2', respectively, for the secondary mirror, i.e.,

$$S_2 = f_1' - t$$
$$= f_2'\left(f' + f_1'\right)/f' \tag{6-171a}$$

$$= R_2/\left(1 + e_2\right) \tag{6-171b}$$

and

$$S_2' = f_2'\left(f' + f_1'\right)/f_1' \tag{6-172a}$$

$$= R_2/\left(1 - e_2\right) \ . \tag{6-172b}$$

Equation (6-171a) has been obtained by using Eq. (6-122). It is clear from Eqs. (6-171b) and (6-172b) that S_2 and S_2' represent the conjugate foci of a conic mirror of eccentricity e_2 and vertex radius of curvature R_2.

The magnification $M_2 = -f'/f_1'$ of the image formed by the secondary mirror is positive in the case of a Cassegrain telescope and negative in the case of a Gregorian. Similarly, the pupil magnification m_2 is positive for the Cassegrain telescope and negative for the Gregorian. This may be seen from Eq. (6-121) in which f_2' is negative for a Cassegrain telescope, but positive for a Gregorian. Since $s_2' > 0$ for a Gregorian telescope, its exit pupil is real. The opposite is true for a Cassegrain; its exit pupil is virtual. The exit pupil lies behind (to the left of) the secondary mirror in a Cassegrain telescope, but between the primary and secondary mirrors in a Gregorian. The final image formed by both telescopes is real.

From Eq. (6-134), the radius of curvature R_p of the Petzval image surface is numerically negative for a Cassegrain telescope and positive for a Gregorian; i.e., for a Cassegrain telescope, the Petzval surface is concave (curved toward the primary mirror) and for a Gregorian it is convex (curved away from the primary mirror) to the rays incident on it. The Petzval surface can be flattened by using a planoconcave field lens in the case of a Cassegrain telescope and a planoconvex lens in the case of a Gregorian, with the planar surface facing the image surface as illustrated in Figure 6-18. The focal length and radius of curvature of the curved surface of the field-flattening lens are given by

$$f_l' = R_p/n \tag{6-173a}$$

and

$$R = R_p(n-1)/n \ , \tag{6-173b}$$

respectively. The signs of various quantities for the two types of telescopes are summarized in Table 6-1.

Substituting Eq. (6-170a) into Eqs. (6-165) and (6-166), we obtain the coma and astigmatism coefficients of the Cassegrain and Gregorian telescopes:

$$a_{ccs} = 1/4m_2^3 f'^3 \tag{6-174a}$$

and

$$a_{acs} = -f_1'(f'+t)/2m_2^2 f'^4(f_1'-t)$$
$$= -1/2m_2^3 f'^3 \ . \tag{6-174b}$$

Their field curvature and distortion are given by Eqs. (6-167) and (6-168), respectively.

Table 6-1. Signs of focal lengths, etc. for Cassegrain and Gregorian telescopes.

Quantity	Cassegrain	Gregorian
f_1	−	−
f_2	−	+
f	+	−
M_2	+	−
m_2	+	−
S_2'	+	+
s_2'	−	+
t	−	−
R_p	−	+

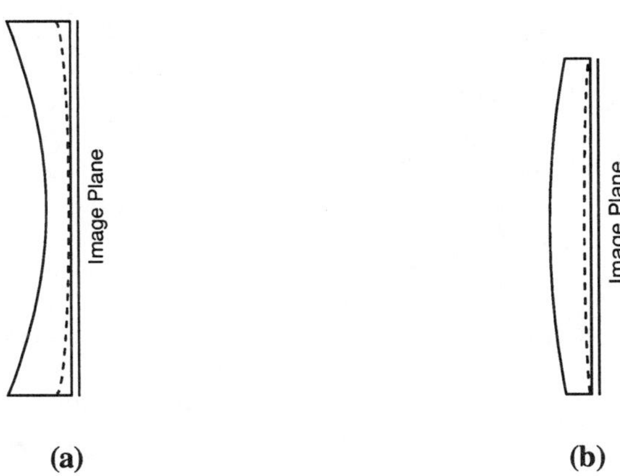

Figure 6-18. Field-flattening lens in (a) Cassegrain and (b) Gregorian telescopes. The Petzval image surface is indicated by the dashed line.

Substituting Eqs. (6-174a) and (6-174b) into Eq. (6-163), we can write the coma and astigmatism aberrations in the form

$$W_{c/a}(r_2,\theta_2;h_2') = \frac{1}{4m_2^3 f'^3} h_2' r_2^3 \cos\theta_2 - \frac{1}{2m_2^3 f'^3} h_2'^2 r_2^2 \cos^2\theta_2 \ . \tag{6-175}$$

Or, substituting Eqs. (6-118) and (6-148), we may write the aberrations at the entrance pupil *EnP*:

$$\boxed{W_{c/a}(r_1,\theta_1;\beta) = \frac{1}{4 f'^2} \beta r_1^3 \cos\theta_1 - \frac{1}{2m_2 f'} \beta^2 r_1^2 \cos^2\theta_1 \ .} \tag{6-176}$$

Comparing Eq. (6-176) with Eq. (6-65), we note that the coma of these two-mirror telescopes is the same as that of a paraboloidal mirror (discussed in Section 6.5) with the same diameter as the primary mirror and a focal length equal to that of the telescope.

However, their astigmatism is larger by a factor of $1/m_2$ (since $|m_2| < 1$ for these telescopes). The field of view of these telescopes is limited by their coma. If the paraboloidal mirror and the telescopes have the same focal ratio, then the angular sizes of the coma ray spots are the same for both, but the angular size of the astigmatism spot is larger by a factor of $1/|m_2|$ for the telescopes compared to that for the paraboloidal mirror.

6.8.8 Aplanatic Cassegrain and Gregorian Telescopes

In the aplanatic Cassegrain and Gregorian telescopes, the eccentricities of the two mirrors are such that spherical aberration and coma are both zero. According to Eqs. (6-164) and (6-165), the eccentricities for zero coefficients must be given by

$$e_1^2 = 1 + 2f_1'^2 \left(f_1' - t\right)/tf'^2 \tag{6-177}$$

and

$$e_2^2 = \left(\frac{f' - f_1'}{f' + f_1'}\right)^2 + \frac{2f'f_1'^3}{t\left(f' + f_1'\right)^3} \quad . \tag{6-178}$$

The corresponding astigmatism according to Eq. (6-166) is given by

$$a_{acs} = -f_1'(2f' + t)/4m_2^2 f'^4\left(f_1' - t\right) \quad , \tag{6-179}$$

which sets the limit on an acceptable field of view of the aplanatic telescopes.

Since $e_1 \neq 0$, the primary mirror does not form an aberration-free image of an on-axis point object. Moreover, since $f_1' - t < 0$ and $f' > |f_1'|$ in a Cassegrain telescope, both e_1 and e_2 are greater than one. Hence, in an aplanatic Cassegrain telescope, both mirrors are hyperboloidal. Such a telescope is called a *Ritchey–Chrétien telescope*. The *Hubble space telescope* is an example of such a telescope. In the aplanatic Gregorian telescope, $f_1' - t > 0$ and $f < 0$; both e_1 and e_2 are less than one and the mirrors are ellipsoidal.

6.8.9 Afocal Telescope

Consider an afocal telescope consisting of a pair of confocal mirrors; i.e., one for which

$$f' = \infty \quad , \tag{6-180}$$

or, following Eq. (6-115),

$$t = f_1' - f_2' \quad . \tag{6-181}$$

Equation (6-181) implies that the mirrors are confocal. Letting $f' = \infty$ in Eqs. (6-164) through (6-166) and using Eq. (6-118), we obtain

$$a_{scs} = \frac{1}{32\, m_2^4\, f_1'^3}\left[1 - e_1^2 + \frac{f_1' - t}{f_1'}\left(e_2^2 - 1\right)\right] \quad , \tag{6-182}$$

$$a_{ccs} h_2' = -\frac{\beta t}{8\, m_2^3\, f_1'^3}\left(e_2^2 - 1\right) \quad , \tag{6-183}$$

and

$$a_{acs} h_2'^2 = \frac{\beta^2 t^2}{8\, m_2^2 \left(f_1' - t\right) f_1'^2}\left(e_2^2 - 1\right) \quad . \tag{6-184}$$

If we let $e_1 = 1$ and $e_2 = 1$, in Eqs. (6-182) through (6-184), we find that the spherical aberration, coma, and astigmatism are all zero. Thus, a telescope consisting of two confocal paraboloidal mirrors is an anastigmat. This is the Mersenne telescope considered earlier in Section 6.7 as a beam expander. Its field curvature and distortion are given by Eq. (6-112). Note that the distortion term in that equation is for an aperture stop located at M_2. Of course, since the system is an anastigmat, its field curvature is independent of the position of its aperture stop.

6.8.10 Couder Anastigmatic Telescopes

If in addition to the eccentricities of the primary and secondary mirror given by Eqs. (6-177) and (6-178), respectively, the spacing between them is given by

$$t = -2f' \quad , \tag{6-185}$$

then, according to Eq. (6-179), astigmatism is also zero. Hence, the name anastigmatic telescope. Note that since t is negative, f' must be positive.

Substituting Eq. (6-117c) and (6-185) into Eq. (6-114c), we obtain

$$S_2' = -f_1' M_2 (1 - 2M_2) \quad . \tag{6-186}$$

Now, for a real final image, $S_2' > 0$. Hence, for a concave primary mirror, i.e., for $f_1' < 0$, we find that $0 < M_2 < 0.5$. Using Eq. (6-185), Eq. (6-186) may also be written

$$S_2' = -t\left[(1/2) - M_2\right] \tag{6-187a}$$

or

$$0 < S_2' < -t/2 \tag{6-187b}$$

for the range of M_2 values under consideration. Thus, the final image lies between the two mirrors but closer to the mirror M_2, as shown in Figure 6-19. Substituting Eq. (6-117c) into Eq. (6-114d), we obtain

$$S_2' = f_2'(1 - M_2) \quad , \tag{6-188}$$

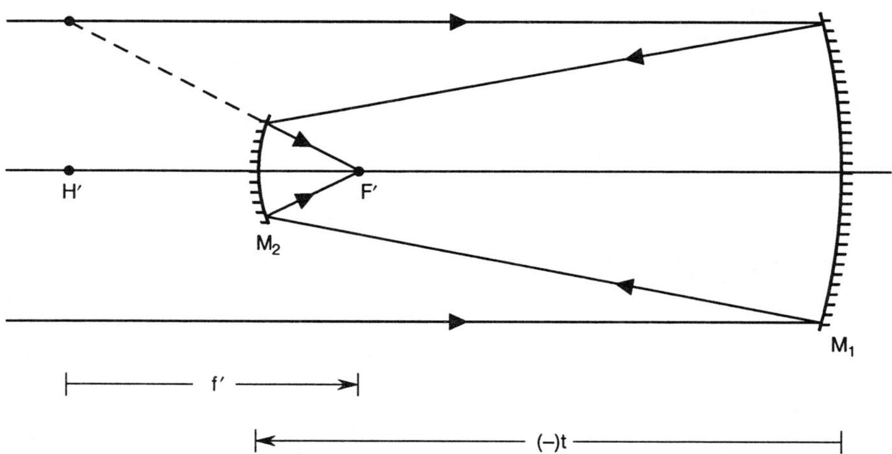

Figure 6-19. Couder anastigmatic telescope with a concave primary mirror. H' **and** F' **are the principal and focal points of the telescope, respectively.**

which shows that for the range of M_2 values under consideration, $S_2' > 0$ if $f_2' > 0$. Thus, the secondary mirror is also concave.

If the primary mirror is convex, i.e., if f_1' is positive, then $S_2' > 0$ if $M_2 < 0$. When $-0.5 < M_2 < 0$, then $|t/2| < S_2' < |t|$, i.e., the image lies between the two mirrors but closer to M_1. The secondary mirror is again concave according to Eq. (6-188). The central portion of the incident beam is blocked from reaching the focal plane directly as indicated in Figure 6-20 by a thick vertical bar. Although $S_2' > |t|$ according to Eq. (6-187a) for $M_2 < -0.5$, the light reflected by the secondary mirror is blocked by the primary mirror unless $M_2 < -1$. An example of such a telescope is the concentric Schwarzschild telescope consisting of two spherical mirrors, which is discussed next.

6.8.11 Schwarzschild Telescope

A special case of the anastigmatic telescopes is the *Schwarzschild telescope* consisting of two concentric spherical mirrors, i.e., one for which

$$e_1 = 0 \ , \tag{6-189}$$

$$e_2 = 0 \ , \tag{6-190}$$

and

$$t = 2\left(f_1' - f_2'\right) \ . \tag{6-191}$$

Substituting for f_2 from Eqs. (6-122) into Eq. (6-191) and solving for t, we obtain

$$t = 2f_1'^2 \big/ \left(f_1' - f'\right) \ . \tag{6-192}$$

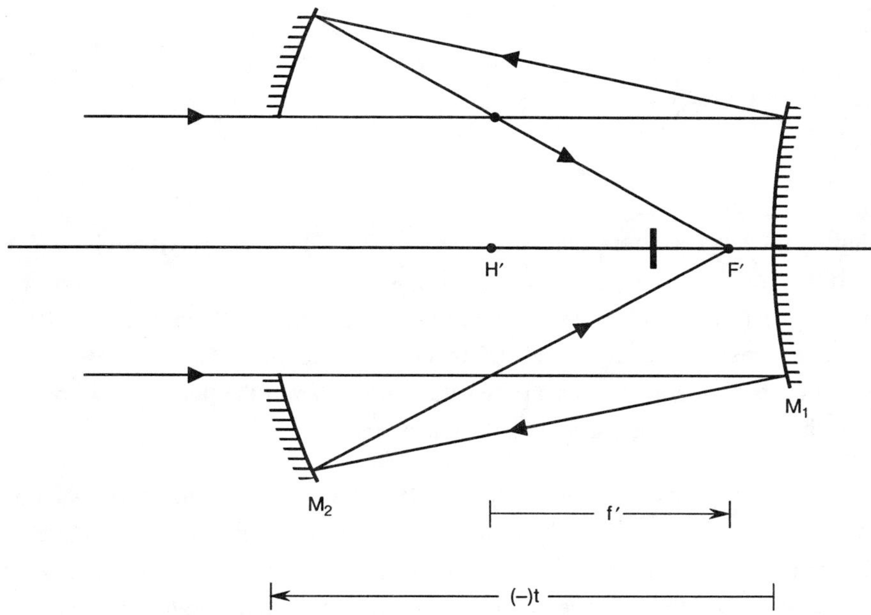

Figure 6-20. Couder anastigmatic telescope with a convex primary mirror. *H'* and *F'* **are the principal and focal points of the telescope, respectively.**

Substituting Eqs. (6-189), (6-190), and (6-192) into Eq. (6-163), we obtain

$$a_{scs} = \frac{1}{32m_2^4 f_1'^3}\left[1 - \frac{(f' - f_1')(f' + f_1')^2}{f'^3}\right] \ .$$

(6-193)

Now a_{scs} is zero if

$$f' = \frac{1}{2}(1 \mp \sqrt{5})f_1'$$

(6-194a)

$$= -0.618 f_1' \text{ or } 1.618 f_1'$$

(6-194b)

corresponding to

$$M_2 = 0.618 \text{ or } -1.618 \ .$$

(6-195)

Substituting Eq. (6-194a) into Eq. (6-192), we obtain

$$t = -2f' \ .$$

(6-196)

Comparing Eqs. (6-191) and (6-196), we obtain

$$f' = f_2' - f_1' \ .$$

(6-197)

Comparing Eqs. (6-194a) and (6-197), we find that the ratio of the radii of curvature of the two surfaces is given by

$$\frac{R_2}{R_1} = \frac{f_2'}{f_1'}$$

$$= \frac{1}{2}\left(3 \mp \sqrt{5}\right) \qquad (6\text{-}198a)$$

$$= 0.382 \text{ or } 2.618 \quad . \qquad (6\text{-}198b)$$

Substituting for t and f' from Eqs. (6-192) and (6-194a), respectively, into Eqs. (6-165) and (6-166), we find that $a_{ccs} = 0 = a_{acs}$. Thus, a telescope consisting of two concentric spherical mirrors such that the ratio of their radii of curvature is given by Eq. (6-198) (so that its focal length is equal to the difference of the focal lengths of the mirrors) is anastigmatic. Its field curvature as an aberration disappears if the image is observed on a spherical surface that is concentric with the mirrors.

For a concave primary mirror, a value of $M_2 = 0.618$ yields a negative value of S_2' according to Eq. (6-186). Hence, the final image is virtual, as illustrated in Figure 6-21, which is not a practical solution. For a convex primary mirror, $M_2 = -0.618$ yields a positive value of S_2'; a real image is obtained in this case, as illustrated in Figure 6-22, Note that the diameter of the secondary mirror is quite large in this case (6.236 times the diameter of the primary mirror or aperture stop for the on-axis beam).

6.8.12 Dall-Kirkham Telescope

Because of the difficulty of a fabricating a convex hyperboloidal secondary mirror of a classical or an aplanatic Cassegrain telescope, many amateur telescopes consist of a spherical secondary mirror as in the *Dall-Kirkham telescope*. With $e_2 = 0$, the spherical aberration of such telescopes is made zero according to Eq. (6-164) by letting

$$e_1^2 = 1 - \frac{f_1' - t}{f_1' f'^3}\left(f' + f_1'\right)\left(f' - f_1'\right)^2 \quad . \qquad (6\text{-}199)$$

From Eq. (6-114c), the working distance ℓ between the primary mirror and the final image plane is given by

$$\begin{aligned}
\ell - t &= S_2' \\
&= -M_2\left(f_1' - t\right) \quad .
\end{aligned} \qquad (6\text{-}200)$$

Thus,

$$t = \frac{\ell + f_1' M_2}{1 + M_2} \qquad (6\text{-}201a)$$

or

$$f_1' - t = -\frac{\ell - f_1'}{1 + M_2} \quad . \qquad (6\text{-}201b)$$

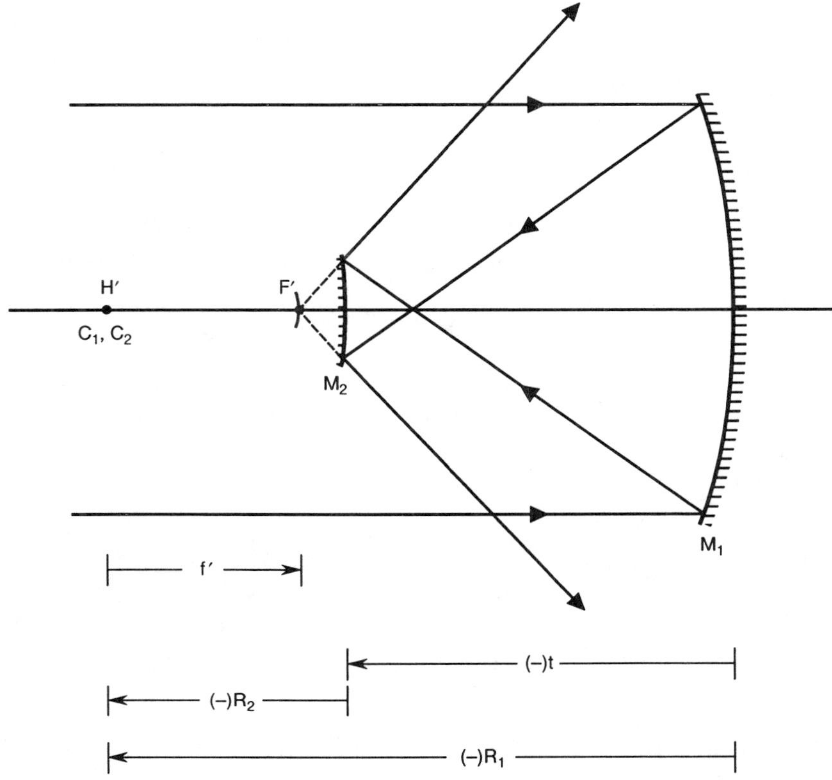

Figure 6-21. Schwarzschild telescope with a concave primary mirror. C_1 and C_2 are the centers of curvature of the mirrors M_1 and M_2, respectively. H' is the principal point of the system. It coincides with C_1 and C_2. The final image is virtual.

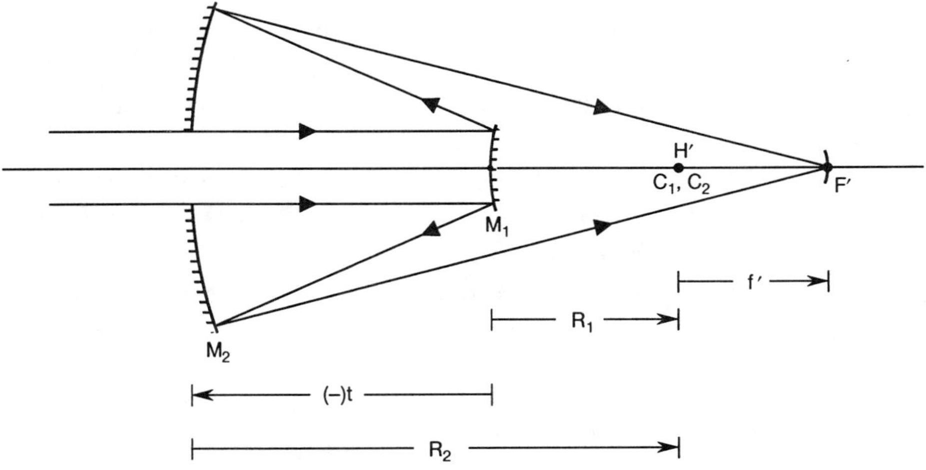

Figure 6-22. Schwarzschild telescope with a convex primary mirror. C_1 and C_2 are the centers of curvature of the mirrors M_1 and M_2, respectively. H' is the principal point of the system. It coincides with C_1 and C_2. The final image is real.

Substituting Eq. (6-201b) into Eq. (6-199), we obtain

$$e_1^2 = 1 + \left(1 - \frac{\ell}{f_1'}\right)\frac{1 - M_2^2}{M_2^3} \tag{6-202}$$

$$< 1 \ ,$$

since $f_1' < 0$ and $M_2 > 1$ for the Cassegrain telescope. Thus, the primary mirror is ellipsoidal. Substituting Eq. (6-201a) into Eq. (6-165), we obtain the coma coefficient:

$$a_{ccs} = \frac{1}{4 m_2^3 f'^3}\left[1 - \left(M_2 + \frac{\ell}{f_1'}\right)\frac{1 - M_2^2}{2M_2}\right] \tag{6-203a}$$

$$\simeq \frac{1 + M_2^2}{8 m_2^3 f'^3} \ . \tag{6-203b}$$

Comparing Eqs. (6-174a) and (6-203b), we note that the coma of a Dall-Kirkham telescope is $\left(1 + M_2^2\right)/2$ times the coma of a classical Cassegrain telescope for $\ell \ll \left|f_1'\right|$, with a correspondingly smaller field of view. However, it is relatively easy to fabricate and test. Moreover, it is less sensitive to misalignments as discussed in Section 7.3; hence it is easy to assemble.

6.9 ASTRONOMICAL TELESCOPES USING ASPHERIC PLATES

6.9.1 Introduction

In the discussion of a Schmidt camera in Section 6.6.2, we showed how the spherical aberration of a spherical mirror is compensated for with a (Schmidt) corrector plate. The plate is placed in the plane of the aperture stop which, in turn, lies at the center of curvature of the mirror. We now consider the use of *aspheric corrector plates* in two-mirror astronomical telescopes such as a Cassegrain or a Ritchey-Chrétien. In astronomical telescopes, the aperture stop generally lies at the primary mirror. The aspheric plate is placed in front of the primary mirror in the collimated object radiation, or in the converging beam forming the final image. The plate is radially symmetric, with its thickness varying typically as r^4, where r is the radial distance from its center, although it may contain higher-order terms such as r^6, r^8, etc. Since the plate does not lie at the aperture stop (unlike the plate in a Schmidt camera), it introduces not only spherical aberration, but other primary aberrations as well. When placed in a converging or a diverging beam, the plane-parallel component of the plate also introduces image displacement and aberrations, as discussed in Section 5.12. These effects are not accounted for in the following discussion.

6.9.2 Aspheric Plate in a Diverging Object Beam

Consider a plate of refractive index n placed at a (numerically negative) distance d_o from the entrance pupil *EnP* of an imaging system as shown in Figure 6-23. We start with an off-axis point object P lying at a height h from the optical axis and at a (numerically

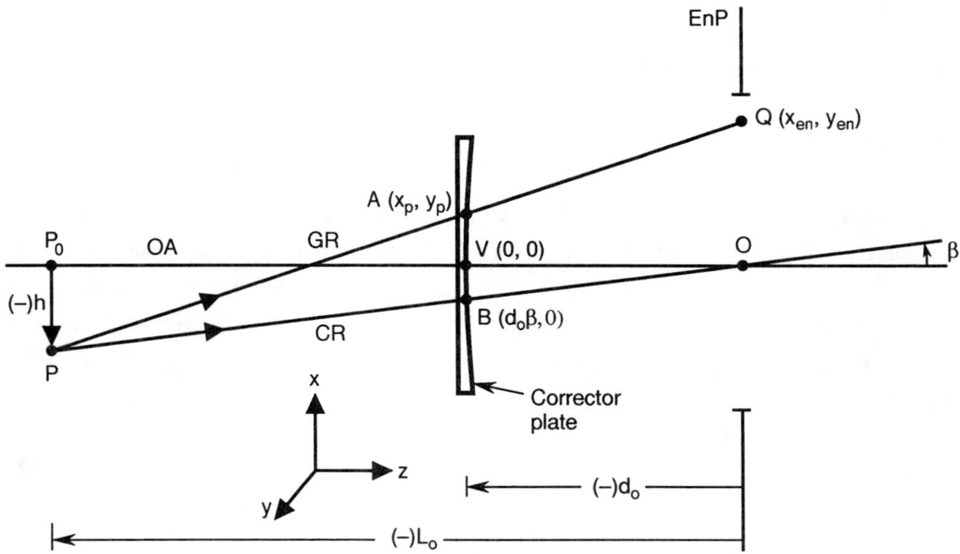

Figure 6-23. Aspheric plate in a divergent object beam incident on a telescope.

negative) distance L_o from the entrance pupil. In the case of an astronomical telescope, since the aperture stop and, therefore, the entrance pupil lie at its primary mirror, d_o would be the distance of the plate from the primary mirror and L_o would equal $-\infty$.

Let the thickness of the plate be given by

$$t\!\left(r_p\right) = t_0 + a_4 r_p^4 \quad , \tag{6-204}$$

where t_0 is a constant, a_4 is the *aspheric coefficient* of its thickness variation, and r_p is the radial distance of a point A on the plate from its center V which lies on the optical axis of the system. The change in the optical path length of an object ray passing through the plate at a distance r_p making a small angle with the optical axis is given by

$$W\!\left(r_p\right) = (n-1)\!\left(t_0 + a_4 r_p^4\right) \quad . \tag{6-205}$$

The wave aberration introduced by the plate to an object ray PA passing through a point Q in the plane of the entrance pupil is given by the difference in the change in its optical path length and that of the chief ray PO passing the center O of the entrance pupil, i.e.,

$$W_p(Q) = W(A) - W(B)$$

$$= a_4(n-1)\!\left(VA^4 - VB^4\right) \quad , \tag{6-206}$$

where B is the point on the plate where the chief ray PO intersects it. For simplicity, we let

$$a_4' = a_4(n-1) \tag{6-207}$$

and write Eq. (6-206) in the form

$$W_p(Q) = a_4'\left(VA^4 - VB^4\right) \ . \tag{6-208}$$

Let (x_{en}, y_{en}) and (x_p, y_p) be the coordinates of points Q and A, respectively. It is evident from Figure 6-23 that the coordinates of the point B are $(d_o\beta, 0)$, where $\beta = h/L_o$ is the angle the chief ray makes with the optical axis in the object space. (As usual, we have assumed that the tangential plane P_0PQ is the zx plane.) From approximately similar triangles QOP and ABP, we note that the coordinates of the points Q and A are related to each other according to

$$(x_{en}, y_{en}) = (x_p - d_o\beta, y_p)\big/\eta_o \ , \tag{6-209}$$

or

$$(x_p, y_p) = (n_o x_{en} + d_o, n_o y_{en}) \ , \tag{6-210}$$

where

$$\eta_o = (L_o - d_o)\big/L_o \ . \tag{6-211}$$

Substituting

$$VA^4 = \left(x_p^2 + y_p^2\right)^2$$

and

$$VB^4 = d_o^4 \beta^4$$

into Eq. (6-208) and substituting Eq. (6-210) in the result obtained, we obtain the primary aberration function contributed by the plate:

$$W_p(x_{en}, y_{en}; \beta) = a_4'\bigg[\eta_o^4\left(x_{en}^2 + y_{en}^2\right)^2 + 4\eta_o^3 d_o\beta x_{en}\left(x_{en}^2 + y_{en}^2\right)$$

$$+ 4\eta_o^2 d_o^2\beta^2 x_{en}^2 + 2\eta_o^2 d_o^2\beta^2\left(x_{en}^2 + y_{en}^2\right) + 4\eta_o d_o^3\beta^3 x_{en}\bigg] \ . \tag{6-212}$$

If (r_{en}, θ_{en}) are the polar coordinates of the point Q, then letting

$$(x_{en}, y_{en}) = r_{en}\left(\cos\theta_{en}, \sin\theta_{en}\right) \ , \tag{6-213}$$

we can write the aberration function in polar coordinates:

$$W_p(r_{en}, \theta_{en}; \beta) = a_4'\bigg(\eta_o^4 r_{en}^4 + 4\eta_o d_o\beta r_{en}^3 \cos\theta_{en} + 4\eta_o^2 d_o^2\beta^2 r_{en}^2 \cos^2\theta_{en}$$

$$+ 2\eta_o^2 d_o^2\beta^2 r_{en}^2 + 4\eta_o d_o^3\beta^3 r_{en}\cos\theta_{en}\bigg) \ . \tag{6-214}$$

Moreover, if (r, θ) are the polar coordinates of the point where the ray PQ intersects the plane of the exit pupil of the system, then by substituting

$$r(\cos\theta, \sin\theta) = mr_{en}(\cos\theta_{en}, \sin\theta_{en}) \tag{6-215}$$

into Eq. (6-214), where m is the magnification of the pupil, we can write the aberration function referred to the exit pupil. If the object lies at infinity, as in the case of astronomical telescopes, then $L_o \to -\infty$ and $\eta_o \to 1$ according to Eq. (6-211).

6.9.3 Aspheric Plate in a Converging Image Beam

We now consider, as illustrated in Figure 6-24, an aspheric plate of refractive index n placed at a distance d_i from the plane of the exit pupil ExP of a system forming the Gaussian image at P' at a height h'. Let the thickness of the plate be represented by Eq. (6-204). The wave aberration introduced by the plate to an image ray passing through a point Q in the plane of the exit pupil is given by

$$W_p(Q) = a_4'\left(VA^4 - VB^4\right) \; , \tag{6-216}$$

where A and B are the points on the plate at which the image ray QP' and the chief ray OP' intersect it. From approximately similar triangles QOP' and ABP', the coordinates (x, y) and (x_p, y_p) of the points Q and A, are related to each other according to

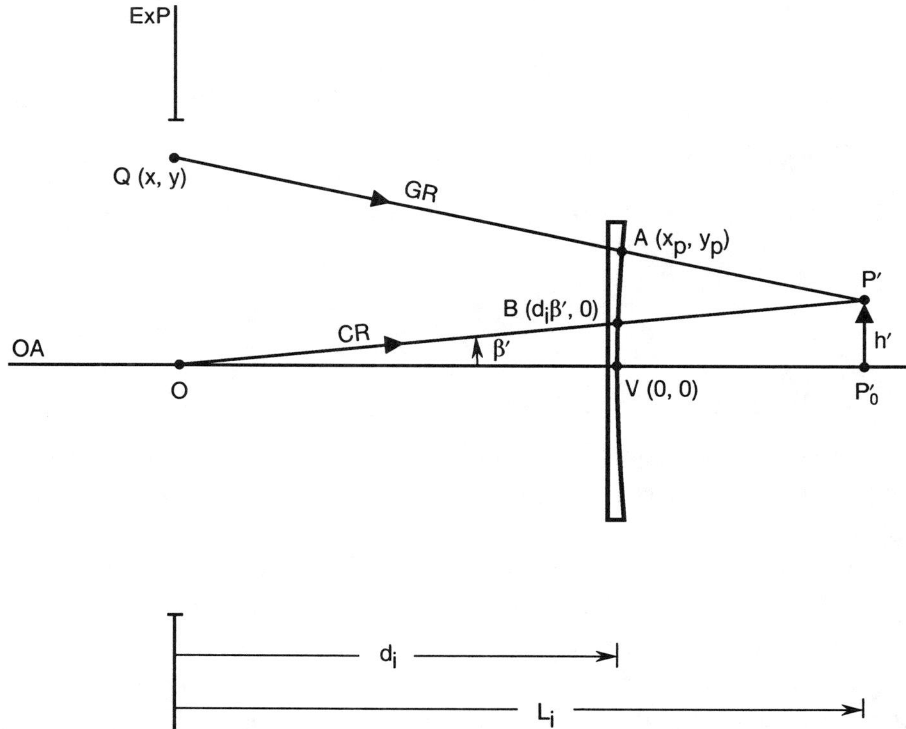

Figure 6-24. Aspheric plate in a converging beam forming the image.

$$(x, y) = \left(x_p - d_i\beta', y_p\right)\big/\eta_i \quad , \tag{6-217}$$

or

$$\left(x_p, y_p\right) = \left(\eta_i\, x + d_i\beta', \eta_i y\right) \quad , \tag{6-218}$$

where

$$\eta_i = \left(L_i - d_i\right)\big/L_i \quad , \tag{6-219}$$

$$\beta' = h'/L_i \tag{6-220}$$

is the angle the chief ray makes with the optical axis in the image space, and L_i is the distance of the image plane from the exit pupil. Substituting

$$VA^4 = \left(x_p^2 + y_p^2\right)^2$$

and

$$VB^4 = d_i^4\beta'^4$$

into Eq. (6-210), we obtain the primary aberration function contributed by the plate:

$$W_p(x, y; \beta') = a_4'\left[\eta_i^4\left(x^2 + y^2\right)^2 + 4\eta_i^3 d_i\,\beta'x\left(x^2 + y^2\right) + 4\eta_i^2 d_i^2\beta'^2 x^2\right.$$

$$\left. + 2\eta_i^2 d_i^2\beta'^2\left(x^2 + y^2\right) + 4\eta_i d_i^3\beta'^3 x\right] \quad . \tag{6-221}$$

In polar coordinates, Eq. (6-221) may be written

$$W_p(r, \theta; \beta') = a_4'\left(\eta_i^4 r^4 + 4\eta_i^3 d_i\,\beta'r^3\cos\theta + 4\eta_i^2 d_i^2\beta'^2 r^2\cos^2\theta\right.$$

$$\left. + 2\eta_i^2 d_i^2\beta'^2 r^2 + 4\eta_i d_i^3\beta'^3 r\cos\theta\right) \quad . \tag{6-222}$$

6.9.4 Aspheric Plate and a Conic Mirror

As an application of an aspheric plate, we first consider a system with a single conic mirror of eccentricity e imaging an object lying at infinity at an angle β from its optical axis, as illustrated in Figure 6-25. We assume that the aperture stop of the system is located at the mirror so that its entrance and exit pupils are also located there. The image is formed in the focal plane of the mirror at a distance f' from it, where f' is its focal length. For example, an aspheric plate may be used with the primary mirror of a two-mirror telescope, in which case it is called a *prime-focus corrector plate*. The primary aberrations introduced by the mirror in the plane of the exit pupil [see Eqs. (6-116) and (6-145)] are given by

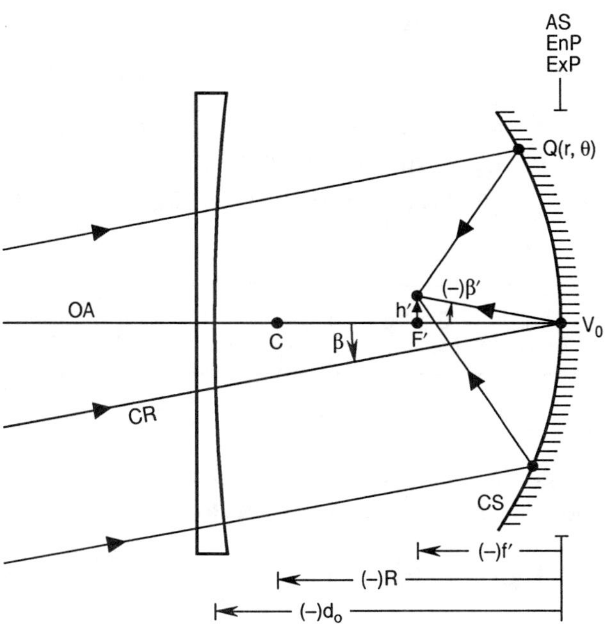

Figure 6-25. Aspheric plate in the object beam and a conic mirror with $e < 1$. Note that $\beta' = -\beta$.

$$W_m\,(r,\theta;\beta) \;=\; \frac{1-e^2}{32f'^3}\,r^4 \;+\; \frac{1}{4f'^2}\,\beta r^3\cos\theta \;+\; \frac{1}{2f'}\,\beta^2 r^2\cos^2\theta \quad . \tag{6-223}$$

If a plate of aspheric coefficient a_4' is placed at a distance d_o from the mirror in the object beam, the aberration introduced by it is given by Eq. (6-214) with $\eta_o = 1$, i.e.,

$$W_p\,(r,\theta;\beta) \;=\; a_4'\Big(r^4 + 4d_o\beta r^3\cos\theta + 4d_o^2\beta^2 r^2\cos^2\theta$$

$$\qquad\qquad + \, 2d_o^2\beta^2 r^2 + 4d_o^3\beta^3 r\cos\theta\Big) \quad . \tag{6-224}$$

Combining the aberrations introduced by the mirror and the plate, we obtain the aberration of the system

$$W_s(r,\theta;\beta) \;=\; W_m(r,\theta;\beta) + W_p(r,\theta;\beta)$$

$$= \left(\frac{1-e^2}{32f'^3} + a_4'\right)r^4 + \left(\frac{1}{4f'^2} + 4a_4'd_o\right)\beta r^3\cos\theta$$

$$+ \left(\frac{1}{2f'} + 4a_4'd_o^2\right)\beta^2 r^2\cos^2\theta + 2a_4'd_o^2\beta^2 r^2 + 4a_4'd_o^3\beta^3 r\cos\theta \quad . \tag{6-225}$$

We note that spherical aberration and coma vanish if

$$a_4' \;=\; -\frac{1-e^2}{32f'^3} \tag{6-226}$$

and

$$d_o = \frac{2f'}{1-e^2} \quad . \tag{6-227}$$

These two parameters determine the shape and the location of the plate. Since d_o is numerically negative, it follows from Eq. (6-227) that e and d_o must satisfy the inequalities $e < 1$ and $|d_o| > |R|$. When Eqs. (6-226) and (6-227) are satisfied, Eq. (6-225) reduces to

$$\boxed{W_s(r,\theta;\beta) = -\frac{e^2}{2f'\left(1-e^2\right)}\beta^2 r^2 \cos^2\theta - \frac{1}{4f'\left(1-e^2\right)}\beta^2 r^2 - \frac{1}{\left(1-e^2\right)^2}\beta^3 r\cos\theta \quad .}$$

$$\tag{6-228}$$

Spherical aberration, coma, and astigmatism are all zero if $e = 0$, $a_4' = -1/32f'^3$, and $d_o = 2f'$. The aberration function in this case reduces to

$$W_s(r,\theta;\beta) = -\left(1/4f'\right)\beta^2 r^2 - \beta^3 r\cos\theta \quad . \tag{6-229}$$

This is the case in a *Schmidt camera*, i.e., a spherical mirror with an aspheric plate placed at its center of curvature. The first term on the right-hand side of Eq. (6-229) represents field curvature. The second term representing distortion reduces to zero if the aperture stop is placed at the center of curvature of the mirror as in Figure 6-9 instead of at the mirror as in Figure 6-25. (see Problem 6.13).

6.9.5 Aspheric Plate and a Two-Mirror Telescope

Combining the aberration function of an aspheric plate with that of a two-mirror telescope, we obtain the aberration of a plate-telescope system. Such an aspheric plate can be used in the collimated light incident on a Cassegrain telescope in a configuration called a *Schmidt-Cassegrain telescope*.[15,16] The aspheric plate can be designed so that a spherical primary mirror may be used, which is easier to fabricate (instead of a paraboloidal mirror as in a classical Cassegrain telescope). The secondary mirror may be ellipsoidal. Spherical aberration and coma of the Schmidt–Cassegrain telescope can be made zero while its astigmatism can be made smaller than that of a corresponding classical Cassegrain telescope. It should be noted that the diameter of the aspheric plate is approximately equal to the diameter of the primary mirror.

If the aspheric plate is placed at a distance d_i from a mirror in the converging image beam, as illustrated in Figure 6-26, then the aberration of the system is given by the sum of the aberrations given by Eqs. (6-222) and (6-223), i.e.,

$$W_s(r,\theta;\beta) = \left(\frac{1-e^2}{32f'^3} + a_4'\eta_i^4\right)r^4 + \left(\frac{1}{4f'^2} - 4a_4'\eta_i^3 d_i\right)\beta r^3\cos\theta$$

$$+ \left(\frac{1}{2f'} + 4a_4'\eta_i^2 d_i^2\right)\beta^2 r^2\cos^2\theta + 2a_4'\eta_i^2 d_i^2\beta^2 r^2 - 4a_4'\eta_i d_i^4\beta^3 r\cos\theta, \tag{6-230}$$

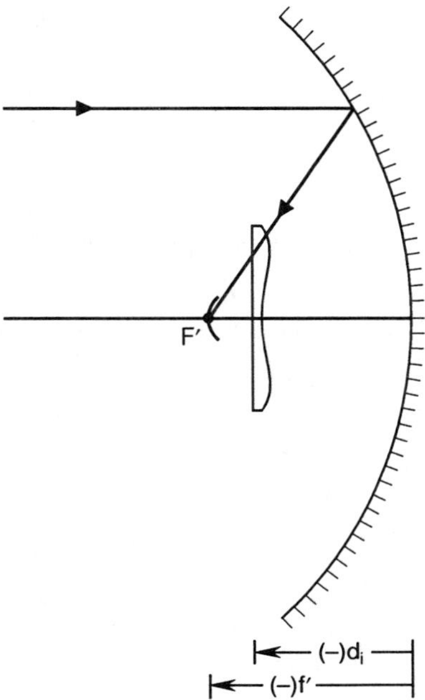

Figure 6-26. Aspheric plate in a converging beam formed by a hyperboloidal $(e > 1)$ mirror.

where (since $L_i \equiv f'$)

$$\eta_i = \left(f' - d_i\right)/f' \tag{6-231}$$

and we have substituted $\beta' = -\beta$.

For a given value of e, there are two free parameters, namely, a'_4 and d_i. Hence, two aberrations can be made zero by a suitable choice of their values. Spherical aberration and coma are zero when

$$d_i = 2f'\left/\left(1 + e^2\right)\right. \tag{6-232}$$

or

$$\eta_i = -\frac{1 - e^2}{1 + e^2} \tag{6-233}$$

and

$$a'_4 = -\frac{\left(1 + e^2\right)^4}{32 f'^3 \left(1 - e^2\right)^3} \quad . \tag{6-234}$$

In that case, the aberration function reduces to

$$W_s(r,\theta;\beta) = \frac{-e^2}{2f'\left(1-e^2\right)}\beta^2 r^2 \cos^2\theta - \frac{1}{4f'\left(1-e^2\right)}\beta^2 r^2 - \frac{1}{\left(1-e^2\right)^2}\beta^3 r\cos\theta \ , \quad (6\text{-}235)$$

which is the same as Eq. (6-228).

Note that $\left|d_i\right|$ must be less than $\left|f'\right|$ for a plate lying in a converging beam. Hence, both spherical aberration and coma of an ellipsoidal $(e<1)$ mirror cannot be corrected with a single plate. For $e>1$, we also note that a_4' is negative (unlike the Schmidt plate, for which it is positive) so that the plate is turned down at the edge as indicated in Figure 6-26, where we have also included an r^2 term to reduce the chromatic aberrations introduced by it (compare it with the plate shown in Figure 6-10). Moreover, as e increases (beyond a value of unity), $\left|d_i\right|$ decreases, a_4' and the aberration coefficients decrease numerically. Hence, the size of the plate and/or the field of view of the system for a given image quality increases. We also note that for $e>1$, the coefficient of the field curvature term in Eq. (6-235) is numerically negative. As discussed in Section 6.4.2, the Petzval curvature of the mirror is zero since the aperture stop is located at the mirror. The effect of the field curvature term as an aberration is reduced to zero if the image is observed on a spherical surface curved as shown in Figure 6-26 with a radius of curvature $f'\left(1-e^2\right)$.

In a Ritchey-Chrétien telescope, both the primary and the secondary mirrors are hyperboloidal. The image of an on-axis point object is aberrated by spherical aberration. It is possible to design a plate which, when placed in the converging beam forming this image, makes it unaberrated. Similarly, the field of view of a Ritchey-Chrétien telescope is limited by its astigmatism.[17] An aspheric plate placed in the converging beam formingthe final image can cancel the telescope astigmatism. However, it will introduce spherical aberration and coma also. If the eccentricities of the two mirrors are adjusted, a plate can be designed so that all three aberrations are zero. As in the case of a Schmidt camera, while the aspheric plate eliminates some aberrations, it also introduces chromatic aberrations. Ultimately, the performance of a telescope will be limited by the higher-order aberrations.

REFERENCES

1. L.C. Epstein, "An all-reflection Schmidt telescope for space research," Sky and Telescope, April 1967, pp. 204–207.

2. L. Epstein, "Improved geometry for the all-reflecting Schmidt telescope," *Appl. Opt.* **12**, 926 (1973).

3. D. Korsch, "Reflective Schmidt corrector," *Appl. Opt.* **13**, 2005–2006 (1974).

4. E. H. Linfoot, *Recent Advances in Optics*, Clarendon, Oxford, p. 190 (1955).

5. R. J. Lurie, "Anastigmatic catadioptric telescopes," *J. Opt. Soc. Am.* **65**, 261–266 (1975).

6. A. Bouwers, *Achievements in Optics*, Chapter 1, Elsevier, Amsterdam (1949).

7. D. D. Maksutov, "New catadioptric meniscus systems," *J. Opt. Soc. Am.* **34**, 270–284 (1944).

8. S. C. B. Gascoigne, "Recent advances in astronomical optics," *Appl. Opt.* **12**, 1419–1429 (1973); also, "Some recent advances in the optics of large telescopes," *Quart. J. Roy. Astron. Soc.* **9**, 98–115 (1968).

9. W. B. Wetherell and M. P. Rimmer, "General analysis of aplanatic Cassegrain, Gregorian, and Schwarzchild telescopes," *Appl. Opt.* **11**, 2817–2832 (1972).

10. R. R. Willey, Jr., "Cassegrain-type telescopes," *Sky and Telescope*, **21**, 191–193 (1962).

11. C. L. Wyman and D. Korsch, "Aplanatic two-mirror telescopes: A systematic study. 1: Cassegrain configuration," *Appl. Opt.* **13**, 2064–2066 (1974), "Systematic study of aplanatic two-mirror telescopes. 2: The Gregorian configuration," *Appl. Opt.* **13**, 2402–2404 (1974), "Aplanatic two-mirror telescopes: a systematic study. 3: The Schwarzschild configuration," *Appl. Opt.* **14**, 992–995 (1975).

12. S. Rosin, "Inverse Cassegrain systems," *Appl. Opt.* **7**, 1483–1497 (1968).

13. R. Gelles, "Unobscured-aperture two-mirror systems," *J. Opt. Soc. Am.* **65**, 1141–1143 (1975).

14. D. Korsch, "Closed-form solutions for imaging systems, corrected for third-order aberrations," *J. Opt. Soc. Am.* **63**, 667–672 (1973).

15. A. S. DeVany, "Schmidt–Cassegrain telescope system with a flat field," *Appl. Opt.* **4**, 1353 (1965); "Schmidt–Cassegrain telescope system with a flat field II," *Appl. Opt.* **6**, 976 (1967).

16. R. D. Sigler, "Family of compact Schmidt–Cassegrain telescope designs," *Appl. Opt.* **13**, 1765–1766 (1974).

17. S. Rosin, "Corrected Cassegrain system," *Appl. Opt.* **3**, 151–152 (1964); "Ritchey-Chrétien corrector system," *Appl. Opt.* **5**, 675–576 (1966).

PROBLEMS

6.1   (a) Consider a concave mirror of radius of curvature R imaging an axial point object lying at infinity. By considering the difference between the optical path lengths of a ray of zone r and the chief ray, show that the spherical aberration of the image is given by $W(r) = nr^4/4R^3$. (b) Repeat the problem for a convex mirror.

6.2   Consider a *spherical mirror* of diameter 4 cm and a radius of curvature of 10 cm imaging an object 2 cm high lying below the optical axis at a distance of 15 cm from it. Let the aperture stop be located at the mirror. Determine the peak values of the primary aberrations for the off-axis point at the tip of the object if the mirror is (a) concave, and (b) convex. (c) Repeat problems (a) and (b) for an object lying at infinity at an angle of 2 milliradians from the optical axis. (d) Repeat problem (c) for a concave *paraboloidal mirror* having the same vertex radius of curvature as the spherical mirror.

6.3   Show that the aberration function for a spherical mirror with its aperture stop located at its center of curvature given by Eq. (6-58) can be obtained from its aberration function given by Eq. (6-46) when the aperture stop lies at the mirror by using Eqs. (5-146) through (5-151).

6.4   Consider a *paraboloidal mirror* imaging an object lying at infinity. Its spherical aberration is zero regardless of the position of its aperture stop. (a) Show that the peak value of its coma is independent of the position of its aperture stop. (b) Determine the position of its aperture stop so that its astigmatism is zero. (c) For the position of its aperture stop obtained in (b), calculate the peak value of coma for an object at 2 milliradians from the optical axis, if the diameter of the aperture stop is 1 cm and the focal length of the mirror is 10 cm. (d) Determine the corresponding field curvature and distortion coefficients. (e) What is the focal ratio of the image-forming light cone?

6.5   Consider the *Mangin mirror* of Problem 1.5 imaging an object lying at infinity. For an aperture stop located at the mirror, show that its primary aberration coefficients are given by

$$a_s = -\frac{n-1}{4n^2}\left[\frac{1}{R_1^3} - \frac{n+3}{2R_1^2 f_s'} + \frac{4n+5}{4f_s'^2} - \frac{4n-3}{8(n-1)f_s'^3}\right] \quad,$$

$$a_c = \frac{1}{4n^2 R_1 f_s'^3}\left[2(n^2-1)f_s' - (2n^2-1)R_1\right] \quad,$$

$$a_a = \frac{1}{2f_s'^3} \quad,$$

$$a_d = \frac{1}{4n^2 R_1 f_s'^3}(n^2-1)(R_1 - f_s') \quad,$$

and

$$a_t = 0 \quad.$$

Also show that the radius of curvature of its Petzval surface is given by

$$R_p^{-1} = \frac{1}{n^2 R_1 f_s'}\left[R_1 + 2\left(n^2 - 1\right)f_s'\right] \ .$$

Determine the peak values of the aberrations for $n = 1.5$, $R_1 = f_s' = 1$ m, and $D = 10$cm for an object at a field angle of 5 milliradians. Compare them with the corresponding aberrations of a thin lens designed for minimum spherical aberration, a spherical mirror, and a paraboloidal mirror each with the same focal length and diameter as the Mangin mirror.

For additional information on the aberrations of a Mangin mirror, refer to: M. J. Reidl, "The Mangin mirror and its primary aberrations," *Appl. Opt.* **13**, 1690-1694, (1974), and R. Gelles, "Aberrations of the Mangin mirror," *Opt. Eng.* **24**, 322-325 (1985).

6.6   Consider a spherical concave mirror of diameter 4 cm and a a radius of curvature 10 cm. (a) Determine the thickness profile of a *Schmidt plate* of refractive index $n = 1.5$ for use in monochromatic light. (b) Repeat problem (a) for white light operation if $\Delta n = 0.025$. (c) Determine the position and size of the white-light image. (d) Determine the focal length of a lens that eliminates the field curvature when placed at the image plane. (e) Calculate the primary aberrations introduced by the lens component of the white-light Schmidt plate.

6.7   Show that the primary aberration function of a *conic mirror* of eccentricity e and vertex radius of curvature R, with its aperture stop located at its conic focus near its vertex, imaging an object lying at infinity, is given by

$$W_c(r_{en}, \theta_{en}; h') = \frac{1}{R^3}\left[\frac{1-e^2}{4}r_{en}^4 - 2eh'r_{en}^3\cos\theta_{en} - 2h'^2 r_{en}^2 + \frac{8e}{(1+e)^2}h'^3 r_{en}\cos\theta_{en}\right],$$

where (r_{en}, θ_{en}) are the coordinates of a point in the plane of the entrance pupil and h' is the image height. Note that astigmatism is zero for any conic mirror, and coma and distortion are zero for a spherical mirror as in a Schmidt camera.

6.8   From Eqs. (6-167) and (6-168), show that the field curvature and distortion aberrations of an afocal telescope discussed in Section 6.8.9 are given by Eq. (6-112).

6.9   Consider a *beam expander* consisting of two confocal paraboloidal mirrors expanding a beam of diameter 10 cm to a beam of diameter 100 cm. Let the focal ratio of the larger of the two mirrors be 2. (a) Determine the point at which a parallel beam incident on it at an angle of 1 degree from its optical axis is focused. (b) Determine the direction of the expanded beam.

6.10   The primary aberration function of a single-mirror system can be obtained from that of a two-mirror system by letting one of the two be a plane mirror. Show that, for example, if we let the secondary mirror be plane, the aberration function given by Eq. (6-163) reduces to that for a primary mirror.

6.11   The *Hubble space telescope* is a Ritchey-Chrétien telescope with a focal ratio of 24. Its primary mirror has a diameter of 2.4 m and a focal ratio of 2.3. The spacing between its two mirrors is 4.905 m. (a) Calculate its working distance. (b) Determine the eccentricities of the mirrors. (c) Determine the location and size of the exit pupil of the system. Also determine the location of its principal and focal points. (d) Calculate the peak values of its primary aberrations for an object at infinity at an angle of 2 milliradians from its optical axis. (e) Determine the diameters of the secondary mirror and the hole in the primary mirror for a field of view of ± 5 milliradians.

6.12   Consider a *spherical mirror* imaging an object lying at infinity. Determine the location of its aperture stop such that its (a) tangential image surface is planar, (b) the sagittal image surface is planar, and (c) the best-image surface is planar.

6.13   Show that the distortion term in Eq. (6-229) for a *Schmidt camera* reduces to zero when the aperture stop located at the mirror as in Figure 6-25 is moved to the center of curvature of the mirror as in Figure 6-9.

CHAPTER 7

CALCULATION OF PRIMARY ABERRATIONS: PERTURBED OPTICAL SYSTEMS

7.1	**Introduction**	**437**
7.2	**Aberrations of a Misaligned Surface**	**438**
	7.2.1 Decentered Surface	438
	7.2.2 Tilted Surface	442
	7.2.3 Despaced Surface	444
7.3	**Aberrations of Perturbed Two-Mirror Telescopes**	**445**
	7.3.1 Decentered Secondary Mirror	445
	7.3.2 Tilted Secondary Mirror	447
	7.3.3 Decentered and Tilted Secondary Mirror	448
	7.3.4 Despaced Secondary Mirror	451
7.4	**Fabrication Errors**	**454**
	7.4.1 Refracting Surface	454
	7.4.2 Reflecting Surface	456
References		**458**
Problems		**459**

Chapter 7

Calculation of Primary Aberrations: Perturbed Optical Systems

7.1 INTRODUCTION

The image quality of an optical system is limited not only by its inherent design aberrations, discussed in Chapters 5 and 6 for rotationally symmetric systems, but also by the fabrication and assembly errors of its elements. New aberrations arise when its elements are misaligned with respect to each other owing to lack of the rotational symmetry of the perturbed system. The *misalignment* of an element may be the decentering of its vertex and/or tilting of its optical axis. The *decenter* of an element usually refers to a misposition of its vertex in a plane normal to its intended optical axis. The decenter along its optical axis is usually called *despace* in that the spacing between it and its adjacent element is incorrect. It should be evident that when one or more elements of a system are decentered or tilted, it loses its rotational symmetry since it no longer has a common optical axis. However, when the elements are only despaced, the system retains its common optical axis and, therefore, its rotational symmetry.

In this chapter, we discuss how to determine the primary aberrations of a perturbed optical system assuming that they are known for the unperturbed system. The first-order effect of a decenter or a tilt of a surface of a system is to produce a transverse displacement of the image formed by the unperturbed system. Its second-order effect is to introduce some new aberrations. It is shown that a small decenter or a tilt does not change the primary spherical aberration of a system. However, if the spherical aberration of the unperturbed system is not zero, it introduces coma that is independent of the image height but depends on the pupil coordinates in the same manner as the primary coma. Since it exists for an on-axis point object, it is called *axial coma*. The other primary aberrations generate aberrations in addition to their own kind in pupil coordinates. For example, coma of the unperturbed system produces coma, astigmatism, and field curvature when it is perturbed. The degree of a new aberration in the image height is one less than the degree of the corresponding aberration of the unperturbed system. Thus, the additional coma is independent of the image height, astigmatism varies linearly with it, and distortion varies quadratically. A despace error displaces the image and the exit pupil (unless it is also the aperture stop) longitudinally and changes the values of the image distance and the distance of the image from the exit pupil. Accordingly, it changes the value of the aberrations introduced by the despaced element. In a multisurface system, the positions of the image and exit pupil change for each surface that follows the despaced surface. The change in the aberrations introduced by each surface can be calculated in a similar manner.

The general equations for the aberrations introduced by a misaligned surface are derived and the results are applied to two-mirror telescopes discussed in Section 6.8. It is shown that a combination of the decenter and tilt of the secondary mirror with respect to the primary mirror introduces no axial coma if the optical axes of the two mirrors intersect at a point called the *neutral point*. The neutral point lies at the center of curvature of the secondary mirror in the case of a Dall-Kirkham telescope, and at a point between the vertex of the secondary mirror and the focus of the primary mirror in the case of a Ritchey-Chrétien telescope. Several papers[1-6] which discuss aberrations of perturbed optical systems are listed under the references. Aberrations of misaligned two-mirror telescopes are also discussed by Schroeder[7] and Wilson.[8]

In calculating the aberrations of a system due to a misalignment of its elements, it is assumed that the elements have their prescribed shapes. In practice, when the elements are fabricated, their shapes will deviate slightly from their prescribed shapes. Such deviations, called *figure errors*, vary randomly across the surface of an element and they introduce *random aberrations* or *wavefront errors*. Relationships between the figure errors of a refracting or a reflecting surface and the aberrations introduced by them are given in the last section of this chapter. It is shown that for comparable figure errors, the wavefront errors introduced by a reflecting surface can be much larger than those introduced by a refracting element of low refractive index even though the latter has two surfaces contributing to the errors.

7.2 ABERRATIONS OF A MISALIGNED SURFACE

7.2.1 Decentered Surface

First, we consider the aberrations introduced by a *decenter* of the surface of a system. Thus, we suppose that an optical surface of the system has been laterally displaced from its optically correct position, as indicated in Figure 7-1 . In the perturbed position, its axis is still parallel to the optical axis of the unperturbed system. Let the displacement be along the x axis with a value of Δ. In its unperturbed position, let the heights of its object and image points P and P' from its optical axis VC be h and h', where V is the vertex and C is the center of curvature of the surface, respectively. The two heights are related to each other according to

$$h' = Mh \quad ,$$
(7-1)

where M is the (transverse) magnification of the image. In the perturbed position, the object and image heights from the new optical axis $V_p C_p$ become

$$h_p = h - \Delta$$
(7-2)

and

$$
\begin{aligned}
h'_p &= Mh_p \\
&= h' - M\Delta \quad ,
\end{aligned}
$$
(7-3)

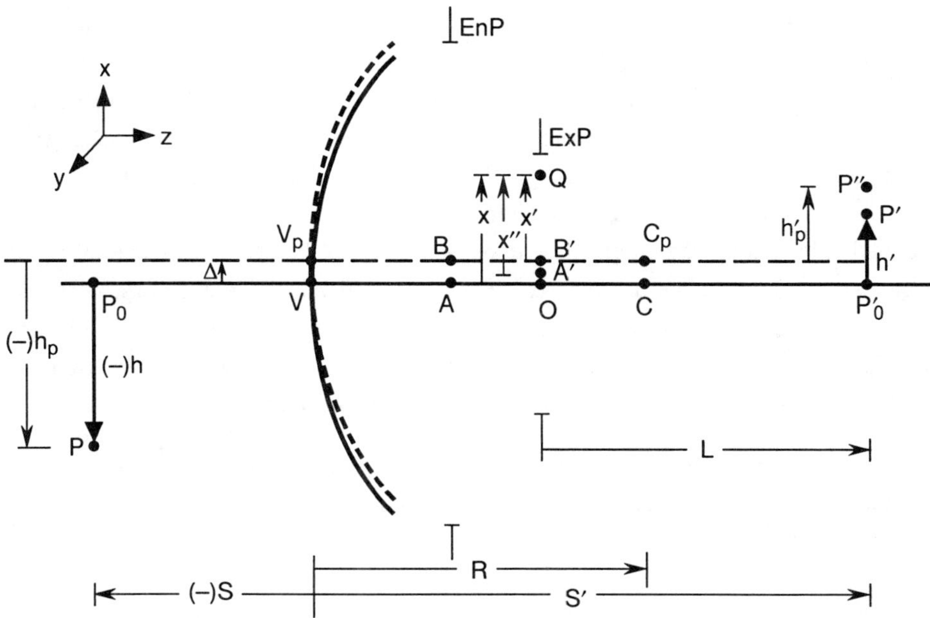

Figure 7-1. Decentered surface. In the unperturbed state, the vertex center of curvature of the surface shown by the solid curve lies at C. The point object P is at a (numerically negative) height h from its optical axis VC. Its Gaussian image P' is at a height h'. The exit pupil ExP is the image of the entrance pupil EnP. The center A of the entrance pupil is imaged at O, the center of the exit pupil. When the surface is decentered by an amount Δ along the x axis indicated by the dashed surface, its center of curvature moves to C_p and the image of P is displaced to P''. The new object and image heights are h_p and h'_p, respectively. The image of the center A of the entrance pupil now lies at A'. B and B' are the points where the new optical axis intersects the entrance and exit pupils, respectively. It is assumed here that the entrance pupil is the exit pupil of a preceding imaging element.

respectively. Note that in Figure 7-1, h and M are numerically negative, and we have assumed that the displacement Δ of the surface, which is positive, is in the tangential (i.e., zx) plane. The image point for the decentered surface lies at P''. The image displacement, which is also along the x axis, is given by

$$P'P'' = h'_p + \Delta - h'$$
$$= (1 - M)\Delta \quad , \tag{7-4a}$$

or

$$\boxed{P'P'' = (1 - M)\Delta_{c/d}} \quad . \tag{7-4b}$$

where $\Delta_{c/d} = \Delta$ is the displacement of the center of curvature of the surface due to its decenter.

Let B and B' be conjugate axial points for the perturbed surface, where its optical axis intersects its entrance and exit pupils EnP and ExP, respectively. If the primary aberrations contributed by the surface under consideration are known for an image height h' (with respect to a reference sphere centered at P' and passing through O), they can be immediately written for an image height h'_p (with respect to a reference sphere centered at P'' and passing through B') by simply replacing h' by h'_p. The aberrations thus obtained at its exit pupil are defined with respect to B' as the origin. However, the center of the exit pupil for the perturbed surface lies at A', which is the image of the center A of the entrance pupil. The transformation of the aberration function as a result of a change in the origin of the aberration coordinate system from B' to A' gives the aberrations with respect to the center A' of the new exit pupil for the image point P''.

Let the contribution to the primary aberration function of the system by the surface under consideration in the unperturbed state be given by

$$W(x, y; h') = a_s\left(x^2 + y^2\right)^2 + a_c h' x\left(x^2 + y^2\right) + a_a h'^2 x^2$$

$$+ a_d h'^2\left(x^2 + y^2\right) + a_t h'^3 x \quad , \tag{7-5}$$

where a_i's are the coefficients of the primary aberrations and (x, y) are the coordinates of a pupil point Q with O as the origin. In the perturbed state, the aberration function is similarly given by

$$W(x', y'; h'_p) = a_s\left(x'^2 + y'^2\right)^2 + a_c h'_p x'\left(x'^2 + y'^2\right) + a_a h'^2_p x'^2$$

$$+ a_d h'^2_p\left(x'^2 + y'^2\right) + a_t h'^3_p x' \quad , \tag{7-6}$$

where (x', y') are the coordinates of the pupil point Q with B' as the origin. Let (x'', y'') be the coordinates of the pupil point Q in a coordinate system with A' as the origin. In this coordinate system, the coordinates of B' are $(m\Delta, 0)$, where m is the magnification of the pupil. It is evident from Figure 7-1 that

$$(x', y') = (x'' - m\Delta, y'') \quad . \tag{7-7}$$

Substituting Eq. (7-7) into Eq. (7-6), we obtain the aberration function for the perturbed surface with respect to A' as the origin:

$$W_{dec}(x'', y''; h'_p) = a_s\left[(x'' - m\Delta)^2 + y''^2\right]^2 + a_c h'_p(x'' - m\Delta)\left[(x'' - m\Delta)^2 + y''^2\right]$$

$$+ a_a h'^2_p(x'' - m\Delta)^2 + a_d h'^2_p\left[(x'' - m\Delta)^2 + y''^2\right]$$

$$+ a_t h'^3_p(x'' - m\Delta) \quad . \tag{7-8}$$

Equation (7-8) describes the contribution to the primary aberration function of the system by the surface under consideration in its perturbed state. We emphasize that it gives the aberration at a point Q with respect to a reference sphere centered at P'' and passing through A'. Each aberration term on its right-hand side can be written in terms of its value for the unperturbed state plus some additional terms. An aberration of a certain order in pupil coordinates contributes all aberrations of lower order as a result of the perturbation. For example, spherical aberration contributes coma, astigmatism, field curvature, and distortion. Similarly, coma contributes astigmatism, field curvature, and distortion, and so on. However, some of these terms depend on Δ^2 and Δ^3, which may be neglected for small values of Δ. (There are terms in Δ^4 also which represent the optical path difference between the rays passing through points A' and O and may be ignored.) Neglecting such terms, the change in the aberration function due to a decenter of the surface may be written

$$\delta W_{dec}(x, y; h') = W_{dec}(x, y; h'_p) - W(x, y; h') \quad ,$$

or

$$\boxed{\begin{aligned} \delta W_{dec}(x, y; h') = &- (Ma_c + 4ma_s)\, \Delta\, x(x^2 + y^2) - 2(Ma_a + ma_c)\Delta\, h' x^2 \\ &- (2Ma_d + ma_c)\Delta\, h'(x^2 + y^2) - [3Ma_t + 2m(a_a + a_d)]\Delta\, h'^2 x \quad , \end{aligned}} \tag{7-9}$$

where $W_{dec}(x, y; h'_p)$ is the aberration of the decentered surface given by Eq. (7-8) with (x'', y'') replaced by (x, y). Equation (7-9) describes the additional aberration at a point Q due to a decenter of the surface, where we have let (x, y) be the coordinates of Q with respect to A' as the origin for convenience. We have also substituted for h'_p in terms of h' according to Eq. (7-3).

It is evident that there is no change in the contribution to spherical aberration of the system by a decenter of its surface. The first term on the right-hand side of Eq. (7–9) depends on the pupil coordinates, as does the primary coma. However, unlike primary coma, it does not depend on the image height h', i.e., it is constant across the entire image of an extended object, including the axial image point P'_0. Hence, it is called *axial coma*. Its coefficient depends upon both coma and spherical aberration of the unperturbed system. Its value is not zero unless both a_c and a_s are zero or $a_c = -4(m/M)a_s$. Similar conclusions can be drawn from the other terms in Eq. (7-9). In terms of their dependence on pupil coordinates, the second term is astigmatism, third is field curvature, and the last is distortion. However, astigmatism and field curvature introduced both vary as h', and the distortion introduced varies as h'^2. Thus, the degree or the power with which each aberration term introduced varies with h' is one less than that for the corresponding terms for an aligned system. Accordingly, the degree of each aberration term introduced in the image (or object) and pupil coordinates is three, i.e., one less than the nominal four for a primary aberration. We also note that except for spherical aberration, each primary aberration introduces additional aberration of its own kind as well. For example, coma introduces additional coma, astigmatism introduces additional astigmatism, etc. In general, if the primary aberrations of an unperturbed system are zero, then a small decenter of its surface does not introduce any additional aberrations.

In a multisurface system, the perturbation of a surface affects not only its aberration contribution but also those of the surfaces that follow it. As the locations of the image point and the center of the exit pupil for the perturbed surface change, the locations of the point object and the center of the entrance pupil for the next surface also change (even if the next surface is not perturbed) thereby changing its contribution to the aberration of the system. The aberrations of the following surfaces can be calculated in a similar manner. The observations made above about the dependence on the image height of the additional aberrations introduced in a system by a decenter of one or more of its elements do not change.

7.2.2   Tilted Surface

Now we consider primary aberrations introduced by the *tilt* of an optical surface of a system from its nominal orientation. We assume that the surface has been rotated by a small angle β about its vertex in the tangential plane, as illustrated in Figure 7-2. In the unperturbed position, the point object P and its Gaussian image P' are at heights h and h' from the optical axis VC. When the surface is tilted, the Gaussian image of the point object P is displaced to P''. With respect to the tilted optical axis of the surface, the heights of its object point P and image point P'' are given by

$$h_p = h - \beta S \tag{7-10}$$

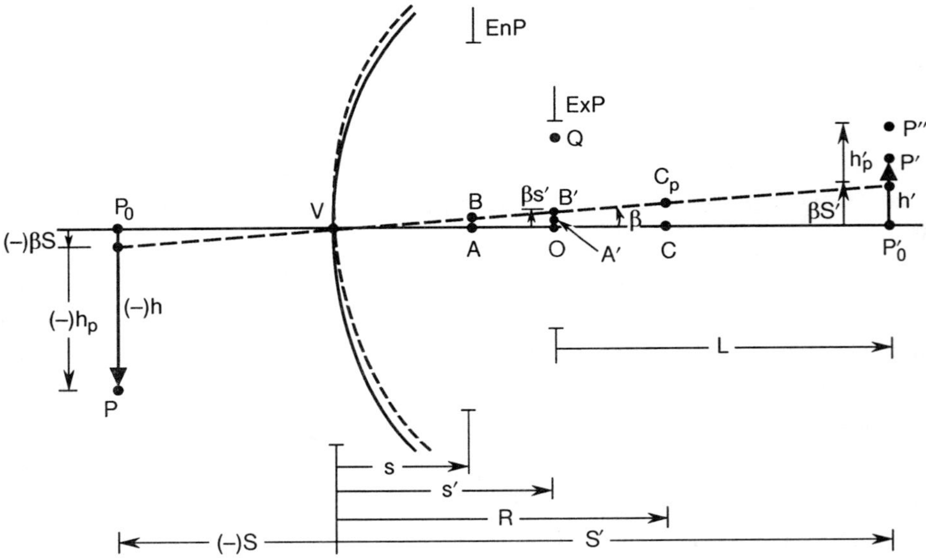

Figure 7-2. Tilted surface. When the surface is tilted by an angle β, indicated by the dashed surface, its vertex center of curvature C moves to C_p. The heights of the object P and image P' change from h to h_p and from h' to h'_p, respectively. The image for the tilted surface is located at P''. The center of the entrance pupil lies at A and its image by the tilted surface lies at A'.

and

$$h'_p = Mh_p$$

$$= h' - M\beta S \quad . \tag{7-11}$$

Note that since h is numerically negative in the figure, $h - \beta S$ is a numerically smaller height than h.

The image displacement, which is along the x axis, as in the case of a decentered surface, is given by

$$P'P'' = h'_p - (h' - \beta S')$$

$$= (S' - MS)\beta \quad . \tag{7-12a}$$

Substituting for S in terms of S' from Eq. (5-9c) for the image magnification and S' in terms of R from Eq. (5-5) for imaging, we find that

$$P'P'' = (1 - M)\beta R \quad , \tag{7-12b}$$

or

$$\boxed{P'P'' = (1 - M)\Delta_{c/t}} \quad , \tag{7-12c}$$

where $\Delta_{c/t} = \beta R$ is the displacement of the center of curvature of the surface due to its tilt.

Given the primary aberrations contributed by the surface for an image height h' (with respect to a reference sphere centered at P'), they can be obtained for an image height h'_p (with respect to a reference sphere centered at P'') by replacing h' by h'_p. The aberrations thus obtained at its exit pupil are defined with respect to an origin at B', where B' is the image of a point B where the perturbed optical axis intersects the entrance pupil. B and B' are axial conjugate points for the perturbed surface. However, the center of the exit pupil for the purturbed surface lies at A', which is the image of the center A of the entrance pupil.

Once again we assume that the primary aberration function for the unperturbed surface is given by Eq. (7-5). The aberration function for the tilted surface with B' as the origin is given by Eq. (7-6), where h'_p is given by Eq. (7-11). Now, the coordinates of B' with respect to the origin at A' are given by $(ms\beta, 0)$ where s is the distance of the entrance pupil from the surface. Let the coordinates of a pupil point Q with respect to A' and B' as the origins be (x'', y'') and (x', y'), respectively. They are related to each other according to

$$(x', y') = (x'' - ms\beta, y'') \quad . \tag{7-13}$$

Substituting Eq. (7-13) into Eq. (7-6), we obtain the aberration function for the tilted surface with respect to A' as origin:

$$W_{tilt}\left(x'', y''; h_p'\right) = a_s\left[\left(x'' - ms\beta\right)^2 + y''^2\right]^2 + a_c h_p'\left(x'' - ms\beta\right)\left[\left(x'' - ms\beta\right)^2 + y''^2\right]$$

$$+ a_a h_p'^2\left(x'' - ms\beta\right)^2 + a_d h_p'^2\left[\left(x'' - ms\beta\right)^2 + y''^2\right]$$

$$+ a_t h_p'^3\left(x'' - ms\beta\right) \quad . \tag{7-14}$$

The change in the aberration function due to a tilt of the surface may be written

$$\delta W_{tilt}\left(x, y; h'\right) = W_{tilt}\left(x, y; h_p'\right) - W\left(x, y; h'\right) \quad , \tag{7-15}$$

where $W_{tilt}\left(x, y; h_p'\right)$ is the aberration of the tilted surface given by Eq. (7-14) with $\left(x'', y''\right)$ replaced by $\left(x, y\right)$. Substituting Eqs. (7-5), (7-11), and (7-14) into Eq. (7-15) and neglecting terms in β of powers higher than one for small values of β, we obtain

$$\boxed{\begin{aligned}\delta W_{tilt}\left(x, y; h'\right) = {}&-\left(MSa_c + 4msa_s\right)\beta x\left(x^2 + y^2\right) - 2\left(MSa_a + msa_c\right)\beta h' x^2 \\ &-\left(2MSa_d + msa_c\right)\beta h'\left(x^2 + y^2\right) - \left[3MSa_t + 2ms\left(a_a + a_d\right)\right]\beta h'^2 x\end{aligned}} \tag{7-16}$$

Comparing Eqs. (7-16) and (7-9), we find that the comments made following the latter are applicable here as well. Thus, for example, a surface tilt does not introduce spherical aberration; coma introduced is independent of image height h', i.e., it is axial axial coma; astigmatism and field curvature vary as h', and distortion varies as h'^2. Incidentally, Eq. (7-16) may be obtained from Eq. (7-9) by replacing $M\Delta$ by $MS\beta$ and $m\Delta$ by $ms\beta$.

7.2.3 Despaced Surface

When an optical surface of an imaging system is displaced longitudinally, i.e., along the common optical axis, the distance of its object point from it changes and, therefore, the distance of its image point also changes. However, the heights of the object and image points do not change. Similarly, the distances of its entrance and exit pupils also change, and, of course, the distance between the exit pupil and the image point also changes. But the centers of the pupils still lie on the optical axis. For a longitudinal movement Δ of the surface, the image and the exit pupil move by $\left(1 - n'M^2/n\right)\Delta$ and $\left(1 - n'm^2/n\right)\Delta$, respectively, where n and n'. are the refractive indices of the object and image spaces of the surface. Thus, the distances S' and L of the image from the surface and from the exit pupil become $S' - \left(n'/n\right)M^2\Delta$ and $L - \left(n'/n\right)\left(M^2 - m^2\right)\Delta$, respectively. Substituting these new values of S' and L in the equations for the aberrations of the unperturbed surface such as Eq. (5-85), we obtain the aberrations of the longitudinally displaced or *despaced surface*.

In a multisurface system, as one surface is displaced, the distances of the object and the entrance pupil for each of the surfaces that follow the one that is displaced also change. The aberration contribution of each surface can be calculated in a manner similar to that for the displaced surface. In a two-surface system, it is not essential to calculate the new aberrations for both surfaces separately. Instead, it is the spacing between the two surfaces that determines the aberrations of the system. Thus, the effect of a change in the spacing can be determined from the system aberration.

7.3 ABERRATIONS OF PERTURBED TWO-MIRROR TELESCOPES

In Section 6.8, we derived the aberrations of properly aligned two-mirror telescopes, i.e., those for which the two mirrors have a common optical axis with the appropriate spacing between them. Now we discuss how their aberrations change as one mirror is decentered, tilted, or despaced with respect to the other. As in Section 6.8, the object lies at infinity and the aperture stop of the system is located at the primary mirror. For the purpose of analysis, it is convenient to assume that the primary mirror is fixed and the secondary mirror is misaligned with respect to it. A decenter or a tilt of the secondary mirror displaces the image laterally, but its despace produces a longitudinal image displacement. Moreover, aberrations are introduced including those that are absent in a properly aligned telescope. Thus, for example, a properly aligned Cassegrain or Gregorian telescope does not suffer from spherical aberration. However, this aberration is introduced if the telescope is despaced. Similarly, by definition, spherical aberration and coma are absent in the aplanatic version of these telescopes. But axial coma is introduced if the secondary mirror is decentered or tilted, and regular coma is introduced if the mirror is despaced. A decenter or a tilt of the secondary mirror does not introduce any spherical aberration. It is shown that the lateral image displacement is proportional to the separation of the axes of the two mirrors in a transverse plane passing through the center of curvature of the secondary mirror. Similarly, axial coma is proportional to their separation in a transverse plane passing through a point called the *neutral point*. Thus, no axial coma is introduced for a combination of decenter and tilt so that the axes of the two mirrors intersect at the neutral point.

7.3.1 Decentered Secondary Mirror

Figure 7-3a shows a properly aligned two-mirror telescope. When the secondary mirror is decentered by a small amount Δ along the x axis, as in Figure 7-3b, the image is displaced by an amount $(1 - M_2)\Delta$. According to Eq. (7-9), the axial coma introduced as a result of the mirror decenter is given by

$$\delta W_{cd}(r_2,\theta_2;h_2') = -\left(M_2 a_{cc2} + 4m_2 a_{sc2}\right)\Delta\, r_2^3 \cos\theta_2 \quad , \tag{7-17}$$

where M_2 and m_2 are the magnifications of the image and exit pupil given by Eqs. (6-117c) and (6-123), and a_{cc2} and a_{sc2} are the coma and spherical aberration coefficients of the secondary mirror given by Eqs. (6-159) and (6-157), respectively. Substituting for the magnifications and the aberration coefficients, we may write the coma coefficient due to decenter in the form (note that the terms containing t cancel out):

$$\begin{aligned}
a_{cd} &= -\left(M_2\, a_{cc2} + 4m_2\, a_{sc2}\right)\Delta \\[2mm]
&= \left\{\frac{M_2}{4m_2^3 f'^3}\left(M_2^2 - 1\right) + \frac{(M_2 - 1)^3}{8m_2^3 f'^3}\left[e_2^2 - \left(\frac{M_2 + 1}{M_2 - 1}\right)^2\right]\right\}\Delta \quad ,
\end{aligned}$$

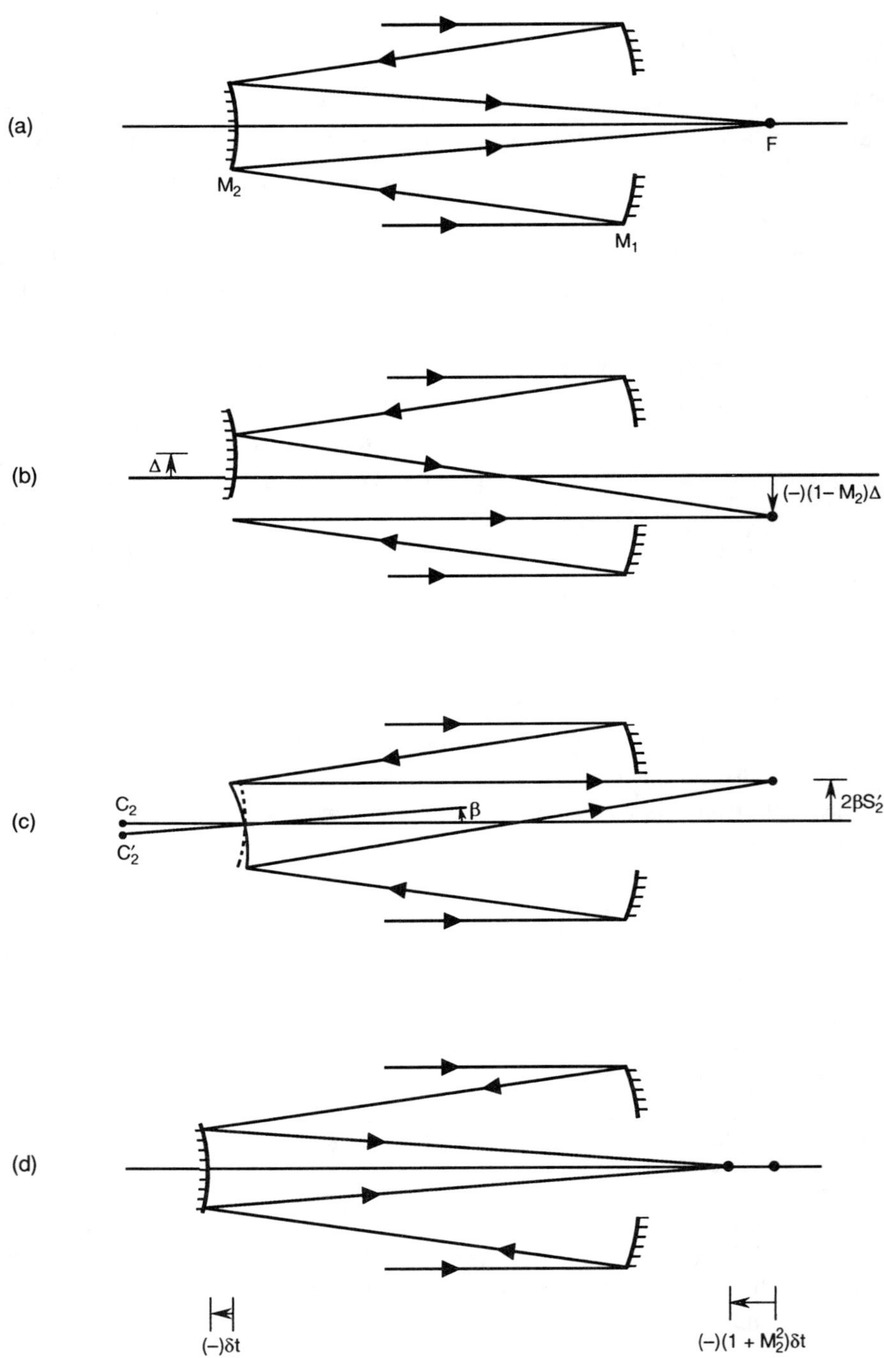

Figure 7-3. Misalignments of a two-mirror telescope. (a) Aligned telescope. (b) Secondary mirror decentered along x axis by Δ. (c) Secondary mirror tilted in zx plane by an angle β so that its center of curvature is displaced from C_2 to C_2'. (d) Secondary mirror despaced by δt .

or

$$a_{cd} = \frac{(M_2 - 1)^3}{8m_2^3 f'^3}\left(e_2^2 + \frac{M_2 + 1}{M_2 - 1}\right)\Delta \quad .$$

(7-18)

Note that the aberration coefficient depends on the eccentricity e_2 of the secondary mirror.

Substituting the value of e_2 for the various telescopes into Eq. (7-18), we can obtain the results for the specific cases. For example, substituting for e_2 from Eq. (6-170) we obtain for the classical Cassegrain and Gregorian telescopes

$$(a_{cd})_{cl} = \frac{M_2(M_2^2 - 1)\Delta}{4m_2^3 f'^3} \quad .$$

(7-19)

Similarly, using Eq. (6-178) for the aplanatic telescopes, we obtain

$$(a_{cd})_{ap} = \frac{M_2[M_2^2 - 1 + (f/t)]\Delta}{4m_2^3 f'^3} \quad .$$

(7-20)

Comparing Eqs. (7-19) and (7-20), we note that an aplanatic telescope is somewhat more sensitive to decenter than the classical. Since the secondary mirror in a Dall-Kirkham telescope is spherical $(e_2 = 0)$, Eq. (7-18) reduces to

$$(a_{cd})_{D-K} = \frac{(M_2 - 1)(M_2^2 - 1)\Delta}{8m_2^3 f'^3} \quad .$$

(7-21)

For an afocal telescope, $M_2 \to \infty$ and Eq. (7-18) reduces to

$$(a_{cd})_{af} = -\frac{(1 + e_2^2)\Delta}{8m_2^3 f_1'^3} \quad ,$$

(7-22)

which, in turn, reduces for the beam expander (Mersenne telescope) of Section 6.7 to

$$(a_{cd})_{be} = -\frac{\Delta}{4m_2^3 f_1'^3} \quad .$$

(7-23)

7.3.2 Tilted Secondary Mirror

When the secondary mirror is tilted with respect to the primary mirror by an angle β as in Figure 7-3c, the image is displaced by an amount $2S_2'\beta$. According to Eq. (7-16), the axial coma introduced as a result of the mirror tilt is given by

$$\delta W_{ct}(r_2, \theta_2; h_2') = -(M_2 S_2 a_{cc2} + 4m_2 s_2 a_{sc2})\beta r_2^3 \cos\theta_2 \quad ,$$

(7-24)

where

$$S_2 = f_1' - t \tag{7-25a}$$

and

$$s_2 = -t \tag{7-25b}$$

are the distances of the primary image and the entrance pupil from the secondary mirror, respectively. Substituting Eqs. (7-25) and (7-26) into Eq. (7-24), the coma coefficient due to tilt may be written

$$a_{ct} = -\left[M_2\left(f_1'-t\right)a_{cc2} - 4m_2 t a_{sc2}\right]\beta \quad . \tag{7-26}$$

Substituting for the coma and spherical aberration coefficients from Eqs. (6-159) and (6-157) respectively, we obtain

$$a_{ct} = -\frac{M_2\left(f_1'-t\right)}{4m_2^3 f'^3}\left(M_2^2 - 1\right)\beta \quad , \tag{7-27a}$$

or

$$\boxed{a_{ct} = -\frac{f_2'}{4m_2^3 f'^3}\left(M_2 - 1\right)^2\left(M_2 + 1\right)\beta \quad .} \tag{7-27b}$$

We note that the coma introduced by a tilt of the secondary mirror does not depend on its eccentricity. For the afocal telescope, $M_2 \to \infty$ and Eq. (7-27b) reduces to

$$\left(a_{ct}\right)_{af} = \frac{f_2'\beta}{8m_2^3 f_1'^3} \quad . \tag{7-28}$$

7.3.3 Decentered and Tilted Secondary Mirror

If the decenter and tilt of the secondary mirror are such that the axes of the two mirrors are coplanar (as considered above), the total axial coma coefficient due to both perturbations is given by the sum of their separate contributions as given by Eqs. (7-18) and (7-28); i.e.,

$$a_{c/dt} = a_{cd} + a_{ct} \quad ,$$

or

$$\boxed{a_{c/dt} = \frac{\left(M_2 - 1\right)^2}{4m_2^3 f'^3}\left[\frac{M_2 - 1}{2}\left(e_2^2 + \frac{M_2 + 1}{M_2 - 1}\right)\Delta - f_2'\left(M_2 + 1\right)\beta\right] \quad .} \tag{7-29}$$

The combination of tilt and decenter that does not introduce any axial coma is given by

$$\beta = \left(1 + e_2^2\,\frac{M_2 - 1}{M_2 + 1}\right)\frac{\Delta}{2f_2'} \quad . \tag{7-30}$$

If we write Eq. (7-30) in the form

$$\beta = \Delta / d \tag{7-31}$$

where

$$d = \frac{2f_2'}{1 + e_2^2 \dfrac{M_2 - 1}{M_2 + 1}} , \tag{7-32}$$

we note that, for zero axial coma, the tilt angle β may be defined by a decenter Δ at a distance d from the vertex of the secondary mirror. A combination of tilt and decenter giving zero axial coma is equivalent to rotating the axis of the secondary mirror about a fixed point on the axis of the primary mirror. This point lying at a distance d from the vertex of the secondary mirror is called the *neutral point*. Thus, no axial coma is introduced when the axes of the two mirrors intersect at this point. If the decenter at a distance d due to tilt is not equal to $-\Delta$, there will be axial coma. The axial coma is proportional to the separation of the axes of the mirrors in a transverse plane passing through the neutral point (see Figure 7-4). Thus, if Δ_d and $\Delta_t = \beta d$ are the decenters in the plane of the neutral point due to surface decenter and tilt, respectively, the net decenter in this plane is given by

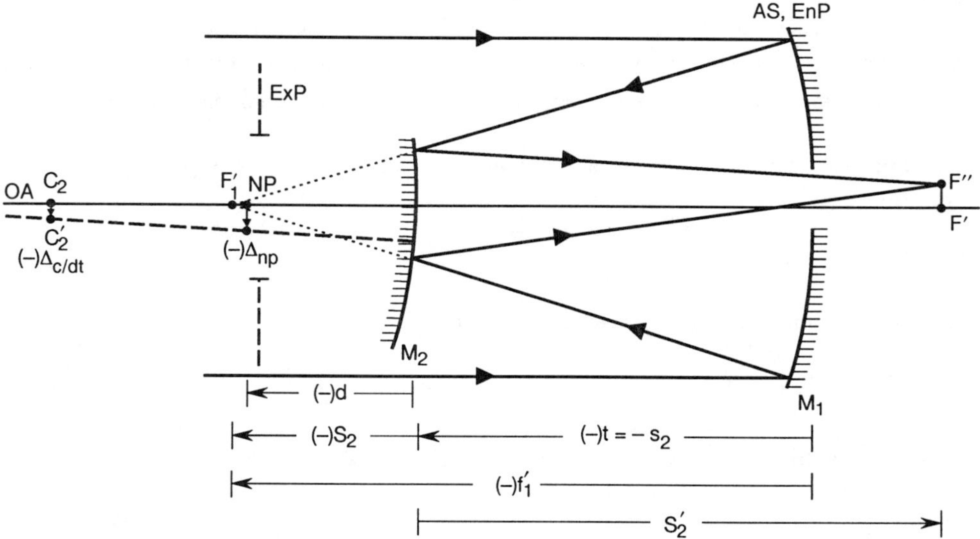

Figure 7-4. Cassegrain telescope with a decentered and tilted secondary mirror. The image displacement $F'F''$ is proportional to the transverse displacement C_2C_2' of the center of the curvature of the secondary mirror. The axial coma is proportional to the separation of the axes of the primary and secondary mirrors in the plane of the neutral point NP. The size of the secondary mirror is exaggerated in the figure for convenience.

$$\Delta_{np} = \Delta_d + \Delta_t \tag{7-33a}$$

$$= \Delta_d + \beta d \tag{7-33b}$$

and the coefficient of axial coma is given by

$$\boxed{a_{c/dt} = \frac{(M_2 - 1)^3}{8M_2^3 f'^3}\left(e_2^2 + \frac{M_2 + 1}{M_2 - 1}\right)\Delta_{np}} \quad . \tag{7-34}$$

Adding the displacements of the center of curvature of the secondary mirror due to the surface decenter and tilt, we may write its total displacement as

$$\Delta_{c/dt} = \Delta_{c/d} + \Delta_{c/t} \tag{7-35a}$$

$$= \Delta_d + \beta R_2 \quad , \tag{7-35b}$$

where R_2 is the vertex radius of curvature of the mirror. From Eqs. (7-4b) and (7-12b), the total image displacement is given by

$$\boxed{F'F'' = (1 - M_2)\Delta_{c/dt}} \quad . \tag{7-36}$$

We note that the image displacement caused by a misalignment of the secondary mirror is zero if the misalignment is only a rotation of the mirror about its center of curvature. This is understandable since the image location depends on the vertex radius of curvature of a mirror, which is not changed by a rotation about its center of curvature. The image displacement is proportional to the separation of the axes of the two mirrors in a transverse plane passing through the center of curvature of the secondary mirror.

For zero axial coma, the surface decenter and tilt are related to each other according to

$$\Delta_d = -\beta d \quad . \tag{7-37}$$

The corresponding image displacement is given by

$$F'F'' = (1 - M_2)(R_2 - d)\beta \quad . \tag{7-38}$$

Thus, by rotating the secondary mirror about the neutral point, a moving object can be tracked without introducing axial coma.

In the case of a Dall-Kirkham telescope, the secondary mirror is spherical; i.e., $e_2 = 0$, and Eq. (7-32) yields $d = 2f_2'$. Thus, the neutral point lies at the center of curvature of the secondary mirror. If this mirror is decentered and tilted so that its center of curvature is not displaced from the optical axis of the primary mirror, then no axial coma is introduced, and, of course, there is no image displacement. In the case of a classical Cassegrain telescope, substituting the values of e_2 and M_2 given by Eqs. (6-170a) and (6-117c) into Eq. (7-32), we obtain

$$d = f_2'(f' + f_1')/f' \ .$$

(7-39)

Substituting for f_2' from Eq. (6-122), we find that

$$d = f_1' - t \ .$$

(7-40)

Thus, the neutral point lies at the focus of the primary mirror. The value of e_2 for an aplanatic Cassegrain (or a Ritchey-Chrétien) telescope is given by Eq. (6-178). For the Hubble telescope, the neutral point lies between the center of the exit pupil and the focus of the primary mirror, as shown in Figure 7-4 (see Problem 7.3). In this figure, for an oblate spheroid secondary mirror $\left(e_2^2 < 0\right)$, the neutral point lies to the left of its center of curvature until it approaches $-\infty$ for $e_2^2 = -(M_2 + 1)/(M_2 - 1)$. According to Eq. (7-18), the axial coma due to a surface decenter is zero for this value of e_2. As e_2^2 decreases further, the neutral point moves to the right of the secondary mirror.

7.3.4 Despaced Secondary Mirror

The effect of a longitudinal displacement of the secondary mirror relative to the primary mirror is to change the spacing t between them. Since the distance of the image formed by the primary mirror from the secondary mirror changes, i.e., since the object distance for the secondary mirror changes, the distance of the (final) image formed by it also changes. Thus, the final image is displaced resulting in a longitudinal defocus. If the secondary mirror moves by an amount δt, as indicated in Figure 7-3d, the image formed by it moves by an amount $\left(1 + M_2^2\right)\delta t$. Hence, for a fixed observation plane, there is a longitudinal defocus of $\left(1 + M_2^2\right)\delta t$. Besides defocus, since the aberrations of the unperturbed system depend on t (see Section 6.8.6), additional aberrations are also introduced.

Let us consider the spherical aberration a_{scs} of a telescope given by Eq. (6-164). Its change with a small change δt in the spacing t may be obtained by taking a derivative of a_{scs} with respect to t. It is convenient to write the aberration in terms of M_2 in the form

$$a_{scs} = \frac{1}{32 m_2^4 f_1'^3} \left\{ 1 - e_1^2 + \frac{f_2'}{f_1'}\left(\frac{M_2 - 1}{M_2}\right)^4 \left[e_2^2 - \left(\frac{M_2 + 1}{M_2 + 1}\right)^2 \right] \right\} \ .$$

(7-41)

The dependence of the aberration on t lies in the dependence of f' and, therefore, $M_2 = -f'/f_1'$ on t, where

$$\frac{\partial M_2}{\partial t} = -\frac{M_2^2}{f_2'}$$

(7-42a)

$$= -\frac{M_2(M_2 - 1)}{f_1' - t} \ .$$

(7-42b)

Differentiatig Eq. (7-41), we obtain

$$\frac{\partial a_{scs}}{\partial t} = \frac{f_2'}{32 m_2^4 f_1'^4} \frac{\partial}{\partial t}\left\{\left(\frac{M_2-1}{M_2}\right)^4\left[e_2^2-\left(\frac{M_2+1}{M_2-1}\right)^2\right]\right\} \quad , \tag{7-43}$$

or

$$\delta a_{scs} = -\frac{1}{8 m_2^4 M_2^3 f_1'^4}\left\{(M_2-1)^3\left[e_2^2-\left(\frac{M_2+1}{M_2-1}\right)^2\right]+M_2(M_2^2-1)\right\}\delta t \quad . \tag{7-44}$$

Substituting the value of e_2 from Eq. (6-170) for the classical Cassegrain and Gregorian telescopes, Eq. (7-44) reduces to

$$(\delta a_{scs})_{cl} = -\frac{1}{8 m_2^4 M_2^2 f_1'^4}(M_2^2-1)\delta t \quad . \tag{7-45}$$

Similarly, for the aplanatic telescopes, substituting the value of e_2 from Eq. (6-178), Eq. (7-44) reduces to

$$(\delta a_{scs})_{ap} = -\frac{1}{8 m_2^4 M_2^3 f_1'^4}\left[M_2(M_2^2-1)-\frac{2f'}{t}\right]\delta t \quad . \tag{7-46}$$

Equations (7-45) and (7-46) give the spherical aberration of the despaced telescopes, which is zero otherwise. It can be seen that the aplanatic telescopes are somewhat more sensitive to despacing than the classical ones. In the case of a Dall-Kirkham telescope since the secondary mirror is spherical $(e_2 = 0)$, Eq. (7-44) reduces to

$$(\delta a_{scs})_{D-K} = -\frac{M_2^2-1}{8 m_2^4 M_2^3 f_1'^4}\delta t \quad . \tag{7-47}$$

The spherical aberration introduced by despacing of an afocal telescope is obtained by letting $M_2 \to \infty$. Thus, Eq. (7-44) yields

$$(\delta a_{scs})_{af} = -\frac{e_2^2}{8 m_2^4 f_1'^4}\delta t \quad . \tag{7-48}$$

For the beam expander (Mersenne telescope) discussed in Section 6.7, $e_2 = 1$ and Eq. (7-48) reduces to

$$(\delta a_{scs})_{be} = -\frac{\delta t}{8 m_2^4 f_1'^4} \quad . \tag{7-49}$$

In order that a telescope despacing yield zero spherical aberration, the value of e_2, according to Eq. (7-44), must be given by

$$e_2 = \frac{(M_2+1)^{1/2}}{M_2-1} \quad . \tag{7-50}$$

For other primary aberrations for an object with a field angle β_o, the variation of the final image height $h_2' = \beta_o f' = -M_2\beta_o f_1'$ [see Eqs. (6-117c) and (6-118)] with respect to t must also be taken into account. For example, the additional coma introduced by a despace error can be obtained by taking the derivative of $h_2' a_{ccs}$, where a_{ccs} is given by Eq. (6-165). Thus,

$$h_2' a_{ccs} = \frac{\beta_o}{4m_2^3 M_2^2 f_1'^2}\left\{1 - \frac{t(M_2-1)^3}{2f_1'M_2}\left[e_2^2 - \left(\frac{M_2+1}{M_2-1}\right)^2\right]\right\} \quad , \tag{7-51}$$

and

$$\frac{\partial\left(h_2' a_{ccs}\right)}{\partial t} = \frac{\beta_o}{8m_2^3 M_2^3 f_1'^3\left(f_1'-t\right)}$$

$$\times\left\{4M_2(M_2-1)f_1' + 4tM_2\left(M_2^2-1\right) - \left(f_1'-4t\right)(M_2-1)^3\left[e_2^2 - \left(\frac{M_2+1}{M_2-1}\right)^2\right]\right\} \quad . \tag{7-52}$$

For the classical Cassegrain and Gregorian telescopes, Eq. (7-52) reduces to

$$\left[\delta\left(h_2' a_{ccs}\right)\right]_{cl} = \frac{\beta_o}{2m_2^3 M_2^2 f_1'^3\left(f_1'-t\right)}\left[(M_2-1)f_1' + t\left(M_2^2-1\right)\right]\delta t \quad , \tag{7-53}$$

while for the aplanatic telescopes, it reduces to

$$\left[\delta\left(h_2' a_{ccs}\right)\right]_{ap} = \frac{\beta_o}{4m_2^3 M_2^2 f_1'^3\left(f_1'-t\right)}\left\{f_1'\left[2(M_2+1) - \left(f_1'/t\right)\right] + 2t\left(M_2^2-1\right)\right\}\delta t \quad . \tag{7-54}$$

For the classical telescopes, Eq. (7-53) gives the coma due to despace error in addition to the coma given by Eq. (6-174a) for the properly spaced telescope. Similarly, an improperly spaced aplanatic telescope is not aplanatic; its coma is given given by Eq. (7-54).

For an afocal telescope, $M_2 \to \infty$ and Eq. (7-52) reduces to

$$\left[\delta\left(h_2' a_{ccs}\right)\right]_{af} = \frac{\beta_o}{8m_2^3 f_1'^3\left(f_1'-t\right)}\left[f_1' - e_2^2\left(f_1'-4t\right)\right]\delta t \quad , \tag{7-55}$$

which, in turn, reduces for the beam expander (Mersenne telescope) to

$$\left[\delta\left(h_2' a_{ccs}\right)\right]_{be} = -\frac{\beta_o t}{2m_2^3 f_1'^3\left(f_1'-t\right)}\delta t \quad . \tag{7-56}$$

The results for an aplanatic telescope obtained here are applied to the Hubble telescope in Problem 7.3.

7.4 FABRICATION ERRORS

So far, in calculating the aberrations of a system, we have assumed that its surfaces, whether misaligned or not, have their *prescribed shapes*. The aberrations of a properly aligned system when its elements have their prescribed shapes are called its *design aberrations*. Any misalignments of its elements introduce additional aberrations. In practice, when the elements of a system are fabricated, their exact shapes will deviate slightly from their prescribed shapes. These *fabrication* or *manufacturing errors* are generally referred to as the *surface* or *figure errors*. They are typically random in that if an element is fabricated in large quantities, its errors will vary randomly from one sample to another. However, these errors have certain statistical properties that depend on the fabrication process. For example, the width (*correlation length*) of the polishing irregularities of an element depends on the size of the tool used to polish it.

In this section, we derive a relationship between the figure errors of a surface and the corresponding changes in the optical path lengths of the rays, called *wavefront errors*. Both refracting and reflecting surfaces are considered. The relationships obtained are applicable not only to figure errors but to surface misalignments as well.[3-6]

7.4.1 Refracting Surface

Consider an optical system imaging a point object P_0 at P_0', as indicated in Figure 7-5. A typical ray from P_0 is shown taking the path $P_0 ABCD P_0'$. If one or more of the surfaces of the system does not have its prescribed shape, the optical path length of the ray will change from its design value. Figure 7-5 illustrates how its optical path changes when the second surface separating media of refractive indices n and n' differs from its prescribed shape. The actual shape and the corresponding ray path are shown by dashed lines.

The paths $BB'C'D'$ and $BGCD$ are parts of the new and original ray paths between the surface that has been perturbed and the element DD' of the wavefront W in the final image space, where G is the point of intersection of the perpendicular from B' with the original ray. The perturbed surface may be regarded as deriving from the original surface by a small "figuring" BH measured along the unperturbed surface normal at the point B.

Now for small values of BH, the rays GCD and $B'C'D'$ are *neighboring rays*. Moreover, $B'G$ and $D'D$ are perpendicular to the ray GCD. Therefore, it follows from the discussion in Section 1.2.3 that the optical path lengths $[GCD]$ and $[B'C'D']$ are equal. Hence, the change in the optical path length of the ray due to the perturbation may be written

$$\delta W = \left[P_0 AB'C'D'\right] - \left[P_0 ABGCD\right]$$

$$= \left[BB'\right] - \left[BG\right]$$
$$= nBB' - n'BG$$
$$= nBB' - n'BB'\cos(\theta - \theta') \quad , \tag{7-57}$$

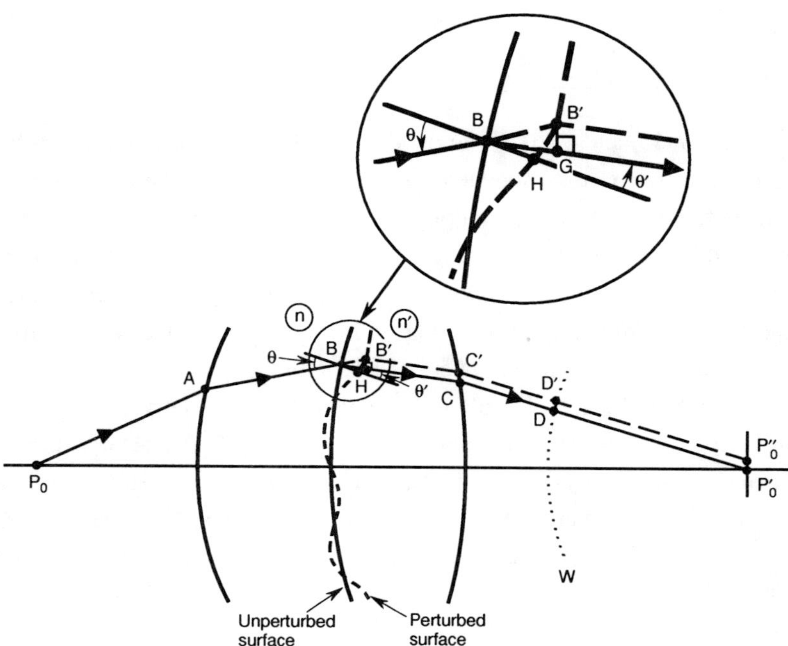

Figure 7-5. Imaging in the presence of a refracting surface perturbation. P_0 is a point object and P_0' is its Gaussian image. In the absence of a perturbation, an object ray $P_0 A$ incident on the first surface takes the path $P_0 ABCD$. Its path changes to $P_0 AB'C'D'$ when the surface is perturbed.

where θ and θ' are the angles of incidence and refraction of the ray at the point B, respectively. Noting that

$$BH = BB' \cos\theta \tag{7-58}$$

and from Snell's law,

$$n\sin\theta = n'\sin\theta' \quad, \tag{7-59}$$

Eq. (7-57) reduces to

$$\delta W = BH(n\cos\theta - n'\cos\theta') \quad. \tag{7-60}$$

In 3D, letting $\hat{i}$, $\hat{i}'$, and $\hat{g}$ be the unit vectors along the incident ray, refracted ray, and the surface normal at the point B, respectively, Eq. (7-60) may be written

$$\boxed{\delta W = \delta\vec{r} \cdot \hat{g}\left(n\hat{i} - n'\hat{i}'\right) \cdot \hat{g} \quad,} \tag{7-61}$$

where

$$\delta\vec{r} = \delta B\hat{i} \tag{7-62}$$

is the displacement vector of the surface along the ray incident at the point B and

$$\delta B \equiv BB' \tag{7-63}$$

is the corresponding displacement.

From Eqs. (7-60) and (7-61), we note that the change in the optical path length of a ray, or the wavefront error associated with it, depends upon the deviation BH of the surface (from the prescribed shape) along the surface normal at the point of incidence B of the ray and its angles of incidence and refraction associated with the unperturbed surface. Thus, it is not essential to know the true path of a ray for the perturbed surface to determine the wavefront error associated with it. If other surfaces are perturbed, the wavefront errors for them can be calculated in a similar manner. For example, under normal incidence, a surface of a plane-parallel plate of refractive index n introduces wavefront errors that are $(n-1)$ times its corresponding figure errors. Because of the random nature of the fabrication errors, the figure errors of its two surfaces will be added as a root sum square to determine their tolerances. If the standard deviation of the figure errors of a surface is σ_F, the standard deviation of the total wavefront error contributed by the plate will be

$$\boxed{\sigma_W = \sqrt{2}(n-1)\sigma_F} \tag{7-64}$$

7.4.2 Reflecting Surface

The relationship between the figure errors of a reflecting surface and the wavefront errors introduced by them can be obtained in a manner similar to that for a refracting surface. From Figure 7-6, we note that the change in the optical path length of a ray incident at a point B on the unperturbed surface at an angle of incidence θ is given by

$$\begin{aligned}
\delta W &= BB' - BG \\
&= BB'\left[1 - \cos(\pi - 2\theta)\right] \\
&= BB'(1 + \cos 2\theta) \\
&= 2BB'\cos^2\theta \\
&= 2BH\cos\theta \quad ,
\end{aligned} \tag{7-65}$$

where BB' is the displacement of the surface along the incident ray, BH is the corresponding displacement along the surface normal, and the angle of reflection $\theta' = -\theta$. In 3D, Eq. (7-65) may be written

$$\delta W = 2\delta\vec{r}\cdot\hat{g}\left(\hat{i}\cdot\hat{g}\right) \quad , \tag{7-66}$$

where $\delta\vec{r}$ is the displacement vector of the surface along the incident ray, $\hat{i}$ is a unit vector along this ray, and $\hat{g}$ is a unit vector along the normal to the surface at the point of incidence of the ray. We note that Eqs. (7-65) and (7-66) for a reflecting surface can be obtained from the corresponding Eqs. (7-60) and (7-61) for a refracting surface by letting $n = 1$, $n' = -1$, and $\theta' = -\theta$. We also note that the maximum value of the wavefront error is two times the corresponding figure error along the surface normal. Thus, if σ_F

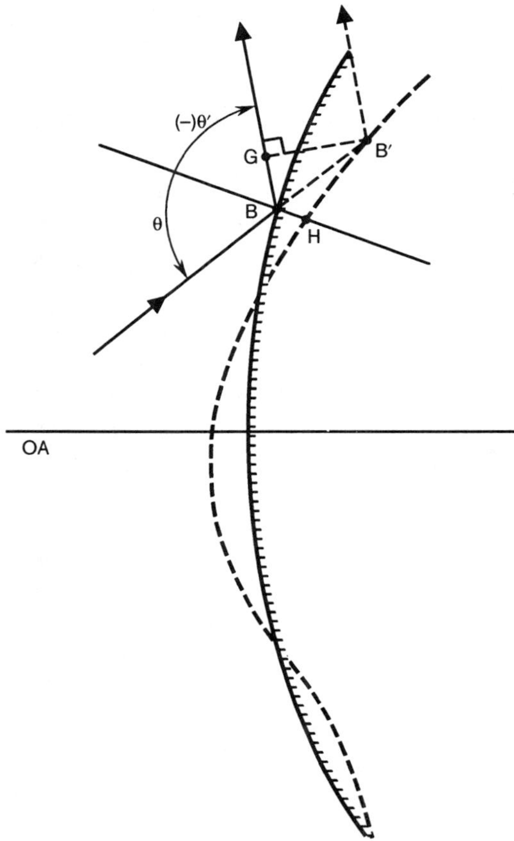

Figure 7-6. Change in the optical path length of a ray due to perturbations of a reflecting surface. The perturbed surface is indicated by the dashed line. *BH* is the normal to the unperturbed surface at the point of incidence *B* of the ray.

is the standard deviation of the figure errors of a reflecting surface, the maximum wavefront error introduced by it will be given by

$$\boxed{\sigma_W = 2\sigma_F} \quad . \tag{7-67}$$

Comparing Eqs. (7-64) and (7-67), we find that the figure errors of a reflecting surface contribute a much larger wavefront error than those of a refracting element of low refractive index. For example, a reflecting surface contributes a maximum wavefront error that is 2.8 times the corresponding maximum error introduced by the two surfaces of a refracting element with $n = 1.5$. For equal surface figure errors, the wavefront errors introduced by a reflecting surface and a refracting element are equal when $n = 2.4$. Of course, if the refractive index of an element is very high, it will contribute larger wavefront errors than a comparable reflecting surface.

REFERENCES

1.   R. Gelles, "Off-center aberrations in nonaligned systems," *J. Opt. Soc. Am.* **68**, 1250–1254 (1978).

2.   P. L. Ruben, "Aberrations arising from decenterations and tilts," *J. Opt. Soc. Am.* **54**, 45–52 (1964).

3.   W. B. Wetherell and M. P. Rimmer, "General analysis of aplanatic Cassegrain, Gregorian, and Schwarzschild telescopes," *Appl. Opt.* **11**, 2817–2832 (1972).

4.   M. Rimmer, "Analysis of perturbed lens systems," *Appl. Opt.* **9**, 533–537 (1970).

5.   H. H. Hopkins and H. J. Tiziani, "A theoretical and experimental study of lens centering errors and their influence on optical image quality," *British J. Appl. Phys.* **17**, 33–54 (1966).

6.   G. Catalan, "Intrinsic and induced aberration sensitivity to surface tilt," *Appl. Opt.* **27**, 22–23 (1988).

7.   D. J. Schroeder, *Astronomical Optics*, Section 6.III, Academic Press, New York (1987).

8.   R. N. Wilson, *Reflecting Telescope Optics* I, Sections 3.7 and 3.8, Springer, New York (1996).

PROBLEMS

7.1 Determine the aberrations of a Schmidt camera in which the axis of the *Schmidt plate* is displaced from the axis of the mirror by an amount Δx. Apply these results to Problem 6.4 if $\Delta x = 0.15$ mm.

7.2 Consider a beam of light incident parallel to the axis of a *beam expander* consisting of two confocal paraboloidal mirrors with focal lengths $f_i, i = 1$ and 2. (a) Determine the change in mirror spacing required to focus the beam at a distance $L \gg f_i$ from the beam expander. (b) Assuming the configuration of problem (a), determine the output beam direction and the longitudinal defocus when the mirrors are misaligned by small amounts $(\alpha_i, \beta_i, \gamma_i)$ and (x_i, y_i, z_i).

7.3 Consider the *Hubble space telescope* described in Problem 6.10 (a) Determine the decenter tolerance to give axial coma with a peak value of $\lambda/10$ at $\lambda = 0.53\,\mu\text{m}$. Calculate the corresponding image displacement. (b) Determine the neutral point of the telescope. (c) Determine the despace tolerance for a $\lambda/10$ peak spherical aberration. What is the corresponding image defocus aberration due to the image displacement? For an object lying at 2 milliradians from the optical axis, calculate the corresponding coma also.

7.4 Consider a *Schmidt-Cassegrain telescope*. Determine the tolerance on the *figure errors* of the corrector plate and the two mirrors so that the standard deviation of the total wavefront error contributed by them is $\lambda/10$ at $\lambda = 0.5\,\mu\text{m}$. Assume that the refractive index of the plate is 1.5.

Bibliography

M. Born and E. Wolf, *Principles of Optics*, Pergamon, New York (1985).

H. A. Buchdahl, *Optical Aberration Coefficients*, Oxford, London (1954); reprinted with Buchdhal's ressearch papers on aberrations by Dover, New York (1968).

A. E. Conrady, *Applied Optics and Optical Design*, Parts I and II, Oxford, London, (1929); Reprinted by Dover, New York (1957).

A. Cox, *A System of Optical Design*, Focal, London (1964).

E. Hecht and A. Zajac, *Optics*, Addition-Wesley, Reading, Massachusetts (1973).

H. H. Hopkins, *Wave Theory of Aberrations*, Oxford, London (1950).

F. A. Jenkins and H. E. White, *Fundamentals of Optics*, McGraw-Hill, New York, 4th ed. (1976).

R, Kingslake, *Lens Design Fundamentals*, Academic Press, New York (1978).

R. Kingslake, *Optical System Design,* Academic Press, New York (1983).

M. V. Klein, *Optics*, Wiley, New York (1970).

M. V. Klein and T. E. Furtak, *Optics*, Wiley, New York (1988).

D. Korsch, *Reflective Optics*, Academic Press, Sand Diego (1991).

E. H. Linfoot, *Recent Advances in Optics*, Clarendon, Oxford (1955).

V. N. Mahajan, *Aberration Theory Made Simple*, SPIE Press, Bellingham, Washington (1991).

D. Malacara and Z. Malacara, *Handbook of Lens Design*, Dekkar, New York (1994).

L. C. Martin and W. T. Welford, *Technical Optics*, Vol. I, 2nd ed., Pitman, London, (1966).

W. R. McCluney, *Introduction to Radiometry and Photometry*, Artech, Norwood, Massachusetts (1994).

P. Mouroulis and J. Macdonald, *Geometrical Optics and Optical Design*, Oxford, New York (1997).

D. C. O'Shea, *Elements of Modern Optical Design*, Wiley (1985).

H. Rutten and M. Van Venrooij, *Telescope Optics*, Willmann-Bell, Richmond, Virginia (1988).

D. J. Schroeder, *Astronomical Optics*, Academic Press, New York (1987).

R. R. Shannon, *The Art and Science of Optical Design*, Cambridge University Press, New York (1997).

G. G. Slyusarev, *Aberration and Optical Design Theory*, 2nd ed., Hilger, Bristol (1984).

W. J. Smith, *Modern Optical Engineering*, 2nd ed., McGraw-Hill, New York (1990).

A. Walther, *The Ray and Wave Theory of Lenses*, Cambridge University Press, New York (1995).

W. T. Welford, *Aberrations of the Symmetrical Optical System*, Academic Press, New York (1974).

R. N. Wilson, *Reflecting Telescope Optics* I, Springer, New York (1996).

Index

A

Abbe number ...327

aberration
chromatic ...322
definition ...143
defocus ...148
geometrical ...143
interferogram ...169
invariance ...356
order ...157
primary ...157
ray ...143, 207
Schwarzschild ...158
secondary ...158
tolerance ...242
transverse ...143
wave ...143

aberration balancing
astigmatism ...224
coma ...223
definition ...205
spherical ...216

achromatic systems
aplanatic doublet ...347
doublet ...248, 323, 340
meniscus corrector ...398
meniscus lens ...397

afocal system
beam expander ...402
for telephoto lens ...42
for wide angle lens ...43
reflecting telescope ...416
refracting telescope ...38, 86, 98

Airy pattern ...5

Amici lens
Gaussian properties ...87
aberrations ...362

anastigmatic system
beam expander ...402
field-flattening lens ...314
Schmidt camera ...387
telescope
Couder ...417
Schwarzschild ...420

angular aperture ...119, 121

angular demagnification ...400
angular field of view ...91
angular magnification ...15, 29, 48
aperture stop ...91, 93
aplanatic system ...265
lens ...309, 311, 362
Cassegrain telescope ...416
cemented doublet ...347
Gregorian telescope ...416

aplanatic
conjugates ...181
planes ...266
points
spherical mirror ...379
spherical refracting surface ...265
thin lens ...310

areal magnification ...124
aspheric corrector plates ...367, 422

aspheric plate
and a conic mirror ...426
and a two-mirror telescope ...428
in a converging image beam ...425
in a diverging object beam ...422

aspheric surface ...280
astigmatic focal line ...176, 226

astigmatism
Cassegrain and Gregorian telescope
aplanatic ...416
classical ...414
definition ...157
interferogram ...177
Mangin mirror ...433
shape ...174
refracting surface
spherical ...265
conic ...279
plane-parallel plate ...321
paraboloidal mirror ...384
spherical mirror ...372
thin lens ...304
thin lenses in contact ...361

astronomical telescope ...86, 402
atmospheric coherence length ...176
atmospheric turbulence ...165
auxiliary axis ...187, 253

axial color 325
    see longitudinal chromatic aberration also
axial coma
    definition 437
    refracting surface
        due to decenter 441
        due to tilt 444
    two-mirror telescope
        decentered mirror 446
        tilted mirror 447
        decentered and tilted mirror 448, 449, 450

B

back focal distance 61
barrel distortion 125, 235, 297
beam expander 367, 398, 434
Bouwers-Maksutov camera 69, 125, 367, 377, 394

C

cardinal points 4, 31
Cartesian pair
    definition 12
    glass sphere 361
    refracting surface 269, 361
    reflecting surface 369
Cartesian surface
    definition 12
    reflecting 369
    refracting 361
Cassegrain telescope 67, 367, 402, 413, 449, 450
catadioptric system
    Bouwers-Maksutov camera 394
    Mangin mirror
        see Mangin mirror
    Schmidt camera 385
    thin lens-mirror combination
        focal length 67
        Petzval curvature 376
catoptric system 367
cemented doublet 345
centered system 14
centrally obscured beam 67
centroid
    definition 209

for coma 223
chief ray 91, 94, 205
chromatic aberrations
    as wave aberrations 397
    Bouwers-Maksutov camera 397
    concentric lens 364
    definition 323
        longitudinal 323
        transverse 323, 325
    doublet 340, 363
    general system 331, 336
    plane-parallel plate 334, 363
    refracting surface 323
    thick lens 364
    thin lens 327
circle of least confusion
    astigmatic 226
    spherical 213
classical aberrations 163
cold stop 100
coma
    astronomical telescopes 412
    Cassegrain and Gregorian telescopes 414, 416
    definition 157
    despaced mirror 451
    interferogram 177
    Mangin mirror 433
    paraboloidal mirror 384
    plane-parallel plate 321
    refracting surface
        spherical 265
        conic 280
    shape 174
    spherical mirror 372
    symmetric system 348
    thin lens 304
compound lens 31
concave mirror 46, 376
concentric lens
    aberrations 364
    focal length 87
confocal paraboloidal mirrors 398
conic constant 275
conic mirror 402, 433
conic of revolution 274, 367
conic reflecting surface 367
conic refracting surface 247, 271

conjugate matrix 73
conjugate points 17
conjugate-shift equations 355
contact lens 86
contact magnifiers 87, 269
converging mirror 46
convex mirror 376
corrector plates
   see aspheric corrector plates
correlation length 450, 453
cosine law of intensity 91
cosine law of irradiance 101
cosine-fourth law of irradiance by an
   extended source 108
cosine-third law of irradiance
   by a point source 103
Couder telescope..............................417

D

Dall-Kirkham telescope 438, 449, 450
decenter 437, 438
decentered surface 438
defocus wave aberration 141, 149
design aberrations 450, 453
despace 437
despaced surface
   reflecting 451
   refracting 444
diffraction 5, 217
diopter 35
dispersive constant 328
distortion
   astronomical telescope 412
   field flattener....................................317
   for uniform image irradiance 125
   image of a square 234
   image of a square grid.....................235
   perturbed surface
      decented 441
      tilted 444
   plane-parallel plate 322
   reflecting surface 374
   refracting surface 261, 266
   spherical mirror..............................378
   symmetric system...........................348
   thin lens...............................306, 314
distortion wave aberration 233, 260
diverging mirror...................................46

E

eccentricity 270, 367
effective aperture stop 97
effective entrance pupil 96, 97
ellipse 271
ellipsoid 273, 274
entrance pupil 91, 94
entrance window 100
equiconvex lens 309
even aberration 206, 220
exact ray tracing 3
exit pupil 91, 94
exit window 91, 100
exitance 105
extended object....................................93
extended source 91
eye
   astigmatism....................................233
   cardinal points 85
   nearsighted 86
   spectral response 126, 127

F

fabrication errors 454
Fermat's principle 3, 5, 9, 278
field curvature 158, 172, 255
   refracting surface 261, 265, 280
   thin lens 305
   beam exapnder 402
   telescope 412
   aspheric plate 428
field flattener 248, 314, 376
field stop 91, 98
field-flattening lens 314, 394, 414
figure errors 438, 454, 456, 457
finite ray tracing 3
f-number 120, 397
focal length 25, 44, 57, 65
focal planes 31
focal points 31, 57, 336
focal ratio 114, 150
focusing power 69
fourth-order wave aberrations 255
fringe 175

G

Galilean telescope 86
Gaussian approximation 3, 14

Gaussian image 14
Gaussian imaging equation 17, 52
Gaussian optics 3
Gaussian reference sphere 143
generalized Lagrange invariant...........135
geometrical focus 271, 367
geometrical optics................................. 3
geometrical ray aberration 143
geometrical path length.................. 5, 142
geometrical point-spread function 206
glass hemisphere 88
glass sphere 87, 361
Gregorian telescope
                ....67, 367, 402, 413, 416

H

Hamilton's point characteristic
        function 13, 134, 178
Herschel condition 183, 359
Hubble space telescope 416, 434
Huygens eyepiece 86, 342
hyperbola 272
hyperboloid 274, 369

I

image magnifications......................... 281
image-space and object-space
        focal lengths 18
image-space focal distance 59, 67
image-space principal plane 31
imaging system 3
immersed detectors 88
intensity.. 91, 100
interference pattern 172
interferogram 175
inverse square 91
inverse-square law of irradiance 101
irradiance 91

J

Jacobian 134, 207

L

Lagrange invariant
        afocal system 39
        generalized 135
        general system 35
        reflecting surface 49
        refracting surface 21
        thin lens 30
        two-ray...70
Lagrange invariance 15, 35, 285
Lambert's cosine law of intensity 105
Lambertian disc 108
Lambertian source 91, 105
lateral aberrations 158
lateral color 326, 336
lateral spherical aberration 158, 394
lens bending 303, 307
lensmaker's formula 26
line-of-sight error 242
linear coma 178, 184
linear obscuration 394
longitudinal astigmatism 226
longitudinal chromatic aberration
        achromatic doublet 340
        definition 325
        general system 332, 338
        plane-parallel plate 335
        refracting surface 325
        thin lens 327
longitudinal defocus 141, 149
longitudinal magnification 21, 256
longitudinal spherical
        aberration 189, 212
lower marginal ray 97
lower rim ray 97
Lyot stop 100

M

Malus-Dupin theorem 3, 11
Mangin mirror
        aberrations
                chromatic 364
                primary 433
        focal length 84
manufacturing errors 450, 453
marginal focus 175
marginal image plane 175
marginal image points 212
marginal ray 91, 94
        lower 97
        upper...97
matrix approach.................................73
mean intensity 100

meridional plane 14, 143, 205
microscope objective 269
minimum root-mean-square
    radius 205, 236, 237
minimum-aberration-variance
    plane 176
misalignment..................................... 437
    decenter 437, 438
    despace 437, 444, 451
    tilt 442, 447, 448

N

negative lenses 302
neighboring ray 11, 185, 451, 454
neutral point 438, 445, 448
neutral zone 392
Newtonian imaging equation
    ....24, 31, 38, 52
nodal planes 31
nodal points 36
numerical aperture 119, 122

O

object imagery 350
object-space focal distance 59, 67
object-space focal point 31
object-space principal plane 31
oblate ellipse 271
oblique spherical aberration 158, 394
obscuration ratio 137, 394
odd aberration 206, 220
offense against the sine
    condition 178, 188, 191
oil immersion microscope 362
optical axis 14
optical path 5, 142, 300
optical sine theorem 186, 188
optical wavefront 11
optimum defocus 216
orthogonal aberrations 165
orthonormal Zernike aberrations 165

P

parabola 273
parabolic image......................................235
paraboloid 273, 274, 359
paraboloidal mirror 5, 384, 433
parallel beam 362

paraxial image plane 175
paraxial ray tracing 3
paraxial refracting surface 22
peak aberration coefficients 288
peak value 160, 189
perfect imaging 178
perfect image 142, 145
perturbed optical system 437
Petzval image point 247, 255
Petzval image surface
    definition 231, 258
    general formula 283
    mirror 375
    Schmidt camera 376, 387
    thin lens 306
    telescope 376, 408
Petzval sum 288
photometry 126
pincushion distortion 235, 297
piston aberration 157
plane-parallel plate
    aberrations
        primary 322
        chromatic 334, 363
        cardinal points 85
point-spread function 205
position factor..................................303
power-series coefficients 168
power-series expansion 152, 156, 160
primary aberrations
    concentric lens 364
    definitions 157
    field flattener 317
    in terms of Seidel sums 288
    in terms of Seidel coefficients 288
    Mangin mirror 432
    plane-parallel plate 322
    reflecting surface
        conic 372
        spherical 378
        parapoloidal 384
    refracting surface
        spherical
            stop at the surface.................260
            stop not at the surface 265
        conic 280
    thin lens 302
prime focus, ...403

principal ray 57, 94
principal planes 31
principal points 336
projected area 101
prolate ellipse 271
pupil aberrations 350
pupil distortion.....................118
pupil imagery 350

R

radial image 225
radiance 91, 104
radiance theorem 116, 134
radiometry 91
   of point object imaging 112
   of extended object imaging.............114
random aberrations 438, 454
ray aberrations 12, 143, 207
ray angular magnification 20
ray fan 205
ray spot diagram 143, 206
ray tracing 52
rectilinear propagation 8
reduced power-series expansion 160
reflection ray-tracing equation 54, 64
refraction matrix 74
relative aperture 120
rim ray.....................97
Ritchey-Chrétien telescope 430, 438, 451
root mean square radius
   astigmatism 229
   coma 223
   general 209
   minimum 236, 237
   spherical 210, 214, 216
rotational invariants 152, 153
rotationally symmetric system
   ....14, 141, 152, 427

S

sag of a conic surface 275
sagittal image 172, 189, 225
sagittal coma 189, 219
sagittal plane 143, 205
sagittal ray fan 205, 206
sagittal rays 147
Schmidt camera 125, 385, 428, 430
Schmidt plate 323, 433

Schmidt-Cassegrain telescope 428
Schott glass.....................330
Schwarzschild aberrations.....................158
Schwarzschild telescope
   aberrations 418
   Gaussian properties 136
secondary aberrations
   ....156, 158, 161, 394
secondary spectrum 345
Seidel aberrations 157, 247, 256
Seidel coefficients 248, 288
Seidel sums.....................248, 288
shape factor 303
sign convention 14
sine condition
   ....178, 181, 182, 186, 191, 359, 361
sixth-order astigmatism 394
sixth-order spherical aberration 394
skew rays 14
Smith-Helmholtz invariant 21
Snell's law 9
speed of a lens 120
spherical aberration
   astronomical telescope.....................412
   circle of least confusion 213
   definition 157
   interferogram 177
   plane-parallel plate 322
   rms radius 216
   refracting surface
      spherical.....................265
      conic 280
   shape 174
   spherical mirror 372
spherical mirror 42, 247, 249, 377
spherochromatism 213, 323, 387
spot diagram 205, 236
spot radius 214
spot size 205
standard deviation 168
stop-shift equations
   primary aberrations 294
   chromatic aberrations 327
surface coefficients 281
symmetrical principle 248, 348
system matrix 73, 76
   afocal system 88
   reversed system 88

symmetrical system 88

T

tangential plane 14, 143, 205
tangential image 172, 225, 297
tangential image surface 227
tangential coma 219
tangential ray fan 205, 206
tangential rays 147
telecentric system 98
telecentric stop 98
telephoto lens 86, 87
telephoto system 41
telescope
   astronomical 65, 86, 392
   Cassegrain
      aplanatic 416, 447, 452, 453
      classical 413, 447, 452, 453
   Couder.................... 417, 447, 452, 453
   Dall-Kirkham 420, 447, 452
   Galilean 38, 39, 86
   Gregorian
      aplanatic 416, 447, 452, 453
      classical 413, 447, 452, 453
   Hubble............................416, 434, 459
   Keplerian................................38, 39
   Mersenne......................398, 447, 452
   Ritchey Chrétien 416, 430, 438, 451
   Schwarzschild 136, 418
tertiary aberrations 142, 152, 155, 156
thick lens 31, 59, 85
thin lens
   aberrations
      with spherical surfaces 303-305
      with conic surfaces....................312
   aplanatic 310
   field flattener....................................314
   focal length....................................25
   imaging equation 25, 31
   magnification 28, 299
   Petzval surface 306
thin-lens doublet 343, 363
third-order ray aberrations 256
tilted surface 432
transfer matrix 74
transfer ray-tracing equation 53
transverse chromatic aberration
   achromatic doublet......................340
   definition 325
   general system 333, 339
   refracting surface 325
   thin lens 327
transverse magnification .. 19, 34, 47, 252
transverse ray aberration 143, 207
two thin lenses 57
two-ray Lagrange invariant 70
Twyman-Green interferometer 173

U

uniform diffuser 105
unobscured two-mirror system 403
upper marginal ray 97
upper rim ray 97

V

variance 166, 216
vertex radius of curvature 271, 368
vignetting 91, 93, 96, 136
vignetting diagram 98, 136
virtual image.................................17
virtual path 251, 301, 369

W

wave aberration
   definition 12, 143, 145
   relationship with ray aberration 147
   due to defocus................................149
   due to Petzval curvature 283
wavefront 142
wavefront errors 438, 454
   reflecting surface 457
   refracting surface 456
wavefront tilt 141
wavefront tilt aberration 141, 150
wide-angle lens 87
wings 394
working distance 137, 405, 420

Z

Zernike aberrations.............................165
Zernike annular polynomials 168, 200
Zernike circle polynomials
      .... 142, 152, 163, 235
Zernike coefficients 165, 168
zonal rays 95

ABOUT THE AUTHOR

 **Virendra N. Mahajan** was born in Vihari, Pakistan, and educated in India and the United States. He received his Ph.D. degree in Optical Sciences from the Optical Sciences Center, University of Arizona, in 1974. He spent nine years at the Charles Stark Draper Laboratory in Cambridge, Massachusetts, where he started and headed an optics group. Since 1983, he has been at The Aerospace Corporation in El Segundo, California, where he is Systems Director working on a space-based surveillance system. He is also an adjunct professor at the University of Southern California in the electrical engineering-electrophysics department, where he teaches a graduate course in optical imaging and aberrations. He was a visiting professor at the Indian Institute of Technology, New Delhi, under a grant from the United Nations Development Program during spring 1990. He has taught short courses on aberration theory at the National Central University, Chung Li, Taiwan, and at the annual meetings of the Optical Society of America and SPIE. He has published numerous papers on diffraction, aberration theory, adaptive optics, and acousto-optics. He is a member of the Optical Society of America and past chairman of its Astronomical, Aeronautical, and Space Optics technical group. He is also a member of the SPIE and has participated in its Education Committee. He is the author of *Aberration Theory Made Simple* published by SPIE in 1991 under their Tutorial Text Series, and the editor of *Selected Papers on Effects of Aberrations in Optical Imaging* published by SPIE in 1994 under their Milestone Series.